ASTROCYTES WIRING the BRAIN

FRONTIERS IN NEUROSCIENCE

Series Editors
Sidney A. Simon, Ph.D.
Miguel A.L. Nicolelis, M.D., Ph.D.

Published Titles

New Concepts in Cerebral Ischemia
Rick C. S. Lin, Ph.D., Professor of Anatomy, University of Mississippi Medical Center, Jackson, Mississippi

DNA Arrays: Technologies and Experimental Strategies
Elena Grigorenko, Ph.D., Technology Development Group, Millennium Pharmaceuticals, Cambridge, Massachusetts

Methods for Alcohol-Related Neuroscience Research
Yuan Liu, Ph.D., National Institute of Neurological Disorders and Stroke, National Institutes of Health, Bethesda, Maryland
David M. Lovinger, Ph.D., Laboratory of Integrative Neuroscience, NIAAA, Nashville, Tennessee

Primate Audition: Behavior and Neurobiology
Asif A. Ghazanfar, Ph.D., Princeton University, Princeton, New Jersey

Methods in Drug Abuse Research: Cellular and Circuit Level Analyses
Barry D. Waterhouse, Ph.D., MCP-Hahnemann University, Philadelphia, Pennsylvania

Functional and Neural Mechanisms of Interval Timing
Warren H. Meck, Ph.D., Professor of Psychology, Duke University, Durham, North Carolina

Biomedical Imaging in Experimental Neuroscience
Nick Van Bruggen, Ph.D., Department of Neuroscience Genentech, Inc.
Timothy P.L. Roberts, Ph.D., Associate Professor, University of Toronto, Canada

The Primate Visual System
John H. Kaas, Department of Psychology, Vanderbilt University, Nashville, Tennessee
Christine Collins, Department of Psychology, Vanderbilt University, Nashville, Tennessee

Neurosteroid Effects in the Central Nervous System
Sheryl S. Smith, Ph.D., Department of Physiology, SUNY Health Science Center, Brooklyn, New York

Modern Neurosurgery: Clinical Translation of Neuroscience Advances
Dennis A. Turner, Department of Surgery, Division of Neurosurgery, Duke University Medical Center, Durham, North Carolina

Sleep: Circuits and Functions
Pierre-Hervé Luppi, Université Claude Bernard, Lyon, France

Methods in Insect Sensory Neuroscience
Thomas A. Christensen, Arizona Research Laboratories, Division of Neurobiology, University of Arizona, Tuscon, Arizona

Motor Cortex in Voluntary Movements
Alexa Riehle, INCM-CNRS, Marseille, France
Eilon Vaadia, The Hebrew University, Jerusalem, Israel

Neural Plasticity in Adult Somatic Sensory-Motor Systems
Ford F. Ebner, Vanderbilt University, Nashville, Tennessee

Advances in Vagal Afferent Neurobiology
Bradley J. Undem, Johns Hopkins Asthma Center, Baltimore, Maryland
Daniel Weinreich, University of Maryland, Baltimore, Maryland

The Dynamic Synapse: Molecular Methods in Ionotropic Receptor Biology
Josef T. Kittler, University College, London, England
Stephen J. Moss, University College, London, England

Animal Models of Cognitive Impairment
Edward D. Levin, Duke University Medical Center, Durham, North Carolina
Jerry J. Buccafusco, Medical College of Georgia, Augusta, Georgia

The Role of the Nucleus of the Solitary Tract in Gustatory Processing
Robert M. Bradley, University of Michigan, Ann Arbor, Michigan

Brain Aging: Models, Methods, and Mechanisms
David R. Riddle, Wake Forest University, Winston-Salem, North Carolina

Neural Plasticity and Memory: From Genes to Brain Imaging
Frederico Bermudez-Rattoni, National University of Mexico, Mexico City, Mexico

Serotonin Receptors in Neurobiology
Amitabha Chattopadhyay, Center for Cellular and Molecular Biology, Hyderabad, India

TRP Ion Channel Function in Sensory Transduction and Cellular Signaling Cascades
Wolfgang B. Liedtke, M.D., Ph.D., Duke University Medical Center, Durham, North Carolina
Stefan Heller, Ph.D., Stanford University School of Medicine, Stanford, California

Methods for Neural Ensemble Recordings, Second Edition
Miguel A.L. Nicolelis, M.D., Ph.D., Professor of Neurobiology and Biomedical Engineering,
    Duke University Medical Center, Durham, North Carolina

Biology of the NMDA Receptor
Antonius M. VanDongen, Duke University Medical Center, Durham, North Carolina

Methods of Behavioral Analysis in Neuroscience
Jerry J. Buccafusco, Ph.D., Alzheimer's Research Center, Professor of Pharmacology and Toxicology,
    Professor of Psychiatry and Health Behavior, Medical College of Georgia,
    Augusta, Georgia

In Vivo Optical Imaging of Brain Function, Second Edition
Ron Frostig, Ph.D., Professor, Department of Neurobiology, University of California,
    Irvine, California

Fat Detection: Taste, Texture, and Post Ingestive Effects
Jean-Pierre Montmayeur, Ph.D., Centre National de la Recherche Scientifique, Dijon, France
Johannes le Coutre, Ph.D., Nestlé Research Center, Lausanne, Switzerland

The Neurobiology of Olfaction
Anna Menini, Ph.D., Neurobiology Sector International School for Advanced Studies, (S.I.S.S.A.),
    Trieste, Italy

Neuroproteomics
Oscar Alzate, Ph.D., Department of Cell and Developmental Biology,
    University of North Carolina, Chapel Hill, North Carolina

Translational Pain Research: From Mouse to Man
Lawrence Kruger, Ph.D., Department of Neurobiology, UCLA School of Medicine, Los Angeles,
    California
Alan R. Light, Ph.D., Department of Anesthesiology, University of Utah, Salt Lake City, Utah

Advances in the Neuroscience of Addiction
Cynthia M. Kuhn, Duke University Medical Center, Durham, North Carolina
George F. Koob, The Scripps Research Institute, La Jolla, California

Neurobiology of Huntington's Disease: Applications to Drug Discovery
Donald C. Lo, Duke University Medical Center, Durham, North Carolina
Robert E. Hughes, Buck Institute for Age Research, Novato, California

Neurobiology of Sensation and Reward
Jay A. Gottfried, Northwestern University, Chicago, Illinois

The Neural Bases of Multisensory Processes
Micah M. Murray, CIBM, Lausanne, Switzerland
Mark T. Wallace, Vanderbilt Brain Institute, Nashville, Tennessee

Neurobiology of Depression
Francisco López-Muñoz, University of Alcalá, Madrid, Spain
Cecilio Álamo, University of Alcalá, Madrid, Spain

Astrocytes: Wiring the Brain
Eliana Scemes, Albert Einstein College of Medicine, Bronx, New York
David C. Spray, Albert Einstein College of Medicine, Bronx, New York

ASTROCYTES WIRING the BRAIN

Edited by
Eliana Scemes

Albert Einstein College of Medicine
Bronx, New York

David C. Spray

Albert Einstein College of Medicine
Bronx, New York

CRC Press
Taylor & Francis Group
Boca Raton London New York

CRC Press is an imprint of the
Taylor & Francis Group, an **informa** business

CRC Press
Taylor & Francis Group
6000 Broken Sound Parkway NW, Suite 300
Boca Raton, FL 33487-2742

First issued in paperback 2019

ISBN-13: 978-1-4398-2584-6 (hbk)
ISBN-13: 978-1-138-37430-0 (pbk)

Library of Congress Cataloging-in-Publication Data

Astrocytes : wiring the brain / editors, Eliana Scemes, David C. Spray.
    p. ; cm. -- (Frontiers in neuroscience)
  Includes bibliographical references and index.
  ISBN 978-1-4398-2584-6 (hardcover : alk. paper)
  I. Scemes, Eliana. II. Spray, David C. III. Series: Frontiers in neuroscience (Boca Raton, Fla.)

[DNLM: 1. Astrocytes--physiology. 2. Central Nervous System--physiology. 3. Central Nervous System Diseases--physiopathology. 4. Mental Disorders--physiopathology. WL 101]
612.8'2--dc23   2011036046

Visit the Taylor & Francis Web site at
http://www.taylorandfrancis.com

and the CRC Press Web site at
http://www.crcpress.com

Contents

SECTION I   Introduction to Cluster A: Functional Biomarkers at the Interface

SECTION II   Introduction to Cluster B: Establishment of Functional Units: Multimodality at the Tips

SECTION III   CNS Pathology: Disruption of Astrocyte Connectivity: Introduction

Series Preface

FRONTIERS IN NEUROSCIENCE

The Frontiers in Neuroscience Series presents the insights of experts on emerging experimental technologies and theoretical concepts that are, or will be at the vanguard of Neuroscience.

The books cover new and exciting multidisciplinary areas of brain research, and describe breakthroughs in fields like visual, gustatory, auditory, olfactory neuroscience, as well as aging biomedical imaging. Recent books cover the rapidly evolving fields of multisensory processing, depression and different aspects of reward.

Each book is edited by experts and consists of chapters written by leaders in a particular field. Books are richly illustrated and contain comprehensive bibliographies. Chapters provide substantial background material relevant to the particular subject.

The goal is for these books to be the references every neuroscientist uses in order to acquaint themselves with new information and methodologies in brain research. We view our task as series editors to produce outstanding products that contribute to the broad field of Neuroscience. Now that the chapters are available on line the effort put in by us, the publisher, the book editors, and individual authors will contribute to the further development of brain research To the extent that you learn from these books we will have succeeded.

Sidney A. Simon, Ph.D.
Miguel A.L. Nicolelis, M.D., Ph.D.

Preface: The Astrocyte: A Polarized Cell

Eliana Scemes and David C. Spray
Dominick P. Purpura Department of Neuroscience,
Albert Einstein College of Medicine, Bronx, New York

The goal of this book has been to compile an overview of the most current findings on the diverse roles played by astrocytes in central nervous system (CNS) function and dysfunction, emphasizing how the connections that the astrocyte makes with other cells of the brain are essential for a variety of important neural functions. A key concept that we have highlighted throughout (wiring the brain) is that astrocytes are morphologically highly elaborated cells, establishing associations through their fine processes with practically all cellular types in the CNS. Such close morphological associations, together with the fact that astrocytes express a multitude of ion channels, transporters, and membrane receptors, endow these cells with the unique capability to sense and influence diverse CNS functions.

A view of the astrocyte that is now becoming classical is that one or more of its processes surrounds synaptic areas of neurons (providing the third element of the so-called tripartite synapse) and another process projecting toward the vasculature to provide a bidirectional pathway for nutrient delivery from the blood supply and metabolite clearance toward the vasculature. Moreover, astrocytes are highly polarized, being specialized for interactions with each neural cell type. In contrast to cell polarity present in epithelial cells, where basolateral and apical domains are separated by junctions with fence and barrier properties, the polarity of astrocytes is similar to interneurons, where distinct processes mediate input and output functions and possess the proteins specifying either pre- or postsynaptic functions. Like neurons, astrocytes integrate information and are believed able to relay it through intracellular calcium wave signaling.

We have divided this book into three sections, with the individual sections dealing with biomarkers that define astrocytes, connections of astrocytes to other neural cells, and damage to astrocytes in disease. Specifically, chapters in Section I identify major astrocyte biomarkers and how they define the different connectivity domains. Chapters in Section II describe the role of these connections, how their function can be manipulated under physiological conditions, and how dysfunction of the connectivity leads to aberrant brain performance. Chapters in Section III examine the alterations of glia that have been observed in specific diseases of the brain (epilepsy, autoimmune encephalitis (MS), Alzheimer's, autism, and major depression) and attempts to identify key mechanisms responsible for these alterations.

As research becomes more and more translational, we expect that compilations such as this will be increasingly useful in defining both the state of what is known with regard to a specific field and what is unknown with regard to causes of pathological conditions and the possibility for novel therapeutic strategies.

Acknowledgments

We gratefully acknowledge the editorial assistance of Ms. Frances Andrade and the support over the years from the National Institutes of Health for our studies of astrocytes (NS041023 and NS052245 to E.S. and NS041282 to D.C.S.).

Contributors

Celia F. Brosnan
Dominick P. Purpura Department of
    Neuroscience
Department of Pathology
Albert Einstein College of Medicine
Bronx, New York

Emeline Camand
Institut Pasteur
Cell Polarity and Migration Group and
    CNRS URA 2582
Paris, France

Salvatore Carbonetto
Centre for Research in Neuroscience
McGill University Health Centre
Montreal, Quebec, Canada

Monica J. Carson
Division of Biomedical Sciences
Center for Glial-Neuronal Interactions
University of California, Riverside
Riverside, California

Kimberly G. V. Davidson
Department of Biomedical Sciences
Colorado State University
Fort Collins, Colorado

Rolf Dermietzel
Department of Neuroanatomy and
    Molecular Brain Research
University of Bochum
Bochum, Germany

Nilufer Esen
Department of Neurology
University of Michigan Medical School
Ann Arbor, Michigan

Sandrine Etienne-Manneville
Institut Pasteur
Cell Polarity and Migration Group and
    CNRS URA 2582
Paris, France

S. Hossein Fatemi
Departments of Psychiatry,
    Pharmacology, and
    Neuroscience
University of Minnesota Medical School
Minneapolis, Minnesota

Timothy D. Folsom
Department of Psychiatry
University of Minnesota Medical School
Minneapolis, Minnesota

Christian Giaume
CIRB, UMR CNRS 7241/
    INSERM U1050
Collège de France, Paris, France

Tammy Kielian
Department of Pathology and
    Microbiology
University of Nebraska Medical
    Center
Omaha, Nebraska

Harold K. Kimelberg
St. Augustine, Florida

Rachel E. Kneeland
Department of Psychiatry
University of Minnesota Medical School
Minneapolis, Minnesota

Stephanie B. Liesch
Department of Psychiatry
University of Minnesota Medical School
Minneapolis, Minnesota

Sarah E. Lutz
Dominick P. Purpura Department of
    Neuroscience
Albert Einstein College of Medicine
Bronx, New York

Carlos Matute
Department of Neurosciences
University of the Basque Country
    UPV/EHU
Leioa, Spain

and

CIBERNED and Neurotek
    UPV/EHU
Leioa, Spain

Aye-Mu Myint
Klinikum der Universität München
Klinik und Poliklinik für Psychiatrie
    und Psychotherapie
Ludwig-Maximilians Universität
    München
Munich, Germany

Grazia Paola Nicchia
Dipartimento di Fisiologia Generale ed
    Ambientale
Facoltà di Scienze Biotecnologiche
Università degli Studi di Bari
Bari, Italy

Cedric S. Raine
Dominick P. Purpura Department of
    Neuroscience
Department of Pathology and
    Department of Neurology
Albert Einstein College of Medicine
Bronx, New York

John E. Rash
Department of Biomedical Sciences
    and Program in Molecular,
    Cellular, and Integrative
    Neurosciences
Colorado State University
Fort Collins, Colorado

José Julio Rodríguez Arellano
IKERBASQUE
Basque Foundation for Science
Bilbao, Spain

and

Department of Neurosciences
University of the Basque Country UPV/
    EHU
Leioa, Spain

and

Institute of Experimental Medicine
ASCR
Prague, Czech Republic

Daniela Rossi
Laboratory for Research on
    Neurodegenerative Disorders
Salvatore Maugeri Foundation
Pavia, Italy

Eliana Scemes
Dominick P. Purpura Department of
    Neuroscience
Albert Einstein College of Medicine
Bronx, New York

Markus J. Schwarz
Klinikum der Universität München
Klinik und Poliklinik für Psychiatrie
    und Psychotherapie
Ludwig-Maximilians Universität
    München
Munich, Germany

Gerald Seifert
Institute of Cellular Neurosciences
University of Bonn
Bonn, Germany

David C. Spray
Dominick P. Purpura Department of
    Neuroscience
Albert Einstein College of Medicine
Bronx, New York

Christian Steinhäuser
Institute of Cellular Neurosciences
University of Bonn
Bonn, Germany

Alexei Verkhratsky
Faculty of Life Sciences
University of Manchester
Manchester, United Kingdom
IKERBASQUE, Basque Foundation
    for Science,
Bilbao, Spain
Department of Neurosciences
University of the Basque Country
Leioa, Spain
Institute of Experimental Medicine
ASCR, Prague, Czech Republic

Andrea Volterra
Department of Cell Biology and
    Morphology
University of Lausanne
Lausanne, Switzerland

Emma H. Wilson
Division of Biomedical Sciences
Center for Glial-Neuronal Interactions
University of California, Riverside
Riverside, California

Section I

Introduction to Cluster A: Functional Biomarkers at the Interface

The intent of the chapters in this first section is to define glial cells, and specifically to identify the biomarkers that characterize the glial cells and their interactions with other neural components. Recent transcriptomic profiling has elegantly delineated gene expression patterns of neurons, astrocytes, and oligodendrocytes, providing catalogs of genes whose expression is enriched in one neural cell population or another and during development (Cahoy et al., 2008). One surprising finding was that the genes showing astrocyte-enriched expression encoded numerous proteins without previous linkage to astrocyte functions (such as the aldehyde dehydrogenase Aldh1L1, the Transcription factors Rfx4, Pbixp1, and Gli3, as well as phagocytotic genes of the CED family); another surprise was the astrocytic expression of genes associated with neuropsychiatric disorders (e.g., *Npas3*, *Mlc1*, *Lgi1/4*, and *Gpr56*). Together with the cell-type-specific proteins and genes that have been made available through the Allen Brain Atlas, this characterization of neural cell autonomous expression data sets will likely revolutionize such tools for studying specific cell types as antibodies, transgenic targeting, and may ultimately allow even more complete profiling of the domains of astrocytes.

The first chapter in this section (Chapter 1), by Harold Kimelberg, characterizes protoplasmic astrocytes in terms of the morphology of their processes and whether the processes are perisynaptic or perivascular. The author raises the question of whether morphological distinctions between the processes are major contributors to spatial buffering, Ca^{2+} diffusion, blood flow, and integration of neuronal activity.

Chapter 2, by Eliana Scemes and Grazia Paola Nicchia, describes the distribution of three classes of biomarkers that identify astrocyte microdomains: AQP-4, gap junction proteins, and purinergic receptors. Functions of the gap junctions and purinergic receptors are considered in potassium buffering, calcium signaling, and metabolic cooperation. AQP-4 in perivascular domains controls water flux and contributes to shape and migration disorders when it is disrupted (such conditions include inflammation, edema). The involvement of astrocyte connexin mutations in genetic diseases is also discussed.

Chapter 3, by Salvatore Carbonetto, Emeline Camand, and Sandrine Etienne-Manneville, deals with the issue of how astrocyte polarization is determined, in terms of receptors on the astrocyte surface responding to signals from the extracellular matrix, contribution of cell adhesive junctions with neighboring cells, and interaction of junctional molecules with the cytoskeleton. The polarity of the astrocyte is defined as the contact of its processes with different targets, including meninges, vessels, and neuronal processes; the polarity of astrocytes is very dependent on physical properties of their environment. It is argued that understanding such contact-dependent signaling will facilitate knowledge about effects on astrocyte polarization.

The fourth and final chapter in this section (Chapter 4), by Nilufer Esen and Tammy Kielian, deals with the role of pattern recognition receptors of astrocytes establishing contact with invading cells/species and thus playing a role in central nervous system (CNS) immunity and inflammation. Besides having molecules that establish contact with other cells, astrocytes also have pattern recognition (Toll-like) receptors (TLRs) for pathogen-induced signals that can release both protective and apoptosis-inducing molecules, including cytokines, chemokines, and growth factors, that result in the activation of large signaling pathways. Pathogens enter the CNS by disrupting the blood–brain barrier. Since astrocytes are the most numerous cell type surrounding the vessel wall and their endfeet are found at this strategic site, astrocytes provide a first line of defense against CNS inflammation. In particular, activation of TLR3 in astrocytes induces chemokine expression, decreases glutamate uptake, and decreases expression of the astrocyte gap junction protein Cx43. Moreover, TLR3 may be a mechanism by which astrocytes influence neurogenesis.

REFERENCE

Cahoy, J. D., B. Emery, A. Kaushal, L. C. Foo, J. L. Zamanian, K. S. Christopherson, Y. Xing, J. L. Lubischer, P. A. Krieg, S. A. Krupenko, W. J. Thompson, and B. A. Barres. 2008. A transcriptome database for astrocytes, neurons, and oligodendrocytes: a new resource for understanding brain development and function. *J Neurosci* 28(1):264–78.

1 Mature Protoplasmic Mammalian Astrocytes

Morphology, Interrelationships, and Implications for Function

Harold K. Kimelberg
St. Augustine, Florida

CONTENTS

What passes for knowledge are most often beliefs which we hold more or less tenaciously without any clear awareness as to what precisely we are claiming to know. Even when our beliefs happen to be true, our lack of precision and our ignorance of the grounds upon which these beliefs could be based permit us to hold other beliefs which are contradictory. (Stebbing, 1954, ix).

1.1 MATURE MAMMALIAN ASTROCYTES

1.1.1 CLASSIFICATION

There are two central questions regarding mature astrocytes. What cells should comprise this class and what do they do? The cells of the nervous system, collectively termed neural cells, are traditionally divided into two main groups, neurons and glia. These are the excitable and nonexcitable cells of the nervous system, respectively. Glia is a Greek-derived word best translated as "glue" and was coined by the pathologist Rudolph Virchow, who was an early and highly respected contributor to cell theory and pathology (Wikipedia, 2010). Virchow actually used the term *neuroglia*, but it is now shortened to *glia*. He also used the German word *nervenkitt*, which Somjen translates as "soft putty" (Somjen, 1988). Regardless of whether we now say "mere glue," to Virchow it meant a connective tissue that surrounded the neurons and was responsible for the inflammatory processes that can affect the ventricular cavities of the brain (Somjen, 1988). This view, naturally enough since neurobiology is an empirical science, has long been superseded by later, more accurate microscopic findings, and the term *glia* is now so archaic that it would best be jettisoned as it serves to obfuscate rather than clarify, or is self-servingly quoted to support some novel, more complex function as being a needed advance on the still-held idea that glia are mere glue, which anyway is an erroneous view of what Virchow thought he had shown 150 years ago. The glia do not even form a single group or class. For example the major types of glia in terms of their mRNA expression are as different from each other as any one is from neurons (Cahoy et al., 2008).

The astroglia, frequently termed astrocytes, are, with the oligodendroglia, classified as macroglia distinct from the other class of glia called microglia. There are other nonexcitable cell types, such as ependymal cells, which line the ventricles. We also now have proliferating stem cells, some of which have been defined as astroglia, which are found in adult brains (Ihrie and Alvarez-Buylla, 2008). Also there are the interesting nonexcitable cells previously referred to as smooth protoplasmic astrocytes and then oligodendrocyte precursor cells (OPCs), and now usually referred to as NG2 cells after their specific staining for this chondroitin sulfate proteoglycan (Nishiyama et al., 2009). Taxonomy is a challenging task and especially so when it is applied to entities such as cells that require microscopy and specialized techniques to observe them, which may be too selective or nonselective and can alter some of the cells' characteristic features. It also depends, like all science, on unpredicted new data. However, the ability to precisely identify what one is studying within the context of a current state of knowledge is obviously vital to avoid the problem of talking at cross purposes when attempting to describe the subject's properties and functions and discuss them with others.

I only deal with adult astroglia cells of the adult mammalian nervous system, which I have defined elsewhere as protoplasmic and fibrous astrocytes, and the Bergmann and Muller glia (Kimelberg, 2010). Others may have different ideas about what should be included as astrocytes. For example, a very recent review (Matyash and Kettenmann, 2010) emphasizes a more extensive heterogeneity for astrocytes but bases much of this on findings in different preparations, including cultured astrocytes and astrocyte precursors from younger animals, as well as mRNA and promoter-driven gene expression. While this may be a call for much future work, it is not yet applicable for a current analysis of function, so perhaps we should simply collect more and more data, as in Bacon's scientific method, until unifying themes emerge. My more restricted definition at least clarifies mature mammalian astroglia for this chapter and perhaps will be useful for the future studies and discussions of others. In practical terms this chapter only deals with protoplasmic astrocytes because these have been most comprehensively studied, and most of the work is for the rodent hippocampus.

My restriction also ignores the role of astroglia in development. Precursors that are on their way to developing into mature cells are not the same as the final cell type. We do like to have single, simple words to distinguish between entities in science, as in other human activities, but the underlying realities are always about shades of differences. Thus the term *astrocyte*, as it has been defined, has got into the same problems as glia. How to define and identify immature astrocytes I will leave to others who are more expert and energetic in these matters than me. So the rest of this chapter is all about mature astrocytes, and mainly protoplasmic astrocytes from rodent hippocampus. How far this is applicable to protoplasmic astrocytes from other regions and to the three other types of astrocytes I mentioned and the nine astrocyte groups that others suggest (Matyash and Kettenmann, 2010) will only be resolved by actually studying them.

The problem now becomes easier but not easy! How are we to define mature protoplasmic astrocytes?

1. Morphology. This got its start with the noble metal staining techniques of Golgi (silver) and Ramón y Cajal (gold). This began in the 1870s and was a major advance on the staining techniques that Virchow used. His "connective tissue" was now seen to be actual cells that differed in their morphology and interrelationships from neurons.

2. Markers. These are proteins that are held to be specifically expressed by particular classes of cells. This has led to some clarification, but also a lot of confusion. We have, of course, known for some time that proteins are synthesized according to the specific sequences of bases in DNA that are transcribed into mRNA, which is the direct template for protein assembly. But all cells in an individual organism have the same DNA, and we now know that the secret of what proteins are made with synthesized is due to the complex control of gene expression at the DNA or RNA levels. All this is now elementary and fundamental, although the details are highly complex. Yet it is surprising how professional biologists seem to gloss over this when they define cell types on the basis of the immutability of protein markers.

3. Dynamic properties. These would be any physiological property, be it a metabolic, transport, or electrophysiological process. These share with morphology an improvement on markers in that they are due to the expression of a number of genes and it is known that multiple characteristics are less

likely to be misleading than one characteristic alone, according to Bayesian probability theory (Kimelberg, 2004). However, 1 and 3 are less easy to specify, so being human, we often take the easy route and use 2.

1.2 MORPHOLOGY

Figure 1.1 shows old and new views of what the four types of astrocytes I mentioned above look like. You will notice that the protoplasmic (Figure 1.1A,C,D) and fibrous (Figure 1.1C) astrocytes look, perhaps, like stars, while the Muller (Figure 1.1B) and Bergmann glia (Figure 1.1C), being columnar, do not. Also, it is reassuring that the older Golgi-stained cells closely resemble the cells visualized by newer filling methods, as in Figure 1.1D for protoplasmic astrocytes and Figure 1.1B for living

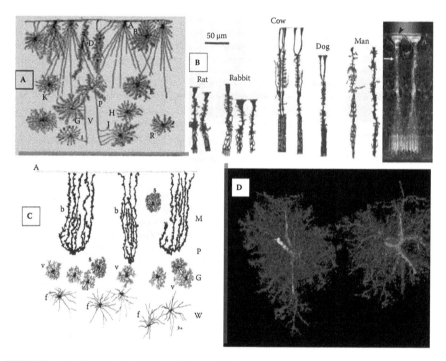

FIGURE 1.1 (See color insert.) (A) Golgi-stained astrocytes from a 2-month-old human infant in the molecular layer and layers II and III of the cerebral cortex. A–D are cells in the first cortical lamina and E–H are cells in the second and third lamina. I and J are cells with endfeet contacting blood vessels. V, blood vessel. (From Ramón y Cajal, *Trab. Lab. Invest. Biol. Univ. Madrid*, 11, 255–315, 1913.) (B) Muller cells in different species as indicated. Furthest right, Muller cells as light guides. (Courtesy of S. Skatchov, from Franze et al., *Proc. Natl. Acad. Sci. USA*, 104, 8287–8292, 2007, cover page of that issue.) (C) Different astrocyte cells in mature mammalian cerebellum by Golgi staining (Golgi, 1885). Working down from the cerebellar surface A at the top, represented by dotted line: M, molecular; P, Purkinje; G, granule cell; and W, white matter layers. b, Bergmann glia; s, protoplasmic astrocytes; v, velate astrocytes; and f, fibrous astrocytes (From Kimelberg, 2010, Figure 1 on p. 83.) (D) Dye-filled astrocytes of rat hippocampus. (From Eric Bushong, personal communication.)

Muller cells as visualized by confocal reflection microscopy. Note, however, that for the dye-filled cells in Figure 1.1D the immunostaining for the astrocyte-specific protein glial-derived neurotrophic factor (GFAP) shows only a small part of the entire cell: the cell soma and the most proximal, widest parts of the processes. This is also shown later in Figure 1.4.

Now we know that the primary five to eight per cell processes emanate from the cell body (Nedergaard et al., 2003) of protoplasmic astrocytes:

1. They subdivide progressively to form finer and finer processes of up to 100,000 per astrocyte.
2. Most of these processes interact with synapses and can be termed perisynaptic astrocytic processes (PSAPs).
3. Fewer and thicker ones contact capillaries, arterioles, and venules, and can be termed perivascular astrocytic processes (PVAPs).
4. The remaining processes (average of 11; Xu et al., 2009) contact the processes of their nearest-neighbor astrocytes and form gap junctions with these to form the astrocyte syncytium.

This has led to several ideas and findings:

1. PSAPs take up products of neuronal activity as first speculated some time ago by Lugaro (1907). We now know that these products at the neuronal articulation (i.e., the synapse) are most importantly neurotransmitters. The major uptake system is for glutamate, consistent with the major glutamate-metabolizing enzyme glutamine synthetase and the major forms of the glutamate transporter family, GLT-1 and GLAST, being found only in astrocytes, and that ~90% of all synapses are glutamatergic. GABA uptake is also significant.
2. Currently it is being proposed that PSAPs also modulate synaptic activity by releasing glutamate, ATP, and D-serine (discussed below) to modulate transmitter release from the presynaptic terminal by acting on the extra-cleft neuronal receptors that do this type of thing, such as the adenosine receptor, since ATP is rapidly converted to adenosine in the extra cellular space (ECS).
3. PVAPs, mainly around arterioles, transduce neuronal activity sensed at the synapses by metabotropic receptors for glutamate on PSAPs, by a secondary increase in astrocytic intracellular Ca^{2+}, to release substances at the PVAPs, such as eicosanoids, to alter arteriolar diameter and therefore blood flow. These and the pericapillary endfeet can also modulate inflow and exit of substances to and from the brain.

A recent novel observation was that bushy astrocytes occupy their own domains, with only up to 10% overlap of their processes, as shown in the rat hippocampus (Bushong et al., 2002; Ogata and Kosaka, 2002). This may be a space-conserving process but should also have functional implications. One of these, I have suggested (Kimelberg, 2007, 2010), is that the up to 100,000 PSAPs function independently since the spread of information from several PSAPs to other sites through the processes will be random and governed by basic biophysical principles, such as diffusion, which will be inversely proportional to the cross-sectional area of the processes. However, the processes to the vasculature are bigger and fewer. This fits into an idea where one of the main functions of astrocytes is to transmit changes in synaptic activity to a signal to the vasculature to increase blood flow when neuronal synaptic activity, manifested as increased glutamate spillover, is increased. This is a type of wiring, but does each astrocyte form a separate functional unit for its own population of neurons, or more strictly its own population of synapses, or does it spread to other astrocytes? This is discussed below and shown diagrammatically in Figure 1.2, where it is contrasted to the classical polarization of the neuronal input-output elements. The intracellular signal that spreads among astrocytes is an increase in astrocytic Ca^{2+}, and the output can be a variety of agents that act on the vasculature (Takano et al., 2006; Gordon et al., 2008). There is good evidence for prostaglandins, but other compounds speculated on have not been conclusively ruled in or out, such as K^+, H^+, and adenosine (Iadecola and Nedergaard, 2007).

Recent observations that could be interpreted as supporting this view of vascular control as a primary role of astrocytes are studies in living ferrets that showed that Ca^{2+} increases in astrocyte cell bodies mapped with that of neurons upon activation of the ferret visual cortex according to the pinwheel orientation responding to different directions of the stimuli, with a few seconds delay for the astrocyte responses. It also correlated with increased blood volume as measured by the intrinsic optical signal (Schummers et al., 2008). Further, a small reduction in the neuronal response led to large reductions

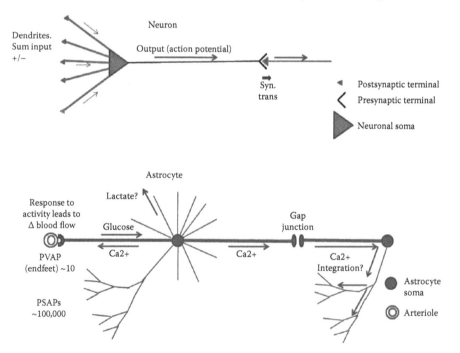

FIGURE 1.2 Input-output of neurons and astrocytes compared. See section 1.4.5, Control of Blood Flow.

in the astrocytic Ca^{2+} response, and "each astrocyte interacts quasi-independently with a small number of neurons surrounding it" (Schummers et al., 2008, 1640). Inhibition of the astrocyte response, which was shown to be achieved with glutamate transport blockers, inhibited the vascular and some of the neuronal responses. The last correlation could be interpreted as the astrocytes driving the neuronal responses, or alternatively causing the neuronal responses to quickly fade because the increased vascular response does not follow. The latter seems more consistent with the temporal delay.

The idea of the protoplasmic astrocytes controlling blood flow can be understood in terms of the Krogh tissue cylinder, which emphasizes that the ~50 μm average intercapillary distance results in exponentially decreasing levels of glucose and oxygen toward a midpoint inverse watershed point (Lubbers, 1977) (see Figure 1.3). This fits with both the astrocyte domain concept and its dimensions and the mean intercapillary distances in grey matter (Lubbers, 1977; Auen et al., 1979), so that the inverse watershed point roughly corresponds to the astrocyte soma. Figure 1.4 shows this for rat hippocampus where the capillaries outlined by AQP-4 can be seen by inspection to be separated from each other by distances of around 50 μm.

However, this Krogh cylinder does not fit control of blood flow, which mainly occurs at the levels of the arterioles whose walls contain smooth muscle (Iadecola and Nedergaard, 2007). These are outlined in Figure 1.4 by astrocytic processes that

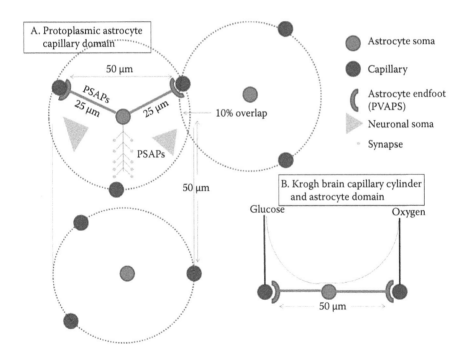

FIGURE 1.3 **(See color insert.)** (A) Domain concept of protoplasmic astrocyte related to distances between capillaries in the grey matter. (B) The Krogh cylinder referring to the exponential decline of oxygen and glucose with increasing distance from the capillary creating a low mid-distance level (see Lubbers, 1977), which would correspond approximately to the position of the astrocyte soma.

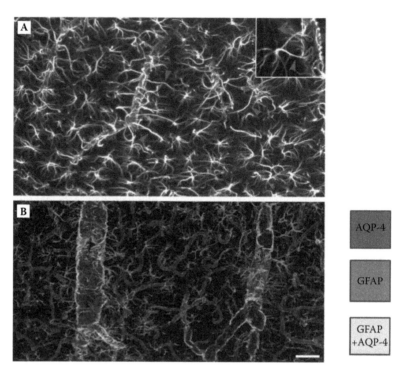

FIGURE 1.4   (See color insert.) AQP-4 and GFAP labeling of astrocytes in rat cortex. (A) GFAP immunolabeling of astrocytes. Individual astrocytes appear somewhat star shaped and are uniformly distributed, with little overlap with neighboring astrocytes. Astrocytic processes to blood vessels differ from other processes by being straight, unbranched, and of wider diameter (red arrowheads). The surfaces of large- to medium-size vessels were densely covered by GFAP+ astrocytic endfeet. Inset, an astrocyte with two vascular processes. (B) Double immunolabeling for AQP-4 (red) and GFAP (green). The larger and even smaller arterioles (precapillary sphincters; see text) are surrounded by endfeet that are positive for both aquaporin-4 and GFAP, possibly because the larger size of the processes leading to the arterioles allows GFAP filaments to localize there. AQP-4 is in the endfeet membranes. The capillaries are only AQP-4 positive, being covered by astrocytic processes, but these are GFAP negative, possibly because they are too small for the GFAP filament bundles to penetrate. Scale bar = 60 μm. (From Simard et al., *J. Neurosci.*, 23, 9254–9262, 2003. With permission.)

express membrane-bound AQP-4 and also show GFAP staining, and are 300–400 μm apart. Presumably these processes are large enough that the GFAP(+) intermediate filaments can be present, and this illustrates the omnipresent possibility that the molecular makeup, and therefore functions, of processes could vary. The increased Ca^{2+} that sums in many PSAPs due to glutamate spillover from many synapses activating astrocytic mGluRs would then diffuse through the respective processes linking these PSAPs to the astrocyte soma, and from there to other soma and their processes, and so on, so that a sufficiently large enough field of astrocytes is involved that some of them will have processes contacting arterioles, to release vasoactive agents to increase arteriolar diameter and blood flow. Since each astrocyte occupies a volume of diameter ~50 μm, this would be six to eight astrocytes at a minimum.

Recent work has, however, proposed that the pericytes in the capillary wall provide some control of capillary diameter (Hamilton et al., 2010), so that the control of blood flow by astrocytes could be at a more localized level.

A more complex informational integration role via effects at some synapses specifically affecting other synapses by messages conveyed through a single astrocyte or the astrocyte syncytium seems to founder on the factual rock that the processes leading to the PSAPs have not been shown to polarize in the way that dendrites and axons do (Figure 1.2).

All this refers only to the bushy protoplasmic astrocytes of the grey matter. To what extent the less exuberant process-bearing fibrous astrocyte of the white matter and the two elongate cells of the mammalian CNS, which I also classify as astrocytes, share these properties has yet to be resolved.

1.3   ASTROCYTES AS MODIFIED EPITHELIAL CELLS

Developmentally, the astrocyte can be viewed as a modified epithelial cell derived from the columnar epithelial cells that form the original neural ectoderm. Epithelial cells are characterized by asymmetric arrangement of ion channels and transporters for ions and other substances on the two faces of the epithelial cell that face two compartments. Short circuiting of the gradients between the two compartments produced by this arrangement is prevented by tight junctions between the epithelial cells. In terms of K^+, H^+, Cl^-, and HCO_3^- transport, I proposed some 30 years or so ago a polarized model for this (Figure 1.5).

One critical feature of an epithelial barrier that has been lost in the astrocytes is the tight junctions, so the astrocytes do not develop compartmental gradients but facilitate the movement of ions to and from the neurons and sequester ions such as H^+ and HCO_3^- to control the pH of the ECS.

1.4   THE ASTROCYTE SYNCYTIUM AND SIGNAL SPREAD VIA GAP JUNCTIONS

1.4.1   Magnitude of the Syncytium and Its Molecular Bases

As noted, one of the dominant characteristics of mature astrocytes is that they are linked by gap junctions to form extensive networks of cells that can be easily visualized by the diffusion of dye injected into one cell. This should not be taken as saying that astrocytes are the only neural cells linked in this way, but their linkage is by far the most extensive, and an average of 100 cells can be seen to be linked by the dye diffusion method (see Figure 1.6A).

These gap junctions are formed by connexin proteins that for mature mammalian astrocytes are both Cx 43 and 30. Several other chapters in this book discuss these in detail, as befits the editors having contributed much to this area. The dogma is that hemichannels are inserted into nonjunctional membrane and, then, diffused to the junctional domain from which they are removed by internalization, and there is no reuse after junctions are formed. If the hemichannels remain in the open state for any reasonable time, this will be deleterious to the cells (Spray et al., 2006).

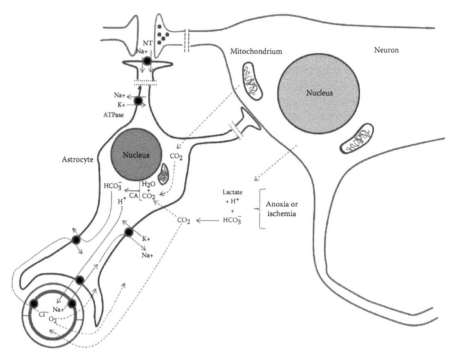

FIGURE 1.5 Model of selective localization of pH and ion transporters in a mature astrocyte linking blood vessels to synaptic and neuronal activity. Starting at the top and proceeding clockwise, the following transporters are depicted by circles with arrows to show direction of transport: neurotransmitter (NT) + Na^+ cotransporter; next, circle ($Na^+ + K^+$) pump and then a Na^+/H^+ and Cl^-/HCO^-_3 exchanger, followed by another depiction of the ($Na^+ + K^+$) pump. C.A. is carbonic anhydrase 2, and O_2 and CO_2 are shown as diffusing freely across membranes. Na^+/H^+ and Cl^-/HCO^-_3 exchangers are also shown located in the endothelial cells of a capillary (red). Under conditions of hypoxia or ischemia a net efflux of lactate + H^+ is shown coming from a neuron. We now know these to be obligatorily cotransported via a transporter (see text). Gap in astrocyte process leading to synapse indicates indeterminate distance. The increase in $H^+ + HCO^-_3$ derived from the increased CO_2, a reaction greatly accelerated in the astrocyte by C.A., was proposed to exchange for Na^+ and Cl^-, respectively, leading to cell swelling. This would therefore increase under anoxic or ischemic conditions. (Redrawn from Kimelberg, in *Neural Trauma*, ed. Popp AJ, Bourke RS, Nelson LR, Kimelberg HK, 137–153, Raven Press, New York, 1979.)

1.4.2 Functional Significances of Domain
and Syncytial Characteristics

The specificity of the neuronal unit is that the many dendritic inputs are summed to cause the firing of an action potential along the axon. So far, no similar polarity has been described for the numerous astrocytic processes so that messages from some processes cannot *specifically* travel to others to affect some integrative role between synapses. Thus, any summing, say of intracellular Ca^{2+} increases, would be random and not directed (see Figure 1.2).

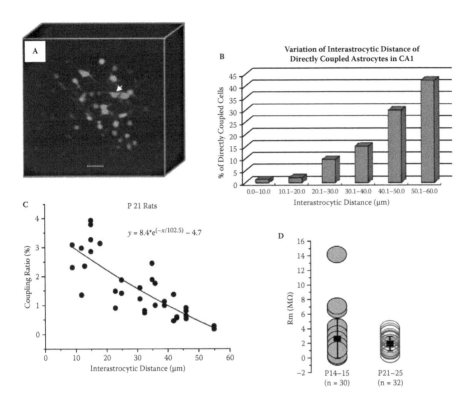

FIGURE 1.6 Transfer of electrical current between nearest-neighbor astrocytes. (A) Typical astrocyte syncytium revealed by injection of biocytin into the astrocyte marked with a white arrow in a P21 rat hippocampal slice, which showed a passive I-V relation indicative of mature astrocytes. Mean number of cells in a syncytium was 79. 3D confocal image. Scale bar = 20 μm. (B) Percentage of directly coupled astrocytes in the different interastrocytic soma distance ranges shown. (C) Exponential decline of coupling ratio (I injected into cell/I measured in nearest-neighbor cell × 100%) with distance. Neighboring cells whole cell patch clamped and amount of current needed to maintain Vc in second cell after injection of current into the other cell were measured. This approach avoids problems due to Rm ≪ Rs. (D) Variation of membrane resistance (Rm) in younger and older rats. Rm measured using membrane test protocol in PClamp 9.2 program, which corrects for high Rs. (From Xu et al., *Glia*, 58(4), 481–493, 2009. With permission.)

As already noted, the only way the astrocyte conforms to an input-output model is that there are many processes that subdivide into smaller and smaller processes that end as PSAPs, and a few larger processes that end on blood vessels as PVAPs (Figures 1.2 and 1.3). If sufficient PSAPs by the mostly glutamatergic synapses are activated, then a signal, presumably Ca^{2+}, will spread with an output selected by diameter based on ease of diffusion to the arterioles to affect a vasodilation. One caveat to this model is whether the intracellular Ca^{2+} "wave" is also regenerative by diffusion of IP_3 to effect continued release, but are there Ca^{2+} stores in especially the smaller-diameter processes?

What substances normally travel through the gap junctions, and does this affect the domain concept of astrocytes and functional implications that might flow therefrom? The syncytium has been and is still viewed as something that links individual astrocytes to achieve complex functions. The earliest of these was the potassium spatial buffering idea (Kuffler et al., 1966), and the latest a multisynaptic integrative function (Volterra and Meldolesi, 2005; Perea et al., 2009). After 50 years the first has not been resolved and the latter is a much more recent proposal that is still in its very early stages of formulation and experimentation. I now discuss both of these topics, as well as the issue of what signals (Ca^{2+}) or energy substrates and metabolites (as in the astrocyte-neuron lactate-shuttle hypothesis [ANLSH]) to pass through the syncytium. I also present alternative viewpoints that the gap junctions function as boundary markers for the individual astrocyte domains and that the astrocytic compartments linked by gap junctions provide a larger reservoir than can be afforded by one cell for dilution of substances taken up from the ECS that the astrocyte is regulating.

1.4.3 K⁺ SPATIAL BUFFERING

This was an ingenious proposal put forward by the group that first studied astrocyte-like cells electrophysiologically in an amphibian optic nerve, chosen for easier experimental control than the mammalian brain (Kuffler et al., 1966). Accounts of its history and attempts to support it experimentally have been repeated many times as it has remained, together with glutamate and GABA uptake, one of the two canonical functions of astrocytes. The fundamental finding was that the glia membrane potential was determined essentially exclusively by the transmembrane K^+ gradient, and this was later shown to apply to mature mammalian astrocytes in vivo (Somjen, 1975). Thus, a localized increase in $[K^+]_o$ would lead to a current inward flow (i.e., positive current) carried by K^+ at this site to outward flow, also carried by K^+, at some distant site.

There are several aspects of this model that remain unclear. First and foremost, we now know for sure that the resting membrane potential (Rm) of mature mammalian astrocytes in situ (so far studied for some hippocampal regions) is very low (Zhou et al., 2009). This means the length constant of the membrane is small and the current will not flow before the potential difference that drives the current is dissipated. How far, and is it far enough to be useful? However, the small voltage differences applied to give small currents to measure Rm are by patch electrodes located on the soma. The resistance of the processes could be much higher, but current flow is always through the path of least resistance. Thus one might have reasonable current flow across a length of processes to a capillary, but this would only be around 30 μm (see Figure 1.3). To travel farther it would always have to go through the low Rm soma.

Recently, we have measured transfer of current between astrocytes in rat hippocampal slices by dual-voltage clamp measurements and shown that transfer of current from one astrocyte to another single astrocyte is around 1%–2% of the injected current (Xu et al., 2009). This applies only to nearest-neighbor astrocytes, and there were an average of 11 of these (Figure 1.6). Current was never measured in cells farther away. So up to 20% of any current generated by a K⁺-dependent depolarization in one astrocyte cell body could be transferred to the nearest-neighbor astrocytes. $[K^+]_o$ increases would be mainly at the synapses, so the current transfer would then start at the PSAPs. Our measurements were, of necessity, made from one astrocyte soma to the next, and we could reasonably expect that this would be along only one process linked by gap junctions to a second process belonging to the next astrocyte, or a distance of around 50 μm. This might be of use physiologically for synaptic activities. Here small changes in $[K^+]_o$ occur on a millisecond timescale, and it could be of use if current carried by K^+ flows a few tens of microns away. However, the boundary of the increased $[K^+]_o$ will be an exponentially changing diffusional boundary. If the calculations are to be made, we will have to use an average value for $[K^+]_o$ at the 50% value from the peak. Faced with all this, we might expect a transporter-mediated uptake of K^+ plus an anion to be more effective and easier to understand. It could be local or widespread, depending only on the sites of the increases in $[K^+]_o$.

The one case where K^+ spatial buffering has been reasonably well shown to occur is in the elongate Muller cell, where the more epithelial shape of the cell is complemented by a characteristic epithelial asymmetric distribution of K^+ channels with regions of low K^+ density, and therefore high Rm, intervening between regions of higher K^+ channel density (Newman et al., 1984). Kir4.1 is one of the main K^+ channels, but also the recently identified two pore K^+ channels are present (Eaton et al., 2004). Such channels have also been identified in protoplasmic astrocytes of the rat hippocampus (Zhou et al., 2009), and it would be of interest to see how their localizations compare to the abundant Kir4.1 channels predominantly localized at the process endings (Butt and Kalsi, 2006).

1.4.4 Ca²⁺ Diffusion and Signaling

In situ in slices, but also seen in a few in vivo studies, spontaneous Ca^{2+} increases brain occur randomly in astrocytes in localized regions within the cell. Under intense applied neuronal stimulation the Ca^{2+} increase fills the entire cell but spreads only very slowly to the neighboring gap-junction-linked astrocytes (Fiacco and McCarthy, 2006). This is very different from the first observations in the GFAP(+) monolayer primary cultures, where stimulation of one cell, usually by poking it with an electrode, led to a wave of Ca^{2+} spread to many other cells (Cornell-Bell et al., 1990). Now it appears that the preferred route of spread is not, as one might expect, diffusion of Ca^{2+} or the second messenger, IP3, which releases Ca^{2+} from the intracellular stores, through the syncytium via gap junctions, but release of ATP, which either as itself or as adenosine activates metabotropic purinergic and adenosine receptors on the next cell (Nedergaard et al., 2003; Fiacco and McCarthy, 2006). This would appear to reduce the specificity and precision of the signal, as the ECS is full of molecular obstacles, such as abundant exo- and ecto-ATPases.

An intercellular Ca^{2+} spread (i.e., wave) is more likely to occur in astrocytes in primary culture because morphologically they are very different from astrocytes in situ. They form contact-inhibited monolayers where essentially all the entire perimeter of a single cell forms gap junctions with the neighboring cells. There are also many differences in gene expression (i.e., mRNA levels) between such cultures and astrocytes freshly isolated from the brain (Cahoy et al., 2008), and these should lead to protein differences and differences in properties dependent on these in situ vs. in culture. For example, in relation to gap junction properties the expression of Cx43 (Gja1) and Cx30 (Gjb6) mRNAs is 3- and 83-fold higher, respectively, for isolated astrocytes than for cultured astrocytes (Cahoy et al., 2008). Thus astrocytes in primary culture can only be partial and random predictors of astrocyte properties in situ, and are best avoided to determine astroglial properties, unless it is known that the specific property of the astrocyte occurs in situ, and the cultures robustly express that property (Kimelberg, 2010).

1.4.5 CONTROL OF BLOOD FLOW

As already noted in Section 1.2, if the Ca^{2+} increases seen in astrocytes in situ due to stimulation of their various metabotropic receptors are localized to individual astrocytes, this imposes certain constraints on astrocytic control of blood flow. Each astrocyte domain interacts at the ends of the finest astrocytic processes with up to 100,000 synapses, but only around 100 processes, or 3 orders of magnitude fewer, contact blood vessels, of which most will be capillaries. However, the control of blood flow is generally held to be at the precapillary arterioles (precapillary sphincter), which contain smooth muscle, while the capillaries only contain pericytes (Iadecola and Nedergaard, 2007). Whether these confer on capillaries the ability to maintain a tone that alters capillary diameter has been long debated and is still unresolved (Krueger and Bechmann, 2010; Hamilton et al., 2010; Dermietzel and Spray, this book). In Figure 1.4 we see that the larger-diameter vessels, presumably arterioles, and profusely covered with AQP-4(+) and GFAP(+) astrocytic foot processes, are 375 μm apart, providing enough space for seven or eight astrocytes to be interposed. Thus a signal restricted to one astrocyte will only have a one in seven chance of being next to an arteriole. There are two intermediate-sized vessels, also APQ4 and GFAP(+), that look like precapillary arterioles that approach to about 150 μm at the bottom of Figure 1.4B, so there may be a higher chance. If pericytes in capillaries can contract, then control would be at the level of a single astrocyte. Otherwise it seems that synaptic signals sufficient to generate an increase in blood flow will need to cause sufficient increases in astrocytic $[Ca^{2+}]_i$ over a synaptic field that may need to encompass up to seven or eight astrocytes to change blood flow. This model actually seems to be supported by the important in vivo studies of Schummers et al. (2008), discussed in Section 1.2.

1.4.6 LACTATE DIFFUSION

This is a specific model for how PVAPs (and this will apply mainly to capillaries, being the most numerous blood vessels) take up substrates from the blood and redistribute them throughout the brain via the other processes. It was first proposed by Golgi that such astrocytic PVAPs had nutritive functions of this nature based on the fact that his *reazione nera*, for the first time, showed the complete morphology of neurons and neuroglia (i.e., astroglia), and that some of their processes surrounded blood vessels (Kettenmann and Ransom, 2005).

In the astrocyte-neuron lactate-shuttle hypothesis (ANLSH) it is proposed that the most abundant energy-supplying substrate, glucose, passes first from the blood to the brain through the PVAPs. It is then converted to lactate, which exits the other astrocyte processes to be taken up by neurons, where it is converted to pyruvate, which is oxidatively metabolized to yield energy (Magistretti et al., 1999; Pellerin and Magistretti, 2004).

To determine whether the ANLSH works as proposed, we might first look at the molecular anatomy for the constituents of the system. However, such data provide only circumstantial evidence. The monocarboxylate lactate transporter (MCT) system for efflux, which preferentially transports lactate anion plus H^+ out of a cell, should be close to sites of uptake on neurons for maximum efficiency. Alternatively, the released lactate could be released from the entire and extensive astrocytic perimeter. For the human neocortical protoplasmic astrocyte soma this can be estimated based on a volume of 70,000 μm^3 per soma (Oberheim et al., 2006), giving a surface area of 8,210 μm^2. Due to the extensive process arborization of the astrocyte processes, the actual surface area could be ~100 times this or ~800,000 μm^2. This is a large number, but we do not know whether much of this membrane area lacks transporters and channels serving just to link the business ends of the processes to the indispensable cell soma. As noted already, the spatial segregation of astrocyte membrane proteins to different parts of the astrocyte processes, more than their identities, is difficult to resolve but essential for understanding function.

It has been reported that the pyruvate dehydrogenase isoform that preferentially converts pyruvate to lactate predominates in astrocytes, whereas the one that converts lactate to pyruvate is more abundant in neurons. As glycolysis proceeds in the cytoplasm and the pyruvate is shuttled into mitochondria, free diffusion of these products within the cytoplasm until they find the appropriate transporter might be a reasonable model. The average relative concentration of lactate and pyruvate will then determine what is moved and where according to the basic principles for transporters. For maximum efficiency of the proposed sequence of the ANSLH,

$$\text{glucose} \rightarrow \text{endothelial cells} \rightarrow \text{astrocytes} \rightarrow \text{lactate} \rightarrow \text{neuron} \qquad (1.1)$$

the glucose transporter should be preferentially localized to the PVAP membranes. The glucose transporter isoform (gene symbol SLC2), present in all neural cells, is the higher-affinity GLUT3, while the lower-affinity GLUT1 is localized on the endothelial cells to effect transport of glucose across the blood–brain barrier (BBB).

There are several problems with the ANSLH proceeding according to the obligatory sequence (1.1) shown above. First and foremost, both neurons and astrocytes express GLUT3, although whether there are specific plasma membrane sites whose locations fit the hypothesis has not yet been determined. At a minimum therefore, the sequence will not operate as an obligatory one but will be determined by metabolic conditions. It fits the Krogh cylinder (Figure 1.3B), but note that the low-watershed level is for oxygen as well as glucose, and the metabolism of lactate by neurons needs to be oxidative for (1.1) to be useful.

1.4.7 INTEGRATION OF NEURONAL ACTIVITY

This is a novel idea first proposed some 15 years ago (see recent review by Halassa and Haydon, 2010). It proposes release of transmitters, so far glutamate, ATP, and D-serine, by PSAPs through regulated vesicular release, as a further control of synaptic transmission. It has been termed the tripartite synapse, but this does not seem a very felicitous term, as the existence of PSAPs has been recognized for 100 years (Lugaro, 1907), and the term could as well be applied to support roles as to the more controlling role envisioned. I suggest the term *astrocytic synaptic modulation hypothesis*, with the acronym ASMH.

Synaptic transmission is initiated by release of transmitter from presynaptic nerve terminals by exocytosis of some of the transmitter-filled vesicles present there in high density. It occurs upon influx of Ca^{2+} by activation of voltage-sensitive Ca^{2+} channels in the memebrane due to the sudden depolarization caused by the arrival of an action potential. Classically, such release was proposed to be altered by transmitter diffusing from the cleft acting back on neuronal presynaptic receptors, often termed autoreceptors. Their activation can dampen further an excessive presynaptic release of transmitter. The ASMH 1 adds a further modulatory role by releasing transmitter present in the astrocyte controlled by an increase in Ca^{2+} due to activation of astrocytic metabotropic receptors, either by transmitter again diffusing from the synaptic cleft or by, for example, catecholamines released by nearby monoaminergic axons (Winder et al., 1996). The first could act as a booster or damper for the purely neuronal process, while the second suggests a way whereby the negative feedback process related solely to events at the synapse could be affected by other neuronal activities.

The ASMH clearly has merit in terms of suggesting additional control of synaptic activity, but it needs to be proposed in a precise way that can be tested and needs to fit with broader views of neural activity. There are a number of problems. For example, why if astrocytes have specific Ca^{2+}-vesicular-dependent release of transmitters, has a high density of vesicles, as seen in nerve endings, never been seen in the perisynaptic astrocytic processes (PSAPs)? Even if regulated exocytotis from PSAPs is demonstrated in vivo, what does this, by itself, imply? If the criterion is a fusion event, how does one distinguish between exocytosis for release of transmitter and insertion of a channel or transporter for release?

Should astrocytes be termed excitable because they have metabotropic receptors that, as in all cells, increase their intracellular $[Ca^{2+}]$? This confuses the term *excitability*, which has a precise meaning in neuroscience.

The proponents of the new tripartite concept also recognize that "astrocytes have important roles in key aspects of brain development and function, such as homeostasis of the extracellular milieu," but also propose that "they integrate and process synaptic information" (Perea et al., 2009). How does a single cell type do both? What will determine whether the perisynaptic astrocytic processes (PSAPs) take up glutamate released at the synapse or release it onto the synapse due to stimulation of astrocytic mGluRs under physiological conditions? How does a system that responds automatically to any changes in a feedback manner, which is how most biologists think of biological systems, including the brain, operate, distinguish these opposing things to lead to useful activities? It would seem that a plausible mechanism as to how astrocytes do all this should be proposed, and a clear problem that this mechanism clarifies be identified before we can take the idea seriously.

There are also some technical problems with the experiments that have been designed to test the hypothesis, and whose results are held to support the idea. For example, recent papers from the McCarthy laboratory (Fiacco et al., 2007, 2009b; Agulhon et al., 2008), who supplied the transgenic animals used in some of the studies interpreted as supporting the ASMH, do not reproduce, using other transgenic mice where astrocyte specificity is similarly produced, some of the critical published results that are interpreted as supporting the information processing hypothesis. It has been suggested that technical problems with the transgenic animals (Perea et al., 2009), which are very real (Fiacco et al., 2009a), could "account for these negative results." But then the same reservations need to be applied to all results obtained from transgenic mice, including those that are supportive. In addition to the problems with transgenic mice, some of the other manipulations, such as uncaging Ca^{2+} in the cell, are likely to lead to nonphysiologically large increases in $[Ca^{2+}]_i$, as McCarthy and colleagues (Agulhon et al., 2008) have pointed out.

In contrast to these critical comments on the ASMH, vesicular release of D-serine from astrocytes has recently been shown to be obligatory for induction of long term potentintion (LTP) in hippocampal slices (Henneberger et al., 2010). This finally demonstrates a function for the important observation, first made in 1995 (Schell et al., 1995), that D-serine was highly concentrated in astrocytes and could be released, albeit in primary cultures, by adding glutamate. This process can be most simply interpreted as follows: overflow of glutamate at active synapses, caused by the excitation required for LTP, stimulates mGluR receptors on astrocytes to increase astrocytic $[Ca^{2+}]_i$. This means that only activity above a level that will cause the requisite overflow of synaptically released glutamate will induce LTP. The study by Henneberger and collaborators also showed that under the intense stimulation conditions used for inducing LTP, it was still limited to the synaptic area covered by one astrocytic domain. This could only be done by observing the area where the induction of LTP could be reversed when a Ca^{2+} buffer or the exocytosis inhibitor tetanus toxin was injected into one astrocyte, as the intense stimulation of the Schaffer collaterals used to induce LTP activates fields of synapses that encompass many astrocytic domains.

Another feedforward system was recently reported for glutamate synapses in the hypothalamic paraventricular nucleus (Gordon et al., 2009), this time involving release of ATP from astrocytes. For this and the D-serine system the astrocytic release is acting as a booster for synaptic activity, but since this potentially "creates a feedforward loop prone to instability" (Gordon et al., 2009), there needs to be a negative feedback process also present to dampen the excitation. These are not yet identified. The latter finding is also at variance with reported inhibition of synaptic activity by astrocytic release of ATP that is then rapidly converted to adenosine (Halassa et al., 2009). Possibly, there are different effects at different synapses, which will then take forever to describe. The basic implication is that PSAPs can act as boosters or dampers at individual synapses. Whether this also becomes a permanent aspect of any neuronal circuit is yet to be determined.

1.5 CONCLUSIONS

So what is the most instructive way to view astrocyte functions to guide future studies? My view is obviously that they have support functions that are simpler to understand, and which themselves are better supported by the data, a little of which has been summarized and discussed in this chapter. But this is opinion, for we have no way of precisely concluding from it, or any other empirical database, a single incontrovertible conclusion. This inherent philosophical problem is exacerbated when we select only aspects of the database. We can recognize this problem in science. A scientist wishing to take a position will buttress it with relevant quotes of data from the literature and, much as a lawyer or, heaven forbid, a politician, or anyone else, will support a position with selected evidence. Scientists will also do experiments designed to support a hypothesis. But what about all the other evidence not quoted or the innumerable number of other experiments that could have been done?

This applies to things we cannot yet measure, such as the activities localized to the ~100,000 finest perisynaptic astrocytic processes (PSAPs) of the protoplasmic astrocytes. We can therefore say with Galileo, "Where the senses fail us reason must step in" (Galilei, 2010). There is then the separate problem of a lack of knowledge of all the data relevant to the question being asked. The ease of retrieval, as well as the amount of knowledge to be retrieved, has been greatly increased by the computer revolution, but such retrieval does not speak to reliability and relevance. Here the retriever needs the knowledge to select.

We ultimately rely in science on the testing of an idea, i.e., performing an experiment to support its validity. Indeed this is the powerful and distinguishing feature of science, the now dominant form of reasoning for understanding causes applied throughout most of the world. As Newton wrote concerning his new explanation of the visible spectrum:

> For the best and safest method of philosophizing seems to be, first to inquire diligently into the properties of things, and establishing those properties by experiments and then to proceed slowly to explanations of them. For hypotheses should be subservient only in explaining the properties of things, but not answered in determining them; unless so far as they may furnish experiments. For if the possibility of hypotheses is to be the test of the truth and reality of things, I see not how certainty can be obtained in any science. (Christianson, 1984, 165)

Here is a cogent description of the modern scientific method by its primary inventor. The language is arcane, so it may be in order to give explanations of what the most recondite phrases may mean. *Philosophizing* clearly means what we now term scientific inquiry. I think the term *measurements* should be used instead of *experiments* in the same sentence, for what is being done is to first describe properties—in modern parlance, establishing a database. Experiment (from the Latin word *experimentum*, a "trial or test" applies when one makes a measurement to see if a single prediction or, more globally, a mechanistic hypothesis remains possible because the measurements (i.e., experiments) show the practical result conforms to what we expect to happen. In this and the third sentence Newton additionally gets to the core of any objective inquiry, often now termed scientific, as in "what does the science show us?" It is simply that our measurements and observations should be made uninfluenced by prior bias as to how things work, and we should be willing to accept the death of a beautiful hypothesis by an ugly fact. However, the scientific method cannot be used like some engineering blueprint, and it is the unpredictable combination of these two selected views (i.e., the quotations from Galileo and Newton), which also contradict Galileo's precise measurements of things, like gravitational acceleration and Newton's first insightful hypothesis that gravity was universal and the earth's gravity controlled the motions of the moon, which reciprocally were responsible for tides, that led to groundbreaking advances, that often-quoted criteria for a truly excellent NIH grant, which of course is hardly, if ever, achieved, if indeed, given the foregoing, it can ever be discerned ahead of time. Further, NIH grants are now required, by program officers or the peer reviewers (I don't know which), to have a hypothesis to be tested, and these grants have therefore now devolved into an experimental program for a particular hypothesis. This is all well and good, but is bad news for those "cutting-edge areas" where the database is meager. Further, according to the scientific method, the experimental program should be stopped at the first experiment that shows the hypothesis is untenable, i.e., cannot be reasonably modified to accommodate contradictory data. The grant funds can then be better used to make other observations or test other ideas, but these will not have been "peer reviewed."

Therefore, what we can conclude as to what the accepted view of astrocyte function settles down to, as distinct from the confusion of the present, is something we can only now guess at. The astrocyte field has been and is replete with many ideas as to functions inferred from their appearances (used in a general sense to include all appearances: morphological, physiological, and biochemical), but very few have been *clearly* established when put to the test. This is a confusing aspect of astrocyte studies since, although not an overly large amount, considerable effort on these questions has been, and is being, expended. Is it that the most supported hypothesis of astrocyte function, namely, homeostasis of the neuronal environment, i.e., a vital support role, is too pedestrian for modern tastes? Is it that our modern complex methods of measurement and experimentation are so far too unclear, precisely because they are new and not time-tested, as to what they really show? Are we ignoring the obvious to choose the more obscure, so that we are too biased in what we test? These coupled with an apparent reluctance to always keep foremost in our minds that our experiments can never logically refute anything—we can rarely say that a hypothesis is clearly disproved and never that it is proved—make us accept things we should not accept, except with strong reservations, and cause all this confusion. The brain, in being unique among (human) organs in not being a machine, still defies our understanding in regard to its higher functions. Applying Ockham's razor, as I have done before (Kimelberg, 2007, 2010), the simplest, well-supported view of information flow and processing in the human brain is that they reside in the codes of the neuronal circuits, whose exquisite complexity and specificity require a corresponding exquisite control of the neuronal environment, i.e., the extracellular space, to which the plasma membrane transporter and channel activities of the mature astrocytes critically contribute.

REFERENCES

Agulhon C, Petravicz J, McMullen AB, Sweger EJ, Minton SK, Taves SR, Casper KB, Fiacco TA, McCarthy KD. (2008). What is the role of astrocyte calcium in neurophysiology? *Neuron* 59: 932–946.

Auen EL, Bourke RS, Barron KD, Filippo San BD, Waldman JB. (1979). Alterations in cat cerebrocortical capillary morphometrical parameters following K+ induced cerebrocortical swelling. *Acta Neuropathol* 47: 175–181.

Bushong EA, Martone ME, Jones YZ, Ellisman MH. (2002). Protoplasmic astrocytes in CA1 stratum radiatum occupy separate anatomical domains. *J Neurosci* 22: 183–192.

Butt AM, Kalsi A. (2006). Inwardly rectifying potassium channels (Kir) in central nervous system glia: a special role for Kir4.1 in glial functions. *J Cell Mol Med* 10: 33–44.

Cahoy JD, Emery B, Kaushal A, Foo LC, Zamanian JL, Christopherson KS, Xing Y, Lubischer JL, Krieg PA, Krupenko SA, Thompson WJ, Barres BA. (2008). A transcriptome database for astrocytes, neurons, and oligodendrocytes: a new resource for understanding brain development and function. *J Neurosci* 28: 264–278.

Christianson GE. (1984). *In the presence of the creator: Isaac Newton and his times.* New York: The Free Press.

Cornell-Bell AH, Finkbeiner SM, Cooper MS, Smith SJ. (1990). Glutamate induces calcium in cultured astrocytes: long-range glial signaling. *Science* 247: 470–473.

Eaton MJ, Veh RW, Makarov F, Shuba YM, Reichenbach A, Skatchkov SN. (2004). Tandem-pore K(+) channels display an uneven distribution in amphibian retina. *NeuroReport* 15: 321–324.

Fiacco TA, Agulhon C, McCarthy KD. (2009b). Sorting out astrocyte physiology from pharmacology. *Annu Rev Pharmacol Toxicol* 49: 151–174.

Fiacco TA, Agulhon C, Taves SR, Petravicz J, Casper KB, Dong X, Chen J, McCarthy KD. (2007). Selective stimulation of astrocyte calcium *in situ* does not affect neuronal excitatory synaptic activity. *Neuron* 54: 611–626.

Fiacco T, Casper K, Sweger E, Agulhon C, Taves S, Kurtzer-Minton S, McCarthy KD. (2009a). Molecular approaches for studying astrocytes. In *Astrocytes in (patho)physiology of the nervous system*, Parpura V, Haydon PG, eds., 383–405. New York: Springer.

Fiacco TA, McCarthy KD. (2006). Astrocyte calcium elevations: properties, propagation, and effects on brain signaling. *Glia* 54: 676–690.

Franze K, Grosche J, Skatchkov SN, Schinkinger S, Foja C, Schild D, Uckermann O, Travis K, Reichenbach A, Guck J. (2007). Muller cells are living optical fibers in the vertebrate retina. *Proc Natl Acad Sci USA* 104: 8287–8292.

Galilei G. (2010). http://thinkexist.com/quotes/galileo_galilei/.

Golgi C. (1885). Sulla fina anatomia degli organi centrali del sisterma nervoso. *Riv Sper Fremiat Med Leg Alienazione Ment* 11: 72–123.

Gordon GR, Choi HB, Rungta RL, Ellis-Davies GC, MacVicar BA. (2008). Brain metabolism dictates the polarity of astrocyte control over arterioles. *Nature* 456: 745–749.

Gordon GR, Iremonger KJ, Kantevari S, Ellis-Davies GC, MacVicar BA, Bains JS. (2009). Astrocyte-mediated distributed plasticity at hypothalamic glutamate synapses. *Neuron* 64: 391–403.

Halassa MM, Fellin T, Haydon PG. (2009). Tripartite synapses: roles for astrocytic purines in the control of synaptic physiology and behavior. *Neuropharmacology* 57: 343–346.

Halassa MM, Haydon PG. (2010). Integrated brain circuits: astrocytic networks modulate neuronal activity and behavior. *Annu Rev Physiol* 72: 335–355.

Hamilton NB, Atwell D, Hall CN. (2010). Pericyte-mediated regulation of capillary diameter: a component of neurovascular coupling in health and disease. *Front Neuroenergetics* 2: ii, 5.

Henneberger C, Papouin T, Oliet SH, Rusakov DA. (2010). Long-term potentiation depends on release of D-serine from astrocytes. *Nature* 463: 232–236.

Iadecola C, Nedergaard M. (2007). Glial regulation of the cerebral microvasculature. *Nat Neurosci* 10: 1369–1376.

Ihrie RA, Alvarez-Buylla A. (2008). Neural stem cells disguised as astrocytes. In *Astrocytes in (patho)physiology of the nervous system*, Parpura V, Haydon PG, eds., 27–47. New York: Springer.

Kettenmann H, Ransom B. (2005). The concept of neuroglia: a historical perspective. In *Neuroglia*, Kettenmann H, Ransom B, eds., 1–16. New York: Oxford University Press.

Kimelberg HK. (1979). Glial enzymes and ion transport in brain swelling. In *Neural trauma*, Popp AJ, Bourke RS, Nelson LR, Kimelberg HK, eds., 137–153. New York: Raven Press.

Kimelberg HK. (2004). The problem of astrocyte identity. *Neurochem Int* 45: 191–202.

Kimelberg HK. (2007). Supportive or information-processing functions of the mature protoplasmic astrocyte in the mammalian CNS? A critical appraisal. *Neuron Glia Biol* 3: 181–189.

Kimelberg HK. (2010). Functions of mature mammalian astrocytes: a current view. *The Neuroscientist* 16: 79–106.

Krueger M, Bechmann I. (2010). CNS pericytes: concepts, misconceptions, and a way out. *Glia* 58: 1–10.

Kuffler SW, Nicholls JG, Orkand RK. (1966). Physiological properties of glial cells in the central nervous system of amphibia. *J Neurophysiol* 29: 768–787.

Lubbers DW. (1977). Quantitative measurement and description of oxygen supply to the tissue. *Oxy Physiol Funct* 254–276.

Lugaro E. (1907). Sulle Funzioni Della Nevroglia. *Riv D Pat Nerv Ment* 12: 225–233.

Magistretti PJ, Pellerin L, Rothman DL, Shulman RG. (1999). Energy on demand. *Science* 283: 496–497.

Matyash V, Kettenmann H. (2010). Heterogeneity in astrocyte morphology and physiology. *Brain Res Rev.* 63: 2–10.

Nedergaard M, Ransom B, Goldman SA. (2003). New roles for astrocytes: redefining the functional architecture of the brain. *Trends Neurosci* 26: 523–530.

Newman EA, Frambach DA, Odette LL. (1984). Control of extracellular potassium levels by retinal glial cell K+ siphoning. *Science* 225: 1174–1175.

Nishiyama A, Komitova M, Suzuki R, Zhu X. (2009). Polydendrocytes (NG2 cells): multifunctional cells with lineage plasticity. *Nat Rev Neurosci* 10: 9–22.

Oberheim NA, Wang X, Goldman S, Nedergaard M. (2006). Astrocytic complexity distinguishes the human brain. *Trends Neurosci* 29: 547–553.

Ogata K, Kosaka T. (2002). Structural and quantitative analysis of astrocytes in the mouse hippocampus. *Neuroscience* 113: 221–233.

Pellerin L, Magistretti PJ. (2004). Neuroenergetics: calling upon astrocytes to satisfy hungry neurons. *Neuroscientist* 10: 53–62.

Perea G, Navarrete M, Araque A. (2009). Tripartite synapses: astrocytes process and control synaptic information. *Trends Neurosci* 32: 421–431.

Ramón y Cajal S. (1913). Contribucion al conocimento de la neuroglia del cerebro humano. *Trab Lab Invest Biol Univ Madrid* 11: 255–315.

Schell MJ, Molliver ME, Snyder SH. (1995). D-serine, an endogenous synaptic modulator: localization to astrocytes and glutamate-stimulated release. *Proc Natl Acad Sci USA* 92: 3948–3952.

Schummers J, Yu H, Sur M. (2008). Tuned responses of astrocytes and their influence on hemodynamic signals in the visual cortex. *Science* 320: 1638–1643.

Simard M, Arcuino G, Takano T, Liu QS, Nedergaard M. (2003). Signaling at the gliovascular interface. *J Neurosci* 23: 9254–9262.

Somjen GG. (1975). Electrophysiology of neuroglia. *Ann Rev Physiol* 37: 163–190.

Somjen GG. (1988). Nervenkitt: notes on the history of the concept of neuroglia. *Glia* 1: 2–9.

Spray DC, Ye ZC, Ransom BR. (2006). Functional connexin "hemichannels": a critical appraisal. *Glia* 54: 758–773.

Stebbing LS. (1954). *Logic in practice*. London: Methuen & Co. Ltd.

Takano T, Tian GF, Peng W, Lou N, Libionka W, Han X, Nedergaard M. (2006). Astrocyte-mediated control of cerebral blood flow. *Nat Neurosci* 9: 260–267.

Volterra A, Meldolesi J. (2005). Astrocytes, from brain glue to communication elements: the revolution continues. *Nat Rev Neurosci* 6: 626–640.

Wikipedia. (2010). Rudolf Virchow. http://en.wikipedia.org/wiki/Wikipedia.

Winder DG, Ritch PS, Gereau RW, Conn PJ. (1996). Novel glial-neuronal signalling by coactivation of metabotropic glutamate and b-adrenergic receptors in rat hippocampus. *J Physiol (Lond)* 494: 743–755.

Xu G, Wang W, Kimelberg HK, Zhou M. (2009). Electrical coupling of astrocytes in rat hippocampal slices under physiological and simulated ischemic conditions. *Glia* 58(4): 481–493.

Zhou M, Xu G, Xie M, Zhang X, Schools GP, Ma L, Kimelberg HK, Chen H. (2009). TWIK-1 and TREK-1 are potassium channels contributing significantly to astrocyte passive conductance in rat hippocampal slices. *J Neurosci* 29: 8551–8564.

2 Biomarkers of Astrocyte Microdomains

Focus on Gap Junctions, Purinergic Receptors, and Aquaporins

Eliana Scemes
Dominick P. Purpura Department of Neuroscience,
Albert Einstein College of Medicine, Bronx, New York

Grazia Paola Nicchia
Dipartimento di Fisiologia Generale ed
Ambientale, Facoltà di Scienze Biotecnologiche,
Università degli Studi di Bari, Bari, Italy

CONTENTS

2.1 THE MULTIDOMAINS OF ASTROCYTES

Astrocytes are highly elaborated and ramified cells and likely the only neural cell that simultaneously establishes close morphological associations with most cellular types. Their fine processes are in close proximity to astrocytes themselves, oligodendrocytes, neurons, and brain endothelial cells (Figure 2.1). Such close morphological associations, together with the fact that astrocytes express a multitude of ion channels, transporters, and membrane receptors, endows these cells with the unique capability to sense and influence diverse central nervous system (CNS) functions.

Distinct astrocyte processes are enriched with particular sets of marker molecules (Figure 2.1); for instance, the perisynaptic processes are extensively labeled with transmitter-related molecules such as glutamate and Gamma aminobutyric acid (GABA) transporters and glutamatergic, purinergic, and adrenergic receptors (Wolff and Chao 2004), while at the interglial and perivascular processes, gap junctions and aquaporins are the main biomarkers, respectively (Dermietzel 1974, Nielsen et al. 1997, Nagelhus et al. 1998, Rash and Yasumura 1999).

In this chapter we will focus on three groups of proteins that are abundantly expressed in astrocytes: the gap junctions, the purinergic P2 receptors, and the aquaporins.

2.2 GAP JUNCTION EXPRESSION AND FUNCTION

2.2.1 GAP JUNCTION FAMILY OF PROTEINS: CONNEXINS, INNEXINS, AND PANNEXINS

Gap junctions are morphological structures found at appositional membranes of cells characterized by the presence of hexagonal arrays of particles interposed by 2–4 nm intermembrane gaps. These structures were first seen in electron microscope studies in which lanthanum was used as a marker of extracellular space (Revel and Karnovsky 1967). In 1968, Revel coined the term *gap* junctions to these structures,

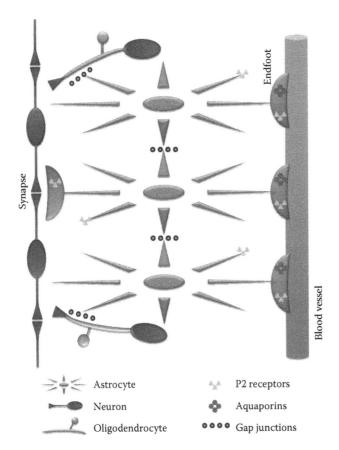

FIGURE 2.1 Astrocyte domains and biomarkers. Schematics of astrocyte domains in close morphological association with blood vessels, neurons, and oligodendrocytes. Gap junctions are shown at the glia-glia and glia-vasculature interfaces, aquaporin-4 is shown at the astrocyte endfeet surrounding the blood vessels, and P2 receptors are indicated at the perisynaptic and vasculature domains.

and later, in 1970, McNutt and Weinstein (1970) confirmed by freeze-fracture the presence of particles arranged as hexameric arrays within the gap junctions. The term *connexin* was proposed by Goodenough (1974) for the proteins composing the gap junctions that were purified from mouse liver. The first evidence for the formation of intercellular channels by gap junction proteins came from experiments in which induction of electrical coupling and the formation of membrane junctions were obtained following the incorporation into communication-deficient cells of liposomes containing mRNA extracted from myometrium expressing abundant gap junctions (Dahl et al. 1981).

It is now well established that direct communication between the cytosolic compartments of two or more adjoining cells is possible due to the presence of intercellular gap junction channels. Gap junctions are present in all tissues with the exception of sperm, erythrocytes, and mature skeletal muscle.

2.2.1.1 Connexins

There are about 20 gap junction (GJ) genes in humans and rodents (see Willecke et al. 2002) that encode the connexin (Cx) proteins. Individual connexin types are expressed in overlapping patterns in tissues, and different connexins are often coexpressed in the same cell type. For example, connexin43 (Cx43) is highly expressed in the heart, myometrium, and brain, and astrocytes predominantly express Cx43 (Figure 2.2A), but also Cx30 and Cx26, and oligodendrocytes express Cx32, Cx47, and Cx29, although the last of these probably does not form functional gap junction channels (Theis et al. 2005, Nagy and Rash 2003, Altevogt and Paul 2004).

Six connexin subunits oligomerize in the Golgi–endoplasmic reticulum (ER) to form the connexons that contribute one side of the gap junction channel when joined across the extracellular gap to a connexon from an adjacent cell (reviewed in Evans et al. 2006).

Analyses of membrane topology and hydrophobicity plots indicate that all connexins are tetraspan membrane proteins (segments M1, M2, M3, and M4) with intracellular C- and N-termini and two extracellular (E1 and E2) and one intracellular (CL) loop (Figure 2.2B). At the amino acid level, connexins share about 50% sequence identity. The transmembrane and extracellular residues are the most similar, while the cytoplasmic domains are the most divergent. The most divergent cytoplasmic domain of connexins in terms of its amino acid sequence and length is the carboxyl terminal. The extracellular loops are structurally conserved, with cysteine residues identically positioned in all connexins (three cysteines in each loop). These loops provide a high-affinity

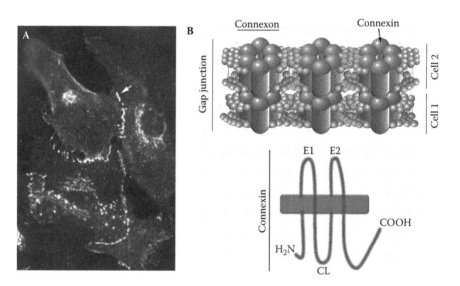

FIGURE 2.2 Astrocyte gap junction. (A) Epifluorescence images of astrocytes in culture showing Cx43 immunoreactivity. Arrow indicates Cx43 gap junctions between two astrocytes. (B) Schematic drawing of gap junction channels that are formed at the appositional membrane of two neighboring cells (cells 1 and 2) by the docking of two connexons (hemichannels). Each connexon is formed by six subunits, the connexins, which are tetraspan proteins, with cytoplasmic N- and C-termini, two extracellular loops (E1 and E2), and a cytoplasmic loop (CL).

intercellular interaction between connexons formed by a single connexin type (so-called homomeric, homotypic gap junctions) and also between many pairs of connexons formed by different connexin types (so-called heterotypic channels, with connexons formed by individual or multiple connexin types). Given the high affinity of the paired connexons, as evidenced by the requirement for hypertonic urea and ethylene glycol-bis (2-amino-ethylether)-N,N,N′,N′N-tetra-acetic acid (EGTA) treatment for appreciable dissociation in biochemical experiments (Ghoshroy et al. 1995), the turnover of gap junctions involves the incorporation of the neighbor connexon into one cell of the pair rather than splitting the connexin subunits (Larsen et al. 1979, Mazet et al. 1985).

The diameter of a gap junction channel is large, and all types of connexin channels allow direct communication of virtually all molecules with molecular weights less than 1,000 Da. This size limit allows diffusion through gap junctions of many molecules that function as second messengers, such as Ca^{2+}, inositol trisphosphate (IP_3), and 3′-5′-cyclic Adenosine monophosphate (AMP) and also such important metabolites as glucose, Adenosine-5′-triphosphate (ATP) and their degraded by-products. In addition to these moderately small molecules, there is recent evidence that larger linear molecules, such as polypeptides and RNAi, might "wiggle through" these channels, thereby directly controlling gene expression or antigen presentation (Valiunas et al. 2005, Neijssen et al. 2005).

Although gap junction channels are not absolutely selective for ions on the basis of their charges, certain types of connexins may provide some degree of selectivity with regard to whether anions or cations permeate the intercellular channels. Studies using dyes of different diameters and charge indicated that gap junctions differ in terms of their permeabilities, some being more permeable to cations than anions, such as the case of Cx45 ($P_{anion}:P_{cation}$ 1:5); others more permeable to anions, such as Cx32 ($P_{anion}: P_{cation}$ 2:1); and still others equally selective to cations and anions, such as Cx43 (Kanaporis et al. 2010; reviewed in Scemes and Spray 2008).

Despite the broad range of permeability, gap junctions impose a resistance to permeation, establishing gradients throughout a tissue. This resistance enables a high cytosolic concentration of gap-junction-permeant molecule in one cell to dissipate only gradually in time and space through the coupled network. Spatial gradients may be established either through generation of these molecules or in their uptake from or release to the extracellular space in a subpopulation of cells. Fluorescent dyes, most notably the highly charged anionic dye Lucifer yellow, have been extensively employed to determine the presence of gap junctional communication between cells, and the extent to which dye diffuses among coupled cells used to evaluate the degree of the so-called dye-coupling strength.

Gap junction channels are voltage gated. However, differently from other plasma membrane voltage-gated ion channels, they are insensitive to changes in plasma membrane potential (Vm), but are sensitive to transjunctional voltage (Vj). Maximal conductance of all gap junction channels is seen at zero transmembrane potential, and minimal conductance when transjunctional potential deviates from 0 mV, the degree of which is particular for each connexin type. This means that in the absence of a difference in membrane potential of two coupled cells ($V_{m1} = V_{m2}$; Vj = 0 mV), the gap junction channel will be fully open. Closure of astrocyte gap junction channels formed by Cx43 and Cx30 occurs when Vj > 70 and 50 mV, respectively (Beltramello et al. 2003, Bukauskas and Verselis 2004). Thus, under normal physiological conditions, the astrocyte gap junction channels are fully open, allowing for the direct transfer of ions and small molecules driven by their concentration gradient.

Besides being gated by voltage, all gap junction channels can be closed by acidification. Cx43 gap junction is closed by minor changes in intracellular pH since it has an apparent pKa of about 6.8. Differently from the voltage gating that drives the channels to a subconductance state, acidification leads to complete closure of gap junction channels. Such high sensitivity to intracellular acidification, as opposed to that of voltage gating, indicates that intracellular acidification is most likely one of the prevalent mechanisms that modulates gap junctional communication under patho(physio)logical conditions, such as in ischemic conditions.

2.2.1.1.1   Glial Gap Junctions

Extensive immunostaining for Cx43 has been reported within the domains of a single astrocyte both in brain sections and in cultured cells (Rohlmann and Wolff 1996, Wolff et al. 1998). These domains were termed autocellular zones by Rohlmann and Wolff (1996), consistent with their observation that the abundant localization of Cx43 within the autocellular space is due in part to small astrocytic processes forming autaptic gap junctions onto other fine processes or onto major branches of the same astrocyte. Such an arrangement is similar in principle to the autaptic or reflexive gap junctions formed between cytoplasm-containing regions squeezed off by compact myelin in Schmidt-Lantermann incisures and paranodal loops of myelinating Schwann cells and in paranodal regions of oligodendrocytes (Spray and Dermietzel 1995, Scherer et al. 1999). The function of such junctions is presumed to lie in nutrient delivery from the nucleus to the innermost regions of the Schwann cell (or alternatively, for exchange of signaling molecules generated at the contacts between axons and the innermost Schwann cell membrane with the Schwann cell body), thereby shunting the tortuous route created by as many as 100 wraps of the Schwann cell around the axon (Balice-Gordon et al. 1998). A somewhat similar role might be played by reflexive gap junctions in astrocytes, where changes in coupling strength might isolate or integrate microdomains, as could occur where fine astrocytic processes surround synapses. For example, functional microdomains have been demonstrated in Bergmann glia, to which the spread of Ca^{+2} elevations is regionally limited in response to neuronal activity (Grosche et al. 1999).

Although processes of neighboring astrocytes were not found to penetrate the outer processes of adjacent astrocytes, oligodendrocyte processes freely intermingled within the autocellular zone. Astrocytes and oligodendrocytes are coupled by gap junctions (Dermietzel 1998, Nagy and Rash 2000, Massa and Mugnaini 1985). Direct oligodendrocyte-astrocyte gap junctions have long been known from thin-section and freeze-fracture studies (Massa and Mugnaini 1985), and their abundance has been confirmed recently using freeze-fracture immunolabeling (FRIL) (Nagy and Rash 2000), giving rise to the concept of global functional intercellular communication throughout the brain, the so-called panglial syncytium (Rash et al. 1997).

A number of different types of gap junction proteins (connexins) connect astrocytes and oligodendrocytes among themselves and between one another. The gap junction gene family currently has about 20 identified members in mammals, more than half of which are found in the brain (Willecke et al. 2002, Spray et al. 2004). The primary gap junction protein in astrocytes from neonate mice is Cx43. Although Cx26, Cx30, Cx40, Cx45, and Cx46 have also been reported to be expressed in astrocytes in vivo or in culture, total junctional conductance between cortical astrocytes from Cx43-null mice is 95% less than between astrocytes from wild-type siblings (Scemes et al. 1998), indicating that this gap junction protein ordinarily supplies the vast majority of the intercellular communication. During CNS development, however, the levels of Cx43 seem to decrease while those of Cx30 increase (Cotrina et al. 2001). Interestingly, using a conditional Cx43-null mice (GFAP-Cre:Cx43$^{f/f}$) a substantial increase in Cx30 expression levels was found, such that the degree of coupling was reduced by 50% in GFAP-Cre:Cx43$^{f/f}$ astrocytes (Theis et al. 2003).

In oligodendrocytes, Cx32 and Cx45 were reported initially (Dermietzel et al. 1997, Kunzelmann et al. 1997), although it now appears that the Cx45 probes used in those studies may have cross-reacted with the newly identified Cx47. Although Cx47 was initially described as neuronal in its distribution (Teubner et al. 2001), it now appears to be mainly (or exclusively) expressed in myelinating cells of the CNS and peripheral nervous system (PNS) (Magnotti et al. 2011). Because the most abundant astrocytic connexin, Cx43, does not form functional channels when paired with cells expressing Cx32 (White and Bruzzone 1996), it seems likely that astrocytic-oligodendrocytic coupling is established by heterotypic gap junctions, with Cx43 on the astrocyte side and Cx47 contributed by the oligodendrocytes (Magnotti et al. 2011). Such connections between oligodendrocytes and astrocytes may explain in part the abundance of Cx43 staining within the autocellular astrocytic zones, although it is certainly likely that the demonstrated autaptic contacts between the fine processes of a single astrocyte also contribute to the Cx43 staining in these zones.

2.2.1.2   Innexins, Pannexins, and Hemichannels

The proteins forming gap junction channels in invertebrates are the innexins (Phelan 2005, Hua et al. 2003); they share no sequence homology and differ topologically from connexins in that they have longer extracellular loops, generally a larger cytoplasmic hinge region, and only two cysteines in each extracellular loop. Innexins expressed in the same organism are more closely related to each other than they are to those expressed in different organisms. A database search for innexin homologues in vertebrates identified a group of proteins sharing low but significant sequence homology with innexins (Baranova et al. 2004). Because two innexin-like sequences were found in the human genome using a polymerase chain reaction (PCR) sequence derived from degenerate primers to clone a molluscan gap junction protein (Panchin et al. 2000), the possibility was raised that these new genes corresponded to the missing links between the invertebrate and vertebrate gap junction proteins; thus they were collectively called pannexins. However, because vertebrate pannexin sequences are so different from those of invertebrate innexins, most authors continue to separate the gene families.

There are three identified pannexin (Panx) members: Panx1, Panx2, and Panx3. Pannexins, like connexins and innexins, are tetraspan membrane proteins with cytoplasmic N- and C-termini and two extracellular loops. Recent studies indicate that Panx1, like connexins and innexins, forms hexameric oligomers, while Panx2 forms octameric structures (Boassa et al. 2007, Ambrosi et al. 2010). Panx1 is ubiquitously expressed, Panx2 is exclusively present in CNS neurons, and Panx3 has restricted expression, being found in cartilage, skin, and osteoblasts (Baranova et al. 2004, Panchin 2005).

It is now well accepted that pannexins do not form gap junctions and intercellular channels but instead form plasma membrane channels, at least in the case of Panx1. A likely explanation for the lack of formation of gap junction channels regards the presence of only two cysteine residues in each extracellular loop and the existence of a glycosylation site (N254) in the second extracellular loop (Penuela et al. 2007, 2009, Boassa et al. 2007). Initial studies performed in *Xenopus* oocytes, however, have indicated that, with suitable Panx1 overexpression, gap junction channels may be formed in this preparation (Bruzzone et al. 2003). Presumably in this system, the secondary modifications that serve to separate potentially interacting membrane domains are not fully competent, thus allowing for the docking of two pannexon channels.

Among the three pannexins, Panx1 is the best characterized. Pannexin1 forms large conductance channels (400–500 pS) that are permeable to relatively large molecules, such as ATP. They are voltage-dependent channels that open upon membrane depolarization above 0 mV, are mechanosensitive responding to mechanical stretch of the plasma membrane, and can be activated following purinergic P2 receptor stimulation and elevation of cytosolic calcium levels. Among the P2 receptors, pannexin1 has been shown by coimmunoprecipitation assays to associate with the $P2X_7$ receptor (Pelegrin and Surprenant 2006, Iglesias et al. 2008) and to provide the membrane permeabilization pathway that follows $P2X_7$ receptor activation (Pelegrin and Surprenant 2006, Locovei et al. 2007).

Pannexin1 is present in neurons and glia (oligodendrocytes and astrocytes) (Bruzzone et al. 2003, Huang et al. 2007, Locovei et al. 2007, Iglesias et al. 2009), where it contributes to paracrine signaling. At present, only in vitro studies have investigated the functional role of astrocyte pannexin1. Through the use of shRNA strategy and pharmacological approaches it was shown that astrocyte Panx1 channels contribute to the transmission of calcium waves by providing sites of ATP release (Scemes et al. 2007, Iglesias et al. 2009).

Pannexin1 has also been shown to play important roles under inflammatory conditions. Differently from macrophages and other cells of the immune system, where $P2X_7$ receptor and Panx1 act together to induce inflammasome formation and cytokine release (Pelegrin and Surprenant 2006), in astrocytes and neurons, inflammasome formation can be induced solely by Panx1 activation (Silverman et al. 2009). Coimmunoprecipitation assays using Panx1 antibodies showed that several components of the inflammasome (NLPR1, ASC, caspase-1, caspase-11, XIAP, $P2X_7R$) were pulled down together from whole cell lysates of neurons and astrocytes. Moreover, activation of caspase-1 in astrocytes and neurons was achieved following Panx1 activation by elevated extracellular potassium and was prevented in Panx1 knockdown cells (Silverman et al. 2009).

Hemichannels, or connexons, are half gap junction channels (Figure 2.2A,B) (for reviews see Bennett et al. 2003, John et al. 2003, Contreras et al. 2004, Evans et al. 2006, Goodenough and Paul 2003, Martin and Evans 2004, Parpura et al. 2004, Saez et al. 2005, Verselis et al. 2000). When open, these channels would connect a cell's interior to extracellular space, a profound functional distinction compared to open gap junctions. Because gap junction channels are such large and rather nonselective pores, opening of hemichannels to the extracellular environment would be expected to be catastrophic, causing not only the collapse of ionic gradients necessary for maintenance of resting potential and transport, but also the loss of precious metabolites, energy sources, and diffusible second-messenger molecules. However, if hemichannel openings were brief enough or controlled, these channels could conceivably provide a pathway for release (or uptake) of large molecules and ions. Such a role has been proposed for Cx43 hemichannel opening in release of glutamate and ATP (and certainly other biological active molecules as well) from astrocytes, which could have a major impact on glial-glial and glial-neuronal interactions.

As mentioned above, pannexins do not form gap junction channels, and therefore the term *hemichannel*, although extensively used for pannexin channels, is used here only when referring to connexin channels.

2.2.2 Roles of Gap Junction Channels, Hemichannels, and Pannexin Channels

Gap junctions connect astrocytes into functional networks, and the purpose of this strong astroglial coupling has long been hypothesized to provide for the direct intercellular exchange of ions, nutrients, and signaling molecules, thus contributing to the spatial buffering of extracellular potassium, providing metabolic cooperation among the interconnected cells and for the long-range transmission of calcium signals. Hemichannels and pannexin channels, in contrast, are proposed to provide sites for the release of molecules from cells, thus contributing to autocrine/paracrine signaling.

2.2.2.1 Potassium Buffering

The spatial buffering of the extracellular K^+ concept was first proposed by Dick Orkand, John Nichols, and Steve Kuffler (Orkand et al. 1966, Kuffler et al. 1966). In this model, the removal of excess K^+ released at synaptic sites (source) to regions of low extracellular K^+ (sink) was proposed to be performed by glial cells. Experimental evidence for such enrollment of glial cells came from experiments performed in Mueller cells showing that these modified retinal glial cells transported K^+ from the plexiform layers to the vitreous body, blood vessels, and subretinal space (Newman 1985, Reichenbach et al. 1992, Newman and Reichenbach 1996). Although the Na^+/K^+ ATPase and the KCl transporter may participate in K^+ redistribution (Walz 2000, Chen and Nicholson 2000), two different subtypes of inwardly rectifying K^+ channels expressed in Mueller cells seem to play a role in siphoning of K^+ into glia (Newman 1985, Reichenbach et al. 1992, Kofuji et al. 2002). These are (1) the strongly rectifying K^+ channel (Kir2.1), highly expressed in membrane domains of the Mueller cells extending into the (source) plexiform layers that would mediate the influx of K^+ into the Mueller cell, and (2) the weakly rectifying K^+ channel (Kir4.1), expressed predominantly at (sink) the endfeet located at the vitreous body and blood vessels, which will favor the efflux of K^+ from the Mueller cells (Kofuji et al. 2002).

Differently from the K^+ siphoning model of Mueller cells, gap junctions between CNS astrocytes have been proposed as being essential for spatial buffering (Orkand et al. 1966). In various brain regions, Kir4.1 immunoreactivity was found in about half of the astrocytes, being mainly expressed at regions where the astrocytic endfeet meet the blood vessels (Takumi et al. 1995, Poopalasundaram et al. 2000, Higashi et al. 2001, Schroder et al. 2002, Kofuji et al. 2002), whereas Kir2.1 transcripts were observed in subpopulations of astrocytes in different brain regions (Kofuji et al. 2002, Schroder et al. 2002). It is conceivable that the distribution of these two inward K^+ channels is confined to different subpopulations of astrocytes in the brain, and that gap junction channels, by linking the members into a network, would assemble the different specialized membrane areas. This arrangement would form a multicellular astrocytic network functionally equivalent to a Mueller cell. Although the space constant for spatial buffering is unknown, it has been estimated that the astrocytic syncytium is far more likely to support K^+ transport than the extracellular diffusion (Gardner-Medwin 1983, Gardner-Medwin and Nicholson 1983) and propagates K^+ waves faster than the extracellular space (Amzica et al. 2002). Gap junction channels are permeable to K^+, and thus are likely to provide the astrocytic network with the necessary volume to accommodate the focal influx of high concentrations of K^+. In this regard, it has been shown that exposure of astrocytes to high K^+ solutions increases the coupling strength (Enkvist and McCarthy 1994, De Pina-Benabou et al. 2001), thus expanding the effective volume of the interconnected astrocytes while maintaining a high surface area for K^+ uptake.

Only recently, experimental evidence was provided in support for a role of gap junction in the redistribution of K^+ (Wallraff et al. 2006). Using the double (Cx43 and Cx30) knockout mice, the laminar profile of increased $[K^+]_o$ from the hippocampal *stratum pyramidale* throughout the *stratum radiatum* and *stratum lacunosum moleculare* was investigated electrophysiologically. The authors found no difference in the redistribution of $[K^+]_o$ within the *stratum radiatum,* but altered $[K^+]_o$ redistribution in the *stratum lacunosum moleculare* of the double knockout compared to that of wild-type mice (Wallraff et al. 2006). In view of these results, they proposed an unequal contribution of gap junctions to radial transport of $[K^+]_o$, depending on the size and orientation of astrocytes within the *stratum radiatum* and *stratum lacunosum moleculare*. Because single astrocytes in the stratum radiatum span are much larger areas than those in the *stratum lacunosum moleculare* and are perpendicularly orientated with regard to the *stratum pyramidale* (Nixdorf-Bergweiler et al. 1994, Wallraff et al. 2006), these stratum radiatum astrocytes could move K^+ radially, without the participation of gap junctions. Lacunosum astrocytes, on the other hand, show no preferential orientation, and their processes are less extended compared with radiatum astrocytes (Nixdorf-Bergweiler et al. 1994, Wallraff et al. 2006). Such differences might explain why, in certain astrocyte networks, gap junctions play a role in radial transport of K^+.

It is interesting to note that the so-called shape of astrocyte networks, as measured by the pattern of stained cells following the diffusion of dyes along the interconnected astrocytes, varies in different brain regions (Houades et al. 2006). Such differences, which likely reflect distinct degrees of coupling within a single network of cells, may have significant impact to the overall directionality of transport of ions and signaling molecules within interconnected astrocytes.

2.2.2.2 Metabolic Cooperation

Glucose, together with lactate, ketone bodies, and glutamate/glutamine, is the main source of energy utilized by the brain (Williamson 1982, Lopes-Cardozo and Klein 1985, Vicario et al. 1993). Although glucose is the only blood-borne substrate used by the brain as an energy source, lactate and other nonoxidized products of glucose metabolism are consumed as fuel during elevated brain activity (Dienel and Hertz 2001). A selective transport system localized in the blood–brain barrier (BBB) provides glucose to the cells within the brain. Given that astrocytes are interposed between the capillaries and neuronal elements, and thus are the first cellular elements that glucose entering the brain encounters after crossing the endothelium of the BBB, astrocytes were assigned a nutritive role for neurons. Astrocytes express the glucose transporter Glut1 (Maher et al. 1994) and store glucose in the form of glycogen. Given that astrocytes lack the enzyme that transforms glucose-6-phospate into glucose during glycogen breakdown, the glycolytic product lactate can be delivered from astrocytes to neurons as a source of energy (Giaume et al. 1997). Glutamate, the main excitatory neurotransmitter released by CNS neurons, is taken up by astrocytes and, together with ammonia, is converted into glutamine, which is then delivered to neurons (Broer and Brookes 2001). Although glucose, lactate, glutamate, and glutamine can diffuse through the extracellular space and then be taken up by neuronal cells, it is also likely these fuels travel more efficiently through the astrocytic network, by diffusing through gap junction channels (for review see Dienel and Hertz 2001). Permeability of gap junction channels to 2-deoxyglucose has been demonstrated in cell lines (Pitts and Finbow 1977), in the lens (Goodenough et al. 1980), and between smooth muscle cells (Cole and Garfield 1986, Cole et al. 1985). The passages of glutamine and glutamate, and lactate and glucose through the astrocytic gap junction were shown using the scrape-loading technique applied to cultured astrocytes (Tabernero et al. 1996, Giaume et al. 1997) and in vivo by the use of gap junction channel blockers showing a 50% reduction of (^{14}C)glucose spread through the brain of a conscious rat (Cruz et al. 1999, Dienel et al. 2002).

Recently, astrocyte gap junctional communication has been shown to contribute to the supply of glucose to neurons in an activity-dependent manner. By analyzing the extent of diffusion of a fluorescent glucose molecule 2-(N-(7-nitrobenz-2-oxa-1,3-diazol-4-yl)amino)-2-deoxyglucose (NBDG) through the astrocyte network in hippocampal slices, Rouach's group (Rouach et al. 2008) found that under epileptiform activity of CA1 pyramidal neurons, there was an increased diffusion of 2-NBGD among the astrocytes, while a reduction in the extent of the diffusion of this glucose analogue was seen after blocking neuronal activity with tetrodotoxin TTX. Since the diffusion of the dye Lucifer yellow was unchanged under these same conditions, their results indicate that the altered diffusion of 2-NBGD was not due to changes in the strength of dye coupling but to the altered glucose consumption by the neurons. These findings, together with the one showing that the impairment of 2-NBDG diffusion among the astrocytes lacking both Cx30 and Cx43 influences synaptic activity (Rouach et al. 2008), showed that astrocyte gap junctions are an important route for the transfer of glucose from the vasculature to the neurons that sustain neuronal activity.

2.2.2.3   Intercellular Communication by Calcium Waves

Intercellular calcium wave spread is a phenomenon characterized by an increase in cytosolic calcium levels within one cell that is followed by increased intracellular Ca^{2+} levels in adjacent cells in a wave-like, propagating fashion. Coordinated endogenous neural activity in the form of propagating calcium waves has been observed in developing retina, cortex, and hippocampal slices (Galli et al. 1988, Meister et al. 1991, Feller et al. 1996, Yuste et al. 1992, Kandler and Katz 1998, Garaschuck et al. 2000). Using brain slices from transgenic mice in which astrocytes were labeled by the green fluorescent protein (GFP) expressed under the *glial fibrillary acidic protein* (GFAP) promoter, it was shown that electrical stimulation induced intercellular wave propagation that spread not only through GFP fluorescent cells, but also to other nonastrocytic glial cells over a distance of 100 μm (Schipke et al. 2002). Propagating intercellular calcium waves have been observed to occur not only following mechanical, electrical, and chemical stimulation, but also spontaneously, independent of neuronal activity. In the presence of TTX to block synaptic activity, astrocytes from the ventrobasal thalamic slices of rat were shown to display spontaneous Ca^{2+} oscillations (Parri et al. 2001, Tashiro et al. 2002) that could be propagated to other astrocytes and induce neuronal activity along the wave path (Parri et al. 2001). Modulatory actions of neuronal activity corresponding to calcium wave spread between astrocytes have been shown to be related to the release of glutamate and ATP from these glial cells acting on glutamatergic and purinergic receptors expressed in neuronal plasma membranes (Parpura et al. 1994, Kang et al. 1998, Fields and Stevens-Graham 2002).

Besides this intracellular gap-junction-mediated pathway, an extracellular pathway can contribute to the communication of Ca^{2+} waves, working in parallel with the intercellular pathway. This paracrine route involves release of an extracellular messenger, such as ATP, other adenosine nucleotides, glutamate, or other neurotransmitters or hormones that relay the Ca^{2+} signal through the activation of cell surface membrane receptors. A variety of distinct mechanisms are involved in the release of gliotransmitters from astrocytes, including regulated exocytosis, and ion channels (Deitmer et al. 2006). Among the ion channels, hemichannels have been proposed as sites for ATP and glutamate release from astrocytes (Stout et al. 2002, Kang et al. 2008, Ye et al. 2003). These studies showed that, under conditions in which hemichannels were believed to be open, the compounds known to block gap junction channels reduced the amount of ATP and glutamate released from astrocytes. However, because gap junction channel blockers have targets other than connexins, including the $P2X_7$ receptor-permeabilization pore, pannexin1 channel, and swelling activated anion channels (Bruzzone et al. 2005, Suadicani et al. 2006, Ye et al. 2009), such a pharmacological approach is not an ideal way to identify permeation pathways. Using transgenic mice lacking Cx43, it was shown that astrocytes were still able to take up membrane-impermeable dye and to release ATP, and that this permeation pathway was obliterated following the knockdown of Panx1 from Cx43-deficient astrocytes (Iglesias et al. 2009). Moreover, knockdown of astrocyte Panx1 by an siRNA strategy, but not the deletion of Cx43, prevented the expansion of intercellular calcium waves seen under low extracellular divalent cation concentrations

(Scemes et al. 2007, Suadicani et al. 2006), a condition reported to promote hemi-channel activity (Stout et al. 2002). Thus, it seems more likely that pannexin1 channels and not Cx43 hemichannels are the sites for the release of gliotransmitters from astrocytes that contribute to the transmission of intercellular calcium waves.

The relative contribution of each of the two pathways involved in the transmission of calcium waves among astrocytes is affected by the CNS environment. Under certain pathological conditions, such as inflammation, a switch from a gap-junction-dependent to a gap-junction-independent, purinoceptor-dependent mechanism for calcium wave propagation was observed. Following interleukin 1-beta treatment of human fetal astrocytes, which decreased Cx43 expression levels and upregulated P2 receptors, calcium wave transmission was solely sustained by the extracellular pathway (John et al. 1999). A similar switch from gap-junction-dependent to purinergic-receptor-mediated transmission of calcium waves was reported to occur in cultured spinal cord astrocytes from Cx43 knockout astrocytes and in wild-type cells treated with Cx43 antisense oligonucleotides (Scemes et al. 2000, Suadicani et al. 2003).

Thus, although intercellular calcium waves among astrocytes may be sustained under diverse conditions, changes in the mode by which these waves travels may have distinct consequences, given that activation of each purinergic receptor subtypes is linked to different signaling transduction pathways, as described below.

2.3 PURINERGIC RECEPTORS

2.3.1 The Family of Purine- and Pyrimidine Nucleotide-Sensitive Receptors

Purines (adenosine, Adenosine 5'-diphosphate (ADP), Adenosine-5'-triphosphate (ATP)) and pyrimidines (Uridine 5'-diphosphate (UDP), Uridine 5'-triphosphate (UPT)) Nucleotides as extracellular signaling molecules mediate several biological functions by acting through membrane receptors called purinergic receptors. These

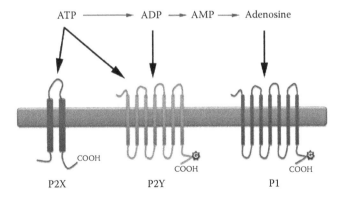

FIGURE 2.3 P2 and P1 receptors. Schematic drawing showing two groups of P2 receptors, the ionotropic P2X and the metabotropic, G-protein-coupled P2Y receptors, and the P1 receptor group comprised by the adenosine receptors. Both P2X and P2Y receptors are activated by ATP, while certain P2Y receptors are also sensitive to the ATP by product ADP. P1 receptors are G-protein-coupled receptors and respond to ATP end product adenosine.

receptors are divided in two main categories, the P1 and P2 receptors (Figure 2.3). P1 receptors are adenosine-sensitive receptors that comprise four subtypes: A1, A2A, A2B, and A3 receptors, all of which are G-protein-coupled seven-transmembrane-domain metabotropic receptors. The P2 receptors, which are mainly sensitive to ATP, are formed by two classes of proteins: ionotropic P2X receptors (ligand-gated cation channels) and the metabotropic P2Y (G-protein-coupled) receptors. There are seven P2X (P2X$_{1-7}$) and eight P2Y (P2Y$_{1,2}$, P2Y$_4$, P2Y$_6$, P2Y$_{11-14}$) receptors.

Astrocytes in vitro and in situ have been reported to express all subtypes of P1 and P2 receptors, except P2Y$_{11}$ receptor (reviewed in Verkhratsky et al. 2009). Adenosine receptors in astrocytes are implicated in the control of carbohydrate metabolism, astrogliosis, and in the release of neuroactive substances. Activation of adenosine A2B receptors in astrocytes mediates rapid glycogenolysis and transcriptional changes in genes encoding proteins regulating energy metabolism (Allaman et al. 2003). Following traumatic brain injury, the high levels of adenosine lead to upregulation of astrocyte intermediate filaments' GFAP and vimentin as well as to the increase in the number of GFAP positive cells and hypertrophy of astrocytes; such astrogliosis is mediated by both A2B and A3 adenosine receptors (Abe and Saito 1998, Abbracchio and Burnstock 1998). On the other hand, physiological levels of adenosine (10 nM), can mediate through activation of A2A receptors the release of glutamate from astrocytes, either inhibiting glutamate transporter GLT-1 or through a GLT-1-independent, PKA-dependent pathway (Nishizaki et al. 2002, Li et al. 2001, Fredholm et al. 2005).

Although all P2X receptor subtypes have been found at the mRNA and protein levels in astrocytes from different CNS regions (Fumagalli et al. 2003, Dixon et al. 2004, Franke et al. 2001, Pannicke et al. 2000, Lalo et al. 2008), there are limited reports on the role played by these ionotropic P2 receptors. This is because no P2X-mediated currents were recorded in astrocytes from slice preparations in response to ATP (Kukley et al. 2001, Jabs et al. 2007). However, in vitro systems revealed that acutely isolated astrocytes from cortex displayed functional heteromeric P2X$_{1/5}$ receptors (Lalo et al. 2008). The presence of P2X$_7$ receptors in the CNS remains controversial, mainly due to questionable specificity of antibodies (Sim et al. 2004), although in situ hybridization and reverse transcriptase (RT)–PCR studies have indicated the presence of P2X$_7$ receptor mRNA in many areas of the brain, and in oligodendrocytes, microglial, and astrocytes (Yu et al. 2008, Sperlagh et al. 2006). As for the other P2X receptors, evidence for functional expression of P2X$_7$ receptors in cultured astrocytes is quite vast, and many studies have indicated that P2X$_7$ receptor plays several roles, mainly related to pathological states involving astrogliosis. For instance, prolonged activation of P2X$_7$ receptor in cultured astrocytes leads to the release of glutamate, ATP, and tumor necrosis factor (TNF)-alpha, stimulates the production of nitric oxide, regulates NFkB signaling, and reduces the expression of glutamine synthetase and aquaporin-4 (Kucher and Neary 2005, Murakami et al. 2003, D'Alimonte et al. 2007, Franke et al. 2004). Although there is a paucity of evidence for the physiological role of P2X$_7$ receptors in brain tissues, it seems more likely that these receptors are related to astrocyte responses to brain injuries. For instance, upregulation of P2X$_7$ receptors in astrocytes occurs following ischemia (Franke et al. 2001) and in hypertrophic astrocytes located at chronic active multiple sclerosis lesions obtained from brain autopsies of human patients (Narcisse et al. 2005).

The P2Y metabotrobic receptors can be subdivided into two groups based on their sequence similarity, the presence of amino acid sequence motifs important for ligand binding, and the type of G-proteins that they are coupled to (reviewed in Abbracchio and Ceruti 2006). The $P2Y_{1,2,4,6,11}$ receptors are coupled to the Gq/G11 proteins that induce calcium release from intracellular stores via the phospholipase C (PLC) and $InsP_3$ pathways, while the $P2Y_{12-14}$ receptors are preferentially coupled to the Gi/o proteins, which inhibit adenylyl cyclase (reviewed in Abbracchio et al. 2009). Pharmacologically, P2Y receptors can be grouped as adenine-nucleotide- (ATP and ADP) preferring receptors ($P2Y_1$ and $P2Y_{11,12,13}$), uracyl nucleotide- (UTP and UDP) selective receptors (human $P2Y_{4,6}$), receptors with mixed selectivity ($P2Y_2$, rodent $P2Y_{11}$), and receptors ($P2Y_{14}$) selective for sugar nucleotides (UDP-glucose and UDP-galactose) (reviewed in Abbracchio and Ceruti 2006 and Verkhratsky et al. 2009).

2.3.2 ROLE OF P2 RECEPTORS IN GLIA-GLIA AND NEURON-GLIA COMMUNICATION

At the transcript level, astrocytes express all P2Y receptor subtypes except $P2Y_{11}$ receptor (reviewed in Verkhratsky et al. 2009); however, only the $P2Y_1$ and $P2Y_2$ receptors are involved in calcium signaling in astrocytes and in the transmission of calcium waves, as described above. In response to ATP, astrocytes release ATP in a calcium-independent manner (Wang et al. 2000, Anderson et al. 2004) either through volume-sensitive chloride channels or hemichannels and pannexin channels (Wang et al. 2000, Stout et al. 2002, Iglesias et al. 2009). In astrocyte progenitors and also mature astrocytes, however, a vesicular Soluble N-ethylmaleimide-sensitive factor (v-SNARE), calcium-dependent release of attachment protein receptor (ATP) was detected, a process that together with the activation of $P2Y_1$ receptors was shown to contribute to the migration of progenitor cells (Scemes et al. 2003, Striedinger et al. 2007) and to the transmission of calcium waves in rat hippocampal astrocytes (Bowser and Khakh 2007).

ATP released from astrocytes not only contributes to the communication with astrocytes themselves and microglia cells (reviewed in Scemes and Giaume 2006, Verderio and Matteoli 2001), but also is used as a source of adenosine, a molecule that modulates synaptic activity (Cunha 2008, Martin et al. 2007). Also, activation of $P2Y_1$ receptors by ATP induces the release of glutamate from astrocytes, which is controlled by TNF-alpha and prostaglandins (Domercq et al. 2006). The release of this excitatory neurotransmitter by astrocytes not only contributes to calcium signaling within the astrocyte network, but also modulates synaptic transmission (Finkbeiner 1992, Parpura et al. 1994).

2.4 AQUAPORINS AT THE PERIVASCULAR DOMAINS

2.4.1 AQUAPORIN FAMILY

Water regulation is important for every cell to maintain intracellular fluid composition and cell volume. Evolutionary processes have developed numerous, very efficient membrane protein channels specific for water passage in almost all living organisms, including plants, insects, amphibia, yeast, and bacteria. These water channels have been named aquaporins (AQPs) and belong to a family of homologous water channels widely distributed in a unique conserved aquaporin fold. The AQP family belongs to the larger membrane integral protein (MIP) family, including over 100 small transmembrane channels allowing passive transport of water, small neutral solutes (carbohydrate, urea, and gas), and perhaps ions. MIP stands for major intrinsic protein, the major component of lens fiber cells in the eye, which is the first cloned protein for this family. Based on their primary amino acid sequence and on the channel properties, the MIP family proteins can be divided into two groups, the first one specific for water transport, and the other one specific for transport for small neutral solutes, such as glycerol. The AQP family is a subgroup of channels further divided into orthodox aquaporins, allowing a selective flux of water molecules, and aquaglyceroporins, allowing the diffusion of water together with glycerol and other small solutes.

In mammals, the first water channel was discovered in human red blood cells and called CHIP28 (channel-forming integral protein of 28 kDa) (Preston et al. 1992). The cloning of this water channel led to the identification of other related water channels by homology cloning. After CHIP28, other water channel proteins were then discovered and named aquaporin, followed by a progressive number based on the order of their discovery. Therefore, MIP itself was renamed AQP0; CHIP28, AQP1; and so on. So far 13 AQPs have been discovered from AQP0 to AQP12, distributed in several organs and tissues. AQP0, AQP1, AQP2, AQP4, AQP5, AQP6, and AQP8 are water selective and therefore considered orthodox AQPs, whereas AQP3, AQP7, AQP9, and AQP10 allow also the passage of small solutes, such as glycerol, and are therefore called aquaglyceroporins. AQP9 is the only promiscuous water channel allowing also the passage of a wide variety of neutral solutes, including carbamides, polyols, purines, and pyrimidines. AQP11 and AQP12 expressed in kidney and pancreas, respectively, belong to a new subfamily called super-aquaporins (Ishibashi 2009), still poorly characterized.

2.4.1.1 The Aquaporin Fold—Structure and Function

Structural and dynamic studies provided key information to understand the function of a protein and to design potential binders and modulators. AQPs are small membrane proteins permeable to water but impermeable to protons. Moreover, they allow bidirectional transport across the lipid bilayer. Hydrophobicity plots for AQPs display six α-helices connected by loops A to E with intracellular N- and C-termini. The homology between the first and second halves of AQPs suggests a sequence derived by an ancient gene duplication event. Each half contains the key signature sequence with two repetitions of the Asn-Pro-Ala (NPA) motif, within loop B and loop E, highly conserved from plants to animals and from prokaryotes to eukaryotes, indicating an important role of the NPA boxes in the channel properties. The first hourglass model (Figure 2.4A) of AQP organization into the membrane (Jung et al. 1994) was hypothesized on the basis of mutagenesis experiments, and it was later grossly confirmed by crystallography studies, which have revealed the atomic AQP1 structure at increasingly higher levels of resolution, now 3.8 Å by electron (Murata et al. 2000) and 2.2 Å by x-ray (Sui et al. 2001) crystallography.

The atomic model of AQP1 has been used as an archetype to study the structure, permeability, and selectivity properties of other water channels. The AQP fold is

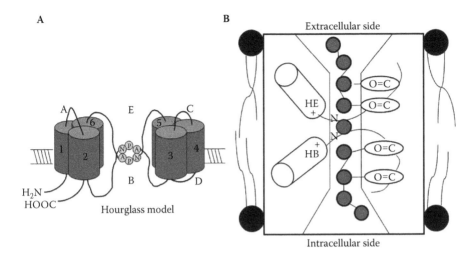

FIGURE 2.4   Aquaporin topology. (A) Hourglass model for AQP1 topology. The hydrophobic loops B and E, carrying the highly conserved NPA motifs, deepened into the center of the membrane layer from the opposite side and are essential for the water flux. (B) Schematic diagram of AQP channel, which is a hydrophilic central hole allowing water molecules to pass in a single line. The water passage is facilitated by hydrogen bonds between water molecules and with main-chain carbonyloxygens, or nitrogens of the NPA motifs.

a right-handed bundle of six α-helices packed in a barrel-like structure having a hydrophilic central hole. A key feature of this structure is that the hydrophobic loops B and E, carrying the NPA motifs, penetrate into the center of the membrane layer from opposite sides, and getting closer to each other, they create a mechanism of proton exclusion that is essential to maintain the proton gradient across the bilayer (Figure 2.4B). At the NPA level, the diameter of the pore becomes very narrow (3 Å) and the amino acids here are hydrophobic, except the two Asn residues. This is an environment that allows water molecules to pass in single file, 2.8 Å being the diameter of a single water molecule. On the other hand, hydrogen bonds between water molecules to the Asn residues facilitate the water molecule entering this constriction, therefore improving the speed of the water passage. While this fold is conserved, differences in the length and amino acid composition of the other loops as well as in the cytoplasmic N- and C-termini play important functional and regulative roles and determine differences in the size of the pore, which in turn affects its selectivity for water. Larger pores in fact allow also the passage of small neutral solutes. The absence of proton and ion permeation through AQPs has been attributed to the presence of steric barriers combined with electrostatic interactions and hydrogen bonds (Verkman and Mitra 2000). Interestingly, it was suggested that AQP1 could transport CO_2, thus playing an important role in gas exchange at the level of pulmonary capillaries and erythrocytes (Nakhoul et al. 1998). However, this possibility was not confirmed in a later study, showing that there was no difference in terms of transport of CO_2 in the lungs of wild-type and AQP1 knockout mice (Yang et al. 2000). Thus, the role of AQPs in gas transport remains an open question.

2.4.1.2 Orthogonal Arrays of Particles—The Super-Molecular Organization of the Major CNS Water Channel Aquaporin-4

AQP monomers contain independent water channels, but they assemble as tetramers in the membrane, probably in order to maximize stability (Figure 2.5A). All AQPs form homo-tetramers of the same isoform in the membrane except the CNS water channel AQP4 (Frigeri et al. 1995a, 1995b, Nielsen et al. 1997), which forms hetero-tetramers of two major polypeptides of 30 and 32 kDa, called M23 and M1, respectively, and named accordingly to their rise from two in-frame translation-initiating methionines (Hasegawa et al. 1994, Jung et al. 1994, Neely et al. 1999).

It is interesting to highlight that a peculiar morphological feature of AQP4 and the lens AQP0 is to form well-organized structures (Figure 2.5B) called orthogonal arrays of particles (OAPs) (Yang et al. 1996, Verbavatz et al. 1997, Rash et al. 1998). OAPs are made of multiple tetramers that further aggregate in the plasma membrane in super-molecular structures, also called square arrays (Nicchia et al. 2010), as visualized by freeze-fracture electron microscopy (Figure 2.5C) (Furman et al. 2003) and by Blue Native/Sodium dodecyl sulfate polyacrylamide gel electrophoresis (SDS)-

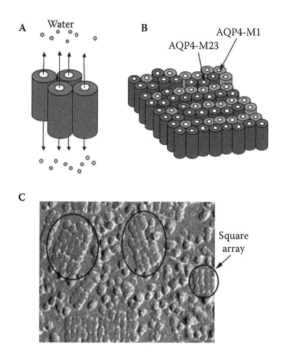

FIGURE 2.5 Aquaporin structure. (A) Schematics illustrating tetrameric organization of AQPs. Each monomer contains an independent water pore (arrows). (B) In the case of orthogonal arrays of particles (OAPs) forming AQPs, multiple tetramers further aggregate in the plasma membrane in well-organized square arrays formed by two different AQP4 isoforms: M23 (dark gray) and M1 (light gray). M23 is the major component of large OAPs and is the isoform that forms OAPs. (C) Electron micrographs of freeze-fracture preparations illustrating the morphological feature of OAPs (circles). (Modified from Furman et al., *Proc. Natl. Acad. Sci. USA* 100 (23), 13609–14, 2003. Copyright 2003 National Academy of Sciences, U.S.A.)

PAGE (Nicchia et al. 2008a). In the case of AQP4, the expression ratio of M1 and M23 isoforms is the major determinant of the size of in vivo OAPs. When the long AQP4-M1 form is expressed alone, no stable OAPs are visible. Formation of stable and large OAPs requires the expression of AQP4-M23 (Yang et al. 1996, Furman et al. 2003, Rossi et al. 2010), which is therefore called the OAP-forming isoform. The coexpression of M1 and M23 leads to OAPs of different sizes, depending on the ratio of the two isoforms. The function OAPs and their different sizes under normal conditions are as yet largely unknown. OAPs of different sizes may be functionally distinct AQP4 macromolecular complexes associated with different proteins (Nicchia et al. 2008c), but whether this aggregation enhances water permeability is still a matter of controversy (Silberstein et al. 2004). Most likely, aggregation of AQP tetramers will confer higher-level plasma membrane stability necessary for AQP4-selective expression in cellular domains not delimited by tight junctions; such is the case for astrocyte foot processes, where in fact the larger OAPs are expressed.

2.4.2 CNS Aquaporins—Control of Water Balance

Water balance is of fundamental importance in brain physiology due to a continuous movement of fluids between the brain compartments. Cellular walls separate, sometimes creating a real barrier, the parenchyma, which is the main compartment, from the vascular and ventricular ones in order to control and protect the CNS milieu. Exchange of water between these compartments is driven by osmotic and hydrostatic forces across the blood–brain and blood-cerebrospinal fluid (CSF) barriers, the ventricular ependymal layer, and glia limitans. The discovery of AQP water channels and their localization in the CNS has clarified the mechanism of water transport associated with fluid shifts (Figure 2.6). The main information on AQP-dependent water flux has come from immunolocalization studies and from functional analysis performed on AQP null mice (Verkman 2009).

The most abundant water channel in the CNS is AQP4, which is very highly concentrated at the astrocyte processes surrounding blood vessels at the blood–brain barrier (BBB) and at those forming the glial limitans layer (Frigeri et al. 1995b, Nielsen et al. 1997). The BBB is a very highly selective barrier between the parenchyma and the vasculature, made up by endothelial cells endowed of tight junctions, normally absent in the capillary outside the brain. The endothelial cells here are completely enveloped by astrocyte foot processes highly enriched in AQP4 and taking part in regulating the blood flow in the brain. Although all AQPs allow bidirectional water flux depending on the osmotic gradient, several studies suggest that in physiological conditions AQP4 could serve at the BBB level for the water to exit in a brain-blood direction. It has been shown that AQP4 expression appears during the first weeks of life (Wen et al. 1999, Nicchia et al. 2004) in parallel with the rapid decline of the extracellular space and with the capability to clear potassium (Connors et al. 1982). Thus, AQP4 during the first weeks of development could be important for the shrinkage of the extracellular space and at the same time to facilitate the clearance of ions such as potassium. This last function is also supported by several studies showing a role for AQP4 in facilitating the water flux coupled with potassium siphoning necessary

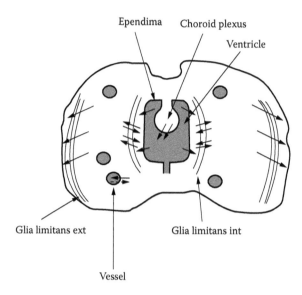

FIGURE 2.6 Aquaporin-mediated water flux in the brain. The small arrows indicate the pathways for AQP-dependent water movements, into and out of the brain, at the blood–brain and CSF-brain barriers. At the blood–brain barrier (vessels), AQP4 is an exit route for water under normal conditions but may also be used as a pathway for water entry under pathological conditions characterized by an altered osmotic equilibrium. At the CSF-brain barriers, AQP1 and AQP4 participate in the production and absorption of the CSF, respectively, that occurs at the choroids plexus, ependymal cell layer, and internal (int) glial limitans. At the external (ext) glia limitans AQP4 may be involved in the movement of water associated with CSF drainage at the subarachnoid space.

to sustain the neuronal activity. A delayed potassium clearance and a consequent increased susceptibility to seizure have been reported for AQP4 KO mice (Binder et al. 2004) and for alpha-syntrophin null mice, which secondarily lack AQP4 (Amiry-Moghaddam et al. 2003a, 2003b), indicating a role in signal transduction (Verkman 2009). It is important to highlight that molecules to be cleared, such as potassium, move together with water through gap junctions from the neuropile toward the perivascular glial domain. Once they arrive there, their extrusion in the blood is facilitated by specific channels that colocalize at the BBB, such as AQP4 and Kir4.1 (Connors et al. 2004). The collaboration between water channels and gap junctions in eliminating molecules from the brain into the blood may explain why AQP4 is often modulated in parallel with connexin43 (Nicchia et al. 2005, Kong et al. 2008, Li et al. 2009, Lichter-Konecki et al. 2008), the main astrocyte gap junction protein.

Another finding supporting AQP4-mediated water exit from the brain to the vascular compartment is the presence of perivascular glial swelling in mice lacking AQP4, such as the alpha-syntrophin null mice (Amiry-Moghaddam et al. 2003a) and the mdx mice (Frigeri et al. 2001), the animal models of Duchenne muscular dystrophy. In this case, the absence of water channels causes a perivascular accumulation of water with consequent glial swelling.

AQPs are also important in water flux associated with CSF production and drainage. The choroid plexus and the arachnoid membrane are endowed with tight junctions and therefore represent the CSF-brain barriers regulating the production and drainage of the CSF, respectively. The choroid plexus is an extension of the pia mater within the ventricles. It is a cauliflower-like structure made of a membrane layer of epithelial cells, folding into several villi around a core of capillaries, in charge of filtering liquid from blood to become CSF at a rate of ~0.5 L a day (Brown et al. 2004). Choroid cells control exchanges between the blood and CSF as they actively transport sodium, chloride, and bicarbonate into the ventricles and water as a consequence of the osmotic gradient. The choroid cell water channel is AQP1, selectively expressed at the apical membrane and involved in CSF production (Nielsen et al. 1993). AQP1 null mice have a lower intracranial pressure compared to wild-types, with ~25% reduced CSF production (Oshio et al. 2005), indicating that AQP1 inhibitors might be used as a novel therapy to reduce CSF production and elevated intracranial pressure in pathologies such as hydrocephalus. The choroid plexus is continuous with the ependymal cell layer separating the ventricular to the parenchyma compartment. AQP4 localizes also at the basolateral membrane of ependymal cells. The precise function of AQP4 here is not very clear. However, when the primary ventricular CSF clearance pathways (the subarachnoid space and the venous sinuses) are blocked, as in hydrocephalus (Bloch et al. 2006), AQP4 is here important in mediating the transparenchymal CSF clearance into the cerebral microvasculature, suggesting that AQP4 induction may be a nonsurgical therapy for hydrocephalus. Different from the choroid plexus, the subarachnoid space has mainly a passive role as a blood-CSF barrier, allowing CSF drainage and preventing the passage of substances from the blood through the arachnoid membrane by the presence of tight junctions. Across the CSF-brain interfaces, AQP4 is also expressed at the multilayer of astrocyte processes forming the external and internal glia limitans (Frigeri et al. 1995a, Nielsen et al. 1997, Rash et al. 1998). The external glia limitans represents a protective surface for the CNS and is a boundary between brain parenchyma and subarachnoid CSF. The internal glia limitans is located under the ependyma and is a boundary between the brain parenchyma and ventricular CSF. AQP4 localization at the astrocyte processes forming the glia limitans is probably important for the movement of water into and out of the brain related with the CSF production and drainage. Besides AQP4 and AQP1, AQP9 is also reported to belong to the CNS AQPs, although its localization and even its expression in the brain are still not universally accepted. It has been reported to localize in the tanycytes where the BBB is absent (Elkjaer et al. 2000, Nicchia et al. 2001) and in some catecholaminergic neurons (Badaut et al. 2004, Mylonakou et al. 2009) even though its expression level is very low.

2.4.3 Changes in Astrocyte Shape and Cell Migration

The analysis of AQP4 knockdown (KD) astrocytes and AQP4 null mice has revealed a new role for AQP4, not related to CNS water balance. AQP4 is important in facilitating changes in cell shape (Nicchia et al. 2005) and cell migration (Saadoun et al. 2005), probably through a functional interaction with actin cytoskeleton (Nicchia et al. 2008b). In astrocyte primary cultures, AQP4 KD by RNA interference induces dramatic morphological changes and actin depolymerization (Nicchia et al. 2005), and cytochalasin-D treatment impairs correct localization of AQP4 if conducted during the first phase of cell adhesion. After brain injury, AQP4 becomes upregulated in reactive astrocytes migrating toward the site of injury to form a glial scar (Lim et al. 2007), and experiments performed using AQP4 null mice have shown that AQP4 deletion slows astrocyte migration in vitro and delays glial scar formation in vivo (Papadopoulos et al. 2008). Moreover, in agreement with in vivo data, reporting an increase in AQP4 during reactive gliosis (Bloch et al. 2005, Papadopoulos and Verkman 2005, Vizuete et al. 1999), AQP4 increases and is correctly delivered to the plasma membrane if a morphological differentiation of astrocyte cultures is induced by cAMP, enabling them to acquire a reactive phenotype.

What is the mechanism proposed for AQP involvement in changes in cell shape occurring during the transformation of astrocytes in reactive cells or during cell migration? Astrocyte swelling induces changes in the actin cytoskeleton and cell morphology (Moran et al. 1996, Allansson et al. 2001), and a transitory functional interaction between AQP4 and the actin cytoskeleton seems important for the change in shape that the cell undergoes during morphological changes, swelling, and migration (Nicchia et al. 2008b). A local swelling is necessary for changes in cell shape or for the formation of the membrane protrusion at the leading edge of the cell. The presence of water channels likely facilitates the water flux, therefore improving the speed of the process. The role of AQP4 and other water channels in cell migration may have therapeutic implications in pathological situations in which morphological and cytoskeletal modifications of astrocytes occur, including glial scarring, glioma infiltration, wound healing, and others.

2.5 DISORDERS AFFECTING THE EXPRESSION AND FUNCTION OF ASTROCYTE BIOMARKERS

2.5.1 INFLAMMATION

It is usually thought that inflammatory demyelinating diseases, such as neuromyelitis optica (NMO) and multiple sclerosis (MS), directly target myelin or oligodendrocytes. However, recent reports suggest that, at least for NMO, demyelination and oligodendrocyte damage are consequences of astrocyte dysfunction. In neuromyelitis optica, the selective binding of NMO-IgG to aquaporin-4 (AQP4) (Lennon et al. 2005) induces the complement-dependent death of astrocytes (Hinson et al. 2008, Vincent et al. 2008, Sabater et al. 2009), which is followed by oligodendrocyte injury and demyelination (Roemer et al. 2007, Bradl and Lassmann 2010). In contrast, in MS patient autopsies, astrocyte loss is minor with no evidence for a disruption of the perivascular limitants. Nevertheless, substantial loss of the gap junction protein Cx43 in spinal cord regions adjacent to infiltrated monocytes was reported in an animal model of MS (Brand-Schieber et al. 2005) and in rat spinal cord lesions induced by NMO-IgG injections (Sharma et al. 2010).

2.5.1.1 AQP4 and Neuromyelitis Optica

Neuromyelitis optica (NMO) is a relapsing inflammatory disease of the CNS often misdiagnosed as multiple sclerosis (MS). NMO can be distinguished from MS by a combination of clinical, neuroimaging, laboratory, and pathological criteria (Pittock 2008, Wingerchuk 2007). The pathological hallmark of NMO is the deposition of typical immunoglobulins, called NMO-IgG, at the perivascular glial domain associated with tissue damage and demyelination. In 2005 it was discovered that AQP4 is the molecular target of the NMO-IgG autoantibodies (Lennon et al. 2005) leading to the classification of NMO as an autoimmune channelopathy. At the moment, the presence of anti-AQP4 antibodies in the serum is considered an important diagnostic and prognostic biomarker of NMO. Several studies indicate that NMO-IgG/anti-AQP4 antibodies have an important role in the development of the pathology even though experiments performed on cell cultures have demonstrated that NMO-IgG binding does not affect AQP4-mediated water transport (Nicchia et al. 2009), indicating that the pathological effect is not due to the inhibition of AQP4 water channel function. As mentioned above, the two major isoforms of AQP4, M1- and M23-AQP4, form in the plasma membrane OAP assemblies whose function is probably to confer a higher level of membrane stability necessary for polarized expression of AQP4 in astrocyte foot processes. Correlations have been reported between OAP alteration and various neuromuscular pathological conditions, such as cerebral ischemia (Neuhaus et al. 1990, Wolburg 1995), epilepsy (Hatton and Ellisman 1984), circulatory arrest (Landis and Reese 1981), and muscular dystrophy (Frigeri et al. 1998, Wakayama et al. 1989). Although all the studies published in the last few years clearly identified AQP4 as the target of NMO-IgG, the field has faced several difficulties in trying to define the precise AQP4 epitope by testing AQP4 linear peptides. The reason for these difficulties was revealed in 2009, when it was demonstrated that the NMO-IgG epitope is intrinsic in AQP4 assemblies into OAPs (Nicchia et al. 2009). Experiments performed using transfected cells, in fact, clearly showed that NMO-IgGs recognize only the OAP-forming isoform M23 and do not recognize AQP4-M1 if expressed without AQP4-M23 (Rossi et al. 2010). Other OAP-forming water channel proteins, such as the lens aquaporin-0 and the insect aquaporin-cic, are not recognized by NMO-IgG, indicating an epitope characteristic of AQP4-OAPs (Nicchia et al. 2009). The discovery that OAPs and not AQP4 have to be considered the real target for NMO-IgG represented an important advance in the field, moving the focus from the effect of NMO-IgG on AQP4 function to the effect on its plasma membrane organization. A detailed analysis of OAP structure and function may aid in the identification of the epitope of the NMO-IgG within the OAPs, which is essential to gain insight into the pathogenesis of the disease and to develop diagnostic and therapeutic tools.

2.5.2 AQP4 AND EDEMA

Cerebral edema is an abnormal accumulation of water into the brain associated with several pathological conditions of the CNS, such as brain tumor, brain abscess, meningitis, and stroke, and also outside the CNS, such as liver failure and sepsis. Brain edema induces an increase in intracranial pressure with consequent brain ischemia, herniation, and death. The role of AQP4 in the pathophysiology of brain edema is widely accepted given that studies have mainly been performed on animal models lacking AQP4, including mice null for AQP4 (Manley et al. 2000), dystrophin (Frigeri et al. 2001, Vajda et al. 2002), and alpha-syntrophin (Amiry-Moghaddam et al. 2003b). Interestingly, AQP4 has been reported to be involved in edema formation, with the absence of AQP4 being deleterious if the edema is cytotoxic and protective if the edema is vasogenic. Cytotoxic edema is also called cellular edema and is associated, for example, with ischemia. It is characterized by cell swelling (mainly astrocyte swelling) due to failure of ATP-dependent transport mechanisms. As a consequence, sodium accumulates within the cells and water follows to maintain osmotic equilibrium. The presence of AQP4 facilitates the entry of water at the perivascular domain, being therefore deleterious in this context. In support of this hypothesis, several authors independently have shown that if cytotoxic edema is induced by water intoxication and focal cerebral ischemia, the brain swelling is significantly reduced in mice lacking AQP4, compared with wild-type (Manley et al. 2000, Vajda et al. 2002, Amiry-Moghaddam et al. 2003a). These studies indicate that AQP4 inhibitors will likely be used to reduce the AQP4-dependent entry of water in cytotoxic brain edema. Vasogenic or noncellular edema, as, for example, the one associated with brain tumor, is characterized by an increase in extracellular fluid volume in the brain parenchyma due to increased permeability of the BBB endothelial cells to macromolecular serum proteins. In this scenario the water entry into the brain through injured tight junction is AQP4 independent. However, because the water transport through AQPs is bidirectional and because the elimination of water in vasogenic edema occurs at the BBB and CFS-brain barrier interfaces, which are highly enriched in AQP4, in this case the presence of AQP4 is beneficial to eliminate the excess water into the blood, the ventricles, and the subarachnoid cerebrospinal fluid (CSF). This hypothesis has been demonstrated by studies showing a worse outcome for AQP4 null mice in which the vasogenic edema has been induced by intraparenchymal fluid infusion, focal cortical freeze injury, and tumor cell implantation (Papadopoulos et al. 2004). In conclusion, in the therapy of brain edema, AQP4 inhibitors would not be beneficial if administered during edema resolution but only if administered at the very early onset of cytotoxic edema. In contrast, AQP4 activators may be used in the resolution phase of brain edema.

2.5.3 GENETIC MUTATIONS

Very little is known about the effects of purinergic receptor mutations on astrocyte physiology, and nothing is known about pannexin mutations in this regard. With respect to P2 receptor mutations, one study reported alterations in the transmission of calcium waves between cultured astrocytes due to a single-nucleotide polymorphism on the $P2X_7$ receptor (Suadicani et al. 2009). In this study it was found that a spontaneous point mutation (P451L) located in the SH3 domain of the $P2X_7$ receptor carboxyl terminal of C57Bl/6 mice interfered with the association of these receptors to pannexin1 channels. Compared to nonmutated receptor, it was found that activation of the P451L $P2X_7$ receptor mutant reduced Panx1 channel current, the amount of ATP released, and the extent to which calcium waves traveled among the astrocyte network (Suadicani et al. 2009).

Although several connexin-related diseases have been documented, there are also no studies reporting alterations in astrocyte connectivity that are related to connexin mutations. Significant impact of astrocyte gap junctions to CNS function, however, is seen after ablation of connexins (reviewed in Sohl et al. 2004, Santiago et al. 2010, Menichella et al. 2003, Lutz et al. 2009).

2.5.3.1 Cx43 Mutations

Because connexin43 is expressed in almost all tissues, genetic mutations affecting the GJA1 gene have a wide phenotypic spectrum of alterations associated with it. For instance, the occulodentaldigital dysplasia (ODDD) syndrome is associated with more than 60 mutations in GJA1, a gene that encodes Cx43. This rare autosomal dominant syndrome is characterized by microphtalmia, microcornea, narrow nose, microdontia, and syndactyly. A fraction of ODDD patients display some neurological symptoms (e.g., white matter changes in the brain, gait disturbances, ataxia), hearing loss, and cardiac arrhythmias (Paznekas et al. 2003, 2009, Judisch et al. 1979). Because the phenotypes seen in ODDD patients do not correlate with those seen in Cx43-null mice it was suggested that loss of function of Cx43 is unlikely to be the mechanism by which the missense mutations cause the human phenotype (Paznekas et al. 2003). Features of Cx43-null mice, including ossification defects of the skull, small mandibles, and blockade of the right ventricle outflow tract (Lecanda et al. 2000, Reaume et al. 1995), are not obvious features of ODDD patients (Paznekas et al. 2003).

Interestingly, however, white matter pathology (vacuolated oligodendrocytes and intramyelin edema) with associated sensorimotor deficits was observed in a double knockout mice (Cx30$^{-/-}$; GFAP:Cre-Cx43$^{f/f}$) lacking the two main astrocyte gap junction proteins (Lutz et al. 2009). The distribution and features of the vacuoles in the myelin of Cx30/Cx43 double knockout mice are similar to those seen when both of the oligodendrocyte connexins (Cx32/Cx47) are knocked out (Menichella et al. 2003, Odermatt et al. 2003). Given that the morphological and behavioral changes seen in Cx30/Cx43 dKO were not present in either of the single knockout mice for Cx43 and Cx30, it was suggested that loss of gap junction function is important for the proper function for astrocyte-oligodendrocyte interaction, and that either one of these astrocyte connexins can compensate for the lack of the other (Lutz et al. 2009).

2.6 SUMMARY

In this chapter, we have described the properties and functions of three astrocyte biomarker families: the gap junctions, purinergic receptors, and aquaporins. We have indicated that although each of these groups of protein families displays particular and independent properties and distinct subcellular distributions, in many instances they work in concert, providing astrocytes with the capability to sense the environment and communicate with other neural cell components. The intercellular gap junction channels, which provide the direct communication between the cytosols of coupled cells, serve as a route for the transport of nutrients, metabolites, and signaling molecules. The ATP-releasing pannexin channels together with purinergic receptors contribute to the paracrine signaling mechanism involved in the transmission of calcium signals between astrocytes and other neural cell types. The aquaporins that participate in the movement of water within brain compartments are also likely to contribute, together with the gap junction channels, to the establishment of concentration gradients that drives the movement of ions and molecules within the astrocyte network.

ACKNOWLEDGMENTS

The work of E. Scemes and G. P. Nicchia presented in this chapter has been supported by NINDS-NIH (NS041023 and NS052245) and by Progetto di Ricerca: "IDEA Giovani Ricercatori" (GRBA085SIS), respectively.

REFERENCES

Abbracchio, M. P., and G. Burnstock. 1998. Purinergic signalling: pathophysiological roles. *Jpn J Pharmacol* 78 (2):113–45.
Abbracchio, M. P., G. Burnstock, A. Verkhratsky, and H. Zimmermann. 2009. Purinergic signalling in the nervous system: an overview. *Trends Neurosci* 32 (1):19–29.
Abbracchio, M. P., and S. Ceruti. 2006. Roles of P2 receptors in glial cells: focus on astrocytes. *Purinergic Signal* 2 (4):595–604.
Abe, K., and H. Saito. 1998. Adenosine stimulates stellation of cultured rat cortical astrocytes. *Brain Res* 804 (1):63–71.
Allaman, I., S. Lengacher, P. J. Magistretti, and L. Pellerin. 2003. A2B receptor activation promotes glycogen synthesis in astrocytes through modulation of gene expression. *Am J Physiol Cell Physiol* 284 (3):C696–704.
Allansson, L., S. Khatibi, T. Olsson, and E. Hansson. 2001. Acute ethanol exposure induces $[Ca^{2+}]_i$ transients, cell swelling and transformation of actin cytoskeleton in astroglial primary cultures. *J Neurochem* 76 (2):472–79.
Altevogt, B. M., and D. L. Paul. 2004. Four classes of intercellular channels between glial cells in the CNS. *J Neurosci* 24 (18):4313–23.
Ambrosi, C., O. Gassmann, J. N. Pranskevich, D. Boassa, A. Smock, J. Wang, G. Dahl, C. Steinem, and G. E. Sosinsky. 2010. Pannexin1 and pannexin2 channels show quaternary similarities to connexons and different oligomerization numbers from each other. *J Biol Chem* 285 (32):24420–31.
Amiry-Moghaddam, M., T. Otsuka, P. D. Hurn, R. J. Traystman, F. M. Haug, S. C. Froehner, M. E. Adams, J. D. Neely, P. Agre, O. P. Ottersen, and A. Bhardwaj. 2003a. An alpha-syntrophin-dependent pool of AQP4 in astroglial end-feet confers bidirectional water flow between blood and brain. *Proc Natl Acad Sci USA* 100 (4):2106–11.
Amiry-Moghaddam, M., A. Williamson, M. Palomba, T. Eid, N. C. de Lanerolle, E. A. Nagelhus, M. E. Adams, S. C. Froehner, P. Agre, and O. P. Ottersen. 2003b. Delayed K$^+$ clearance associated with aquaporin-4 mislocalization: phenotypic defects in brains of alpha-syntrophin-null mice. *Proc Natl Acad Sci USA* 100 (23):13615–20.
Amzica, F., M. Massimini, and A. Manfridi. 2002. Spatial buffering during slow and paroxysmal sleep oscillations in cortical networks of glial cells *in vivo*. *J Neurosci* 22 (3):1042–53.
Anderson, C. M., J. P. Bergher, and R. A. Swanson. 2004. ATP-induced ATP release from astrocytes. *J Neurochem* 88 (1):246–56.

Badaut, J., J. M. Petit, J. F. Brunet, P. J. Magistretti, C. Charriaut-Marlangue, and L. Regli. 2004. Distribution of aquaporin 9 in the adult rat brain: preferential expression in catecholaminergic neurons and in glial cells. *Neuroscience* 128 (1):27–38.

Balice-Gordon, R. J., L. J. Bone, and S. S. Scherer. 1998. Functional gap junctions in the Schwann cell myelin sheath. *J Cell Biol* 142 (4):1095–104.

Baranova, A., D. Ivanov, N. Petrash, A. Pestova, M. Skoblov, I. Kelmanson, D. Shagin, S. Nazarenko, E. Geraymovych, O. Litvin, A. Tiunova, T. L. Born, N. Usman, D. Staroverov, S. Lukyanov, and Y. Panchin. 2004. The mammalian pannexin family is homologous to the invertebrate innexin gap junction proteins. *Genomics* 83 (4):706–16.

Beltramello, M., M. Bicego, V. Piazza, C. D. Ciubotaru, F. Mammano, and P. D'Andrea. 2003. Permeability and gating properties of human connexins 26 and 30 expressed in HeLa cells. *Biochem Biophys Res Commun* 305 (4):1024–33.

Bennett, M. V., J. E. Contreras, F. F. Bukauskas, and J. C. Saez. 2003. New roles for astrocytes: gap junction hemichannels have something to communicate. *Trends Neurosci* 26 (11):610–17.

Binder, D. K., K. Oshio, T. Ma, A. S. Verkman, and G. T. Manley. 2004. Increased seizure threshold in mice lacking aquaporin-4 water channels. *Neuroreport* 15 (2):259–62.

Bloch, O., K. I. Auguste, G. T. Manley, and A. S. Verkman. 2006. Accelerated progression of kaolin-induced hydrocephalus in aquaporin-4-deficient mice. *J Cereb Blood Flow Metab* 26 (12):1527–37.

Bloch, O., M. C. Papadopoulos, G. T. Manley, and A. S. Verkman. 2005. Aquaporin-4 gene deletion in mice increases focal edema associated with staphylococcal brain abscess. *J Neurochem* 95 (1):254–62.

Boassa, D., C. Ambrosi, F. Qiu, G. Dahl, G. Gaietta, and G. Sosinsky. 2007. Pannexin1 channels contain a glycosylation site that targets the hexamer to the plasma membrane. *J Biol Chem* 282 (43):31733–43.

Bowser, D. N., and B. S. Khakh. 2007. Vesicular ATP is the predominant cause of intercellular calcium waves in astrocytes. *J Gen Physiol* 129 (6):485–91.

Bradl, M., and H. Lassmann. 2010. Oligodendrocytes: biology and pathology. *Acta Neuropathol* 119 (1):37–53.

Brand-Schieber, E., P. Werner, D. A. Iacobas, S. Iacobas, M. Beelitz, S. L. Lowery, D. C. Spray, and E. Scemes. 2005. Connexin43, the major gap junction protein of astrocytes, is down-regulated in inflamed white matter in an animal model of multiple sclerosis. *J Neurosci Res* 80 (6):798–808.

Broer, S., and N. Brookes. 2001. Transfer of glutamine between astrocytes and neurons. *J Neurochem* 77 (3):705–19.

Brown, P. D., S. L. Davies, T. Speake, and I. D. Millar. 2004. Molecular mechanisms of cerebrospinal fluid production. *Neuroscience* 129 (4):957–70.

Bruzzone, R., M. T. Barbe, N. J. Jakob, and H. Monyer. 2005. Pharmacological properties of homomeric and heteromeric pannexin hemichannels expressed in *Xenopus* oocytes. *J Neurochem* 92 (5):1033–43.

Bruzzone, R., S. G. Hormuzdi, M. T. Barbe, A. Herb, and H. Monyer. 2003. Pannexins, a family of gap junction proteins expressed in brain. *Proc Natl Acad Sci USA* 100 (23):13644–49.

Bukauskas, F. F., and V. K. Verselis. 2004. Gap junction channel gating. *Biochim Biophys Acta* 1662 (1–2):42–60.

Chen, K. C., and C. Nicholson. 2000. Spatial buffering of potassium ions in brain extracellular space. *Biophys J* 78 (6):2776–97.

Cole, W. C., and R. E. Garfield. 1986. Evidence for physiological regulation of myometrial gap junction permeability. *Am J Physiol* 251 (3 Pt 1):C411–20.

Cole, W. C., R. E. Garfield, and J. S. Kirkaldy. 1985. Gap junctions and direct intercellular communication between rat uterine smooth muscle cells. *Am J Physiol* 249 (1 Pt 1):C20–31.

Connors, B. W., B. R. Ransom, D. M. Kunis, and M. J. Gutnick. 1982. Activity-dependent K+ accumulation in the developing rat optic nerve. *Science* 216 (4552):1341–43.

Connors, N. C., M. E. Adams, S. C. Froehner, and P. Kofuji. 2004. The potassium channel Kir4.1 associates with the dystrophin-glycoprotein complex via alpha-syntrophin in glia. *J Biol Chem* 279 (27):28387–92.

Contreras, J. E., H. A. Sanchez, L. P. Veliz, F. F. Bukauskas, M. V. Bennett, and J. C. Saez. 2004. Role of connexin-based gap junction channels and hemichannels in ischemia-induced cell death in nervous tissue. *Brain Res Brain Res Rev* 47 (1–3):290–303.

Cotrina, M. L., Q. Gao, J. H. Lin, and M. Nedergaard. 2001. Expression and function of astrocytic gap junctions in aging. *Brain Res* 901 (1–2):55–61.

Cruz, N. F., K. Adachi, and G. A. Dienel. 1999. Rapid efflux of lactate from cerebral cortex during K+-induced spreading cortical depression. *J Cereb Blood Flow Metab* 19 (4):380–92.

Cunha, R. A. 2008. Different cellular sources and different roles of adenosine: A1 receptor-mediated inhibition through astrocytic-driven volume transmission and synapse-restricted A2A receptor-mediated facilitation of plasticity. *Neurochem Int* 52 (1–2):65–72.

Dahl, G., R. Azarnia, and R. Werner. 1981. Induction of cell-cell channel formation by mRNA. *Nature* 289 (5799):683–85.

D'Alimonte, I., R. Ciccarelli, P. Di Iorio, E. Nargi, S. Buccella, P. Giuliani, M. P. Rathbone, S. Jiang, F. Caciagli, and P. Ballerini. 2007. Activation of P2X7 receptors stimulates the expression of P2Y2 receptor mRNA in astrocytes cultured from rat brain. *Int J Immunopathol Pharmacol* 20 (2):301–16.

Deitmer, J. W., K. D. McCarthy, E. Scemes, and C. Giaume. 2006. Information processing and transmission in glia: calcium signaling and transmitter release. *Glia* 54 (7):639–41.

De Pina-Benabou, M. H., M. Srinivas, D. C. Spray, and E. Scemes. 2001. Calmodulin kinase pathway mediates the K+-induced increase in gap junctional communication between mouse spinal cord astrocytes. *J Neurosci* 21 (17):6635–43.

Dermietzel, R. 1974. Junctions in the central nervous system of the cat. 3. Gap junctions and membrane-associated orthogonal particle complexes (MOPC) in astrocytic membranes. *Cell Tissue Res* 149 (1):121–35.

Dermietzel, R. 1998. Diversification of gap junction proteins (connexins) in the central nervous system and the concept of functional compartments. *Cell Biol Int* 22 (11–12):719–30.

Dermietzel, R., M. Farooq, J. A. Kessler, H. Althaus, E. L. Hertzberg, and D. C. Spray. 1997. Oligodendrocytes express gap junction proteins connexin32 and connexin45. *Glia* 20 (2):101–14.

Dienel, G. A., and L. Hertz. 2001. Glucose and lactate metabolism during brain activation. *J Neurosci Res* 66 (5):824–38.

Dienel, G. A., R. Y. Wang, and N. F. Cruz. 2002. Generalized sensory stimulation of conscious rats increases labeling of oxidative pathways of glucose metabolism when the brain glucose-oxygen uptake ratio rises. *J Cereb Blood Flow Metab* 22 (12):1490–502.

Dixon, S. J., R. Yu, N. Panupinthu, and J. X. Wilson. 2004. Activation of P2 nucleotide receptors stimulates acid efflux from astrocytes. *Glia* 47 (4):367–76.

Domercq, M., L. Brambilla, E. Pilati, J. Marchaland, A. Volterra, and P. Bezzi. 2006. P2Y1 receptor-evoked glutamate exocytosis from astrocytes: control by tumor necrosis factor-alpha and prostaglandins. *J Biol Chem* 281 (41):30684–96.

Elkjaer, M., Z. Vajda, L. N. Nejsum, T. Kwon, U. B. Jensen, M. Amiry-Moghaddam, J. Frokiaer, and S. Nielsen. 2000. Immunolocalization of AQP9 in liver, epididymis, testis, spleen, and brain. *Biochem Biophys Res Commun* 276 (3):1118–28.

Enkvist, M. O., and K. D. McCarthy. 1994. Astroglial gap junction communication is increased by treatment with either glutamate or high K⁺ concentration. *J Neurochem* 62 (2):489–95.

Evans, W. H., E. De Vuyst, and L. Leybaert. 2006. The gap junction cellular internet: connexin hemichannels enter the signalling limelight. *Biochem J* 397 (1):1–14.

Feller, M. B., D. P. Wellis, D. Stellwagen, F. S. Werblin, and C. J. Shatz. 1996. Requirement for cholinergic synaptic transmission in the propagation of spontaneous retinal waves. *Science* 272 (5265):1182–87.

Fields, R. D., and B. Stevens-Graham. 2002. New insights into neuron-glia communication. *Science* 298 (5593):556–62.

Finkbeiner, S. 1992. Calcium waves in astrocytes-filling in the gaps. *Neuron* 8 (6):1101–8.

Franke, H., J. Grosche, H. Schadlich, U. Krugel, C. Allgaier, and P. Illes. 2001. P2X receptor expression on astrocytes in the nucleus accumbens of rats. *Neuroscience* 108 (3):421–29.

Franke, H., A. Gunther, J. Grosche, R. Schmidt, S. Rossner, R. Reinhardt, H. Faber-Zuschratter, D. Schneider, and P. Illes. 2004. P2X₇ receptor expression after ischemia in the cerebral cortex of rats. *J Neuropathol Exp Neurol* 63 (7):686–99.

Fredholm, B. B., J. F. Chen, R. A. Cunha, P. Svenningsson, and J. M. Vaugeois. 2005. Adenosine and brain function. *Int Rev Neurobiol* 63:191–270.

Frigeri, A., M. A. Gropper, C. W. Turck, and A. S. Verkman. 1995a. Immunolocalization of the mercurial-insensitive water channel and glycerol intrinsic protein in epithelial cell plasma membranes. *Proc Natl Acad Sci USA* 92 (10):4328–31.

Frigeri, A., M. A. Gropper, F. Umenishi, M. Kawashima, D. Brown, and A. S. Verkman. 1995b. Localization of MIWC and GLIP water channel homologs in neuromuscular, epithelial and glandular tissues. *J Cell Sci* 108 (Pt 9):2993–3002.

Frigeri, A., G. P. Nicchia, B. Nico, F. Quondamatteo, R. Herken, L. Roncali, and M. Svelto. 2001. Aquaporin-4 deficiency in skeletal muscle and brain of dystrophic MDX mice. *FASEB J* 15 (1):90–98.

Frigeri, A., G. P. Nicchia, J. M. Verbavatz, G. Valenti, and M. Svelto. 1998. Expression of aquaporin-4 in fast-twitch fibers of mammalian skeletal muscle. *J Clin Invest* 102 (4):695–703.

Fumagalli, M., R. Brambilla, N. D'Ambrosi, C. Volonte, M. Matteoli, C. Verderio, and M. P. Abbracchio. 2003. Nucleotide-mediated calcium signaling in rat cortical astrocytes: role of P2X and P2Y receptors. *Glia* 43 (3):218–03.

Furman, C. S., D. A. Gorelick-Feldman, K. G. Davidson, T. Yasumura, J. D. Neely, P. Agre, and J. E. Rash. 2003. Aquaporin-4 square array assembly: opposing actions of M1 and M23 isoforms. *Proc Natl Acad Sci USA* 100 (23):13609–14.

Galli, C., P. Hannaert, A. S. Diaz, E. Cragoe, Jr., and R. Garay. 1988. A study of the interaction between the Na⁺, K⁺ pump and Na⁺:Ca²⁺ exchange in macrophages and vascular smooth muscle cells. *Am J Hypertens* 1 (3 Pt 3):64S–70S.

Garaschuk, O., Linn, J., Eilers, J., and Konnerth, A. 2000. Large-scale oscillatory calcium waver in the immature cortex. *Nat Neurosci* 3(5): 452–9.

Gardner-Medwin, A. R. 1983. A study of the mechanisms by which potassium moves through brain tissue in the rat. *J Physiol* 335:353–74.

Gardner-Medwin, A. R., and C. Nicholson. 1983. Changes of extracellular potassium activity induced by electric current through brain tissue in the rat. *J Physiol* 335:375–92.

Ghoshroy, S., D. A. Goodenough, and G. E. Sosinsky. 1995. Preparation, characterization, and structure of half gap junctional layers split with urea and EGTA. *J Membr Biol* 146 (1):15–28.

Giaume, C., A. Tabernero, and J. M. Medina. 1997. Metabolic trafficking through astrocytic gap junctions. *Glia* 21 (1):114–23.

Goodenough, D. A. 1974. Bulk isolation of mouse hepatocyte gap junctions. Characterization of the principal protein, connexin. *J Cell Biol* 61 (2):557–63.

Goodenough, D. A., J. S. Dick, 2nd, and J. E. Lyons. 1980. Lens metabolic cooperation: a study of mouse lens transport and permeability visualized with freeze-substitution autoradiography and electron microscopy. *J Cell Biol* 86 (2):576–89.

Goodenough, D. A., and D. L. Paul. 2003. Beyond the gap: functions of unpaired connexon channels. *Nat Rev Mol Cell Biol* 4 (4):285–94.

Grosche, J., V. Matyash, T. Moller, A. Verkhratsky, A. Reichenbach, and H. Kettenmann. 1999. Microdomains for neuron-glia interaction: parallel fiber signaling to Bergmann glial cells. *Nat Neurosci* 2 (2):139–43.

Hasegawa, H., T. Ma, W. Skach, M. A. Matthay, and A. S. Verkman. 1994. Molecular cloning of a mercurial-insensitive water channel expressed in selected water-transporting tissues. *J Biol Chem* 269 (8):5497–500.

Hatton, J. D., and M. H. Ellisman. 1984. Orthogonal arrays are redistributed in the membranes of astroglia from alumina-induced epileptic foci. *Epilepsia* 25 (2):145–51.

Higashi, K., A. Fujita, A. Inanobe, M. Tanemoto, K. Doi, T. Kubo, and Y. Kurachi. 2001. An inwardly rectifying K+ channel, Kir4.1, expressed in astrocytes surrounds synapses and blood vessels in brain. *Am J Physiol Cell Physiol* 281 (3):C922–31.

Hinson, S. R., S. F. Roemer, C. F. Lucchinetti, J. P. Fryer, T. J. Kryzer, J. L. Chamberlain, C. L. Howe, S. J. Pittock, and V. A. Lennon. 2008. Aquaporin-4-binding autoantibodies in patients with neuromyelitis optica impair glutamate transport by down-regulating EAAT2. *J Exp Med* 205 (11):2473–81.

Houades, V., N. Rouach, P. Ezan, F. Kirchhoff, A. Koulakoff, and C. Giaume. 2006. Shapes of astrocyte networks in the juvenile brain. *Neuron Glia Biol* 2 (1):3–14.

Hua, V. B., A. B. Chang, J. H. Tchieu, N. M. Kumar, P. A. Nielsen, and M. H. Saier, Jr. 2003. Sequence and phylogenetic analyses of 4 TMS junctional proteins of animals: connexins, innexins, claudins and occludins. *J Membr Biol* 194 (1):59–76.

Huang, Y., J. B. Grinspan, C. K. Abrams, and S. S. Scherer. 2007. Pannexin1 is expressed by neurons and glia but does not form functional gap junctions. *Glia* 55 (1):46–56.

Iglesias, R., G. Dahl, F. Qiu, D. C. Spray, and E. Scemes. 2009. Pannexin 1: the molecular substrate of astrocyte "hemichannels." *J Neurosci* 29 (21):7092–97.

Iglesias, R., S. Locovei, A. Roque, A. P. Alberto, G. Dahl, D. C. Spray, and E. Scemes. 2008. P2X7 receptor-pannexin1 complex: pharmacology and signaling. *Am J Physiol Cell Physiol* 295 (3):C752–60.

Ishibashi, K. 2009. New members of mammalian aquaporins: AQP10-AQP12. *Handb Exp Pharmacol* 190:251–62.

Jabs, R., K. Matthias, A. Grote, M. Grauer, G. Seifert, and C. Steinhauser. 2007. Lack of P2X receptor mediated currents in astrocytes and GluR type glial cells of the hippocampal CA1 region. *Glia* 55 (16):1648–55.

John, G. R., E. Scemes, S. O. Suadicani, J. S. Liu, P. C. Charles, S. C. Lee, D. C. Spray, and C. F. Brosnan. 1999. IL-1beta differentially regulates calcium wave propagation between primary human fetal astrocytes via pathways involving P2 receptors and gap junction channels. *Proc Natl Acad Sci USA* 96 (20):11613–18.

John, S., D. Cesario, and J. N. Weiss. 2003. Gap junctional hemichannels in the heart. *Acta Physiol Scand* 179 (1):23–31.

Judisch, G. F., A. Martin-Casals, J. W. Hanson, and W. H. Olin. 1979. Oculodentodigital dysplasia. Four new reports and a literature review. *Arch Ophthalmol* 97 (5):878–84.

Jung, J. S., R. V. Bhat, G. M. Preston, W. B. Guggino, J. M. Baraban, and P. Agre. 1994. Molecular characterization of an aquaporin cDNA from brain: candidate osmoreceptor and regulator of water balance. *Proc Natl Acad Sci USA* 91 (26):13052–56.

Kanaporis, G., P. R. Brink, and V. Valiunas. 2011. Gap junction permeability: selectivity for anionic and cationic probes. *Am J Physiol Cell Physiol* 300(3): C600–9.

Kandler, K., and L. C. Katz. 1998. Coordination of neuronal activity in developing visual cortex by gap junction-mediated biochemical communication. *J Neurosci* 18 (4):1419–27.

Kang, J., L. Jiang, S. A. Goldman, and M. Nedergaard. 1998. Astrocyte-mediated potentiation of inhibitory synaptic transmission. *Nat Neurosci* 1 (8):683–92.

Kang, J., N. Kang, D. Lovatt, A. Torres, Z. Zhao, J. Lin, and M. Nedergaard. 2008. Connexin 43 hemichannels are permeable to ATP. *J Neurosci* 28 (18):4702–11.

Kofuji, P., B. Biedermann, V. Siddharthan, M. Raap, I. Iandiev, I. Milenkovic, A. Thomzig, R. W. Veh, A. Bringmann, and A. Reichenbach. 2002. Kir potassium channel subunit expression in retinal glial cells: implications for spatial potassium buffering. *Glia* 39 (3):292–303.

Kong, H., Y. Fan, J. Xie, J. Ding, L. Sha, X. Shi, X. Sun, and G. Hu. 2008. AQP4 knockout impairs proliferation, migration and neuronal differentiation of adult neural stem cells. *J Cell Sci* 121 (Pt 24):4029–36.

Kucher, B. M., and J. T. Neary. 2005. Bi-functional effects of ATP/P2 receptor activation on tumor necrosis factor-alpha release in lipopolysaccharide-stimulated astrocytes. *J Neurochem* 92 (3):525–35.

Kuffler, S. W., J. G. Nicholls, and R. K. Orkand. 1966. Physiological properties of glial cells in the central nervous system of amphibia. *J Neurophysiol* 29 (4):768–87.

Kukley, M., J. A. Barden, C. Steinhauser, and R. Jabs. 2001. Distribution of P2X receptors on astrocytes in juvenile rat hippocampus. *Glia* 36 (1):11–21.

Kunzelmann, P., I. Blumcke, O. Traub, R. Dermietzel, and K. Willecke. 1997. Coexpression of connexin45 and -32 in oligodendrocytes of rat brain. *J Neurocytol* 26 (1):17–22.

Lalo, U., Y. Pankratov, S. P. Wichert, M. J. Rossner, R. A. North, F. Kirchhoff, and A. Verkhratsky. 2008. $P2X_1$ and $P2X_5$ subunits form the functional P2X receptor in mouse cortical astrocytes. *J Neurosci* 28 (21):5473–80.

Landis, D. M., and T. S. Reese. 1981. Astrocyte membrane structure: changes after circulatory arrest. *J Cell Biol* 88 (3):660–63.

Larsen, W. J., H. N. Tung, S. A. Murray, and C. A. Swenson. 1979. Evidence for the participation of actin microfilaments and bristle coats in the internalization of gap junction membrane. *J Cell Biol* 83 (3):576–87.

Lecanda, F., P. M. Warlow, S. Sheikh, F. Furlan, T. H. Steinberg, and R. Civitelli. 2000. Connexin43 deficiency causes delayed ossification, craniofacial abnormalities, and osteoblast dysfunction. *J Cell Biol* 151 (4):931–44.

Lennon, V. A., T. J. Kryzer, S. J. Pittock, A. S. Verkman, and S. R. Hinson. 2005. IgG marker of optic-spinal multiple sclerosis binds to the aquaporin-4 water channel. *J Exp Med* 202 (4):473–77.

Li, X., H. Kong, W. Wu, M. Xiao, X. Sun, and G. Hu. 2009. Aquaporin-4 maintains ependymal integrity in adult mice. *Neuroscience* 162 (1):67–77.

Li, X. X., T. Nomura, H. Aihara, and T. Nishizaki. 2001. Adenosine enhances glial glutamate efflux via A2a adenosine receptors. *Life Sci* 68 (12):1343–50.

Lichter-Konecki, U., J. M. Mangin, H. Gordish-Dressman, E. P. Hoffman, and V. Gallo. 2008. Gene expression profiling of astrocytes from hyperammonemic mice reveals altered pathways for water and potassium homeostasis *in vivo*. *Glia* 56 (4):365–77.

Lim, J. H., H. M. Gibbons, S. J. O'Carroll, P. J. Narayan, R. L. Faull, and M. Dragunow. 2007. Extracellular signal-regulated kinase involvement in human astrocyte migration. *Brain Res* 1164:1–13.

Locovei, S., E. Scemes, F. Qiu, D. C. Spray, and G. Dahl. 2007. Pannexin1 is part of the pore forming unit of the P2X(7) receptor death complex. *FEBS Lett* 581 (3):483–88.

Lopes-Cardozo, M., and W. Klein. 1985. Contribution of acetoacetate to the synthesis of cholesterol and fatty acids in regions of developing rat brain *in vivo*. *Neurochem Int* 7 (4):647–53.

Lutz, S. E., Y. Zhao, M. Gulinello, S. C. Lee, C. S. Raine, and C. F. Brosnan. 2009. Deletion of astrocyte connexins 43 and 30 leads to a dysmyelinating phenotype and hippocampal CA1 vacuolation. *J Neurosci* 29 (24):7743–52.

Magnotti, L. M., D. A. Goodenough, and D. L. Paul. 2011. Functional heterotypic interactions between astrocyte and oligodendrocyte connexins. *Glia* 59 (1):26–34.

Maher, F., S. J. Vannucci, and I. A. Simpson. 1994. Glucose transporter proteins in brain. *FASEB J* 8 (13):1003–11.

Manley, G. T., M. Fujimura, T. Ma, N. Noshita, F. Filiz, A. W. Bollen, P. Chan, and A. S. Verkman. 2000. Aquaporin-4 deletion in mice reduces brain edema after acute water intoxication and ischemic stroke. *Nat Med* 6 (2):159–63.

Martin, E. D., M. Fernandez, G. Perea, O. Pascual, P. G. Haydon, A. Araque, and V. Cena. 2007. Adenosine released by astrocytes contributes to hypoxia-induced modulation of synaptic transmission. *Glia* 55 (1):36–45.

Martin, P. E., and W. H. Evans. 2004. Incorporation of connexins into plasma membranes and gap junctions. *Cardiovasc Res* 62 (2):378–87.

Massa, P. T., and E. Mugnaini. 1985. Cell-cell junctional interactions and characteristic plasma membrane features of cultured rat glial cells. *Neuroscience* 14 (2):695–709.

Mazet, F., B. A. Wittenberg, and D. C. Spray. 1985. Fate of intercellular junctions in isolated adult rat cardiac cells. *Circ Res* 56 (2):195–204.

McNutt, N. S., and R. S. Weinstein. 1970. The ultrastructure of the nexus. A correlated thin-section and freeze-cleave study. *J Cell Biol* 47 (3):666–88.

Meister, M., R. O. Wong, D. A. Baylor, and C. J. Shatz. 1991. Synchronous bursts of action potentials in ganglion cells of the developing mammalian retina. *Science* 252 (5008):939–43.

Menichella, D. M., D. A. Goodenough, E. Sirkowski, S. S. Scherer, and D. L. Paul. 2003. Connexins are critical for normal myelination in the CNS. *J Neurosci* 23 (13):5963–73.

Moran, J., M. Sabanero, I. Meza, and H. Pasantes-Morales. 1996. Changes of actin cytoskeleton during swelling and regulatory volume decrease in cultured astrocytes. *Am J Physiol* 271 (6 Pt 1):C1901–7.

Murakami, K., Y. Nakamura, and Y. Yoneda. 2003. Potentiation by ATP of lipopolysaccharide-stimulated nitric oxide production in cultured astrocytes. *Neuroscience* 117 (1):37–42.

Murata, K., K. Mitsuoka, T. Hirai, T. Walz, P. Agre, J. B. Heymann, A. Engel, and Y. Fujiyoshi. 2000. Structural determinants of water permeation through aquaporin-1. *Nature* 407 (6804):599–605.

Mylonakou, M. N., P. H. Petersen, E. Rinvik, A. Rojek, E. Valdimarsdottir, S. Zelenin, T. Zeuthen, S. Nielsen, O. P. Ottersen, and M. Amiry-Moghaddam. 2009. Analysis of mice with targeted deletion of AQP9 gene provides conclusive evidence for expression of AQP9 in neurons. *J Neurosci Res* 87 (6):1310–22.

Nagelhus, E. A., M. L. Veruki, R. Torp, F. M. Haug, J. H. Laake, S. Nielsen, P. Agre, and O. P. Ottersen. 1998. Aquaporin-4 water channel protein in the rat retina and optic nerve: polarized expression in Muller cells and fibrous astrocytes. *J Neurosci* 18 (7):2506–19.

Nagy, J. I., and J. E. Rash. 2000. Connexins and gap junctions of astrocytes and oligodendrocytes in the CNS. *Brain Res Brain Res Rev* 32 (1):29–44.

Nagy, J. I., and J. E. Rash. 2003. Astrocyte and oligodendrocyte connexins of the glial syncytium in relation to astrocyte anatomical domains and spatial buffering. *Cell Commun Adhes* 10 (4–6):401–6.

Nakhoul, N. L., B. A. Davis, M. F. Romero, and W. F. Boron. 1998. Effect of expressing the water channel aquaporin-1 on the CO_2 permeability of *Xenopus* oocytes. *Am J Physiol* 274 (2 Pt 1):C543–48.

Narcisse, L., E. Scemes, Y. Zhao, S. C. Lee, and C. F. Brosnan. 2005. The cytokine IL-1beta transiently enhances $P2X_7$ receptor expression and function in human astrocytes. *Glia* 49 (2):245–58.

Neely, J. D., B. M. Christensen, S. Nielsen, and P. Agre. 1999. Heterotetrameric composition of aquaporin-4 water channels. *Biochemistry* 38 (34):11156–63.

Neijssen, J., C. Herberts, J. W. Drijfhout, E. Reits, L. Janssen, and J. Neefjes. 2005. Cross-presentation by intercellular peptide transfer through gap junctions. *Nature* 434 (7029):83–88.

Neuhaus, J., E. M. Schmid, and H. Wolburg. 1990. Stability of orthogonal arrays of particles in murine skeletal muscle and astrocytes after circulatory arrest, and human gliomas. *Neurosci Lett* 109 (1–2):163–68.

Newman, E., and A. Reichenbach. 1996. The Muller cell: a functional element of the retina. *Trends Neurosci* 19 (8):307–12.

Newman, E. A. 1985. Membrane physiology of retinal glial (Muller) cells. *J Neurosci* 5 (8):2225–39.

Nicchia, G. P., L. Cogotzi, A. Rossi, D. Basco, A. Brancaccio, M. Svelto, and A. Frigeri. 2008a. Expression of multiple AQP4 pools in the plasma membrane and their association with the dystrophin complex. *J Neurochem* 105 (6):2156–65.

Nicchia, G. P., A. Frigeri, B. Nico, D. Ribatti, and M. Svelto. 2001. Tissue distribution and membrane localization of aquaporin-9 water channel: evidence for sex-linked differences in liver. *J Histochem Cytochem* 49 (12):1547–56.

Nicchia, G. P., M. Mastrototaro, A. Rossi, F. Pisani, C. Tortorella, M. Ruggieri, A. Lia, M. Trojano, A. Frigeri, and M. Svelto. 2009. Aquaporin-4 orthogonal arrays of particles are the target for neuromyelitis optica autoantibodies. *Glia* 57 (13):1363–73.

Nicchia, G. P., B. Nico, L. M. Camassa, M. G. Mola, N. Loh, R. Dermietzel, D. C. Spray, M. Svelto, and A. Frigeri. 2004. The role of aquaporin-4 in the blood-brain barrier development and integrity: studies in animal and cell culture models. *Neuroscience* 129 (4):935–45.

Nicchia, G. P., A. Rossi, M. G. Mola, F. Pisani, C. Stigliano, D. Basco, M. Mastrototaro, M. Svelto, and A. Frigeri. 2010. Higher order structure of aquaporin-4. *Neuroscience* 168 (4):903–14.

Nicchia, G. P., A. Rossi, M. G. Mola, G. Procino, A. Frigeri, and M. Svelto. 2008b. Actin cytoskeleton remodeling governs aquaporin-4 localization in astrocytes. *Glia* 56 (16):1755–66.

Nicchia, G. P., A. Rossi, U. Nudel, M. Svelto, and A. Frigeri. 2008c. Dystrophin-dependent and -independent AQP4 pools are expressed in the mouse brain. *Glia* 56 (8):869–76.

Nicchia, G. P., M. Srinivas, W. Li, C. F. Brosnan, A. Frigeri, and D. C. Spray. 2005. New possible roles for aquaporin-4 in astrocytes: cell cytoskeleton and functional relationship with connexin43. *FASEB J* 19 (12):1674–76.

Nielsen, S., E. A. Nagelhus, M. Amiry-Moghaddam, C. Bourque, P. Agre, and O. P. Ottersen. 1997. Specialized membrane domains for water transport in glial cells: high-resolution immunogold cytochemistry of aquaporin-4 in rat brain. *J Neurosci* 17 (1):171–80.

Nielsen, S., B. L. Smith, E. I. Christensen, M. A. Knepper, and P. Agre. 1993. CHIP28 water channels are localized in constitutively water-permeable segments of the nephron. *J Cell Biol* 120 (2):371–83.

Nishizaki, T., K. Nagai, T. Nomura, H. Tada, T. Kanno, H. Tozaki, X. X. Li, T. Kondoh, N. Kodama, E. Takahashi, N. Sakai, K. Tanaka, and N. Saito. 2002. A new neuromodulatory pathway with a glial contribution mediated via A2a adenosine receptors. *Glia* 39 (2):133–47.

Nixdorf-Bergweiler, B. E., D. Albrecht, and U. Heinemann. 1994. Developmental changes in the number, size, and orientation of GFAP-positive cells in the CA1 region of rat hippocampus. *Glia* 12 (3):180–95.

Odermatt, B., K. Wellershaus, A. Wallraff, G. Seifert, J. Degen, C. Euwens, B. Fuss, H. Bussow, K. Schilling, C. Steinhauser, and K. Willecke. 2003. Connexin 47 (Cx47)-deficient mice with enhanced green fluorescent protein reporter gene reveal predominant oligodendrocytic expression of Cx47 and display vacuolized myelin in the CNS. *J Neurosci* 23 (11):4549–59.

Orkand, R. K., J. G. Nicholls, and S. W. Kuffler. 1966. Effect of nerve impulses on the membrane potential of glial cells in the central nervous system of amphibia. *J Neurophysiol* 29 (4):788–806.

Oshio, K., H. Watanabe, Y. Song, A. S. Verkman, and G. T. Manley. 2005. Reduced cerebrospinal fluid production and intracranial pressure in mice lacking choroid plexus water channel aquaporin-1. *FASEB J* 19 (1):76–78.

Panchin, Y., I. Kelmanson, M. Matz, K. Lukyanov, N. Usman, and S. Lukyanov. 2000. A ubiquitous family of putative gap junction molecules. *Curr Biol* 10 (13):R473–74.

Panchin, Y. V. 2005. Evolution of gap junction proteins—the pannexin alternative. *J Exp Biol* 208 (Pt 8):1415–19.

Pannicke, T., W. Fischer, B. Biedermann, H. Schadlich, J. Grosche, F. Faude, P. Wiedemann, C. Allgaier, P. Illes, G. Burnstock, and A. Reichenbach. 2000. $P2X_7$ receptors in Muller glial cells from the human retina. *J Neurosci* 20 (16):5965–72.

Papadopoulos, M. C., G. T. Manley, S. Krishna, and A. S. Verkman. 2004. Aquaporin-4 facilitates reabsorption of excess fluid in vasogenic brain edema. *FASEB J* 18 (11):1291–93.

Papadopoulos, M. C., S. Saadoun, and A. S. Verkman. 2008. Aquaporins and cell migration. *Pflugers Arch* 456 (4):693–700.

Papadopoulos, M. C., and A. S. Verkman. 2005. Aquaporin-4 gene disruption in mice reduces brain swelling and mortality in pneumococcal meningitis. *J Biol Chem* 280 (14):13906–12.

Parpura, V., T. A. Basarsky, F. Liu, K. Jeftinija, S. Jeftinija, and P. G. Haydon. 1994. Glutamate-mediated astrocyte-neuron signalling. *Nature* 369 (6483):744–47.

Parpura, V., E. Scemes, and D. C. Spray. 2004. Mechanisms of glutamate release from astrocytes: gap junction "hemichannels," purinergic receptors and exocytotic release. *Neurochem Int* 45 (2–3):259–64.

Parri, H. R., T. M. Gould, and V. Crunelli. 2001. Spontaneous astrocytic Ca^{2+} oscillations *in situ* drive NMDAR-mediated neuronal excitation. *Nat Neurosci* 4 (8):803–12.

Paznekas, W. A., S. A. Boyadjiev, R. E. Shapiro, O. Daniels, B. Wollnik, C. E. Keegan, J. W. Innis, M. B. Dinulos, C. Christian, M. C. Hannibal, and E. W. Jabs. 2003. Connexin 43 (GJA1) mutations cause the pleiotropic phenotype of oculodentodigital dysplasia. *Am J Hum Genet* 72 (2):408–18.

Paznekas, W. A., B. Karczeski, S. Vermeer, R. B. Lowry, M. Delatycki, F. Laurence, P. A. Koivisto, L. Van Maldergem, S. A. Boyadjiev, J. N. Bodurtha, and E. W. Jabs. 2009. GJA1 mutations, variants, and connexin 43 dysfunction as it relates to the oculodentodigital dysplasia phenotype. *Hum Mutat* 30 (5):724–33.

Pelegrin, P., and A. Surprenant. 2006. Pannexin-1 mediates large pore formation and interleukin-1beta release by the ATP-gated $P2X_7$ receptor. *EMBO J* 25 (21):5071–82.

Penuela, S., R. Bhalla, X. Q. Gong, K. N. Cowan, S. J. Celetti, B. J. Cowan, D. Bai, Q. Shao, and D. W. Laird. 2007. Pannexin 1 and pannexin 3 are glycoproteins that exhibit many distinct characteristics from the connexin family of gap junction proteins. *J Cell Sci* 120 (Pt 21):3772–83.

Penuela, S., R. Bhalla, K. Nag, and D. W. Laird. 2009. Glycosylation regulates pannexin intermixing and cellular localization. *Mol Biol Cell* 20 (20):4313–23.

Phelan, P. 2005. Innexins: members of an evolutionarily conserved family of gap-junction proteins. *Biochim Biophys Acta* 1711 (2):225–45.

Pittock, S. J. 2008. Neuromyelitis optica: a new perspective. *Semin Neurol* 28 (1):95–104.

Pitts, J. D. and M. E. Finbow. 1997. Junctional permeability and its consequences. In *Intracellular communication*. Ed. W. C. Mello, 61. New York: Plenum.

Poopalasundaram, S., C. Knott, O. G. Shamotienko, P. G. Foran, J. O. Dolly, C. A. Ghiani, V. Gallo, and G. P. Wilkin. 2000. Glial heterogeneity in expression of the inwardly rectifying K^+ channel, Kir4.1, in adult rat CNS. *Glia* 30 (4):362–72.

Preston, G. M., T. P. Carroll, W. B. Guggino, and P. Agre. 1992. Appearance of water channels in *Xenopus* oocytes expressing red cell CHIP28 protein. *Science* 256 (5055):385–87.

Rash, J. E., H. S. Duffy, F. E. Dudek, B. L. Bilhartz, L. R. Whalen, and T. Yasumura. 1997. Grid-mapped freeze-fracture analysis of gap junctions in gray and white matter of adult rat central nervous system, with evidence for a "panglial syncytium" that is not coupled to neurons. *J Comp Neurol* 388 (2):265–92.

Rash, J. E., and T. Yasumura. 1999. Direct immunogold labeling of connexins and aquaporin-4 in freeze-fracture replicas of liver, brain, and spinal cord: factors limiting quantitative analysis. *Cell Tissue Res* 296 (2):307–21.

Rash, J. E., T. Yasumura, C. S. Hudson, P. Agre, and S. Nielsen. 1998. Direct immunogold labeling of aquaporin-4 in square arrays of astrocyte and ependymocyte plasma membranes in rat brain and spinal cord. *Proc Natl Acad Sci USA* 95 (20):11981–86.

Reaume, A. G., P. A. de Sousa, S. Kulkarni, B. L. Langille, D. Zhu, T. C. Davies, S. C. Juneja, G. M. Kidder, and J. Rossant. 1995. Cardiac malformation in neonatal mice lacking connexin43. *Science* 267 (5205):1831–34.

Reichenbach, A., A. Henke, W. Eberhardt, W. Reichelt, and D. Dettmer. 1992. K+ ion regulation in retina. *Can J Physiol Pharmacol* 70 (Suppl):S239–47.

Revel, J. P., and M. J. Karnovsky. 1967. Hexagonal array of subunits in intercellular junctions of the mouse heart and liver. *J Cell Biol* 33 (3):C7–12.

Roemer, S. F., J. E. Parisi, V. A. Lennon, E. E. Benarroch, H. Lassmann, W. Bruck, R. N. Mandler, B. G. Weinshenker, S. J. Pittock, D. M. Wingerchuk, and C. F. Lucchinetti. 2007. Pattern-specific loss of aquaporin-4 immunoreactivity distinguishes neuromyelitis optica from multiple sclerosis. *Brain* 130 (Pt 5):1194–205.

Rohlmann, A., and J. R. Wolff. 1996. Subcellular topography and plasticity of gap junction distribution in astrocytes. In *Gap junctions in the nervous system*, ed. D. C. Spray and R. Dermietzel, 175–192. Austin, TX: Landes.

Rossi, A., F. Pisani, G. P. Nicchia, M. Svelto, and A. Frigeri. 2010. Evidences for a leaky scanning mechanism for the synthesis of the shorter M23 protein isoform of aquaporin-4: implication in orthogonal array formation and neuromyelitis optica antibody interaction. *J Biol Chem* 285 (7):4562–69.

Rouach, N., A. Koulakoff, V. Abudara, K. Willecke, and C. Giaume. 2008. Astroglial metabolic networks sustain hippocampal synaptic transmission. *Science* 322 (5907):1551–55.

Saadoun, S., M. C. Papadopoulos, H. Watanabe, D. Yan, G. T. Manley, and A. S. Verkman. 2005. Involvement of aquaporin-4 in astroglial cell migration and glial scar formation. *J Cell Sci* 118 (Pt 24):5691–98.

Sabater, L., A. Giralt, A. Boronat, K. Hankiewicz, Y. Blanco, S. Llufriu, J. Alberch, F. Graus, and A. Saiz. 2009. Cytotoxic effect of neuromyelitis optica antibody (NMO-IgG) to astrocytes: an *in vitro* study. *J Neuroimmunol* 215 (1–2):31–35.

Saez, J. C., M. A. Retamal, D. Basilio, F. F. Bukauskas, and M. V. Bennett. 2005. Connexin-based gap junction hemichannels: gating mechanisms. *Biochim Biophys Acta* 1711 (2):215–24.

Santiago, M. F., P. Alcami, K. M. Striedinger, D. C. Spray, and E. Scemes. 2010. The carboxyl-terminal domain of connexin43 is a negative modulator of neuronal differentiation. *J Biol Chem* 285 (16):11836–45.

Scemes, E., R. Dermietzel, and D. C. Spray. 1998. Calcium waves between astrocytes from Cx43 knockout mice. *Glia* 24 (1):65–73.

Scemes, E., N. Duval, and P. Meda. 2003. Reduced expression of P2Y$_1$ receptors in connexin43-null mice alters calcium signaling and migration of neural progenitor cells. *J Neurosci* 23 (36):11444–52.

Scemes, E., and C. Giaume. 2006. Astrocyte calcium waves: what they are and what they do. *Glia* 54 (7):716–25.

Scemes, E., and D. C. Spray. 2008. Connexin expression (gap junctions and hemichannels) in astrocytes. In *Astrocytes in (patho)physiology of the nervous system*, ed. V. Parpura and P. G. Haydon, 107–150. New York: Springer.

Scemes, E., S. O. Suadicani, G. Dahl, and D. C. Spray. 2007. Connexin and pannexin mediated cell-cell communication. *Neuron Glia Biol* 3 (3):199–208.

Scemes, E., S. O. Suadicani, and D. C. Spray. 2000. Intercellular communication in spinal cord astrocytes: fine tuning between gap junctions and P2 nucleotide receptors in calcium wave propagation. *J Neurosci* 20 (4):1435–45.

Scherer, S. S., L. J. Bone, S. M. Deschenes, A. Abel, R. J. Balice-Gordon, and K. H. Fischbeck. 1999. The role of the gap junction protein connexin32 in the pathogenesis of X-linked Charcot-Marie-Tooth disease. *Novartis Found Symp* 219:175–85; discussion, 185–87.

Schipke, C. G., C. Boucsein, C. Ohlemeyer, F. Kirchhoff, and H. Kettenmann. 2002. Astrocyte Ca_2^+ waves trigger responses in microglial cells in brain slices. *FASEB J* 16 (2):255–57.

Schroder, W., G. Seifert, K. Huttmann, S. Hinterkeuser, and C. Steinhauser. 2002. AMPA receptor-mediated modulation of inward rectifier K+ channels in astrocytes of mouse hippocampus. *Mol Cell Neurosci* 19 (3):447–58.

Sharma, R., M. T. Fischer, J. Bauer, P. A. Felts, K. J. Smith, T. Misu, K. Fujihara, M. Bradl, and H. Lassmann. 2010. Inflammation induced by innate immunity in the central nervous system leads to primary astrocyte dysfunction followed by demyelination. *Acta Neuropathol* 120 (2):223–36.

Silberstein, C., R. Bouley, Y. Huang, P. Fang, N. Pastor-Soler, D. Brown, and A. N. Van Hoek. 2004. Membrane organization and function of M1 and M23 isoforms of aquaporin-4 in epithelial cells. *Am J Physiol Renal Physiol* 287 (3):F501–11.

Silverman, W. R., J. P. de Rivero Vaccari, S. Locovei, F. Qiu, S. K. Carlsson, E. Scemes, R. W. Keane, and G. Dahl. 2009. The pannexin 1 channel activates the inflammasome in neurons and astrocytes. *J Biol Chem* 284 (27):18143–51.

Sim, J. A., M. T. Young, H. Y. Sung, R. A. North, and A. Surprenant. 2004. Reanalysis of $P2X_7$ receptor expression in rodent brain. *J Neurosci* 24 (28):6307–14.

Sohl, G., B. Odermatt, S. Maxeiner, J. Degen, and K. Willecke. 2004. New insights into the expression and function of neural connexins with transgenic mouse mutants. *Brain Res Brain Res Rev* 47 (1–3):245–59.

Sperlagh, B., E. S. Vizi, K. Wirkner, and P. Illes. 2006. $P2X_7$ receptors in the nervous system. *Prog Neurobiol* 78 (6):327–46.

Spray, D. C., and R. Dermietzel. 1995. X-linked dominant Charcot-Marie-Tooth disease and other potential gap-junction diseases of the nervous system. *Trends Neurosci* 18 (6):256–62.

Spray, D. C., E. Scemes, R. Rozental, and R. Dermietzel. 2004. Cell-cell communication: an overview emphasizing gap junctions. In *An introduction to cellular and molecular neuroscience*, ed. J. Byrne and J. Roberts, 431–58. New York: Academic Press.

Stout, C. E., J. L. Costantin, C. C. Naus, and A. C. Charles. 2002. Intercellular calcium signaling in astrocytes via ATP release through connexin hemichannels. *J Biol Chem* 277 (12):10482–88.

Striedinger, K., P. Meda, and E. Scemes. 2007. Exocytosis of ATP from astrocyte progenitors modulates spontaneous Ca^{2+} oscillations and cell migration. *Glia* 55 (6):652–62.

Suadicani, S. O., C. F. Brosnan, and E. Scemes. 2006. $P2X_7$ receptors mediate ATP release and amplification of astrocytic intercellular Ca^{2+} signaling. *J Neurosci* 26 (5):1378–85.

Suadicani, S. O., M. H. De Pina-Benabou, M. Urban-Maldonado, D. C. Spray, and E. Scemes. 2003. Acute downregulation of Cx43 alters P2Y receptor expression levels in mouse spinal cord astrocytes. *Glia* 42 (2):160–71.

Suadicani, S. O., R. Iglesias, D. C. Spray, and E. Scemes. 2009. Point mutation in the mouse P2X$_7$ receptor affects intercellular calcium waves in astrocytes. *ASN Neuro* 1 (1):ii: e00005.

Sui, H., B. G. Han, J. K. Lee, P. Walian, and B. K. Jap. 2001. Structural basis of water-specific transport through the AQP1 water channel. *Nature* 414 (6866):872–78.

Tabernero, A., C. Vicario, and J. M. Medina. 1996. Lactate spares glucose as a metabolic fuel in neurons and astrocytes from primary culture. *Neurosci Res* 26 (4):369–76.

Takumi, T., T. Ishii, Y. Horio, K. Morishige, N. Takahashi, M. Yamada, T. Yamashita, H. Kiyama, K. Sohmiya, S. Nakanishi, et al. 1995. A novel ATP-dependent inward rectifier potassium channel expressed predominantly in glial cells. *J Biol Chem* 270 (27):16339–46.

Tashiro, A., J. Goldberg, and R. Yuste. 2002. Calcium oscillations in neocortical astrocytes under epileptiform conditions. *J Neurobiol* 50 (1):45–55.

Teubner, B., B. Odermatt, M. Guldenagel, G. Sohl, J. Degen, F. Bukauskas, J. Kronengold, V. K. Verselis, Y. T. Jung, C. A. Kozak, K. Schilling, and K. Willecke. 2001. Functional expression of the new gap junction gene connexin47 transcribed in mouse brain and spinal cord neurons. *J Neurosci* 21 (4):1117–26.

Theis, M., R. Jauch, L. Zhuo, D. Speidel, A. Wallraff, B. Doring, C. Frisch, G. Sohl, B. Teubner, C. Euwens, J. Huston, C. Steinhauser, A. Messing, U. Heinemann, and K. Willecke. 2003. Accelerated hippocampal spreading depression and enhanced locomotory activity in mice with astrocyte-directed inactivation of connexin43. *J Neurosci* 23 (3):766–76.

Theis, M., G. Sohl, J. Eiberger, and K. Willecke. 2005. Emerging complexities in identity and function of glial connexins. *Trends Neurosci* 28 (4):188–95.

Vajda, Z., M. Pedersen, E. M. Fuchtbauer, K. Wertz, H. Stodkilde-Jorgensen, E. Sulyok, T. Doczi, J. D. Neely, P. Agre, J. Frokiaer, and S. Nielsen. 2002. Delayed onset of brain edema and mislocalization of aquaporin-4 in dystrophin-null transgenic mice. *Proc Natl Acad Sci USA* 99 (20):13131–36.

Valiunas, V., J. F. Bechberger, C. C. Naus, P. R. Brink, and G. S. Goldberg. 2005. Nontransformed cells can normalize gap junctional communication with transformed cells. *Biochem Biophys Res Commun* 333 (1):174–79.

Verbavatz, J. M., T. Ma, R. Gobin, and A. S. Verkman. 1997. Absence of orthogonal arrays in kidney, brain and muscle from transgenic knockout mice lacking water channel aquaporin-4. *J Cell Sci* 110 (Pt 22):2855–60.

Verderio, C., and M. Matteoli. 2001. ATP mediates calcium signaling between astrocytes and microglial cells: modulation by IFN-gamma. *J Immunol* 166 (10):6383–91.

Verkhratsky, A., O. A. Krishtal, and G. Burnstock. 2009. Purinoceptors on neuroglia. *Mol Neurobiol* 39 (3):190–208.

Verkman, A. S. 2009. Knock-out models reveal new aquaporin functions. *Handb Exp Pharmacol* 190:359–81.

Verkman, A. S., and A. K. Mitra. 2000. Structure and function of aquaporin water channels. *Am J Physiol Renal Physiol* 278 (1):F13–28.

Verselis, V. K., E. B. Trexler, and F. F. Bukauskas. 2000. Connexin hemichannels and cell-cell channels: comparison of properties. *Braz J Med Biol Res* 33 (4):379–89.

Vicario, C., A. Tabernero, and J. M. Medina. 1993. Regulation of lactate metabolism by albumin in rat neurons and astrocytes from primary culture. *Pediatr Res* 34 (6):709–15.

Vincent, T., P. Saikali, R. Cayrol, A. D. Roth, A. Bar-Or, A. Prat, and J. P. Antel. 2008. Functional consequences of neuromyelitis optica-IgG astrocyte interactions on blood-brain barrier permeability and granulocyte recruitment. *J Immunol* 181 (8):5730–37.

Vizuete, M. L., J. L. Venero, C. Vargas, A. A. Ilundain, M. Echevarria, A. Machado, and J. Cano. 1999. Differential upregulation of aquaporin-4 mRNA expression in reactive astrocytes after brain injury: potential role in brain edema. *Neurobiol Dis* 6 (4):245–58.

Wakayama, Y., T. Jimi, N. Misugi, T. Kumagai, S. Miyake, S. Shibuya, and T. Miike. 1989. Dystrophin immunostaining and freeze-fracture studies of muscles of patients with early stage amyotrophic lateral sclerosis and Duchenne muscular dystrophy. *J Neurol Sci* 91 (1–2):191–205.

Wallraff, A., R. Kohling, U. Heinemann, M. Theis, K. Willecke, and C. Steinhauser. 2006. The impact of astrocytic gap junctional coupling on potassium buffering in the hippocampus. *J Neurosci* 26 (20):5438–47.

Walz, W. 2000. Role of astrocytes in the clearance of excess extracellular potassium. *Neurochem Int* 36 (4–5):291–300.

Wang, Z., P. G. Haydon, and E. S. Yeung. 2000. Direct observation of calcium-independent intercellular ATP signaling in astrocytes. *Anal Chem* 72 (9):2001–7.

Wen, H., E. A. Nagelhus, M. Amiry-Moghaddam, P. Agre, O. P. Ottersen, and S. Nielsen. 1999. Ontogeny of water transport in rat brain: postnatal expression of the aquaporin-4 water channel. *Eur J Neurosci* 11 (3):935–45.

White, T. W., and R. Bruzzone. 1996. Multiple connexin proteins in single intercellular channels: connexin compatibility and functional consequences. *J Bioenerg Biomembr* 28 (4):339–50.

Willecke, K., J. Eiberger, J. Degen, D. Eckardt, A. Romualdi, M. Guldenagel, U. Deutsch, and G. Sohl. 2002. Structural and functional diversity of connexin genes in the mouse and human genome. *Biol Chem* 383 (5):725–37.

Williamson, D. H. 1982. The production and utilization of ketone bodies in the neonate. In *Biochemical development of the fetus and neonate*, ed. C. T. Jones, 621–50. Amsterdan: Elsevier.

Wingerchuk, D. M. 2007. Neuromyelitis optica: new findings on pathogenesis. *Int Rev Neurobiol* 79:665–88.

Wolburg, H. 1995. Orthogonal arrays of intramembranous particles: a review with special reference to astrocytes. *J Hirnforsch* 36 (2):239–58.

Wolff, J. R., and T. I. Chao. 2004. Cytoarchitectonics of non-neuronal cells in the nervous system. In *Non-neuronal cells of the nervous system: function and distinctions*, ed. L. Hertz, 1–51. Amsterdam: Elsevier.

Wolff, J. R., K. Stuke, M. Missler, H. Tytko, P. Schwarz, A. Rohlmann, and T. I. Chao. 1998. Autocellular coupling by gap junctions in cultured astrocytes: a new view on cellular autoregulation during process formation. *Glia* 24 (1):121–40.

Yang, B., D. Brown, and A. S. Verkman. 1996. The mercurial insensitive water channel (AQP-4) forms orthogonal arrays in stably transfected Chinese hamster ovary cells. *J Biol Chem* 271 (9):4577–80.

Yang, B., N. Fukuda, A. van Hoek, M. A. Matthay, T. Ma, and A. S. Verkman. 2000. Carbon dioxide permeability of aquaporin-1 measured in erythrocytes and lung of aquaporin-1 null mice and in reconstituted proteoliposomes. *J Biol Chem* 275 (4):2686–92.

Ye, Z. C., N. Oberheim, H. Kettenmann, and B. R. Ransom. 2009. Pharmacological "cross-inhibition" of connexin hemichannels and swelling activated anion channels. *Glia* 57 (3):258–69.

Ye, Z. C., M. S. Wyeth, S. Baltan-Tekkok, and B. R. Ransom. 2003. Functional hemichannels in astrocytes: a novel mechanism of glutamate release. *J Neurosci* 23 (9):3588–96.

Yu, Y., S. Ugawa, T. Ueda, Y. Ishida, K. Inoue, A. Kyaw Nyunt, A. Umemura, M. Mase, K. Yamada, and S. Shimada. 2008. Cellular localization of $P2X_7$ receptor mRNA in the rat brain. *Brain Res* 1194:45–55.

Yuste, R., A. Peinado, and L. C. Katz. 1992. Neuronal domains in developing neocortex. *Science* 257 (5070):665–69.

3 Adhesion Molecules and Their Function in Astrocyte Polarity

Salvatore Carbonetto
Centre for Research in Neuroscience, McGill University
Health Centre, Montreal, Quebec, Canada

Emeline Camand and
Sandrine Etienne-Manneville
Institut Pasteur, Cell Polarity and Migration
Group and CNRS URA 2582, Paris, France

CONTENTS

3.1 INTRODUCTION

In their classic textbook, Kuffler and Nichols (1976) summarized the then current hypotheses on neuroglial function as (1) structural support of the nervous system, (2) electrical isolation of neurons, (3) buffering of ionic currents resulting from neural activity, and (4) uptake of neurotransmitters. Since that time, the repertoire of astroglial functions has since expanded dramatically. Interneuronal synapses are now often referred to as tripartite synapses (Halassa et al., 2009; Halassa and Haydon, 2010) because they are thought to include glial processes as a third, nonneuronal, element. These astrocyte processes may modulate synaptic transmission through the release of neurotransmitters (Henneberger et al., 2010; Trudel and Bourque, 2010; Agulhon et al., 2010; Hamilton and Attwell, 2010). Another new function attributed to astroglia occurs during development where radial glia not only guide neurons migrating to their proper positions in the brain, but also give rise to neuronal precursors (Kriegstein and Alvarez-Buylla, 2009). Astrocytes have also been implicated in neuroinflammatory responses that occur in several diseases and in neurotrauma (see Chapter 6 by Rossi and Volterra).

Astrocyte function relies heavily on their abilities to respond to and modify their environment. In this chapter we describe the major adhesion molecules that allow astrocytes to interact with the extracellular matrix (ECM) as well as with neighboring cells via cell adhesion. Adhesion molecules are generally classified into four major families, including the integrins, the dystroglycan (DG), the cadherins, and the immunoglobulin superfamily of adhesion molecules. An emerging principle is that cell adhesion molecules (CAMs), which typically lack the ability for signal transduction, bind with low affinity to their ligands. As a result of this low-affinity interaction they must form islands containing a high density of adhesion molecules to mediate effective adhesion. Possibly as an evolutionary by-product, the intracellular domains form scaffolds that attract intracellular kinases, guanosine triphosphate (GTP) hydrolases (GTPhases), etc., for intracellular signal transduction (Carbonetto and Lindenbaum, 1995). More recent evidence suggests that the extracellular domains of adhesion molecules also function as scaffolds assembling ECM proteins, which include growth factors such as neuregulin and coreceptors such as perlecan. These scaffolds of adhesion molecules also form cytoskeletal linkages necessary for fundamental cellular events, such as process extension and polarization. The cytoskeletal and signaling aspects are so intimately interwoven as to be described by some as "mechanochemical signaling" (Ingber et al., 1994).

In astroglia, polarity is manifest in the organization of subcellular domains or more extensive polarization of the entire cell. Radial glia in the developing brain are more obviously polarized, and here connexins have been proposed to function not as channels but as CAMs for migrating neurons (Elias et al., 2007). In the hypothalamus of the adult brain astrocyes extend processes in proximity of the supraoptic nucleus and release neurotransmitters onto those neurons (Theodosis et al., 2008; Trudel and Bourque, 2010) (Figure 3.1). In the central nervous system (CNS) astrocytes may send processes over long distances to contact the basement membrane of blood vessels in the brain. ECM receptors are involved in the targeting of aquaporin and potassium channels to astrocytic endfeet (Guadagno and Moukhles, 2004; Noel et al., 2005). In animal models of multiple sclerosis, brain edema is associated with loss of astrocyte polarity and aquaporin-rich plaques on the endfeet (Wolburg-Buchholz et al., 2009). In addition, astrocytes are reported to constrict blood vessels and regulate blood flow (Gordon et al., 2007). It is interesting to view this interface with the vasculature in a broader context (Lee et al., 2009) in which astrocytes connect and

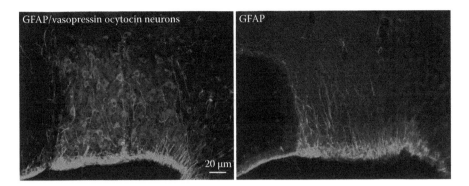

FIGURE 3.1 **(See color insert.)** Polarized astrocytes in the hypothalamus. A coronal section of the ventral part of the adult rat hypothalamus was immunocytochemically labeled with antibodies to glial fibrillary acidic protein (green) to identify astrocytes and oxytocin- and vasopressin-associated neurophysins to identify neurons. Astrocytes (green) send out processes toward the neurons and are able to modify the responsiveness of these neurons (red). The optic tract is at left. A population of astrocytes with cell bodies sitting on the glial limitans send long processes dorsally and ventrally to the supraoptic nucleus, where they surround cell bodies and dendrites of magnocellular neurons. Bar = 20 μm. (Courtesy of Dr. Charles Bourque, McGill University.)

coordinate multiple cell types throughout the brain. Metabolic, osmotic, and other changes initiated in the vasculature may propagate widely through gap junctions among astroglia (Giaume et al., 2010).

Of special interest here is the response of astrocytes to brain trauma. After trauma to the CNS, astrocytes become reactive and are hypertrophic and more extensively polarized. They extend processes (Figure 3.2) toward the region of the wound forming a scar and also secrete extracellular matrix proteins that, with the help of migrating ependymal cells, isolate the wound. The resulting cavity that may be formed following neurotrauma is an impediment to neuronal regeneration, though reactive astroglia may mitigate the secondary effects of trauma (Sofroniew, 2009). It will be important, in future studies, to understand the multiple phases of astrocytic scarring and to be able to manipulate the detrimental and beneficial effects separately in order to enhance those that are salutary and inhibit those that are interfering with neural regeneration.

Our groups have used in vitro models of astrocyte wound healing to study the cellular events and their molecular underpinnings. We have described that process extension is distinct from astrocyte polarization (Etienne-Manneville and Hall, 2001; Peng et al., 2008), and we have identified a number of key molecular interactors that mediate polarization (Etienne-Manneville and Hall, 2003a; Etienne-Manneville et al., 2005; Osmani et al., 2006, 2010). Cell adhesion is essential for the early response to the wound, with respect to both process extension and cell polarization. Here we present the major CAMs expressed in astrocytes and discuss the cellular and molecular bases of their function in astrocyte polarization.

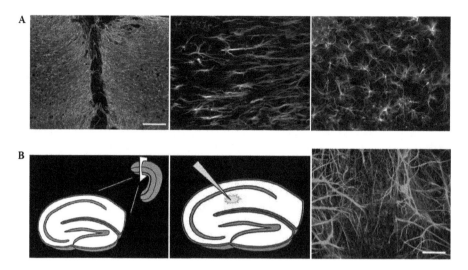

FIGURE 3.2   (See color insert.) Polarized astrocytes orient toward a wound. (A) Traumatic injury of the adult mouse spinal cord was induced using iridectomy scissors. After 8 days, the spinal cord was fixed, removed, and serially sliced. Glial fibrillary acidic protein (GFAP) immunostaining shows that reactive astrocytes of the gray matter present two main morphologies. Astrocytes that directly surround the lesion site (dark area, left panel) most often have extended processes toward the wound (middle panel), whereas the ones that are more distant keep their stellate morphology (right panel). Scale bar = 150 μm in left panel and 40 μm in middle and right panels. (B) Rat hippocampal slices in culture were either wounded (left panel) or α-amino-adipic acid was injected to create a chemical injury in rat hippocampal slices in culture. After 5 days, the slice was fixed and labeled immunocytochemically with antibodies to GFAP that revealed astrocytic processes extending toward the injury (green). The injury site delineated by expression of SPARC (red), a protein secreted by astrocytes as well as the absence of any neurofilament positive axons (blue). Bar = 100 μm. (Courtesy of Drs. Emma V. Jones and Keith Mural, McGill University.)

3.2   ASTROCYTE INTERACTIONS WITH THE EXTRACELLULAR MATRIX

3.2.1   The Extracellular Matrix (ECM) of the Brain

The vast majority of the ECM of the brain consists of a collection of proteoglycans. Because these proteins are decorated with multiple carbohydrate side chains that are negatively charged, they are often described by these glycosaminoglycan chains, e.g., chondroitin sulfate, heparan sulfate proteoglycans, or when more well defined, by their protein cores, such as, e.g., neurocan, brevican, phosphacan (Margolis and Margolis, 1994). The parenchyma of the brain in general and synapses in particular are deficient in basement membranes. This is in contrast to synapses in the peripheral nervous system (Carbonetto and Lindenbaum, 1995), where specialized basement membranes in the cleft have been shown to be critical for synaptic formation and function (Nishimune et al., 2004). Adhesive proteins found in basement membranes, including laminin, nidogen, fibronectin, tenascin, and collagen, are

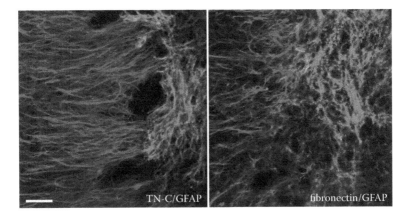

FIGURE 3.3   **(See color insert.)** Astrocytes interact with tenascin-C and fibronectin in the glial scar. Eight days after traumatic injury of the adult mouse spinal cord, the lesion site is invaded with different cell types from nonnervous origin, such as inflammatory cells and fibroblasts. These cells express and secrete extracellular matrix molecules, such as tenascin-C (green, left panel) or fibronectin (green, right panel), organized in fibrillar structures that are in contact with reactive astrocytes (GFAP, red in both pictures) of the glial scar. Tenascin-C in particular is believed to inhibit axon regeneration through the scar and to participate in the formation of a new glia limitans. Scale bar = 80 μm.

largely missing from the parenchyma of adult CNS. During development, more specialized, nonbasement membrane proteins are expressed in the ECM. Reelin, for example, regulates extension of radial glial processes and neuronal migration (Nomura et al., 2008; Hartfuss et al., 2003). Similarly, tenascin-C is expressed in the brain neurogenic zone and is involved in the migration of early postnatal precursors (Steindler et al., 1995; Garcion et al., 2001). Several ECM proteins, including tenascin-C proteins, are upregulated in the brain following a wound, and this overexpression facilitates astrocyte process extension and wound healing (Peng et al., 2008) (Figure 3.3).

3.2.2 EXTRACELLULAR MATRIX RECEPTORS

Integrins and dystroglycan (DG) are major ECM receptors in the nervous system, though novel receptors and channels and new functions for old ones continue to emerge (Elias and Kriegstein, 2008). Both integrins and DG are heterodimers but are otherwise structurally very different. α and β integrin subunits are transmembrane proteins while, αDG is a peripheral membrane protein that is docked to the β subunit, a transmembrane protein. The two integrin subunits form a variety of heterodimers wherein the α subunit will dictate ligand selectivity, as in the $\alpha 1$ family. In contrast, αDG alone binds most of its ligands, not by interaction with the core protein but via O-mannose-linked carbohydrate side chains on this highly glycosylated subunit (Yoshida-Moriguchi et al., 2010). This necessitates a much more tortuous route of transmembrane signaling (see below) as compared with integrins where ligands bind directly to the pocket formed by the two transmembrane subunits.

Nevertheless, integrins and DG participate nonredundantly in similar functions during development and in the adult brain. In the following sections we discuss these aspects of ECM receptors, including their overlapping functions, with a focus on astrocyte polarity and migration.

3.2.2.1 Integrins

There are 24 heterodimeric receptors in the integrin superfamily, with the majority of ECM receptors expressing β1 subunits. These integrins bind to a variety of ECM proteins, including fibronectin, vitronectin, laminin, and various types of collagen, with distinct but overlapping specificities. For instance, five different integrins (α1β1, α2β1, α6β1, α6β4, α7β1) bind to laminin, and certain of these integrins also bind to multiple ECM proteins (e.g., α1β1-integrin binds both collagen IV and laminin; Tawil et al., 1990). The extracellular regions of the two subunits form a pocket for selective ligand binding. Integrin cytoplasmic domains have no intrinsic catalytic activity and function as scaffolds for other structural and signaling proteins (Legate and Fassler, 2009). The heterodimers are inserted into the cell surface in an inactive state that is activated following the binding of talin and kindlin to the intracellular domain of the α1 or α3 subunits (Moser et al., 2009). Kindlin null mice are embryonic lethal, as are integrin α1 null mice, and kindlin is apparently necessary for integrin adhesion. This inside-out signaling can be regulated by parallel pathways, such as ligation of selectins or growth factors or mechanical stress (Miyamoto et al., 1996; Giancotti and Ruoslahti, 1999; Green et al., 2004; Luo et al., 2007; Alon and Ley, 2008; Constantin, 2008). Outside-in signaling follows the binding of ECM proteins to integrins, triggering a conformational change in the extracellular and intracellular domains of α and β subunits (Luo and Springer, 2006). The affinity of individual integrins is low so that the tight cell adhesion only develops following the recruitment of multiple integrin heterodimers to a membrane domain increasing the avidity of that domain for the ECM. This aggregation of integrins is facilitated somewhat by the fact that many ECM proteins are multivalent and, more importantly, by "trapping" of integrins that move into the forming adhesive patch. This is followed by the morphing of integrins into a variety of contacts (Tawil et al., 1990), beginning with a small, immature focal complex and culminating in a mature focal adhesion. Mature focal adhesions are large complexes of integrin-associated proteins that form bridges between the extracellular matrix and the intracellular cytoskeletal network to provide strong mechanical bonds between the cell and its environment. Talin, which binds to integrins early following integrin activation, helps recruit vinculin, which is important in maturation of the adhesion and linkage of focal adhesions to the actin cytoskeleton (Critchley and Gingras, 2008). Other early cytoskeletal accessory proteins include paxillin and tensin, both of which are important in focal adhesion formation/maturation. Integrins are best known for interaction with actin in the lamellipodia and filopodia of migrating cells. In astrocytes responding to a wound, the spectacular microtubule network extends into processes reaching toward the wound. The microtubules emanating from the centrosome, which is polarized by the wound, can be found approaching the edge of the lamellipodium of migrating cells (Etienne-Manneville and Hall, 2001). These microtubules are not part of the focal adhesion per se, but rather are involved in trafficking of proteins at focal adhesions and participate in their disassembly, possibly by facilitating integrin endocytosis and turnover (Small and Kaverina, 2003; Krylyshkina et al., 2003; Palazzo et al., 2004). In keeping with this scenario, microtubules appear to be regulated by signals emanating from the focal adhesion complex (Palazzo et al., 2004).

Astrocytes express several members of the β1-integrin subfamily, including α1β1, α3β1, α5β1, and α6β1 (Tawil et al., 1994; Milner et al., 1999), that can serve as receptors for laminin, collagen, and fibronectin. Astrocytic function varies during development, after trauma, and with disease, as does integrin expression. Qualitative and quantitative changes in integrin expression influence the ability of astrocytes to extend, process, polarize, and migrate within the brain. For instance, αv integrins are expressed on radial glia in the developing brain where they are involved in neuronal migration (Anton et al., 1999). αvβ8 expression promotes astrocyte migration on vitronectin, but its expression decreases during development, whereas αvβ5 expression increases (Milner et al., 2001) and the latter contributes to astrocyte adhesion to vitronectin but not to migration (Milner et al., 1999). The α6β4 integrin, a receptor for laminin, is highly expressed at astrocyte endfeet on blood vessels, but its expression is decreased soon after a stroke, and this decrease also correlates with blood–brain barrier breakdown during cerebral ischemia (Wagner et al., 1997). It appears that α6β4 along with DG is overexpressed during a phase of astrocyte activation and endothelial cell proliferation (Longxuan et al., 2010). This suggests a key role for α6β4 in astrocyte-induced blood–brain barrier function and α6β4 cooperation with DG (discussed below). Finally, αvβ3, like αvβ1 and αvβ6, is not expressed in astrocytes in normal conditions, and increased expression of αvβ3 is associated with invasive gliomas (Gladson et al., 1995). Extrapolating these changes in integrin expression to function is difficult at this point. Recent work on neovascularization of tumors has shown that integrins are necessary for cell polarization and angiogenesis (Hynes, 2002). However, ligand mimetics that were expected to inhibit integrin functions, angiogenesis, and tumor formation have the paradoxical effect of simulating angiogenesis (Reynolds et al., 2009).

3.2.2.2 Dystroglycan

αDG is an ECM receptor and peripheral membrane protein (Smalheiser and Schwartz, 1987; Douville et al., 1988; Ibraghimov-Beskrovnaya et al., 1992; Gee et al., 1993) that is noncovalently bound to βDG, a transmembrane protein (Ibraghimov-Beskrovnaya et al., 1992). DG forms the functional core of a larger complex of cell surface proteins that extends from the ECM to the membrane cytoskeleton (Durbeej and Campbell, 2002; Henry and Campbell, 1996; Lapidos et al., 2004). Carbohydrate side chains on αDG (Ervasti and Campbell, 1993) recognize the G domains of laminin, and similar motifs in agrin, perlecan, as well as the cell adhesion molecule, neurexin (Gee et al., 1994; Peng et al., 1999; Sugita et al., 2001). Biglycan (Bowe et al., 2000) and pikachurin (Sato et al., 2008) also bind to DG.

Despite obvious differences in structure, ligand selectivity, etc., analysis of integrin and DG function has revealed striking overlaps. Dystroglycan and β1-integrins act cooperatively to regulate laminin expression and cell survival

(Li et al., 2003). Similar function of β1-integrins and DG is also suggested by studies in which deletion of the DG or β1-integrin gene in mice results in peri-implantation lethality (Williamson et al., 1997). In muscle, deficiency of these same genes results in a myopathy (β1-integrin) (Schwander et al., 2004) or muscular dystrophy (DG) (Cote et al., 1999). In brain, a cell-specific knockout of the DG or β1-integrin gene results in a phenotype closely resembling type II (cobblestone) lissencephaly (Graus-Porta et al., 2001; Moore et al., 2002) that is characterized by defects in neuronal migration and cortical lamination. In this latter instance, the lissencephaly and resulting mental retardation are thought to devolve from a failure of radial glia to attach to or assemble a basement membrane at the glial-pial interface (Graus-Porta et al., 2001; Moore et al., 2002; Niewmierzycka et al., 2005). This implies that during development DG is targeted to glial endfeet in contact with the ECM at the ventricular and cortical surfaces. At the glial endfeet that contact blood vessels in the adult brain, DG assembles a complex of proteins, including aquaporin, that regulate the extracellular volume of the brain (Guadagno and Moukhles, 2004; Bragg et al., 2006). Like DG, α6β4 integrins are highly expressed at astrocytic endfeet, and loss of these receptors correlates with detachment of astrocyte processes from the vascular basement membrane and disruption of the blood–brain barrier following cerebral ischemia and hypoxia (Wagner et al., 1997; Longxuan et al., 2010).

DG structure and function have been elucidated in most detail in skeletal muscle, and it is instructive to consider some of these issues. In skeletal muscle, DG forms the functional core of a scaffold of proteins linking dystrophin in the submembranous cytoskeleton to the ECM. Genetic evidence demonstrates that DG is critical for the function of dystrophin associated in maintaining muscle integrity (Henry and Campbell, 1996) by buttressing the muscle cell surface against the stresses of contraction (Ervasti, 2003). The underlying cellular and molecular mechanisms, however, are complex. Mutations in genes encoding different dystrophin-associated proteins result in distinct forms of muscular dystrophy. Furthermore, mutations in the major ECM protein laminin result in muscular dystrophy in humans and in mice, but there is no obvious disruption of the plasmalemma (Straub et al., 1997). Conversely, in DG-deficient muscle, severe muscular dystrophy and plasmalemma disruption can occur with no obvious disruption of the basement membrane (Cote et al., 1999). In view of this heterogeneity in muscle, what appears to be a single DG-dystrophin complex is likely to vary in composition and function in other tissues. For example, DG mediates cell survival in muscle cells (Montanaro et al., 1999; Langenbach and Rando, 2002; Li et al., 2002) and cell polarity in noncontractile cells (Mirouse et al., 2009), suggesting involvement of intracellular signaling in addition to its structural function.

DG and its associated proteins have been implicated in establishing specialized domains at neuromuscular junctions (NMJs) (Jacobson et al., 2001; Montanaro et al., 1998). There, it appears to function downstream of the receptor tyrosine kinase muscle specific kinase (MuSK) to stabilize acetylcholine receptors (AChRs) into synaptic plaques that eventually mature into complex postsynaptic densities. In addition, DG mediates assembly of acetylcholinesterase (AchE) into the synaptic basement membrane (Jacobson et al., 2001; Cote et al., 1999). In the CNS, astrocytes may send processes many hundreds of microns to contact the basement membrane of blood vessels in the brain. Aquaporin and potassium channels are targeted to astroglial endfeet, as noted above, and the mechanism by which this targeting occurs is similar to that used by laminin/DG in aggregating acetylcholine receptors in muscle (Guadagno and Moukhles, 2004; Noel et al., 2005).

3.2.3 SIGNALING FROM ECM RECEPTORS AND ITS CONSEQUENCES FOR ASTROCYTE POLARITY

The astrocyte response to a lesion entails the orientation of the centrosome and Golgi apparatus in front of the nucleus toward the portion of the cell facing the wound (Etienne-Manneville and Hall, 2001). Subsequently, microtubules emanating from newly oriented centrosome contribute to the extension of cell processes at the wound side. However, process extension and polarization are distinct events, and this is a clear instance of divergence in the function of DG vs. integrin (Etienne-Manneville and Hall, 2001). DG is mostly involved in process extension, whereas loss of integrin function dramatically affects process extension and cell polarization (Peng et al., 2008).

While integrins launch a variety of signals from focal adhesion complexes via activation of focal adhesion kinase, integrin-linked kinase, and other pathways, the signaling pathways for DG are more obscure. Indeed, there are conceptual as well as empirical challenges to the notion of transmembrane signaling directly through DG. For example, one is faced with the unusual proposition that binding to carbohydrates decorating the core protein alters the conformation of αDG and subsequently βDG to signal intracellularly. Several proteins with the potential for cell signaling interact with βDG either directly, like Grb 2 and caveolin, or indirectly, like neuronal Nitric Oxide Synthase (nNOS) and calmodulin (Lapidos et al., 2004; Yang et al., 1995). nNOS-mediated vasoconstriction is perturbed in αsyntrophin null mice (Thomas et al., 2003). In addition, direct evidence for transmembrane signaling in a physiologically relevant context is just starting to emerge (Mirouse et al., 2009; Spence et al., 2004; Zhou et al., 2006). For these reasons we focus on integrin-mediated signaling from the ECM.

As cells extend processes and establish contacts with the surrounding cerebral parenchyma, new integrins are engaged and lead to important rearrangements of the cytoskeleton, including actin and microtubule networks. The leading edge of migrating cells has structures called lamellipodia and filopodia. The latter are rich in bundles of actin and appear to sense the environment and are especially prominent on neuronal growth cones or precursors to lamellipodia. Lamellipodia have a cortical meshwork of actin and just behind it microtubules. Integrins are best known for regulating the actin cytoskeleton and cell motility (Ridley et al., 2003). Focal adhesions are sufficient to mediate actin polymerization (Butler et al., 2006). In lamellipodia, this involves the protein Arp2/3, which nucleates branched actin filaments in collaboration with the Wiskott-Aldrich syndrome protein (WASP). Thus, focal adhesions form and mature in the region of the lamellipodia where Arp2/3 is recruited by its interaction with the focal adhesion kinase (FAK) (Serrels et al., 2007). In astrocytes responding to a wound, actin remodeling remains essential for process extension and migration, but microtubules are the main regulators of morphological changes and intracellular polarization (Etienne-Manneville, 2004a; Etienne-Manneville and Hall, 2001).

Among the list of hundreds of integrin-associated molecules (Humphries et al., 2009; Zaidel-Bar et al., 2007), focal adhesion kinase (FAK) (del Pozo et al., 2004; Palazzo et al., 2004), integrin-linked kinase (ILK) (Guo et al., 2007), Src, and Cdc42 stand out as key components of the signaling scaffold involved in cell polarization. Here we briefly discuss FAK, ILK, and Cdc42 because of their relevance to cell polarity. Several excellent recent reviews give a more comprehensive treatment of the focal adhesion complex as we know it (Shattil et al., 2010).

FAK can bind directly to β integrin subunits, though it appears that in situ the association occurs through paxillin (Hayashi et al., 2002) (Figure 3.4). FAK is

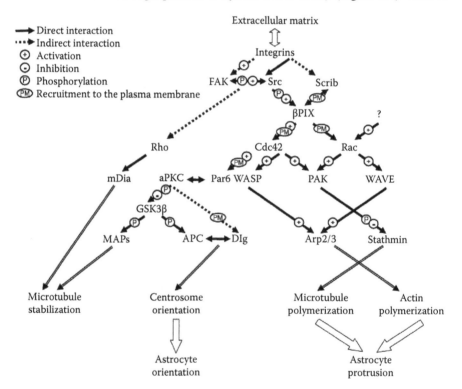

FIGURE 3.4 Integrin signaling promotes cytoskeleton reorganization and astrocyte polarization. This diagram summarizes the signaling pathways that, upon integrin engagement with the extracellular matrix, lead to astrocyte polarization. Direct and indirect molecular interactions and the consequences of these interactions in terms of protein activity or protein localization are indicated by arrows. These signaling pathways ultimately regulate the actin and microtubule cytoskeleton. Cytoskeletal modifications lead to astrocyte polarization with the formation of a membrane protrusion and to the positioning and orientation of this protrusion in the direction of the stimulated integrins.

typically activated by ligand binding resulting in increased kinase activity and FAK autophosphorylation. This in turn leads to the recruitment of Src and Src family kinases, and to increased phosphorylation and binding to paxillin and p130Cas and other focal adhesion proteins (Mitra et al., 2005). In addition, FAK is a structural component of the scaffold binding to a variety of membrane-associated proteins, including growth factor receptors, phosphoinositide 3-kinase (PI3k), Grb 2, and a host of other proteins (Beattie et al., 2010; Mitra et al., 2005; Parsons, 2003; Parsons et al., 2000). FAK appears in more mature focal adhesions and is essential for their turnover (Mitra et al., 2005). In contrast, Src is found constitutively associated with β integrin via its SH3 motif (Arias-Salgado et al., 2005) (Figure 3.4). Src is thought to be a core component of FAK signaling, which includes activation of ERK and Rho GTPases. Cytoskeletal dynamics at focal adhesions are regulated by Rho GTPases, most notably Rac, Cdc42, and RhoA (Etienne-Manneville and Hall, 2002). Rho GTPases are signaling proteins that cycle between an active state bound to Guanosine triphosphate (GTP) and an inactive Guanosine diphosphate (GDP) GDP-bound state. The cycling is in turn regulated by guanine nucleotide exchange factors (GEFs), activating proteins (GAPs), and dissociation inhibitors (GDIs). Jaffe and Hall (2005) described how these factors can regulate cell polarity (Jaffe and Hall, 2005). Though this largely impinges on the actin network, the stability of microtubules in the lamellipodium is regulated by Rho downstream of FAK (Palazzo et al., 2004). Interestingly, this involves a lipid domain at the leading edge. Indeed, lipid rafts have also been implicated in integrin internalization in polarized cells (del Pozo et al., 2004) via Rac and in establishing caveolin 3 signaling complexes that are necessary for Src-integrin association and phosphorylation of focal adhesion proteins (del Pozo and Schwartz, 2007). Thus, internalization may regulate these processes in addition to regulation of microtubule stability.

ILK binds directly to the β1- and β3-integrin subunits and interacts with actin via a heterotrimeric complex with Parvins and Pinch (particularly interesting new cysteine-histidine-rich protein) (Legate et al., 2006). In view of the recent controversy surrounding the function, if any, of the kinase domain in ILK, its structural role may be particularly important (Wickstrom et al., 2010), as, for example, in its role in integrin-actin stabilization on the membrane (Mackinnon et al., 2002; Postel et al., 2008; Zervas et al., 2001). Recent evidence (Fielding et al., 2008; LaFlamme et al., 2008; Reverte et al., 2006) has implicated ILK in the formation of mitotic spindles in dividing cells and possibly in the centrosome cycle during cell division, whereby the centrosome gives rise to two centrioles. This is particularly relevant to the study of cell polarization in wounded astrocytes, where the centrosome in interphase is the major microtubule organizing center (MTOC) and becomes oriented with the Golgi apparatus indicative of cell polarization toward the wound (discussed below).

In addition to ILK and FAK, integrin engagement leads to the recruitment and activation of evolutionary conserved polarity proteins such as Cdc42 (Etienne-Manneville, 2004b). Src can phosphorylate Rho GTPase exchange factors such as βPIX that controls Cdc42 activation (Osmani et al., 2006, 2010) (Figure 3.4). Inhibiting integrin signaling on fibronectin using Argine-Glycine Asparate (RGD) peptides, antibodies to fibronectin, or inhibiting Src prevents Cdc42 activation (Etienne-Manneville and Hall, 2001; Peng et al., 2008). The prolarized recruitment of Cdc42 to the leading edge of migrating cells involves Arf6-dependent vesicular traffic and bPIX interaction with the scaffold protein Scrib (Osmani et al., 2006, 2010). Local activation of Cdc42 is critical in positioning lammellipodia and in controlling the localization of Rac, and promoting both actin and microtubule polymerization. Cdc42 and Rac use WASP and Scar/WAVE, respectively, to regulate actin in the lamellipodium via the Arp2/3 complex. Both can also activate PAK, which has been shown to negatively regulate the microtubule destabilizer Stathmin (Daub et al., 2001; Wittmann et al., 2004). In addition, localized Cdc42 activity determines centrosome and Golgi positioning, microtubule network orientation, and in general, sets up the cell's polarity axis and direction of process extension/migration. This is especially important for astrocytes but occurs in many other cell types (Etienne-Manneville, 2004b; Wittmann and Waterman-Storer, 2001). Indeed, active Cdc42 binds to and activates a protein complex comprised of Par6 and atypical protein kinase C (aPKC) that is key in the regulation of cell polarity in migrating, dividing, or differentiating cells (Etienne-Manneville and Hall, 2003b). aPKC can phosphorylate and inhibit GSK3β kinase (Etienne-Manneville and Hall, 2003a), and thereby leads to the regulation of a large panel of microtubule-associated proteins that regulate the stability and dynamics of microtubules (Etienne-Manneville, 2010). Among these, the tumor suppressor adenomatous polyposis coli (APC) forms clusters at microtubule plus ends at the leading edge of migrating cells (Etienne-Manneville et al., 2005; Manneville, 2010). In addition, aPKC controls microtubule interaction with the actin-rich cell cortex by recruiting the tumor suppressor Dlg to the leading edge of the cell and promoting its interaction. In addition to Cdc42 and Rac, integrin-induced activation of RhoA and its downstream effectors mDia (Diaphanous) also influences microtubule dynamics and promotes stabilization of the microtubules that reach toward the leading edge (Palazzo et al., 2001, 2004). This polarized organization of microtubules driven by integrins is likely to participate in the delivery of membrane vesicles and accumulation of specific transmembrane proteins, including integrins, at the leading edge of astrocytic processes.

3.3 ASTROCYTES INTERACTIONS WITH NEIGHBORING CELLS

The nervous system has been the target of much effort in identifying CAMs and understanding their role in neural development and function in the adult. As a result, we know a good deal about the involvement of the many neuronal CAMs, including cadherins, protocadherins, immunoglobulin family members, L1, neurexins, DsCAM, as well as axonal guidance molecules (netrins, semaphorins, neuropillin, etc.) and antiadhesive molecules (Nogo, MAG, etc.). Our knowledge of the role of CAMs in astrocytes is relatively meager. As noted briefly above, astrocytes are assembled into networks that are coupled electrically and metabolically via gap junctions (Araque and Navarrete, 2010). This network extends from the astroglia vascular interface and is thought to be important in ionic and metabolic regulation in the brain aided by the diffusion of molecules (<1.5 kDa), including signaling intermediates, through gap junctions (Seifert et al., 2006). Furthermore, these networks appear to be linked to neuronal organization in some regions of the cortex (Houades et al., 2008). Interestingly, connexins that form gap junctions have been suggested to function as CAMs to mediate neuronal migration on radial glia (Elias et al., 2007). It would seem parsimonious that they do so in the formation of astrocytic networks, but gap junctions require CAMs to bring together connexin hemichannels expressed in adjacent cells to form an intercellular channel (Chakraborty et al., 2010; Musil and Goodenough, 1990).

CAMs are also necessary for the formation of adherens junctions that classically help to tie together epithelial cells. Recent work (Rieger et al., 2009) indicates that adherens junctions are necessary for cell motility during cerebellar granule migration. This requires cadherins that function not only to establish adhesion among migrating neurons and with neurons and radial glia, but also to polarize the cell and allow nuclear translocation and sustained directional neuronal migration. Nucleokinesis is essential for neuronal migration, and is known to involve the microtubular cytoskeleton and centrosome to polarize the cell and allow directional migration. Dupin et al. (2009) have shown that cadherin can orient the centrosome and polarize rat astrocytes when cultured on patterned substrates. In related studies, tangential migration of cerebellar granule cells (GCs) in mouse is regulated by the transmembrane ligand semaphorin 6A and its receptor plexin-A2 (Kerjan et al., 2005). Again, this appears to affect centrosome-generated cell polarity to position this MTOC in the leading edge of the migrating neuron movements and its stable positioning in front of the nucleus during tangential nucleokinetic migration.

These functions of CAMs and cell polarity are new and exciting and are already being brought to bear on our understanding of astrocyte polarization. Below, we discuss some of the wealth of data on CAM function that one must integrate into future models of formation and modulation of astrocytic networks as well as during astrocyte migration and polarization.

3.3.1 ADHERENS JUNCTION MOLECULES

Two basic adhesive complexes, the cadherin-catenin complex and the nectin-afadin complex form adherens junctions. The nectin-afadin complex is required for the subsequent recruitment of the cadherin-catenin complex, and the association of the two molecular complexes is essential for the formation of the zonula adherens (Takai et al., 2008). Each complex contributes in a different way to the structure and function of adherens junctions, and they cooperate to make a link with the underlying cytoskeleton.

First characterized as calcium-binding molecules, cadherins represent a large family of single-pass transmembrane adhesion molecules (Hyafil et al., 1981; Volk and Geiger, 1984; Damsky et al., 1983; Yoshida-Noro et al., 1984; Hatta et al., 1988). They assume a broad range of functions, including direct cell-cell cohesion and cell signaling. They encompass several different subfamilies, which share sequence similarities: the type I and type II cadherins (both referred to as classical cadherins), the desmosomal cadherins, and the protocadherins. Most cadherin subfamilies are expressed in the CNS, except the desmosomal cadherins that have essentially been found to be limited to epithelial cells so far. Protocadherins are specifically enriched in the CNS and constitute the largest subgroup of the cadherin superfamily (Vanhalst et al., 2005; Frank et al., 2005; Sano et al., 1993; Sago et al., 1995; Hirano et al., 1999; Carroll et al., 2001). Until recently, they had essentially been reported in neurons where they display specific neuronal subpopulation distribution and play fundamental roles during neural development (Frank et al., 2005; Vanhalst et al., 2005; Kohmura et al., 1998). However, it has now been shown that astrocytes express the gamma-protocadherins that are localized at perisynaptic processes and that these proteins contribute to synaptogenesis, at least in culture. As such, gamma-protocadherins are intriguing candidates for CAMs on astrocytes that are intimately involved in the development of the tripartite synapse (Garrett and Weiner, 2009).

Classical cadherins are single-transmembrane-domain proteins that are expressed in a tissue-specific manner during development and in the adult. N-cadherin is mostly expressed in the developing and mature CNS as well as in cardiac and skeletal muscles (Moore and Walsh, 1993; Redies et al., 1993; Miyatani et al., 1989), while E-cadherin, for instance, is rather restricted to epithelial cells (Boller et al., 1985; Shore and Nelson, 1991). The extracellular domains of classical cadherins contain five cadherin repeat domains (EC1–5), characterized by conserved amino acid sequences (Nose et al., 1990; Hatta et al., 1988) that bind calcium and have been shown to protect classical cadherins from protease activity (Shirayoshi et al., 1986; Hyafil et al., 1980; Ringwald et al., 1987; Ozawa et al., 1990a). Calcium binding to cadherins is required for *trans* homophilic cell adhesion (Chitaev and Troyanovsky, 1998). Several regions of the extracellular domain seem to be required for cadherin homophilic interactions, but the most N-terminal EC repeat, which comprises a His-Ala-Val sequence, seems crucial (Blaschuk et al., 1990; Tomschy et al., 1996; Chappuis-Flament et al., 2001; Shan et al., 2004). Classical cadherins also interact in *cis* and are able to form both homophilic and heterophilic dimers (Shan et al., 2000; Tamura et al., 1998). The cytoplasmic tail of classical cadherins is highly conserved and mediates cytoskeleton interaction and signaling events mainly through its association with catenins (Ozawa et al., 1989). β-Catenin and p120 interact directly with cadherins, whereas α-catenin associates with cadherins through a β-catenin interaction (Ozawa et al., 1990b; Aberle et al., 1994; Daniel and Reynolds 1995; Obama and Ozawa, 1997; Huber et al., 1997). The oncogene p120, which was first described as a substrate of Src (Reynolds et al., 1989), associates with the cadherin juxtamembrane region (Yap et al., 1998; Ohkubo and Ozawa, 1999) on the cadherin precursor proteins located in the endoplasmic reticulum prior to cadherin insertion into the plasma membrane. p120 interacts with kinesin and may be involved in cadherin transport to the plasma membrane (Chen et al., 2003; Mary et al., 2002; Yanagisawa et al., 2004). In addition, it regulates cadherin clustering and endocytosis as well as the overall cadherin stability at the plasma membrane (Davis et al., 2003; Reynolds and Carnahan, 2004; Xiao et al., 2007; Thoreson et al., 2000). β-Catenin interacts with the distal part of the cytoplasmic domain of cadherin and also plays crucial roles in the regulation of cadherin-dependent adhesion (Ozawa et al., 1989; Piepenhagen and Nelson, 1993). β-Catenin and α-catenin are recruited to the cadherin cytoplasmic tail when cadherin is phosphorylated by casein kinase I during its transport through the Golgi apparatus. The structure and activity of p120 and β-catenin are regulated by phosphorylation (Piedra et al., 2001) and Fer, a nontransmembrane receptor tyrosine kinase that forms a complex with p120 and β-catenin, functions in catenin phosphorylation (Rosato et al., 1998).

Formation of cadherin-mediated adherens junctions is supported by nectins (Sato et al., 2006; Okamoto et al., 2005; Fukuhara et al., 2003). The superfamily of nectin and nectin-like (Necl) molecules are immunoglobulin-like adhesion molecules that mediate calcium-independent interactions. There are four nectins (nectin 1–4), and each has at least two splice variants (Morrison and Racaniello, 1992; Eberle et al., 1995; Takahashi et al., 1999; Satoh-Horikawa et al., 2000; Reymond et al., 2001). Their pattern of expression is still not fully determined, but nectin 1 is likely expressed in astrocytes (Guzman et al., 2006; Shukla et al., 2006). All the members of the nectin family have three immunoglobulin-like loops in their extracellular domain, a single transmembrane region, and a cytoplasmic tail. Nectins form *cis* dimers, which in turn can form *trans* dimers in a homophilic or heterophilic manner (Satoh-Horikawa et al., 2000; Miyahara et al., 2000). Nectin 1 has been shown to be a substrate for α- and γ-secretase, which cleave it in neurons (Kim et al., 2002, 2010). In addition, five Necls have been identified that can mediate both *trans* homophilic and heterophilic adhesions (Ikeda et al., 2003; Shingai et al., 2003), including those with nectins. Though their roles have not been fully elucidated, several Necls are specifically expressed in the brain (Kakunaga et al., 2005; Niederkofler et al., 2010; Park et al., 2008; Perlin and Talbot, 2007).

3.3.2 THE IMMUNOGLOBULIN SUPERFAMILY OF ADHESION MOLECULES

Neural cell adhesion molecule (NCAM) is a membrane glycoprotein belonging to the immunoglobulin superfamily. It is encoded by a single-copy gene that undergoes alternative splicing (Owens et al., 1987; Hoffman et al., 1982; Murray et al., 1986). This leads to the production of 180, 140, and 120 kDa isoforms that mainly differ in the length of their cytoplasmic domains and their mode of attachment to the membrane (Barthels et al., 1988; Santoni et al., 1987, 1989; Barbas et al., 1988). NCAM-180 and -140 are both transmembrane proteins that differ only in a small portion of their cytoplasmic tails. In contrast, NCAM-120 has no cytoplasmic segment and is linked to the membrane through a glycosylphosphatidylinositol anchor (Hemperly et al., 1986). Some authors have proposed that the different isoforms display specific cell type expression, with NCAM-180 being only expressed in neurons, NCAM-120 only expressed in astrocytes, and NCAM-140 expressed in both cell types (Nybroe et al., 1985; Noble et al., 1985; Ditlevsen et al., 2008). However, this is controversial, possibly because of differences in brain regions assessed (Pollerberg et al., 1985; Sasaki and Endo, 1999). NCAM is also secreted as a result of either the presence of an exon between the two fibronectin type III homology domains (Gower et al., 1988), enzymatic removal of the NCAM-120 by phosphatidylinositol-specific phospholipase C (Sadoul et al., 1986; He et al., 1987), or proteolytic cleavage of the extracellular domain (Olsen et al., 1993; Nybroe et al., 1989).

The extracellular domain of NCAM has five immunoglobulin and two fibronectin type III homology domains. NCAM mediates cell-cell adhesion mainly through calcium-independent homophilic interactions (Rao et al., 1992, 1993; Soroka et al., 2003; Atkins et al., 2004). The crystal structure of NCAM supports several models of *cis* and *trans* homophilic interactions, involving the first three immunoglobulin domains (Soroka et al., 2003; Kasper et al., 2000; Kiselyov et al., 2005). In addition, NCAM can establish heterophilic interactions with other members of the immuno-globulin superfamily, such as L1 (see below) or TAG-1 (Milev et al., 1996; Kadmon et al., 1990). It also functions as an ECM receptor by binding to heparan sulfate, chondroitin sulfate proteoglycans, and collagens (Kallapur and Akeson, 1992; Probstmeier et al., 1989; Kiselyov et al., 1997).

NCAM-mediated cell adhesion can be regulated by insertion of the variable alternative splice exon (VASE) or by attachment to the fifth Ig domain of poly-sialic acid (PSA) negatively charged carbohydrate homopolymers of α-2,8-linked N-acetylneuraminic acid (Small and Akeson, 1990; Reyes et al., 1993; Finne et al., 1983; Hoffman et al., 1982). VASE has an inhibitory effect on neurite growth-promoting activity of NCAM (Doherty et al., 1992).

Addition of PSA to NCAM is mediated by two Golgi-associated polysialyl-transferases (Kitagawa and Paulson, 1994; Nakayama et al., 1995; Kojima et al., 1996; Eckhardt et al., 1995; Yoshida et al., 1995) that exclusively modify NCAM (Tomasiewicz et al., 1993; Ono et al., 1994; Cremer et al., 1994). PSA expression levels on the cell surface are regulated by polysialyltransferases transcription or enzymatic activity (Theodosis et al., 2008), as well as by PSA-NCAM trafficking and localiza-tion at the plasma membrane (Kiss et al., 1994; Minana et al., 2001; Gallagher et al., 2001; Bruses and Rutishauser, 1998). The expression of the two polysialyltransferases is indeed tightly regulated spatially as well as temporally in the developing and adult brain (Wood et al., 1997; Ong et al., 1998). PSA-NCAM is mainly expressed during development, and its expression is associated with periods of myelination in the CNS (Oumesmar et al., 1995). PSA-NCAM expression is strongly decreased in the adult brain, though neurogenic regions, and regions showing long-lasting plasticity may express PSA-NCAM (Alvarez-Buylla and Lois, 1995; Bonfanti and Theodosis, 1994; Nothias et al., 1997; Varea et al., 2005). Most astrocytes do not express PSA, except for those in the hypothalamus (Kiss et al., 1993; Theodosis et al., 1991, 1999) or suprachiasmatic nucleus (Theodosis and Poulain, 1999; Theodosis et al., 1999), where astrocytic process extension is circadian and regulated by osmolality (Trudel and Bourque, 2010; Theodosis et al., 2008).

The effects of PSA addition on NCAM are not fully clear; however, PSA is thought to act through two main mechanisms. First, it is now generally accepted that addition of PSA to NCAM inhibits NCAM homophilic interactions and affects the adhesive properties of other CAMs, including N-cadherin (Fujimoto et al., 2001; Johnson et al., 2005), most likely due to interference by the highly charged carbohydrate side chain. Additionally, PSA may regulate the extracellular volume of the brain (Yang et al., 1992) and may affect extracellular matrix deposition. More recently, it has been shown that PSA can also regulate lateral diffusion of NCAM throughout the cell membrane, and thus constitute another mechanism of PSA action on cell adhe-sion (Conchonaud et al., 2007).

L1 is a transmembrane protein (200 kDa) and member of the immunoglobulin superfamily of CAM that is strongly expressed in the CNS (Kadmon et al., 1998; Faissner et al., 1984; Fushiki and Schachner 1986; Linnemann et al., 1988). Full-length L1 is found only in neurons and not in astrocytes (Horinouchi et al., 2005; Takeda et al., 1996), though a splice variant is specifically found in glial cells such as Schwann cells, astrocytes, and oligodendrocytes (Takeda et al., 1996). The close homologue of L1, CHL1 (Holm et al., 1996), is also found in subpopulations of astrocytes, oligodendrocyte precursors, and Schwann cells (Hillenbrand et al., 1999). CHL1 has three isoforms of 185, 165, or 125 kDa, and primary astrocyte cultures predominantly express the 185 kDa isoform (Hillenbrand et al., 1999). CHL1 expression is increased in reactive astrocytes found at the glial scar, though it inhibits functional recovery after spinal cord injury (Jakovcevski et al., 2007). The extracellular part of L1 comprises six immunoglobulin and five fibronectin type III homology domains. L1 mediates both homophilic and heterophilic cell-cell adhesion, and several ligands for L1 have been discovered, including NCAM and integrins (Miura et al., 1992; Zhao and Siu, 1995; Montgomery et al., 1996; Yip et al., 1998; Gouveia et al., 2008). These interactions have variable effects in the control of neurite outgrowth, neurite fasciculation, neuronal cell differentiation, and migration (Hillenbrand et al., 1999; Jakovcevski et al., 2007; Dihne et al., 2003).

3.3.3 CELL ADHESION MOLECULES IN ASTROCYTE POLARITY AND FUNCTIONS

Transmembrane forms of all CAMs associate with the cytoskeleton. These interactions promote and stabilize the clustering of CAMs, participate in the mechanical coupling between adjacent cells, and regulate the cytoskeleton providing the basis for cell polarity, motility, and migration. β-Catenin mediates cadherin interactions with actin cytoskeleton through its association with α-catenin (Rimm et al., 1995). Although α-catenin can directly interact with actin, recent observations suggest that it cannot simultaneously interact with actin and β-catenin (Drees et al., 2005; Yamada et al., 2005). Nevertheless, α-catenin remains a central player in the interaction between the actin network and cadherin-mediated junctions. α-Catenin interacts with actin-binding proteins such as vinculin, afadin, and also with α-actinin or eplin (Abe and Takeichi, 2008; Kobielak and Fuchs, 2004; Maul et al., 2003). Like cadherins, nectins are connected to the actin cytoskeleton. The major link is directly provided by afadin, a ubiquitously expressed actin filament-binding protein (Takahashi et al., 1999; Ikeda et al., 1999; Miyahara et al., 2000). However, a small isoform of afadin, s-afadin, which lacks the actin-binding domain, is expressed in the brain and may modulate afadin function (Mandai et al., 1997). Most of the nectin isoforms have a four-amino-acid conserved motif at their carboxyl terminus that binds to afadin (Mandai et al., 1997; Takahashi et al., 1999; Reymond et al., 2001). Necls, however, do not have the ability to bind afadin (Shingai et al., 2003). Nectin-bound afadin interacts with α-catenin and contributes to cadherin recruitment to junctional sites. The nectin-afadin complex may also regulate adherens junctions through p120-mediated internalization of cadherins (Hoshino et al., 2005). Like the other CAMs, NCAM is also a major regulator of cytoskeleton dynamics (Ditlevsen et al., 2008). Direct as well as indirect interaction of NCAM with cytoskeleton components has also been shown (Buttner et al., 2003; Pollerberg et al., 1987; Jaffe et al., 1990). In particular, all three NCAM isoforms have been shown to bind spectrin (Leshchyns'ka et al., 2003), which may link NCAM to connexins and thus regroup NCAM-mediated interactions and GAP-43 (Korshunova et al., 2007). Moreover, biochemical evidence suggests that NCAM-140 interacts with β-actin, α-actinin, and tropomyosin (Buttner et al., 2003).

Adherens junctions are best known as the main actors in the transmission of forces between cells, and thus play a key role in the organization of coherent multi-cellular tissues such as epithelia. In addition to their adhesive properties, adherens junctions mediate the regulation of the cytoskeleton to ensure the tissue-specific mechanical properties of the cell. Intercellular adhesions can modulate actin and microtubule organization by controlling small Rho GTPases or their effectors (Etienne-Manneville and Hall, 2002). N-cadherin signaling has been shown to activate RhoA and inhibit Rac1 and Cdc42 (Charrasse et al., 2002). p120 is thought to play a crucial role in cadherin-mediated regulation of GTPases. It sequesters inactive RhoA and prevents RhoA activation (Anastasiadis et al., 2000). When unbound to cadherin, p120 can interact with Vav-2, a Rac, and Cdc42 exchange factor (Noren et al., 2000), suggesting that its recruitment to forming adherens junctions might be important for contact inhibition (Grosheva et al., 2001). In addition, α-catenin interacts with the effectors formins, which promote actin polymerization (Kobielak and Fuchs, 2004), and also inhibits the Arp2/3 com-plex (Drees et al., 2005; Yamada et al., 2005). α-Catenin may provide a more direct and fundamental molecular basis for the mechanosensitivity of adherens junctions (le Duc et al., 2010). Similar to talin in focal adhesions, α-catenin seems to unfold upon stretching and increase vinculin recruitment to adherens junctions (Yonemura et al., 2010). In epithelial cells, adherens junctions form the zonula adherens, a continuous belt-like structure observed in most epithelia (Farquhar and Palade, 1963), while in neurons and glia, adherens junctions are most often discontinuously distributed along the plasma membrane and form spot-like adhe-sions probably associated with weak mechanical forces involved at sites of cell-cell contacts (Liu et al., 2010). This may be related to the softness of the neural tissue. The fact that adherens junctions, like focal adhesions, serve as mechanosensors raises the question of their relative roles in the control of cell polarity and intra-cellular organization. Adherens junctions tend to locally inhibit the formation of focal adhesions and integrin-mediated signaling, and thereby affect actin organi-zation (Borghi et al., 2010; Dupin et al., 2009). By controlling the localization of focal adhesion, anisotropic cadherin-mediated interaction indirectly affects actin-dependent nucleus positioning and intracellular organization (Dupin et al., 2009). NCAM may play a key role in the cross talk between N-cadherin and integrins. In epithelial cells, loss of E-cadherin expression leads to increased NCAM levels (Lehembre et al., 2008), and may thereby promote tumor dissemination (Perl et al., 1999). Interestingly, in epithelial cells, NCAM is also crucial in the assembly of β1-integrin-dependent focal adhesions by inducing FAK phophorylation (Frame and Inman, 2008). Whether the role of NCAM in the cadherin-integrin interplay also occurs in astrocytes remains to be demonstrated.

Interactions with cytoskeleton and cytoskeleton regulators are probably central to the role of adherens junction molecules in the regulation of astrocyte polarity and migration. Our recent observations indicate that asymmetric distribution of adherens junctions can promote astrocyte polarization independently of other extracellular factors (Dupin et al., 2009). Adhesion to a cadherin-coated asymmetric micro-pattern surface induced the orientation of the nucleus-centrosome axis of the cell toward nonadhesive cell edges. This effect does not involve actin remodeling but seems to require the regulation of the microtubule network. Intercellular contacts interact by different means with the microtubule network. Adherens junctions are connected to microtubules that, at least in epithelial cells, are essential for their regulation (Ivanov et al., 2006; Ligon and Holzbaur, 2007). Catenins are likely to play a key role in microtubule recruitment to adherens junctions. β-Catenin regulates adenomatous polyposis coli (APC) association with microtubules, and may thus regulate microtubule dynamics (Votin et al., 2005; Barth et al., 2008). α-Catenin also controls microtubule dynamics (Shtutman et al., 2008). α- and β-catenin can interact with components of the microtubule-associated motor dynein (Ligon et al., 2001; Lien et al., 2008). Dynein, which interacts with catenins, regulates microtubule dynamics at cell-cell contacts, and this involves its interaction with the Par-related polarity protein Par3 and may participate in the orientation of the nucleus-centrosome axis (Schmoranzer et al., 2009). Par3 localizes at cell-cell contacts in astrocytes together with components of adherens junctions (Etienne-Manneville and Hall, 2003b). In epithelial cells, Par3 is recruited to cell-cell junction by the tight junction protein junctional adhesion molecule (JAM) (Ebnet et al., 2001). Interestingly, Par3 also interacts with the cytoplasmic part of nectins (Takekuni et al., 2003). Like E-cadherin, which is often lost in epithelial cancer, changes in N-cadherin expression or N-cadherin localization at the plasma membrane are likely to play a key role in the invasive capacity of astrocyte-derived tumors. Several studies report that N-cadherin expression is low in brain astrocytic tumors or cell lines (Asano et al., 1997, 2000; Hegedus et al., 2006; Foty and Steinberg, 2004). However, other studies reported upregulation of N-cadherin immunoreactivity (Utsuki et al., 2002), or even no modification of protein expression levels (Perego et al., 2002; Shinoura et al., 1995). N-cadherin limits astrocyte migration and is also required for astrocyte polarization during migration (personal unpublished results), as well as in nonmigrating cells (Dupin et al., 2009). Moreover, expression of the endothelial VE-cadherin has been reported in astroglioma (Vitolo et al., 1996), while cadherin-11 appears downregulated (Zhou and Skalli, 2000).

Several studies have revealed that NCAM plays some role in cell polarity during development. NCAM is required for islet cell polarity and normal organogenesis in the pancreas (Esni et al., 1999). NCAM has a polarized distribution in the retinal pigment epithelium, but its function in cell polarity in this context has yet to be demonstrated (Gundersen et al., 1993; McKay et al., 1997). No specialized cellular domain with a high density of NCAM has been described on neurons in vitro (Pollerberg et al., 1985; van den Pol et al., 1986), although it is concentrated at nodes of Ranvier in the sciatic nerve (Rieger et al., 1986). One open question is whether NCAM is involved in the control of astrocyte migration and polarity. NCAM-140 and NCAM-180 interact with α-tubulin and β-tubulin (Buttner et al., 2003). Moreover, they may regulate microtubule stability since they also interact with LANP/pp32, which controls microtubule stability via microtubule-associated protein (MAP) MAP1B. In addition, NCAM-180 interacts with MAP1A. Recruitment of MAPs to NCAM may prevent their association with microtubules and lead to a microtubule destabilizing effect. NCAM-mediated cell adhesion may regulate microtubule dynamics and contribute to the regulation of astrocyte process extension and astrocyte migration. Decrease in NCAM expression has also been associated with CNS invasion by glioma cells (Andersson et al., 1991; Maidment et al., 1997; Gratsa et al., 1997), and inverse correlation was found between NCAM-140 expression and human glioma cell grade (Duenisch et al., 2010). Overexpression of NCAM in tumor astrocytes leads to significant decrease of their invasive behavior (Edvardsen et al., 1994; Owens et al., 1998). On the other hand, PSA has been shown to facilitate glioma invasion (Suzuki et al., 2005) and PSA-NCAM has been recently proposed as a biomarker for the prognosis of patients with glioblastomas (Amoureux et al., 2010). Thus, two independent mechanisms may participate to enhance glioma invasion, one through a decrease in NCAM expression levels and a second by PSA addition on NCAM. These two processes could participate to weaken cell-cell adhesion, and thus facilitate glioma cell migration.

In addition to their role in regulating the cytoskeleton and cell polarity, CAMs are also involved in a large variety of signaling pathways controlling gene transcription and cell proliferation that are likely to play key roles in the control of astrocyte behavior, neuron survival, and regeneration. Both β-catenin and p120 have been reported to be localized in the nucleus (Daniel, 2007). β-Catenin exists as a soluble cytosolic protein that plays a central role in Wnt signaling (Su et al., 1993; Hulsken et al., 1994). p120 modulates gene transcription via its interaction with the transcription factor Kaiso (Daniel and Reynolds, 1999; Kelly et al., 2004). The cytoplasmic tail of NCAM interacts with several signaling proteins, such as Src tyrosine kinases and protein kinase C (PKC), which control mitogen-activated protein kinase (MAPK) signaling and gene transcription (Beggs et al., 1997; Kolkova et al., 2000). Most importantly, cadherins and NCAM share an ability to interact with growth factor receptor tyrosine kinases. NCAM interacts with glial-derived neurotrophic factor (GDNF) receptor (Nielsen et al., 2009). N-cadherin and NCAM interact with the fibroblast growth factor receptor (FGFR) (Kiselyov et al., 2003, 2005; Sanchez-Heras et al., 2006). In the latter instance, this may increase FGFR stability at the plasma membrane and promote tumor cell invasion by increasing FGF interaction with its receptor (Williams et al., 2001; Hazan et al., 2000; Suyama et al., 2002). In neurons, N-cadherin facilitates FGFR dimerization and thus induces neurite extension independent of growth factors (Doherty et al., 2000; Skaper et al., 2001). Like N-cadherin, NCAM on astrocytes stimulates neurite outgrowth (Neugebauer et al., 1988; Tomaselli et al., 1988; Doherty et al., 1990), as well as axon guidance (Matsunaga et al., 1988; Silver and Rutishauser, 1984). NCAM regulates cell motility (Prag et al., 2002). Astrocytes also express NCAM that has been implicated in regulation of neuronal migration, axon guidance and fasciculation, target recognition, and synapse formation (Nybroe et al., 1985; Noble et al., 1985; Maness and Schachner, 2007). N-cadherin has been shown to inhibit oligodendrocytes (Schnadelbach et al., 2000) and Schwann cell migration on astrocyte monolayers (Wilby et al., 1999; Fairless et al., 2005; Lakatos et al., 2000). At the glial scar, N-cadherin is expressed in reactive astrocytes (Vazquez-Chona and Geisert, 1999) that also strongly express PSA-NCAM (Camand et al., 2004; Le Gal La Salle et al., 1992; Oumesmar et al., 1995; Alonso and Privat, 1993; Dusart et al., 1999;

Bonfanti et al., 1996; Nomura et al., 2000) (Figure 3.5). NCAM has been shown to act as a regulator of astrocyte proliferation through glucocorticoid receptor pathways and nuclear factor-kappa B (NF-κB) activation, and may thus contribute to astrocyte proliferation at the lesion site (Crossin et al., 1997; Krushel et al., 1999; Little et al., 2001). Interestingly, PSA-NCAM expression by reactive astrocytes seems to play a permissive role for axon sprouting, although it is not able to overcome the global

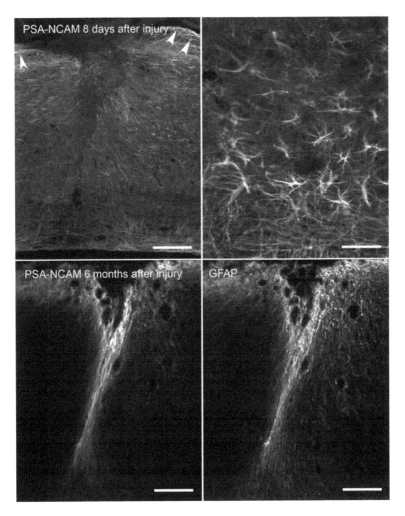

FIGURE 3.5 PSA-NCAM expression by astrocytes of the glial scar. After adult mouse spinal cord traumatic injury, PSA-NCAM immunoreactivity is observed in the glial scar. After 1 week, it is associated with many reactive astrocytes from both gray and white matters, even at sites far from the lesion. PSA-NCAM is also observed in unmyelinated fibers (arrowheads) that have been cut by the lesion. High magnification reveals some punctate astrocytic PSA-NCAM immunostaining. Six months after injury, PSA-NCAM is present in most reactive astrocytes of the glial scar. Scale bar = 150 µm, except in right upper panel, where it = 50 µm.

inhibitory effect of the glial scar on axon regeneration (Camand et al., 2004; Dusart et al., 1999; Morel et al., 2002). Several recent studies reveal that ectopic expression of PSA-NCAM or injection of a PSA mimetic peptide in the glial scar induces sprouting/regeneration of some axonal populations and functional recovery (Zhang et al., 2007a, 2007b; Marino et al., 2009).

3.4 DISCUSSION

Astrocytes are polarized cells that extend processes, interconnecting them with each other, and meninges, as well as blood vessels, to regulate the metabolic state of the brain. In addition, astrocytic processes also participate in synaptic transmission. To accomplish these tasks, astrocytes use multiple CAMs and ECM receptors to interact with the large variety of surrounding molecules and cells. While identifying the repertoire of CAMs and ECM molecules/receptors is still a work in progress, some of the general principles of their action are clear. A common denominator among all CAMs is their ability to connect the extracellular environment to the cytoskeleton, and thereby to influence cell polarity, cell shape, and motility. An emerging theme regards the role of focal adhesions and adherens junctions in mechanosensing. The mechanical properties of the tissue may result from feedback mechanisms between the intrinsic characteristics of the extracellular matrix and the regulation of cell contractility. The general orientation and polarity of cells, and in particular astrocytes, is thus dramatically dependent on the physical properties of their environment. Moreover, CAMs by transmitting specific intracellular signals inform the cells about the chemical nature of their nearby environment. It seems likely from the accumulating data that CAMs and downstream membrane scaffolds, which are used to establish cell polarity, are involved in many cellular interactions in which subcellular domains or frankly polarized astrocytes play a role. Thus, deciphering signaling pathways downstream of CAMs will be helpful in mitigating deleterious effects of the polarized astrocyte or facilitating beneficial ones.

ACKNOWLEDGMENTS

We thank Drs. Charles Bourque, Keith Murai, Emma Thompson, and Isabelle Dusart for providing images of astrocytes. Work on this chapter was supported by grants to S.C. from the Canadian Cancer Society and NSERC and to S.E.-M. from the Centre National de la Recherche Scientific, the Institut Pasteur, the Agence Nationale pour la Recherche. and the cancéropole d'Ile de France. S.E.-M. is member of the EMBO YIP.

REFERENCES

Abe, K., and M. Takeichi. 2008. EPLIN mediates linkage of the cadherin catenin complex to F-actin and stabilizes the circumferential actin belt. *Proc Natl Acad Sci USA* 105 (1):13–19.

Aberle, H., S. Butz, J. Stappert, H. Weissig, R. Kemler, and H. Hoschuetzky. 1994. Assembly of the cadherin-catenin *complex in vitro* with recombinant proteins. *J Cell Sci* 107 (Pt 12):3655–63.

Agulhon, C., T. A. Fiacco, and K. D. McCarthy. 2010. Hippocampal short- and long-term plasticity are not modulated by astrocyte Ca2+ signaling. *Science* 327 (5970):1250–54.

Alon, R., and K. Ley. 2008. Cells on the run: shear-regulated integrin activation in leukocyte rolling and arrest on endothelial cells. *Curr Opin Cell Biol* 20 (5):525–32.

Alonso, G., and A. Privat. 1993. Reactive astrocytes involved in the formation of lesional scars differ in the mediobasal hypothalamus and in other forebrain regions. *J Neurosci Res* 34 (5):523–38.

Alvarez-Buylla, A., and C. Lois. 1995. Neuronal stem cells in the brain of adult vertebrates. *Stem Cells* 13 (3):263–72.

Amoureux, M. C., B. Coulibaly, O. Chinot, A. Loundou, P. Metellus, G. Rougon, and D. Figarella-Branger. 2010. Polysialic acid neural cell adhesion molecule (PSA-NCAM) is an adverse prognosis factor in glioblastoma, and regulates olig2 expression in glioma cell lines. *BMC Cancer* 10:91.

Anastasiadis, P. Z., S. Y. Moon, M. A. Thoreson, D. J. Mariner, H. C. Crawford, Y. Zheng, and A. B. Reynolds. 2000. Inhibition of RhoA by p120 catenin. *Nat Cell Biol* 2 (9):637–44.

Andersson, A. M., N. Moran, H. Gaardsvoll, D. Linnemann, R. Bjerkvig, O. D. Laerum, and E. Bock. 1991. Characterization of NCAM expression and function in BT4C and BT4Cn glioma cells. *Int J Cancer* 47 (1):124–29.

Anton, E. S., J. A. Kreidberg, and P. Rakic. 1999. Distinct functions of alpha3 and alpha(v) integrin receptors in neuronal migration and laminar organization of the cerebral cortex. *Neuron* 22 (2):277–89.

Araque, A., and M. Navarrete. 2010. Glial cells in neuronal network function. *Philos Trans R Soc Lond B Biol Sci* 365 (1551):2375–81.

Arias-Salgado, E. G., S. Lizano, S. J. Shattil, and M. H. Ginsberg. 2005. Specification of the direction of adhesive signaling by the integrin beta cytoplasmic domain. *J Biol Chem* 280 (33):29699–707.

Asano, K., O. Kubo, Y. Tajika, M. C. Huang, K. Takakura, K. Ebina, and S. Suzuki. 1997. Expression and role of cadherins in astrocytic tumors. *Brain Tumor Pathol* 14 (1):27–33.

Asano, K., O. Kubo, Y. Tajika, K. Takakura, and S. Suzuki. 2000. Expression of cadherin and CSF dissemination in malignant astrocytic tumors. *Neurosurg Rev* 23 (1):39–44.

Atkins, A. R., W. J. Gallin, G. C. Owens, G. M. Edelman, and B. A. Cunningham. 2004. Neural cell adhesion molecule (N-CAM) homophilic binding mediated by the two N-terminal Ig domains is influenced by intramolecular domain-domain interactions. *J Biol Chem* 279 (48):49633–43.

Barbas, J. A., J. C. Chaix, M. Steinmetz, and C. Goridis. 1988. Differential splicing and alternative polyadenylation generates distinct NCAM transcripts and proteins in the mouse. *Embo J* 7 (3):625–32.

Barth, A. I., H. Y. Caro-Gonzalez, and W. J. Nelson. 2008. Role of adenomatous polyposis coli (APC) and microtubules in directional cell migration and neuronal polarization. *Semin Cell Dev Biol* 19 (3):245–51.

Barthels, D., G. Vopper, and W. Wille. 1988. NCAM-180, the large isoform of the neural cell adhesion molecule of the mouse, is encoded by an alternatively spliced transcript. *Nucleic Acids Res* 16 (10):4217–25.

Beattie, J., L. McIntosh, and C. F. van der Walle. 2010. Cross-talk between the insulin-like growth factor (IGF) axis and membrane integrins to regulate cell physiology. *J Cell Physiol* 224 (3):605–11.

Beggs, H. E., S. C. Baragona, J. J. Hemperly, and P. F. Maness. 1997. NCAM140 interacts with the focal adhesion kinase p125(fak) and the SRC-related tyrosine kinase p59(fyn). *J Biol Chem* 272 (13):8310–19.

Blaschuk, O. W., R. Sullivan, S. David, and Y. Pouliot. 1990. Identification of a cadherin cell adhesion recognition sequence. *Dev Biol* 139 (1):227–29.

Boller, K., D. Vestweber, and R. Kemler. 1985. Cell-adhesion molecule uvomorulin is localized in the intermediate junctions of adult intestinal epithelial cells. *J Cell Biol* 100 (1):327–32.

Bonfanti, L., A. Merighi, and D. T. Theodosis. 1996. Dorsal rhizotomy induces transient expression of the highly sialylated isoform of the neural cell adhesion molecule in neurons and astrocytes of the adult rat spinal cord. *Neuroscience* 74 (3):619–23.

Bonfanti, L., and D. T. Theodosis. 1994. Expression of polysialylated neural cell adhesion molecule by proliferating cells in the subependymal layer of the adult rat, in its rostral extension and in the olfactory bulb. *Neuroscience* 62 (1):291–305.

Borghi, N., M. Lowndes, V. Maruthamuthu, M. L. Gardel, and W. J. Nelson. 2010. Regulation of cell motile behavior by crosstalk between cadherin- and integrin-mediated adhesions. *Proc Natl Acad Sci USA.* 107(30):13324–29.

Bowe, M. A., D. B. Mendis, and J. R. Fallon. 2000. The small leucine-rich repeat proteoglycan biglycan binds to alpha-dystroglycan and is upregulated in dystrophic muscle. *J Cell Biol* 148 (4):801–10.

Bragg, A. D., M. Amiry-Moghaddam, O. P. Ottersen, M. E. Adams, and S. C. Froehner. 2006. Assembly of a perivascular astrocyte protein scaffold at the mammalian blood-brain barrier is dependent on alpha-syntrophin. *Glia* 53 (8):879–90.

Bruses, J. L., and U. Rutishauser. 1998. Regulation of neural cell adhesion molecule polysialylation: evidence for nontranscriptional control and sensitivity to an intracellular pool of calcium. *J Cell Biol* 140 (5):1177–86.

Butler, B., C. Gao, A. T. Mersich, and S. D. Blystone. 2006. Purified integrin adhesion complexes exhibit actin-polymerization activity. *Curr Biol* 16 (3):242–51.

Buttner, B., C. Kannicht, W. Reutter, and R. Horstkorte. 2003. The neural cell adhesion molecule is associated with major components of the cytoskeleton. *Biochem Biophys Res Commun* 310 (3):967–71.

Camand, E., M. P. Morel, A. Faissner, C. Sotelo, and I. Dusart. 2004. Long-term changes in the molecular composition of the glial scar and progressive increase of serotoninergic fibre sprouting after hemisection of the mouse spinal cord. *Eur J Neurosci* 20 (5):1161–76.

Carbonetto, S., and M. Lindenbaum. 1995. The basement membrane at the neuromuscular junction: a synaptic mediatrix. *Curr Opin Neurobiol* 5 (5):596–605.

Carroll, P., O. Gayet, C. Feuillet, S. Kallenbach, B. de Bovis, K. Dudley, and S. Alonso. 2001. Juxtaposition of CNR protocadherins and reelin expression in the developing spinal cord. *Mol Cell Neurosci* 17 (4):611–23.

Chakraborty, S., S. Mitra, M. M. Falk, et al., 2010. E-cadherin differentially regulates the assembly of connexin43 and connexin32 into gap junctions in human squamous carcinoma cells. *J Biol Chem* 285 (14):10761–76.

Chappuis-Flament, S., E. Wong, L. D. Hicks, C. M. Kay, and B. M. Gumbiner. 2001. Multiple cadherin extracellular repeats mediate homophilic binding and adhesion. *J Cell Biol* 154 (1):231–43.

Charrasse, S., M. Meriane, F. Comunale, A. Blangy, and C. Gauthier-Rouviere. 2002. N-cadherin-dependent cell-cell contact regulates Rho GTPases and beta-catenin localization in mouse C2C12 myoblasts. *J Cell Biol* 158 (5):953–65.

Chen, X., S. Kojima, G. G. Borisy, and K. J. Green. 2003. p120 catenin associates with kinesin and facilitates the transport of cadherin-catenin complexes to intercellular junctions. *J Cell Biol* 163 (3):547–57.

Chitaev, N. A., and S. M. Troyanovsky. 1998. Adhesive but not lateral E-cadherin complexes require calcium and catenins for their formation. *J Cell Biol* 142 (3):837–46.

Conchonaud, F., S. Nicolas, M. C. Amoureux, C. Menager, D. Marguet, P. F. Lenne, G. Rougon, and V. Matarazzo. 2007. Polysialylation increases lateral diffusion of neural cell adhesion molecule in the cell membrane. *J Biol Chem* 282 (36):26266–74.

Constantin, G. 2008. Chemokine signaling and integrin activation in lymphocyte migration into the inflamed brain. *J Neuroimmunol* 198 (1–2):20–26.

Cote, P. D., H. Moukhles, M. Lindenbaum, and S. Carbonetto. 1999. Chimaeric mice deficient in dystroglycans develop muscular dystrophy and have disrupted myoneural synapses. *Nat Genet* 23 (3):338–42.

Cremer, H., R. Lange, A. Christoph, M. Plomann, G. Vopper, J. Roes, R. Brown, S. Baldwin, P. Kraemer, S. Scheff, et al. 1994. Inactivation of the N-CAM gene in mice results in size reduction of the olfactory bulb and deficits in spatial learning. *Nature* 367 (6462):455–59.

Critchley, D. R., and A. R. Gingras. 2008. Talin at a glance. *J Cell Sci* 121 (Pt 9):1345–47.

Crossin, K. L., M. H. Tai, L. A. Krushel, V. P. Mauro, and G. M. Edelman. 1997. Glucocorticoid receptor pathways are involved in the inhibition of astrocyte proliferation. *Proc Natl Acad Sci USA* 94 (6):2687–92.

Damsky, C. H., J. Richa, D. Solter, K. Knudsen, and C. A. Buck. 1983. Identification and purification of a cell surface glycoprotein mediating intercellular adhesion in embryonic and adult tissue. *Cell* 34 (2):455–66.

Daniel, J. M. 2007. Dancing in and out of the nucleus: p120(ctn) and the transcription factor Kaiso. *Biochim Biophys Acta* 1773:59-68.

Daniel, J. M., and A. B. Reynolds. 1995. The tyrosine kinase substrate p120cas binds directly to E-cadherin but not to the adenomatous polyposis coli protein or alpha-catenin. *Mol Cell Biol* 15 (9):4819–24.

Daniel, J. M., and A. B. Reynolds. 1999. The catenin p120(ctn) interacts with Kaiso, a novel BTB/POZ domain zinc finger transcription factor. *Mol Cell Biol* 19 (5):3614–23.

Daub, H., K. Gevaert, J. Vandekerckhove, A. Sobel, and A. Hall. 2001. Rac/Cdc42 and p65PAK regulate the microtubule-destabilizing protein stathmin through phosphorylation at serine 16. *J Biol Chem* 276:1677–80.

Davis, M. A., R. C. Ireton, and A. B. Reynolds. 2003. A core function for p120-catenin in cadherin turnover. *J Cell Biol* 163 (3):525–34.

del Pozo, M. A., N. B. Alderson, W. B. Kiosses, H. H. Chiang, R. G. Anderson, and M. A. Schwartz. 2004. Integrins regulate Rac targeting by internalization of membrane domains. *Science* 303 (5659):839–42.

del Pozo, M. A., and M. A. Schwartz. 2007. Rac, membrane heterogeneity, caveolin and regulation of growth by integrins. *Trends Cell Biol* 17 (5):246–50.

Dihne, M., C. Bernreuther, M. Sibbe, W. Paulus, and M. Schachner. 2003. A new role for the cell adhesion molecule L1 in neural precursor cell proliferation, differentiation, and transmitter-specific subtype generation. *J Neurosci* 23 (16):6638–50.

Ditlevsen, D. K., G. K. Povlsen, V. Berezin, and E. Bock. 2008. NCAM-induced intracellular signaling revisited. *J Neurosci Res* 86 (4):727–43.

Doherty, P., M. Fruns, P. Seaton, G. Dickson, C. H. Barton, T. A. Sears, and F. S. Walsh. 1990. A threshold effect of the major isoforms of NCAM on neurite outgrowth. *Nature* 343 (6257):464–66.

Doherty, P., C. E. Moolenaar, S. V. Ashton, R. J. Michalides, and F. S. Walsh. 1992. The VASE exon downregulates the neurite growth-promoting activity of NCAM 140. *Nature* 356 (6372):791–93.

Doherty, P., G. Williams, and E. J. Williams. 2000. CAMs and axonal growth: a critical evaluation of the role of calcium and the MAPK cascade. *Mol Cell Neurosci* 16 (4):283–95.

Douville, P. J., W. J. Harvey, and S. Carbonetto. 1988. Isolation and partial characterization of high affinity laminin receptors in neural cells. *J Biol Chem* 263 (29):14964–69.

Drees, F., S. Pokutta, S. Yamada, W. J. Nelson, and W. I. Weis. 2005. Alpha-catenin is a molecular switch that binds E-cadherin-beta-catenin and regulates actin-filament assembly. *Cell* 123 (5):903–15.

Duenisch, P., R. Reichart, U. Mueller, M. Brodhun, R. Bjerkvig, B. Romeike, J. Walter, C. Herbold, C. R. Regenbrecht, R. Kalff, and S. A. Kuhn. 2010. Neural cell adhesion molecule isoform 140 declines with rise of WHO grade in human gliomas and serves as indicator for the invasion zone of multiform glioblastomas and brain metastases. *J Cancer Res Clin Oncol.* 137(3): 39999–414.

Dupin, I., E. Camand, and S. Etienne-Manneville. 2009. Classical cadherins control nucleus and centrosome position and cell polarity. *J Cell Biol* 185 (5):779–86.

Durbeej, M., and K. P. Campbell. 2002. Muscular dystrophies involving the dystrophin-glycoprotein complex: an overview of current mouse models. *Curr Opin Genet Dev* 12 (3):349–61.

Dusart, I., M. P. Morel, R. Wehrle, and C. Sotelo. 1999. Late axonal sprouting of injured Purkinje cells and its temporal correlation with permissive changes in the glial scar. *J Comp Neurol* 408 (3):399–418.

Eberle, F., P. Dubreuil, M. G. Mattei, E. Devilard, and M. Lopez. 1995. The human PRR2 gene, related to the human poliovirus receptor gene (PVR), is the true homolog of the murine MPH gene. *Gene* 159 (2):267–72.

Ebnet, K., A. Suzuki, Y. Horikoshi, T. Hirose, M. K. Meyer Zu Bridkwedde, S. Ohno, and D. Vestweber. 2001. The cell polarity protein ASIP/PAR-3 directly associates with junctional adhesion molecule (JAM). *Embo J* 20:3738–48.

Eckhardt, M., M. Muhlenhoff, A. Bethe, J. Koopman, M. Frosch, and R. Gerardy-Schahn. 1995. Molecular characterization of eukaryotic polysialyltransferase-1. *Nature* 373 (6516):715–18.

Edvardsen, K., P. H. Pedersen, R. Bjerkvig, G. G. Hermann, J. Zeuthen, O. D. Laerum, F. S. Walsh, and E. Bock. 1994. Transfection of glioma cells with the neural-cell adhesion molecule NCAM: effect on glioma-cell invasion and growth *in vivo. Int J Cancer* 58 (1):116–22.

Elias, L. A., and A. R. Kriegstein. 2008. Gap junctions: multifaceted regulators of embryonic cortical development. *Trends Neurosci* 31 (5):243–50.

Elias, L. A., D. D. Wang, and A. R. Kriegstein. 2007. Gap junction adhesion is necessary for radial migration in the neocortex. *Nature* 448 (7156):901–7.

Ervasti, J. M. 2003. Costameres: the Achilles' heel of Herculean muscle. *J Biol Chem* 278 (16):13591–94.

Ervasti, J. M., and K. P. Campbell. 1993. A role for the dystrophin-glycoprotein complex as a transmembrane linker between laminin and actin. *J Cell Biol* 122 (4):809–23.

Esni, F., I. B. Taljedal, A. K. Perl, H. Cremer, G. Christofori, and H. Semb. 1999. Neural cell adhesion molecule (N-CAM) is required for cell type segregation and normal ultrastructure in pancreatic islets. *J Cell Biol* 144 (2):325–37.

Etienne-Manneville, S. 2010. From signaling pathways to microtubule dynamics: the key players. *Curr Opin Cell Biol* 22 (1):104–11.

Etienne-Manneville, S. 2004a. Actin and microtubules in cell motility: which one is in control? *Traffic* 5:470–477.

Etienne-Manneville, S. 2004b. Cdc42—the centre of polarity. *J Cell Sci* 117 (Pt 8):1291–300.

Etienne-Manneville, S., and A. Hall. 2001. Integrin-mediated activation of Cdc42 controls cell polarity in migrating astrocytes through PKCzeta. *Cell* 106 (4):489–98.

Etienne-Manneville, S., and A. Hall. 2002. Rho GTPases in cell biology. *Nature* 420:629–635.

Etienne-Manneville, S., and A. Hall. 2003a. Cdc42 regulates GSK-3beta and adenomatous polyposis coli to control cell polarity. *Nature* 421 (6924):753–56.

Etienne-Manneville, S., and A. Hall. 2003b. Cell polarity: Par6, aPKC and cytoskeletal crosstalk. *Curr Opin Cell Biol* 15:67–72.

Etienne-Manneville, S., J. B. Manneville, S. Nicholls, M. A. Ferenczi, and A. Hall. 2005. Cdc42 and Par6/PKCz regulate the spatially localized association of Dlg1 and APC to control cell polarization. *J Cell Biol* 170 (6):895–901.

Fairless, R., M. C. Frame, and S. C. Barnett. 2005. N-cadherin differentially determines Schwann cell and olfactory ensheathing cell adhesion and migration responses upon contact with astrocytes. *Mol Cell Neurosci* 28 (2):253–63.

Faissner, A., J. Kruse, J. Nieke, and M. Schachner. 1984. Expression of neural cell adhesion molecule L1 during development, in neurological mutants and in the peripheral nervous system. *Brain Res* 317 (1):69–82.

Farquhar, M. G., and G. E. Palade. 1963. Junctional complexes in various epithelia. *J Cell Biol* 17:375–412.

Fielding, A. B., I. Dobreva, P. C. McDonald, L. J. Foster, and S. Dedhar. 2008. Integrin-linked kinase localizes to the centrosome and regulates mitotic spindle organization. *J Cell Biol* 180 (4):681–89.

Finne, J., U. Finne, H. Deagostini-Bazin, and C. Goridis. 1983. Occurrence of alpha 2–8 linked polysialosyl units in a neural cell adhesion molecule. *Biochem Biophys Res Commun* 112 (2):482–87.

Foty, R. A., and M. S. Steinberg. 2004. Cadherin-mediated cell-cell adhesion and tissue segregation in relation to malignancy. *Int J Dev Biol* 48 (5–6):397–409.

Frame, M. C., and G. J. Inman. 2008. NCAM is at the heart of reciprocal regulation of E-cadherin- and integrin-mediated adhesions via signaling modulation. *Dev Cell* 15 (4):494–96.

Frank, M., M. Ebert, W. Shan, G. R. Phillips, K. Arndt, D. R. Colman, and R. Kemler. 2005. Differential expression of individual gamma-protocadherins during mouse brain development. *Mol Cell Neurosci* 29 (4):603–16.

Fujimoto, I., J. L. Bruses, and U. Rutishauser. 2001. Regulation of cell adhesion by polysialic acid. Effects on cadherin, immunoglobulin cell adhesion molecule, and integrin function and independence from neural cell adhesion molecule binding or signaling activity. *J Biol Chem* 276 (34):31745–51.

Fukuhara, A., K. Shimizu, T. Kawakatsu, T. Fukuhara, and Y. Takai. 2003. Involvement of nectin-activated Cdc42 small G protein in organization of adherens and tight junctions in Madin-Darby canine kidney cells. *J Biol Chem* 278 (51):51885–93.

Fushiki, S., and M. Schachner. 1986. Immunocytological localization of cell adhesion molecules L1 and N-CAM and the shared carbohydrate epitope L2 during development of the mouse neocortex. *Brain Res* 389 (1–2):153–67.

Gallagher, H. C., K. J. Murphy, A. G. Foley, and C. M. Regan. 2001. Protein kinase C delta regulates neural cell adhesion molecule polysialylation state in the rat brain. *J Neurochem* 77 (2):425–34.

Garcion, E., A. Faissner, and C. ffrench-Constant. 2001. Knockout mice reveal a contribution of the extracellular matrix molecule tenascin-C to neural precursor proliferation and migration. *Development* 128 (13):2485–96.

Garrett, A. M., and J. A. Weiner. 2009. Control of CNS synapse development by {gamma}-protocadherin-mediated astrocyte-neuron contact. *J Neurosci* 29 (38):11723–31.

Gee, S. H., R. W. Blacher, P. J. Douville, P. R. Provost, P. D. Yurchenco, and S. Carbonetto. 1993. Laminin-binding protein 120 from brain is closely related to the dystrophin-associated glycoprotein, dystroglycan, and binds with high affinity to the major heparin binding domain of laminin. *J Biol Chem* 268 (20):14972–80.

Gee, S. H., F. Montanaro, M. H. Lindenbaum, and S. Carbonetto. 1994. Dystroglycan-alpha, a dystrophin-associated glycoprotein, is a functional agrin receptor. *Cell* 77 (5):675–86.

Giancotti, F. G., and E. Ruoslahti. 1999. Integrin signaling. *Science* 285 (5430):1028–32.

Giaume, C., A. Koulakoff, L. Roux, D. Holcman, and N. Rouach. 2010. Astroglial networks: a step further in neuroglial and gliovascular interactions. *Nat Rev Neurosci* 11 (2):87–99.

Gladson, C. L., J. N. Wilcox, L. Sanders, G. Y. Gillespie, and D. A. Cheresh. 1995. Cerebral microenvironment influences expression of the vitronectin gene in astrocytic tumors. *J Cell Sci* 108 (Pt 3):947–56.

Gordon, G. R., S. J. Mulligan, and B. A. MacVicar. 2007. Astrocyte control of the cerebrovasculature. *Glia* 55 (12):1214–21.

Gouveia, R. M., C. M. Gomes, M. Sousa, P. M. Alves, and J. Costa. 2008. Kinetic analysis of L1 homophilic interaction: role of the first four immunoglobulin domains and implications on binding mechanism. *J Biol Chem* 283 (42):28038–47.

Gower, H. J., C. H. Barton, V. L. Elsom, J. Thompson, S. E. Moore, G. Dickson, and F. S. Walsh. 1988. Alternative splicing generates a secreted form of N-CAM in muscle and brain. *Cell* 55 (6):955–64.

Gratsa, A., H. K. Rooprai, J. P. Rogers, K. K. Martin, and G. J. Pilkington. 1997. Correlation of expression of NCAM and GD3 ganglioside to motile behaviour in neoplastic glia. *Anticancer Res* 17 (6B):4111–17.

Graus-Porta, D., S. Blaess, M. Senften, A. Littlewood-Evans, C. Damsky, Z. Huang, P. Orban, R. Klein, J. C. Schittny, and U. Muller. 2001. Beta1-class integrins regulate the development of laminae and folia in the cerebral and cerebellar cortex. *Neuron* 31 (3):367–79.

Green, C. E., D. N. Pearson, R. T. Camphausen, D. E. Staunton, and S. I. Simon. 2004. Shear-dependent capping of L-selectin and P-selectin glycoprotein ligand 1 by E-selectin signals activation of high-avidity beta2-integrin on neutrophils. *J Immunol* 172 (12):7780–90.

Grosheva, I., M. Shtutman, M. Elbaum, and A. D. Bershadsky. 2001. p120 catenin affects cell motility via modulation of activity of Rho-family GTPases: a link between cell-cell contact formation and regulation of cell locomotion. *J Cell Sci* 114 (Pt 4):695–707.

Guadagno, E., and H. Moukhles. 2004. Laminin-induced aggregation of the inwardly rectifying potassium channel, Kir4.1, and the water-permeable channel, AQP4, via a dystroglycan-containing complex in astrocytes. *Glia* 47 (2):138–49.

Gundersen, D., S. K. Powell, and E. Rodriguez-Boulan. 1993. Apical polarization of N-CAM in retinal pigment epithelium is dependent on contact with the neural retina. *J Cell Biol* 121 (2):335–43.

Guo, W., H. Jiang, V. Gray, S. Dedhar, and Y. Rao. 2007. Role of the integrin-linked kinase (ILK) in determining neuronal polarity. *Dev Biol* 306 (2):457–68.

Guzman, G., S. Oh, D. Shukla, H. H. Engelhard, and T. Valyi-Nagy. 2006. Expression of entry receptor nectin-1 of herpes simplex virus 1 and/or herpes simplex virus 2 in normal and neoplastic human nervous system tissues. *Acta Virol* 50 (1):59–66.

Halassa, M. M., T. Fellin, and P. G. Haydon. 2009. Tripartite synapses: roles for astrocytic purines in the control of synaptic physiology and behavior. *Neuropharmacology* 57 (4):343–46.

Halassa, M. M., and P. G. Haydon. 2010. Integrated brain circuits: astrocytic networks modulate neuronal activity and behavior. *Annu Rev Physiol* 72:335–55.

Hamilton, N. B., and D. Attwell. 2010. Do astrocytes really exocytose neurotransmitters? *Nat Rev Neurosci* 11 (4):227–38.

Hartfuss, E., E. Forster, H. H. Bock, M. A. Hack, P. Leprince, J. M. Luque, J. Herz, M. Frotscher, and M. Gotz. 2003. Reelin signaling directly affects radial glia morphology and biochemical maturation. *Development* 130 (19):4597–609.

Hatta, K., A. Nose, A. Nagafuchi, and M. Takeichi. 1988. Cloning and expression of cDNA encoding a neural calcium-dependent cell adhesion molecule: its identity in the cadherin gene family. *J Cell Biol* 106 (3):873–81.

Hayashi, I., K. Vuori, and R. C. Liddington. 2002. The focal adhesion targeting (FAT) region of focal adhesion kinase is a four-helix bundle that binds paxillin. *Nat Struct Biol* 9 (2):101–6.

Hazan, R. B., G. R. Phillips, R. F. Qiao, L. Norton, and S. A. Aaronson. 2000. Exogenous expression of N-cadherin in breast cancer cells induces cell migration, invasion, and metastasis. *J Cell Biol* 148 (4):779–90.

He, H. T., J. Finne, and C. Goridis. 1987. Biosynthesis, membrane association, and release of N-CAM-120, a phosphatidylinositol-linked form of the neural cell adhesion molecule. *J Cell Biol* 105 (6 Pt 1):2489–500.

Hegedus, B., F. Marga, K. Jakab, K. L. Sharpe-Timms, and G. Forgacs. 2006. The interplay of cell-cell and cell-matrix interactions in the invasive properties of brain tumors. *Biophys J* 91 (7):2708–16.

Hemperly, J. J., G. M. Edelman, and B. A. Cunningham. 1986. cDNA clones of the neural cell adhesion molecule (N-CAM) lacking a membrane-spanning region consistent with evidence for membrane attachment via a phosphatidylinositol intermediate. *Proc Natl Acad Sci USA* 83 (24):9822–26.

Henneberger, C., T. Papouin, S. H. Oliet, and D. A. Rusakov. 2010. Long-term potentiation depends on release of D-serine from astrocytes. *Nature* 463 (7278):232–36.

Henry, M. D., and K. P. Campbell. 1996. Dystroglycan: an extracellular matrix receptor linked to the cytoskeleton. *Curr Opin Cell Biol* 8 (5):625–31.

Hillenbrand, R., M. Molthagen, D. Montag, and M. Schachner. 1999. The close homologue of the neural adhesion molecule L1 (CHL1): patterns of expression and promotion of neurite outgrowth by heterophilic interactions. *Eur J Neurosci* 11 (3):813–26.

Hirano, S., Q. Yan, and S. T. Suzuki. 1999. Expression of a novel protocadherin, OL-protocadherin, in a subset of functional systems of the developing mouse brain. *J Neurosci* 19 (3):995–1005.

Hoffman, S., B. C. Sorkin, P. C. White, R. Brackenbury, R. Mailhammer, U. Rutishauser, B. A. Cunningham, and G. M. Edelman. 1982. Chemical characterization of a neural cell adhesion molecule purified from embryonic brain membranes. *J Biol Chem* 257 (13):7720–29.

Holm, J., R. Hillenbrand, V. Steuber, U. Bartsch, M. Moos, H. Lubbert, D. Montag, and M. Schachner. 1996. Structural features of a close homologue of L1 (CHL1) in the mouse: a new member of the L1 family of neural recognition molecules. *Eur J Neurosci* 8 (8):1613–29.

Horinouchi, K., Y. Nakamura, H. Yamanaka, T. Watabe, and S. Shiosaka. 2005. Distribution of L1cam mRNA in the adult mouse brain: *in situ* hybridization and Northern blot analyses. *J Comp Neurol* 482 (4):386–404.

Hoshino, T., T. Sakisaka, T. Baba, T. Yamada, T. Kimura, and Y. Takai. 2005. Regulation of E-cadherin endocytosis by nectin through afadin, Rap1, and p120ctn. *J Biol Chem* 280 (25):24095–103.

Houades, V., A. Koulakoff, P. Ezan, I. Seif, and C. Giaume. 2008. Gap junction-mediated astrocytic networks in the mouse barrel cortex. *J Neurosci* 28 (20):5207–17.

Huber, O., M. Krohn, and R. Kemler. 1997. A specific domain in alpha-catenin mediates binding to beta-catenin or plakoglobin. *J Cell Sci* 110 (Pt 15):1759–65.

Hulsken, J., J. Behrens, and W. Birchmeier. 1994. Tumor-suppressor gene products in cell contacts: the cadherin-APC-armadillo connection. *Curr Opin Cell Biol* 6 (5):711–16.

Humphries, J. D., A. Byron, M. D. Bass, S. E. Craig, J. W. Pinney, D. Knight, and M. J. Humphries. 2009. Proteomic analysis of integrin-associated complexes identifies RCC2 as a dual regulator of Rac1 and Arf6. *Sci Signal* 2 (87):ra51.

Hyafil, F., C. Babinet, and F. Jacob. 1981. Cell-cell interactions in early embryogenesis: a molecular approach to the role of calcium. *Cell* 26 (3 Pt 1):447–54.

Hyafil, F., D. Morello, C. Babinet, and F. Jacob. 1980. A cell surface glycoprotein involved in the compaction of embryonal carcinoma cells and cleavage stage embryos. *Cell* 21 (3):927–34.

Hynes, R. O. 2002. A reevaluation of integrins as regulators of angiogenesis. *Nat Med* 8 (9):918–21.

Ibraghimov-Beskrovnaya, O., J. M. Ervasti, C. J. Leveille, C. A. Slaughter, S. W. Sernett, and K. P. Campbell. 1992. Primary structure of dystrophin-associated glycoproteins linking dystrophin to the extracellular matrix. *Nature* 355 (6362):696–702.

Ikeda, W., S. Kakunaga, S. Itoh, T. Shingai, K. Takekuni, K. Satoh, Y. Inoue, A. Hamaguchi, K. Morimoto, M. Takeuchi, T. Imai, and Y. Takai. 2003. Tage4/Nectin-like molecule-5 heterophilically trans-interacts with cell adhesion molecule Nectin-3 and enhances cell migration. *J Biol Chem* 278 (30):28167–72.

Ikeda, W., H. Nakanishi, J. Miyoshi, K. Mandai, H. Ishizaki, M. Tanaka, A. Togawa, K. Takahashi, H. Nishioka, H. Yoshida, A. Mizoguchi, S. Nishikawa, and Y. Takai. 1999. Afadin: a key molecule essential for structural organization of cell-cell junctions of polarized epithelia during embryogenesis. *J Cell Biol* 146 (5):1117–32.

Ingber, D. E., L. Dike, L. Hansen, S. Karp, H. Liley, A. Maniotis, H. McNamee, D. Mooney, G. Plopper, J. Sims, et al. 1994. Cellular tensegrity: exploring how mechanical changes in the cytoskeleton regulate cell growth, migration, and tissue pattern during morphogenesis. *Int Rev Cytol* 150:173–224.

Ivanov, A. I., I. C. McCall, B. Babbin, S. N. Samarin, A. Nusrat, and C. A. Parkos. 2006. Microtubules regulate disassembly of epithelial apical junctions. *BMC Cell Biol* 7:12.

Jacobson, C., P. D. Cote, S. G. Rossi, R. L. Rotundo, and S. Carbonetto. 2001. The dystroglycan complex is necessary for stabilization of acetylcholine receptor clusters at neuromuscular junctions and formation of the synaptic basement membrane. *J Cell Biol* 152 (3):435–50.

Jaffe, A. B., and A. Hall. 2005. Rho GTPases: biochemistry and biology. *Annu Rev Cell Dev Biol* 21:247–69.

Jaffe, S. H., D. R. Friedlander, F. Matsuzaki, K. L. Crossin, B. A. Cunningham, and G. M. Edelman. 1990. Differential effects of the cytoplasmic domains of cell adhesion molecules on cell aggregation and sorting-out. *Proc Natl Acad Sci USA* 87 (9):3589–93.

Jakovcevski, I., J. Wu, N. Karl, I. Leshchyns'ka, V. Sytnyk, J. Chen, A. Irintchev, and M. Schachner. 2007. Glial scar expression of CHL1, the close homolog of the adhesion molecule L1, limits recovery after spinal cord injury. *J Neurosci* 27 (27):7222–33.

Johnson, C. P., I. Fujimoto, U. Rutishauser, and D. E. Leckband. 2005. Direct evidence that neural cell adhesion molecule (NCAM) polysialylation increases intermembrane repulsion and abrogates adhesion. *J Biol Chem* 280 (1):137–45.

Kadmon, G., A. Kowitz, P. Altevogt, and M. Schachner. 1990. The neural cell adhesion molecule N-CAM enhances L1-dependent cell-cell interactions. *J Cell Biol* 110 (1):193–208.

Kadmon, G., A. M. Montgomery, and P. Altevogt. 1998. L1 makes immunological progress by expanding its relations. *Dev Immunol* 6 (3–4):205–13.

Kakunaga, S., W. Ikeda, S. Itoh, M. Deguchi-Tawarada, T. Ohtsuka, A. Mizoguchi, and Y. Takai. 2005. Nectin-like molecule-1/TSLL1/SynCAM3: a neural tissue-specific immunoglobulin-like cell-cell adhesion molecule localizing at non-junctional contact sites of presynaptic nerve terminals, axons and glia cell processes. *J Cell Sci* 118 (Pt 6):1267–77.

Kallapur, S. G., and R. A. Akeson. 1992. The neural cell adhesion molecule (NCAM) heparin binding domain binds to cell surface heparan sulfate proteoglycans. *J Neurosci Res* 33 (4):538–48.

Kasper, C., H. Rasmussen, J. S. Kastrup, S. Ikemizu, E. Y. Jones, V. Berezin, E. Bock, and I. K. Larsen. 2000. Structural basis of cell-cell adhesion by NCAM. *Nat Struct Biol* 7 (5):389–93.

Kelly, K. F., C. M. Spring, A. A. Otchere, and J. M. Daniel. 2004. NLS-dependent nuclear localization of p120ctn is necessary to relieve Kaiso-mediated transcriptional repression. *J Cell Sci* 117 (Pt 13):2675–86.

Kerjan, G., J. Dolan, C. Haumaitre, S. Schneider-Maunoury, H. Fujisawa, K. J. Mitchell, and A. Chedotal. 2005. The transmembrane semaphorin Sema6A controls cerebellar granule cell migration. *Nat Neurosci* 8 (11):1516–24.

Kim, D. Y., L. A. Ingano, and D. M. Kovacs. 2002. Nectin-1alpha, an immunoglobulin-like receptor involved in the formation of synapses, is a substrate for presenilin/gamma-secretase-like cleavage. *J Biol Chem* 277 (51):49976–81.

Kim, J., C. Lilliehook, A. Dudak, J. Prox, P. Saftig, H. J. Federoff, and S. T. Lim. 2010. Activity-dependent alpha-cleavage of nectin-1 is mediated by a disintegrin and metalloprotease 10 (ADAM10). *J Biol Chem* 285:22919–26.

Kiselyov, V. V., V. Berezin, T. E. Maar, V. Soroka, K. Edvardsen, A. Schousboe, and E. Bock. 1997. The first immunoglobulin-like neural cell adhesion molecule (NCAM) domain is involved in double-reciprocal interaction with the second immunoglobulin-like NCAM domain and in heparin binding. *J Biol Chem* 272 (15):10125–34.

Kiselyov, V. V., G. Skladchikova, A. M. Hinsby, P. H. Jensen, N. Kulahin, V. Soroka, N. Pedersen, V. Tsetlin, F. M. Poulsen, V. Berezin, and E. Bock. 2003. Structural basis for a direct interaction between FGFR1 and NCAM and evidence of a regulatory role of ATP. *Structure* 11 (6):691–701.

Kiselyov, V. V., V. Soroka, V. Berezin, and E. Bock. 2005. Structural biology of NCAM homophilic binding and activation of FGFR. *J Neurochem* 94 (5):1169–79.

Kiss, J. Z., C. Wang, S. Olive, G. Rougon, J. Lang, D. Baetens, D. Harry, and W. F. Pralong. 1994. Activity-dependent mobilization of the adhesion molecule polysialic NCAM to the cell surface of neurons and endocrine cells. *Embo J* 13 (22):5284–92.

Kiss, J. Z., C. Wang, and G. Rougon. 1993. Nerve-dependent expression of high polysialic acid neural cell adhesion molecule in neurohypophysial astrocytes of adult rats. *Neuroscience* 53 (1):213–21.

Kitagawa, H., and J. C. Paulson. 1994. Cloning of a novel alpha 2,3-sialyltransferase that sialylates glycoprotein and glycolipid carbohydrate groups. *J Biol Chem* 269 (2):1394–401.

Kobielak, A., and E. Fuchs. 2004. Alpha-catenin: at the junction of intercellular adhesion and actin dynamics. *Nat Rev Mol Cell Biol* 5 (8):614–25.

Kohmura, N., K. Senzaki, S. Hamada, N. Kai, R. Yasuda, M. Watanabe, H. Ishii, M. Yasuda, M. Mishina, and T. Yagi. 1998. Diversity revealed by a novel family of cadherins expressed in neurons at a synaptic complex. *Neuron* 20 (6):1137–51.

Kojima, N., Y. Tachida, Y. Yoshida, and S. Tsuji. 1996. Characterization of mouse ST8Sia II (STX) as a neural cell adhesion molecule-specific polysialic acid synthase. Requirement of core alpha1,6-linked fucose and a polypeptide chain for polysialylation. *J Biol Chem* 271 (32):19457–63.

Kolkova, K., V. Novitskaya, N. Pedersen, V. Berezin, and E. Bock. 2000. Neural cell adhesion molecule-stimulated neurite outgrowth depends on activation of protein kinase C and the Ras-mitogen-activated protein kinase pathway. *J Neurosci* 20 (6):2238–46.

Korshunova, I., V. Novitskaya, D. Kiryushko, N. Pedersen, K. Kolkova, E. Kropotova, M. Mosevitsky, M. Rayko, J. S. Morrow, I. Ginzburg, V. Berezin, and E. Bock. 2007. GAP-43 regulates NCAM-180-mediated neurite outgrowth. *J Neurochem* 100 (6):1599–612.

Kriegstein, A., and A. Alvarez-Buylla. 2009. The glial nature of embryonic and adult neural stem cells. *Annu Rev Neurosci* 32:149–84.

Krushel, L. A., B. A. Cunningham, G. M. Edelman, and K. L. Crossin. 1999. NF-kappaB activity is induced by neural cell adhesion molecule binding to neurons and astrocytes. *J Biol Chem* 274 (4):2432–39.

Krylyshkina, O., K. I. Anderson, I. Kaverina, I. Upmann, D. J. Manstein, J. V. Small, and D. K. Toomre. 2003. Nanometer targeting of microtubules to focal adhesions. *J Cell Biol* 161 (5):853–59.

Kuffler, S.W., Nichols, J.G., 1976. *From Neuron to Brain.* Sunderland, MA: Sinauer Associates, Inc.

LaFlamme, S. E., B. Nieves, D. Colello, and C. G. Reverte. 2008. Integrins as regulators of the mitotic machinery. *Curr Opin Cell Biol* 20 (5):576–82.

Lakatos, A., R. J. Franklin, and S. C. Barnett. 2000. Olfactory ensheathing cells and Schwann cells differ in their *in vitro* interactions with astrocytes. *Glia* 32 (3):214–25.

Langenbach, K. J., and T. A. Rando. 2002. Inhibition of dystroglycan binding to laminin disrupts the PI3K/AKT pathway and survival signaling in muscle cells. *Muscle Nerve* 26 (5):644–53.

Lapidos, K. A., R. Kakkar, and E. M. McNally. 2004. The dystrophin glycoprotein complex: signaling strength and integrity for the sarcolemma. *Circ Res* 94 (8):1023–31.

le Duc, Q., Q. Shi, I. Blonk, A. Sonnenberg, N. Wang, D. Leckband, and J. de Rooij. 2010. Vinculin potentiates E-cadherin mechanosensing and is recruited to actin-anchored sites within adherens junctions in a myosin II-dependent manner. *J Cell Biol* 189 (7):1107–15.

Le Gal La Salle, G., G. Rougon, and A. Valin. 1992. The embryonic form of neural cell surface molecule (E-NCAM) in the rat hippocampus and its reexpression on glial cells following kainic acid-induced status epilepticus. *J Neurosci* 12 (3):872–82.

Lee, H. S., J. Han, H. J. Bai, and K. W. Kim. 2009. Brain angiogenesis in developmental and pathological processes: regulation, molecular and cellular communication at the neurovascular interface. *Febs J* 276 (17):4622–35.

Legate, K. R., and R. Fassler. 2009. Mechanisms that regulate adaptor binding to beta-integrin cytoplasmic tails. *J Cell Sci* 122 (Pt 2):187–98.

Legate, K. R., E. Montanez, O. Kudlacek, and R. Fassler. 2006. ILK, PINCH and parvin: the tIPP of integrin signalling. *Nat Rev Mol Cell Biol* 7 (1):20–31.

Lehembre, F., M. Yilmaz, A. Wicki, T. Schomber, K. Strittmatter, D. Ziegler, A. Kren, P. Went, P. W. Derksen, A. Berns, J. Jonkers, and G. Christofori. 2008. NCAM-induced focal adhesion assembly: a functional switch upon loss of E-cadherin. *Embo J* 27 (19):2603–15.

Leshchyns'ka, I., V. Sytnyk, J. S. Morrow, and M. Schachner. 2003. Neural cell adhesion molecule (NCAM) association with PKCbeta2 via betaI spectrin is implicated in NCAM-mediated neurite outgrowth. *J Cell Biol* 161 (3):625–39.

Li, S., D. Harrison, S. Carbonetto, R. Fassler, N. Smyth, D. Edgar, and P. D. Yurchenco. 2002. Matrix assembly, regulation, and survival functions of laminin and its receptors in embryonic stem cell differentiation. *J Cell Biol* 157 (7):1279–90.

Li, Z., M. Hannigan, Z. Mo, B. Liu, W. Lu, Y. Wu, A. V. Smrcka, G. Wu, L. Li, M. Liu, C. K. Huang, and D. Wu. 2003. Directional sensing requires G beta gamma-mediated PAK1 and PIX alpha-dependent activation of Cdc42. *Cell* 114 (2):215–27.

Lien, W. H., V. I. Gelfand, and V. Vasioukhin. 2008. Alpha-E-catenin binds to dynamitin and regulates dynactin-mediated intracellular traffic. *J Cell Biol* 183 (6):989–97.

Ligon, L. A., and E. L. Holzbaur. 2007. Microtubules tethered at epithelial cell junctions by dynein facilitate efficient junction assembly. *Traffic* 8 (7):808–19.

Ligon, L. A., S. Karki, M. Tokito, and E. L. Holzbaur. 2001. Dynein binds to beta-catenin and may tether microtubules at adherens junctions. *Nat Cell Biol* 3 (10):913–17.

Linnemann, D., K. Edvardsen, and E. Bock. 1988. Developmental study of the cell adhesion molecule L1. *Dev Neurosci* 10 (1):34–42.

Little, E. B., K. L. Crossin, L. A. Krushel, G. M. Edelman, and B. A. Cunningham. 2001. A short segment within the cytoplasmic domain of the neural cell adhesion molecule (N-CAM) is essential for N-CAM-induced NF-kappa B activity in astrocytes. *Proc Natl Acad Sci USA* 98 (5):2238–43.

Liu, Z., J. L. Tan, D. M. Cohen, M. T. Yang, N. J. Sniadecki, S. A. Ruiz, C. M. Nelson, and C. S. Chen. 2010. Mechanical tugging force regulates the size of cell-cell junctions. *Proc Natl Acad Sci USA* 107 (22):9944–49.

Longxuan, L., J. V. Welser, P. Dore-Duffy, G. J. Del Zoppo, J. C. Lamanna, and R. Milner. 2010. In the hypoxic central nervous system, endothelial cell proliferation is followed by astrocyte activation, proliferation, and increased expression of the alpha6beta4 integrin and dystroglycan. *Glia* 58:1157–67.

Luo, B. H., C. V. Carman, and T. A. Springer. 2007. Structural basis of integrin regulation and signaling. *Annu Rev Immunol* 25:619–47.

Luo, B. H., and T. A. Springer. 2006. Integrin structures and conformational signaling. *Curr Opin Cell Biol* 18 (5):579–86.

Mackinnon, A. C., H. Qadota, K. R. Norman, D. G. Moerman, and B. D. Williams. 2002. C. elegans PAT-4/ILK functions as an adaptor protein within integrin adhesion complexes. *Curr Biol* 12 (10):787–97.

Maidment, S. L., G. J. Rucklidge, H. K. Rooprai, and G. J. Pilkington. 1997. An inverse correlation between expression of NCAM-A and the matrix-metalloproteinases gelatinase-A and gelatinase-B in human glioma cells *in vitro*. *Cancer Lett* 116 (1):71–77.

Mandai, K., H. Nakanishi, A. Satoh, H. Obaishi, M. Wada, H. Nishioka, M. Itoh, A. Mizoguchi, T. Aoki, T. Fujimoto, Y. Matsuda, S. Tsukita, and Y. Takai. 1997. Afadin: a novel actin filament-binding protein with one PDZ domain localized at cadherin-based cell-to-cell adherens junction. *J Cell Biol* 139 (2):517–28.

Maness, P. F., and M. Schachner. 2007. Neural recognition molecules of the immunoglobulin superfamily: signaling transducers of axon guidance and neuronal migration. *Nat Neurosci* 10 (1):19–26.

Manneville, J. B., M. Jehanno, S. Etienne-Manneville. 2010. Dlg1 binds GKAP to control dynein association with microtubules, centrosome positioning, and cell polarity. *J Cell Biol* 191(3):585–98.

Margolis, R. U., and R. K. Margolis. 1994. Aggrecan-versican-neurocan family proteoglycans. *Methods Enzymol* 245:105–26.

Marino, P., J. C. Norreel, M. Schachner, G. Rougon, and M. C. Amoureux. 2009. A polysialic acid mimetic peptide promotes functional recovery in a mouse model of spinal cord injury. *Exp Neurol* 219 (1):163–74.

Mary, S., S. Charrasse, M. Meriane, F. Comunale, P. Travo, A. Blangy, and C. Gauthier-Rouviere. 2002. Biogenesis of N-cadherin-dependent cell-cell contacts in living fibroblasts is a microtubule-dependent kinesin-driven mechanism. *Mol Biol Cell* 13 (1):285–301.

Matsunaga, M., K. Hatta, A. Nagafuchi, and M. Takeichi. 1988. Guidance of optic nerve fibres by N-cadherin adhesion molecules. *Nature* 334 (6177):62–64.

Maul, R. S., Y. Song, K. J. Amann, S. C. Gerbin, T. D. Pollard, and D. D. Chang. 2003. EPLIN regulates actin dynamics by cross-linking and stabilizing filaments. *J Cell Biol* 160 (3):399–407.

McKay, B. S., P. E. Irving, C. M. Skumatz, and J. M. Burke. 1997. Cell-cell adhesion molecules and the development of an epithelial phenotype in cultured human retinal pigment epithelial cells. *Exp Eye Res* 65 (5):661–71.

Milev, P., P. Maurel, M. Haring, R. K. Margolis, and R. U. Margolis. 1996. TAG-1/axonin-1 is a high-affinity ligand of neurocan, phosphacan/protein-tyrosine phosphatase-zeta/beta, and N-CAM. *J Biol Chem* 271 (26):15716–23.

Milner, R., X. Huang, J. Wu, S. Nishimura, R. Pytela, D. Sheppard, and C. ffrench-Constant. 1999. Distinct roles for astrocyte alphavbeta5 and alphavbeta8 integrins in adhesion and migration. *J Cell Sci* 112 (Pt 23):4271–79.

Milner, R., J. B. Relvas, J. Fawcett, and C. ffrench-Constant. 2001. Developmental regulation of alphav integrins produces functional changes in astrocyte behavior. *Mol Cell Neurosci* 18 (1):108–18.

Minana, R., J. M. Duran, M. Tomas, J. Renau-Piqueras, and C. Guerri. 2001. Neural cell adhesion molecule is endocytosed via a clathrin-dependent pathway. *Eur J Neurosci* 13 (4):749–56.

Mirouse, V., C. P. Christoforou, C. Fritsch, D. St. Johnston, and R. P. Ray. 2009. Dystroglycan and perlecan provide a basal cue required for epithelial polarity during energetic stress. *Dev Cell* 16 (1):83–92.

Mitra, S. K., D. A. Hanson, and D. D. Schlaepfer. 2005. Focal adhesion kinase: in command and control of cell motility. *Nat Rev Mol Cell Biol* 6 (1):56–68.

Miura, M., H. Asou, M. Kobayashi, and K. Uyemura. 1992. Functional expression of a full-length cDNA coding for rat neural cell adhesion molecule L1 mediates homophilic intercellular adhesion and migration of cerebellar neurons. *J Biol Chem* 267 (15):10752–58.

Miyahara, M., H. Nakanishi, K. Takahashi, K. Satoh-Horikawa, K. Tachibana, and Y. Takai. 2000. Interaction of nectin with afadin is necessary for its clustering at cell-cell contact sites but not for its *cis* dimerization or *trans* interaction. *J Biol Chem* 275 (1):613–18.

Miyamoto, S., H. Teramoto, J. S. Gutkind, and K. M. Yamada. 1996. Integrins can collaborate with growth factors for phosphorylation of receptor tyrosine kinases and MAP kinase activation: roles of integrin aggregation and occupancy of receptors. *J Cell Biol* 135 (6 Pt 1):1633–42.

Miyatani, S., K. Shimamura, M. Hatta, A. Nagafuchi, A. Nose, M. Matsunaga, K. Hatta, and M. Takeichi. 1989. Neural cadherin: role in selective cell-cell adhesion. *Science* 245 (4918):631–35.

Montanaro, F., S. H. Gee, C. Jacobson, M. H. Lindenbaum, S. C. Froehner, and S. Carbonetto. 1998. Laminin and alpha-dystroglycan mediate acetylcholine receptor aggregation via a MuSK-independent pathway. *J Neurosci* 18 (4):1250–60.

Montanaro, F., M. Lindenbaum, and S. Carbonetto. 1999. alpha-Dystroglycan is a laminin receptor involved in extracellular matrix assembly on myotubes and muscle cell viability. *J Cell Biol* 145 (6):1325–40.

Montgomery, A. M., J. C. Becker, C. H. Siu, V. P. Lemmon, D. A. Cheresh, J. D. Pancook, X. Zhao, and R. A. Reisfeld. 1996. Human neural cell adhesion molecule L1 and rat homologue NILE are ligands for integrin alpha v beta 3. *J Cell Biol* 132 (3):475–85.

Moore, R., and F. S. Walsh. 1993. The cell adhesion molecule M-cadherin is specifically expressed in developing and regenerating, but not denervated skeletal muscle. *Development* 117 (4):1409–20.

Moore, S. A., F. Saito, J. Chen, D. E. Michele, M. D. Henry, A. Messing, R. D. Cohn, S. E. Ross-Barta, S. Westra, R. A. Williamson, T. Hoshi, and K. P. Campbell. 2002. Deletion of brain dystroglycan recapitulates aspects of congenital muscular dystrophy. *Nature* 418 (6896):422–25.

Morel, M. P., I. Dusart, and C. Sotelo. 2002. Sprouting of adult Purkinje cell axons in lesioned mouse cerebellum: "non-permissive" versus "permissive" environment. *J Neurocytol* 31 (8–9):633–47.

Morrison, M. E., and V. R. Racaniello. 1992. Molecular cloning and expression of a murine homolog of the human poliovirus receptor gene. *J Virol* 66 (5):2807–13.

Moser, M., K. R. Legate, R. Zent, and R. Fassler. 2009. The tail of integrins, talin, and kindlins. *Science* 324 (5929):895–99.

Murray, B. A., J. J. Hemperly, E. A. Prediger, G. M. Edelman, and B. A. Cunningham. 1986. Alternatively spliced mRNAs code for different polypeptide chains of the chicken neural cell adhesion molecule (N-CAM). *J Cell Biol* 102 (1):189–93.

Musil, L. S., and D. A. Goodenough. 1990. Gap junctional intercellular communication and the regulation of connexin expression and function. *Curr Opin Cell Biol* 2 (5):875–80.

Nakayama, J., M. N. Fukuda, B. Fredette, B. Ranscht, and M. Fukuda. 1995. Expression cloning of a human polysialyltransferase that forms the polysialylated neural cell adhesion molecule present in embryonic brain. *Proc Natl Acad Sci USA* 92 (15):7031–35.

Neugebauer, K. M., K. J. Tomaselli, J. Lilien, and L. F. Reichardt. 1988. N-cadherin, NCAM, and integrins promote retinal neurite outgrowth on astrocytes *in vitro. J Cell Biol* 107 (3):1177–87.

Niederkofler, V., T. Baeriswyl, R. Ott, and E. T. Stoeckli. 2010. Nectin-like molecules/SynCAMs are required for post-crossing commissural axon guidance. *Development* 137 (3):427–35.

Nielsen, J., K. Gotfryd, S. Li, N. Kulahin, V. Soroka, K. K. Rasmussen, E. Bock, and V. Berezin. 2009. Role of glial cell line-derived neurotrophic factor (GDNF)-neural cell adhesion molecule (NCAM) interactions in induction of neurite outgrowth and identification of a binding site for NCAM in the heel region of GDNF. *J Neurosci* 29 (36):11360–76.

Niewmierzycka, A., J. Mills, R. St.-Arnaud, S. Dedhar, and L. F. Reichardt. 2005. Integrin-linked kinase deletion from mouse cortex results in cortical lamination defects resembling cobblestone lissencephaly. *J Neurosci* 25 (30):7022–31.

Nishimune, H., J. R. Sanes, and S. S. Carlson. 2004. A synaptic laminin-calcium channel interaction organizes active zones in motor nerve terminals. *Nature* 432 (7017):580–87.

Noble, M., M. Albrechtsen, C. Moller, J. Lyles, E. Bock, C. Goridis, M. Watanabe, and U. Rutishauser. 1985. Glial cells express N-CAM/D2-CAM-like polypeptides *in vitro. Nature* 316 (6030):725–28.

Noel, G., M. Belda, E. Guadagno, J. Micoud, N. Klocker, and H. Moukhles. 2005. Dystroglycan and Kir4.1 coclustering in retinal Muller glia is regulated by laminin-1 and requires the PDZ-ligand domain of Kir4.1. *J Neurochem* 94 (3):691–702.

Nomura, T., M. Takahashi, Y. Hara, and N. Osumi. 2008. Patterns of neurogenesis and amplitude of Reelin expression are essential for making a mammalian-type cortex. *PLoS One* 3 (1):e1454.

Nomura, T., T. Yabe, E. S. Rosenthal, M. Krzan, and J. P. Schwartz. 2000. PSA-NCAM distinguishes reactive astrocytes in 6-OHDA-lesioned substantia nigra from those in striatal terminal fields. *J Neurosci Res* 61 (6):588–96.

Noren, N. K., B. P. Liu, K. Burridge, and B. Kreft. 2000. p120 catenin regulates the actin cytoskeleton via Rho family GTPases. *J Cell Biol* 150 (3):567–80.

Nose, A., K. Tsuji, and M. Takeichi. 1990. Localization of specificity determining sites in cadherin cell adhesion molecules. *Cell* 61 (1):147–55.

Nothias, F., P. Vernier, Y. von Boxberg, S. Mirman, and J. D. Vincent. 1997. Modulation of NCAM polysialylation is associated with morphofunctional modifications in the hypothalamo-neurohypophysial system during lactation. *Eur J Neurosci* 9 (8):1553–65.

Nybroe, O., M. Albrechtsen, J. Dahlin, D. Linnemann, J. M. Lyles, C. J. Moller, and E. Bock. 1985. Biosynthesis of the neural cell adhesion molecule: characterization of polypeptide C. *J Cell Biol* 101 (6):2310–15.

Nybroe, O., D. Linnemann, and E. Bock. 1989. Heterogeneity of soluble neural cell adhesion molecule. *J Neurochem* 53 (5):1372–78.

Obama, H., and M. Ozawa. 1997. Identification of the domain of alpha-catenin involved in its association with beta-catenin and plakoglobin (gamma-catenin). *J Biol Chem* 272 (17):11017–20.

Ohkubo, T., and M. Ozawa. 1999. p120(ctn) binds to the membrane-proximal region of the E-cadherin cytoplasmic domain and is involved in modulation of adhesion activity. *J Biol Chem* 274 (30):21409–15.

Okamoto, R., K. Irie, A. Yamada, T. Katata, A. Fukuhara, and Y. Takai. 2005. Recruitment of E-cadherin associated with alpha- and beta-catenins and p120ctn to the nectin-based cell-cell adhesion sites by the action of 12-O-tetradecanoylphorbol-13-acetate in MDCK cells. *Genes Cells* 10 (5):435–45.

Olsen, M., L. Krog, K. Edvardsen, L. T. Skovgaard, and E. Bock. 1993. Intact transmembrane isoforms of the neural cell adhesion molecule are released from the plasma membrane. *Biochem J* 295 (Pt 3):833–40.

Ong, E., J. Nakayama, K. Angata, L. Reyes, T. Katsuyama, Y. Arai, and M. Fukuda. 1998. Developmental regulation of polysialic acid synthesis in mouse directed by two polysia-lyltransferases, PST and STX. *Glycobiology* 8 (4):415–24.

Ono, K., H. Tomasiewicz, T. Magnuson, and U. Rutishauser. 1994. N-CAM mutation inhibits tangential neuronal migration and is phenocopied by enzymatic removal of polysialic acid. *Neuron* 13 (3):595–609.

Osmani, N., F. Peglion, P. Chavrier, and S. Etienne-Manneville. 2010. CdC42 localization and cell polarity depend on membrane traffic. *J Cell Biol* 191(7):1261–69.

Osmani, N., N. Vitale, J. P. Borg, and S. Etienne-Manneville. 2006. Scrib controls Cdc42 localization and activity to promote cell polarization during astrocyte migration. *Curr Biol* 16 (24):2395–405.

Oumesmar, B. N., L. Vignais, E. Duhamel-Clerin, V. Avellana-Adalid, G. Rougon, and A. Baron-Van Evercooren. 1995. Expression of the highly polysialylated neural cell adhe-sion molecule during postnatal myelination and following chemically induced demyeli-nation of the adult mouse spinal cord. *Eur J Neurosci* 7 (3):480–91.

Owens, G. C., G. M. Edelman, and B. A. Cunningham. 1987. Organization of the neural cell adhesion molecule (N-CAM) gene: alternative exon usage as the basis for different membrane-associated domains. *Proc Natl Acad Sci USA* 84 (1):294–98.

Owens, G. C., E. A. Orr, B. K. DeMasters, R. J. Muschel, M. E. Berens, and C. A. Kruse. 1998. Overexpression of a transmembrane isoform of neural cell adhesion molecule alters the invasiveness of rat CNS-1 glioma. *Cancer Res* 58 (9):2020–28.

Ozawa, M., H. Baribault, and R. Kemler. 1989. The cytoplasmic domain of the cell adhesion molecule uvomorulin associates with three independent proteins structurally related in different species. *Embo J* 8 (6):1711–17.

Ozawa, M., J. Engel, and R. Kemler. 1990a. Single amino acid substitutions in one Ca2+ bind-ing site of uvomorulin abolish the adhesive function. *Cell* 63 (5):1033–38.

Ozawa, M., M. Ringwald, and R. Kemler. 1990b. Uvomorulin-catenin complex formation is regulated by a specific domain in the cytoplasmic region of the cell adhesion molecule. *Proc Natl Acad Sci USA* 87 (11):4246–50.

Palazzo, A. F., T. A. Cook, A. S. Alberts, and G. G. Gundersen. 2001. mDia mediates Rho-regulated formation and orientation of stable microtubules. *Nat Cell Biol* 3:723–729.

Palazzo, A. F., C. H. Eng, D. D. Schlaepfer, E. E. Marcantonio, and G. G. Gundersen. 2004. Localized stabilization of microtubules by integrin- and FAK-facilitated Rho signaling. *Science* 303 (5659):836–39.

Park, J., B. Liu, T. Chen, H. Li, X. Hu, J. Gao, Y. Zhu, Q. Zhu, B. Qiang, J. Yuan, X. Peng, and M. Qiu. 2008. Disruption of Nectin-like 1 cell adhesion molecule leads to delayed axonal myelination in the CNS. *J Neurosci* 28 (48):12815–19.

Parsons, J. T. 2003. Focal adhesion kinase: the first ten years. *J Cell Sci* 116 (Pt 8):1409–16.

Parsons, J. T., K. H. Martin, J. K. Slack, J. M. Taylor, and S. A. Weed. 2000. Focal adhe-sion kinase: a regulator of focal adhesion dynamics and cell movement. *Oncogene* 19 (49):5606–13.

Peng, H. B., H. Xie, S. G. Rossi, and R. L. Rotundo. 1999. Acetylcholinesterase clustering at the neuromuscular junction involves perlecan and dystroglycan. *J Cell Biol* 145 (4):911–21.

Peng, H., W. Shah, P. Holland, and S. Carbonetto. 2008. Integrins and dystroglycan regulate astrocyte wound healing: the integrin beta1 subunit is necessary for process extension and orienting the microtubular network. *Dev Neurobiol* 68 (5):559–74.

Perego, C., C. Vanoni, S. Massari, A. Raimondi, S. Pola, M. G. Cattaneo, M. Francolini, L. M. Vicentini, and G. Pietrini. 2002. Invasive behaviour of glioblastoma cell lines is associated with altered organisation of the cadherin-catenin adhesion system. *J Cell Sci* 115 (Pt 16):3331–40.

Perl, A. K., U. Dahl, P. Wilgenbus, H. Cremer, H. Semb, and G. Christofori. 1999. Reduced expression of neural cell adhesion molecule induces metastatic dissemination of pancreatic beta tumor cells. *Nat Med* 5 (3):286–91.

Perlin, J. R., and W. S. Talbot. 2007. Putting the glue in glia: necls mediate Schwann cell axon adhesion. *J Cell Biol* 178 (5):721–23.

Piedra, J., D. Martinez, J. Castano, S. Miravet, M. Dunach, and A. G. de Herreros. 2001. Regulation of beta-catenin structure and activity by tyrosine phosphorylation. *J Biol Chem* 276 (23):20436–43.

Piepenhagen, P. A., and W. J. Nelson. 1993. Defining E-cadherin-associated protein complexes in epithelial cells: plakoglobin, beta- and gamma-catenin are distinct components. *J Cell Sci* 104 (Pt 3):751–62.

Pollerberg, E. G., R. Sadoul, C. Goridis, and M. Schachner. 1985. Selective expression of the 180-kD component of the neural cell adhesion molecule N-CAM during development. *J Cell Biol* 101 (5 Pt 1):1921–29.

Pollerberg, G. E., K. Burridge, K. E. Krebs, S. R. Goodman, and M. Schachner. 1987. The 180-kD component of the neural cell adhesion molecule N-CAM is involved in cell-cell contacts and cytoskeleton-membrane interactions. *Cell Tissue Res* 250 (1):227–36.

Postel, R., P. Vakeel, J. Topczewski, R. Knoll, and J. Bakkers. 2008. Zebrafish integrin-linked kinase is required in skeletal muscles for strengthening the integrin-ECM adhesion complex. *Dev Biol* 318 (1):92–101.

Prag, S., E. A. Lepekhin, K. Kolkova, R. Hartmann-Petersen, A. Kawa, P. S. Walmod, V. Belman, H. C. Gallagher, V. Berezin, E. Bock, and N. Pedersen. 2002. NCAM regulates cell motility. *J Cell Sci* 115 (Pt 2):283–92.

Probstmeier, R., K. Kuhn, and M. Schachner. 1989. Binding properties of the neural cell adhesion molecule to different components of the extracellular matrix. *J Neurochem* 53 (6):1794–801.

Rao, Y., X. F. Wu, J. Gariepy, U. Rutishauser, and C. H. Siu. 1992. Identification of a peptide sequence involved in homophilic binding in the neural cell adhesion molecule NCAM. *J Cell Biol* 118 (4):937–49.

Rao, Y., X. F. Wu, P. Yip, J. Gariepy, and C. H. Siu. 1993. Structural characterization of a homophilic binding site in the neural cell adhesion molecule. *J Biol Chem* 268 (27):20630–38.

Redies, C., K. Engelhart, and M. Takeichi. 1993. Differential expression of N- and R-cadherin in functional neuronal systems and other structures of the developing chicken brain. *J Comp Neurol* 333 (3):398–416.

Reverte, C. G., A. Benware, C. W. Jones, and S. E. LaFlamme. 2006. Perturbing integrin function inhibits microtubule growth from centrosomes, spindle assembly, and cytokinesis. *J Cell Biol* 174 (4):491–97.

Reyes, A. A., S. V. Schulte, S. Small, and R. Akeson. 1993. Distinct NCAM splicing events are differentially regulated during rat brain development. *Brain Res Mol Brain Res* 17 (3–4):201–11.

Reymond, N., S. Fabre, E. Lecocq, J. Adelaide, P. Dubreuil, and M. Lopez. 2001. Nectin4/
    PRR4, a new afadin-associated member of the nectin family that trans-interacts with
    nectin1/PRR1 through V domain interaction. *J Biol Chem* 276 (46):43205–15.
Reynolds, A. B., and R. H. Carnahan. 2004. Regulation of cadherin stability and turnover by
    p120ctn: implications in disease and cancer. *Semin Cell Dev Biol* 15 (6):657–63.
Reynolds, A. B., D. J. Roesel, S. B. Kanner, and J. T. Parsons. 1989. Transformation-specific
    tyrosine phosphorylation of a novel cellular protein in chicken cells expressing onco-
    genic variants of the avian cellular src gene. *Mol Cell Biol* 9 (2):629–38.
Reynolds, A. R., I. R. Hart, A. R. Watson, J. C. Welti, R. G. Silva, S. D. Robinson, G. Da
    Violante, M. Gourlaouen, M. Salih, M. C. Jones, D. T. Jones, G. Saunders, V. Kostourou,
    F. Perron-Sierra, J. C. Norman, G. C. Tucker, and K. M. Hodivala-Dilke. 2009.
    Stimulation of tumor growth and angiogenesis by low concentrations of RGD-mimetic
    integrin inhibitors. *Nat Med* 15 (4):392–400.
Ridley, A. J., M. A. Schwartz, K. Burridge, R. A. Firtel, M. H. Ginsberg, G. Borisy, J. T. Parsons,
    and A. R. Horwitz. 2003. Cell migration: integrating signals from front to back. *Science*
    302 (5651):1704–9.
Rieger, F. et al. 1986. Neuronal cell adhesion molecules and cytotactin are colocalized at the
    node of Ranvier. *J Cell Biol* 103:379-91.
Rieger, S., N. Senghaas, A. Walch, and R. W. Koster. 2009. Cadherin-2 controls directional
    chain migration of cerebellar granule neurons. *PLoS Biol* 7 (11):e1000240.
Rimm, D. L., E. R. Koslov, P. Kebriaei, C. D. Cianci, and J. S. Morrow. 1995. Alpha 1(E)-
    catenin is an actin-binding and-bundling protein mediating the attachment of F-actin to
    the membrane adhesion complex. *Proc Natl Acad Sci USA* 92 (19):8813–17.
Ringwald, M., R. Schuh, D. Vestweber, H. Eistetter, F. Lottspeich, J. Engel, R. Dolz, F. Jahnig,
    J. Epplen, S. Mayer, et al. 1987. The structure of cell adhesion molecule uvomorulin.
    Insights into the molecular mechanism of Ca2+-dependent cell adhesion. *Embo J* 6
    (12):3647–53.
Rosato, R., J. M. Veltmaat, J. Groffen, and N. Heisterkamp. 1998. Involvement of the tyrosine
    kinase fer in cell adhesion. *Mol Cell Biol* 18 (10):5762–70.
Sadoul, K., A. Meyer, M. G. Low, and M. Schachner. 1986. Release of the 120 kDa component
    of the mouse neural cell adhesion molecule N-CAM from cell surfaces by phosphati-
    dylinositol-specific phospholipase C. *Neurosci Lett* 72 (3):341–46.
Sago, H., M. Kitagawa, S. Obata, N. Mori, S. Taketani, J. M. Rochelle, M. F. Seldin,
    M. Davidson, T. St. John, and S. T. Suzuki. 1995. Cloning, expression, and chromo-
    somal localization of a novel cadherin-related protein, protocadherin-3. *Genomics* 29
    (3):631–40.
Sanchez-Heras, E., F. V. Howell, G. Williams, and P. Doherty. 2006. The fibroblast growth
    factor receptor acid box is essential for interactions with N-cadherin and all of the major
    isoforms of neural cell adhesion molecule. *J Biol Chem* 281 (46):35208–16.
Sano, K., H. Tanihara, R. L. Heimark, S. Obata, M. Davidson, T. St. John, S. Taketani, and
    S. Suzuki. 1993. Protocadherins: a large family of cadherin-related molecules in central
    nervous system. *Embo J* 12 (6):2249–56.
Santoni, M. J., D. Barthels, J. A. Barbas, M. R. Hirsch, M. Steinmetz, C. Goridis, and W. Wille.
    1987. Analysis of cDNA clones that code for the transmembrane forms of the mouse
    neural cell adhesion molecule (NCAM) and are generated by alternative RNA splicing.
    Nucleic Acids Res 15 (21):8621–41.
Santoni, M. J., D. Barthels, G. Vopper, A. Boned, C. Goridis, and W. Wille. 1989. Differential
    exon usage involving an unusual splicing mechanism generates at least eight types of
    NCAM cDNA in mouse brain. *Embo J* 8 (2):385–92.
Sasaki, T., and T. Endo. 1999. Evidence for the presence of N-CAM 180 on astrocytes from
    rat cerebellum and differences in glycan structures between N-CAM 120 and N-CAM
    140. *Glia* 28 (3):236–43.

Sato, S., Y. Omori, K. Katoh, M. Kondo, M. Kanagawa, K. Miyata, K. Funabiki, T. Koyasu, N. Kajimura, T. Miyoshi, H. Sawai, K. Kobayashi, A. Tani, T. Toda, J. Usukura, Y. Tano, T. Fujikado, and T. Furukawa. 2008. Pikachurin, a dystroglycan ligand, is essential for photoreceptor ribbon synapse formation. *Nat Neurosci* 11 (8):923–31.

Sato, T., N. Fujita, A. Yamada, T. Ooshio, R. Okamoto, K. Irie, and Y. Takai. 2006. Regulation of the assembly and adhesion activity of E-cadherin by nectin and afadin for the formation of adherens junctions in Madin-Darby canine kidney cells. *J Biol Chem* 281 (8):5288–99.

Satoh-Horikawa, K., H. Nakanishi, K. Takahashi, M. Miyahara, M. Nishimura, K. Tachibana, A. Mizoguchi, and Y. Takai. 2000. Nectin-3, a new member of immunoglobulin-like cell adhesion molecules that shows homophilic and heterophilic cell-cell adhesion activities. *J Biol Chem* 275 (14):10291–99.

Schmoranzer, J., J. P. Fawcett, M. Segura, S. Tan, R. B. Vallee, T. Pawson, and G. G. Gundersen. 2009. Par3 and dynein associate to regulate local microtubule dynamics and centrosome orientation during migration. *Curr Biol* 19 (13):1065–74.

Schnadelbach, O., O. W. Blaschuk, M. Symonds, B. J. Gour, P. Doherty, and J. W. Fawcett. 2000. N-cadherin influences migration of oligodendrocytes on astrocyte monolayers. *Mol Cell Neurosci* 15 (3):288–302.

Schwander, M., R. Shirasaki, S. L. Pfaff, and U. Muller. 2004. Beta1 integrins in muscle, but not in motor neurons, are required for skeletal muscle innervation. *J Neurosci* 24 (37):8181–91.

Seifert, G., K. Schilling, and C. Steinhauser. 2006. Astrocyte dysfunction in neurological disorders: a molecular perspective. *Nat Rev Neurosci* 7 (3):194–206.

Serrels, B., A. Serrels, V. G. Brunton, M. Holt, G. W. McLean, C. H. Gray, G. E. Jones, and M. C. Frame. 2007. Focal adhesion kinase controls actin assembly via a FERM-mediated interaction with the Arp2/3 complex. *Nat Cell Biol* 9 (9):1046–56.

Shan, W. S., H. Tanaka, G. R. Phillips, K. Arndt, M. Yoshida, D. R. Colman, and L. Shapiro. 2000. Functional cis-heterodimers of N- and R-cadherins. *J Cell Biol* 148 (3):579–90.

Shan, W., Y. Yagita, Z. Wang, A. Koch, A. Fex Svenningsen, E. Gruzglin, L. Pedraza, and D. R. Colman. 2004. The minimal essential unit for cadherin-mediated intercellular adhesion comprises extracellular domains 1 and 2. *J Biol Chem* 279 (53):55914–23.

Shattil, S. J., C. Kim, and M. H. Ginsberg. 2010. The final steps of integrin activation: the end game. *Nat Rev Mol Cell Biol* 11 (4):288–300.

Shingai, T., W. Ikeda, S. Kakunaga, K. Morimoto, K. Takekuni, S. Itoh, K. Satoh, M. Takeuchi, T. Imai, M. Monden, and Y. Takai. 2003. Implications of nectin-like molecule-2/IGSF4/RA175/SgIGSF/TSLC1/SynCAM1 in cell-cell adhesion and transmembrane protein localization in epithelial cells. *J Biol Chem* 278 (37):35421–27.

Shinoura, N., N. E. Paradies, R. E. Warnick, H. Chen, J. J. Larson, J. J. Tew, M. Simon, R. A. Lynch, Y. Kanai, S. Hirohashi, et al. 1995. Expression of N-cadherin and alpha-catenin in astrocytomas and glioblastomas. *Br J Cancer* 72 (3):627–33.

Shirayoshi, Y., K. Hatta, M. Hosoda, S. Tsunasawa, F. Sakiyama, and M. Takeichi. 1986. Cadherin cell adhesion molecules with distinct binding specificities share a common structure. *Embo J* 5 (10):2485–88.

Shore, E. M., and W. J. Nelson. 1991. Biosynthesis of the cell adhesion molecule uvomorulin (E-cadherin) in Madin-Darby canine kidney epithelial cells. *J Biol Chem* 266 (29):19672–80.

Shtutman, M., A. Chausovsky, M. Prager-Khoutorsky, N. Schiefermeier, S. Boguslavsky, Z. Kam, E. Fuchs, B. Geiger, G. G. Borisy, and A. D. Bershadsky. 2008. Signaling function of alpha-catenin in microtubule regulation. *Cell Cycle* 7 (15):2377–83.

Shukla, D., P. M. Scanlan, V. Tiwari, V. Sheth, C. Clement, G. Guzman-Hartman, T. S. Dermody, and T. Valyi-Nagy. 2006. Expression of nectin-1 in normal and herpes simplex virus type 1-infected murine brain. *Appl Immunohistochem Mol Morphol* 14 (3):341–47.

Silver, J., and U. Rutishauser. 1984. Guidance of optic axons *in vivo* by a preformed adhesive pathway on neuroepithelial endfeet. *Dev Biol* 106 (2):485–99.

Skaper, S. D., S. E. Moore, and F. S. Walsh. 2001. Cell signalling cascades regulating neuronal growth-promoting and inhibitory cues. *Prog Neurobiol* 65 (6):593–608.

Smalheiser, N. R., and N. B. Schwartz. 1987. Cranin: a laminin-binding protein of cell membranes. *Proc Natl Acad Sci USA* 84 (18):6457–61.

Small, J. V., and I. Kaverina. 2003. Microtubules meet substrate adhesions to arrange cell polarity. *Curr Opin Cell Biol* 15 (1):40–47.

Small, S. J., and R. Akeson. 1990. Expression of the unique NCAM VASE exon is independently regulated in distinct tissues during development. *J Cell Biol* 111 (5 Pt 1):2089–96.

Sofroniew, M. V. 2009. Molecular dissection of reactive astrogliosis and glial scar formation. *Trends Neurosci* 32 (12):638–47.

Soroka, V., K. Kolkova, J. S. Kastrup, K. Diederichs, J. Breed, V. V. Kiselyov, F. M. Poulsen, I. K. Larsen, W. Welte, V. Berezin, E. Bock, and C. Kasper. 2003. Structure and interactions of NCAM Ig1–2-3 suggest a novel zipper mechanism for homophilic adhesion. *Structure* 11 (10):1291–301.

Spence, H. J., Y. J. Chen, C. L. Batchelor, J. R. Higginson, H. Suila, O. Carpen, and S. J. Winder. 2004. Ezrin-dependent regulation of the actin cytoskeleton by beta-dystroglycan. *Hum Mol Genet* 13 (15):1657–68.

Steindler, D. A., D. Settles, H. P. Erickson, E. D. Laywell, A. Yoshiki, A. Faissner, and M. Kusakabe. 1995. Tenascin knockout mice: barrels, boundary molecules, and glial scars. *J Neurosci* 15 (3 Pt 1):1971–83.

Straub, V., J. A. Rafael, J. S. Chamberlain, and K. P. Campbell. 1997. Animal models for muscular dystrophy show different patterns of sarcolemmal disruption. *J Cell Biol* 139 (2):375–85.

Su, L. K., B. Vogelstein, and K. W. Kinzler. 1993. Association of the APC tumor suppressor protein with catenins. *Science* 262 (5140):1734–37.

Sugita, S., F. Saito, J. Tang, J. Satz, K. Campbell, and T. C. Sudhof. 2001. A stoichiometric complex of neurexins and dystroglycan in brain. *J Cell Biol* 154 (2):435–45.

Suyama, K., I. Shapiro, M. Guttman, and R. B. Hazan. 2002. A signaling pathway leading to metastasis is controlled by N-cadherin and the FGF receptor. *Cancer Cell* 2 (4):301–14.

Suzuki, M., M. Suzuki, J. Nakayama, A. Suzuki, K. Angata, S. Chen, K. Sakai, K. Hagihara, Y. Yamaguchi, and M. Fukuda. 2005. Polysialic acid facilitates tumor invasion by glioma cells. *Glycobiology* 15 (9):887–94.

Takahashi, K., H. Nakanishi, M. Miyahara, K. Mandai, K. Satoh, A. Satoh, H. Nishioka, J. Aoki, A. Nomoto, A. Mizoguchi, and Y. Takai. 1999. Nectin/PRR: an immunoglobulin-like cell adhesion molecule recruited to cadherin-based adherens junctions through interaction with Afadin, a PDZ domain-containing protein. *J Cell Biol* 145 (3):539–49.

Takai, Y., W. Ikeda, H. Ogita, and Y. Rikitake. 2008. The immunoglobulin-like cell adhesion molecule nectin and its associated protein afadin. *Annu Rev Cell Dev Biol* 24:309–42.

Takeda, Y., H. Asou, Y. Murakami, M. Miura, M. Kobayashi, and K. Uyemura. 1996. A nonneuronal isoform of cell adhesion molecule L1: tissue-specific expression and functional analysis. *J Neurochem* 66 (6):2338–49.

Takekuni, K., W. Ikeda, T. Fujito, K. Morimoto, M. Takeuchi, M. Monden, and Y. Takai. 2003. Direct binding of cell polarity protein PAR-3 to cell-cell adhesion molecule nectin at neuroepithelial cells of developing mouse. *J Biol Chem* 278 (8):5497–500.

Tamura, K., W. S. Shan, W. A. Hendrickson, D. R. Colman, and L. Shapiro. 1998. Structure-function analysis of cell adhesion by neural (N-) cadherin. *Neuron* 20 (6):1153–63.

Tawil, N. J., M. Houde, R. Blacher, F. Esch, L. F. Reichardt, D. C. Turner, and S. Carbonetto. 1990. Alpha 1 beta 1 integrin heterodimer functions as a dual laminin/collagen receptor in neural cells. *Biochemistry* 29 (27):6540–44.

Tawil, N. J., P. Wilson, and S. Carbonetto. 1994. Expression and distribution of functional integrins in rat CNS glia. *J Neurosci Res* 39 (4):436–47.

Theodosis, D. T., R. Bonhomme, S. Vitiello, G. Rougon, and D. A. Poulain. 1999. Cell surface expression of polysialic acid on NCAM is a prerequisite for activity-dependent morphological neuronal and glial plasticity. *J Neurosci* 19 (23):10228–36.

Theodosis, D. T., and D. A. Poulain. 1999. Contribution of astrocytes to activity-dependent structural plasticity in the adult brain. *Adv Exp Med Biol* 468:175–82.

Theodosis, D. T., D. A. Poulain, and S. H. Oliet. 2008. Activity-dependent structural and functional plasticity of astrocyte-neuron interactions. *Physiol Rev* 88 (3):983–1008.

Theodosis, D. T., G. Rougon, and D. A. Poulain. 1991. Retention of embryonic features by an adult neuronal system capable of plasticity: polysialylated neural cell adhesion molecule in the hypothalamo-neurohypophysial system. *Proc Natl Acad Sci USA* 88 (13):5494–98.

Thomas, G. D., P. W. Shaul, I. S. Yuhanna, S. C. Froehner, and M. E. Adams. 2003. Vasomodulation by skeletal muscle-derived nitric oxide requires alpha-syntrophin-mediated sarcolemmal localization of neuronal nitric oxide synthase. *Circ Res* 92 (5):554–60.

Thoreson, M. A., P. Z. Anastasiadis, J. M. Daniel, R. C. Ireton, M. J. Wheelock, K. R. Johnson, D. K. Hummingbird, and A. B. Reynolds. 2000. Selective uncoupling of p120(ctn) from E-cadherin disrupts strong adhesion. *J Cell Biol* 148 (1):189–202.

Tomaselli, K. J., K. M. Neugebauer, J. L. Bixby, J. Lilien, and L. F. Reichardt. 1988. N-cadherin and integrins: two receptor systems that mediate neuronal process outgrowth on astrocyte surfaces. *Neuron* 1 (1):33–43.

Tomasiewicz, H., K. Ono, D. Yee, C. Thompson, C. Goridis, U. Rutishauser, and T. Magnuson. 1993. Genetic deletion of a neural cell adhesion molecule variant (N-CAM-180) produces distinct defects in the central nervous system. *Neuron* 11 (6):1163–74.

Tomschy, A., C. Fauser, R. Landwehr, and J. Engel. 1996. Homophilic adhesion of E-cadherin occurs by a co-operative two-step interaction of N-terminal domains. *Embo J* 15 (14):3507–14.

Trudel, E., and C. W. Bourque. 2010. Central clock excites vasopressin neurons by waking osmosensory afferents during late sleep. *Nat Neurosci* 13 (4):467–74.

Utsuki, S., Y. Sato, H. Oka, B. Tsuchiya, S. Suzuki, and K. Fujii. 2002. Relationship between the expression of E-, N-cadherins and beta-catenin and tumor grade in astrocytomas. *J Neurooncol* 57 (3):187–92.

van den Pol, A. N., di Porzio, U., and Rutishauser, U. 1986. Growth cone localization of neural cell adhesion molecule on central nervous system neurons in vitro. *J Cell Biol* 102: 2281–94.

Vanhalst, K., P. Kools, K. Staes, F. van Roy, and C. Redies. 2005. delta-Protocadherins: a gene family expressed differentially in the mouse brain. *Cell Mol Life Sci* 62 (11):1247–59.

Varea, E., J. Nacher, J. M. Blasco-Ibanez, M. A. Gomez-Climent, E. Castillo-Gomez, C. Crespo, and F. J. Martinez-Guijarro. 2005. PSA-NCAM expression in the rat medial prefrontal cortex. *Neuroscience* 136 (2):435–43.

Vazquez-Chona, F., and E. E. Geisert, Jr. 1999. N-cadherin at the glial scar in the rat. *Brain Res* 838 (1–2):45–50.

Vitolo, D., P. Paradiso, S. Uccini, L. P. Ruco, and C. D. Baroni. 1996. Expression of adhesion molecules and extracellular matrix proteins in glioblastomas: relation to angiogenesis and spread. *Histopathology* 28 (6):521–28.

Volk, T., and B. Geiger. 1984. A 135-kd membrane protein of intercellular adherens junctions. *Embo J* 3 (10):2249–60.

Votin, V., W. J. Nelson, and A. I. Barth. 2005. Neurite outgrowth involves adenomatous polyposis coli protein and beta-catenin. *J Cell Sci* 118 (Pt 24):5699–708.

Wagner, S., M. Tagaya, J. A. Koziol, V. Quaranta, and G. J. del Zoppo. 1997. Rapid disruption of an astrocyte interaction with the extracellular matrix mediated by integrin alpha 6 beta 4 during focal cerebral ischemia/reperfusion. *Stroke* 28 (4):858–65.

Wickstrom, S. A., A. Lange, E. Montanez, and R. Fassler. 2010. The ILK/PINCH/parvin complex: the kinase is dead, long live the pseudokinase! *Embo J* 29 (2):281–91.

Wilby, M. J., E. M. Muir, J. Fok-Seang, B. J. Gour, O. W. Blaschuk, and J. W. Fawcett. 1999. N-cadherin inhibits Schwann cell migration on astrocytes. *Mol Cell Neurosci* 14 (1):66–84.

Williams, E. J., G. Williams, F. V. Howell, S. D. Skaper, F. S. Walsh, and P. Doherty. 2001. Identification of an N-cadherin motif that can interact with the fibroblast growth factor receptor and is required for axonal growth. *J Biol Chem* 276 (47):43879–86.

Williamson, R. A., M. D. Henry, K. J. Daniels, R. F. Hrstka, J. C. Lee, Y. Sunada, O. Ibraghimov-Beskrovnaya, and K. P. Campbell. 1997. Dystroglycan is essential for early embryonic development: disruption of Reichert's membrane in Dag1-null mice. *Hum Mol Genet* 6 (6):831–41.

Wittmann, T., G. M. Bokoch, and C. M. Waterman-Storer. 2004. Regulation of microtubule destabilizing activity of op18/stathmin downstream of rac1. *J Biol Chem* 279 (7):6196–203.

Wittmann, T., and C. M. Waterman-Storer. 2001. Cell motility: can Rho GTPases and microtubules point the way? *J Cell Sci* 114:3795–803.

Wolburg-Buchholz, K., A. F. Mack, E. Steiner, F. Pfeiffer, B. Engelhardt, and H. Wolburg. 2009. Loss of astrocyte polarity marks blood-brain barrier impairment during experimental autoimmune encephalomyelitis. *Acta Neuropathol* 118 (2):219–33.

Wood, G. K., J. J. Liang, G. Flores, S. Ahmad, R. Quirion, and L. K. Srivastava. 1997. Cloning and *in situ* hybridization analysis of the expression of polysialyltransferase mRNA in the developing and adult rat brain. *Brain Res Mol Brain Res* 51 (1–2):69–81.

Xiao, K., R. G. Oas, C. M. Chiasson, and A. P. Kowalczyk. 2007. Role of p120-catenin in cadherin trafficking. *Biochim Biophys Acta* 1773 (1):8–16.

Yamada, S., S. Pokutta, F. Drees, W. I. Weis, and W. J. Nelson. 2005. Deconstructing the cadherin-catenin-actin complex. *Cell* 123 (5):889–901.

Yanagisawa, M., I. N. Kaverina, A. Wang, Y. Fujita, A. B. Reynolds, and P. Z. Anastasiadis. 2004. A novel interaction between kinesin and p120 modulates p120 localization and function. *J Biol Chem* 279 (10):9512–21.

Yang, B., D. Jung, D. Motto, J. Meyer, G. Koretzky, and K. P. Campbell. 1995. SH3 domain-mediated interaction of dystroglycan and Grb2. *J Biol Chem* 270 (20):11711–14.

Yang, P., X. Yin, and U. Rutishauser. 1992. Intercellular space is affected by the polysialic acid content of NCAM. *J Cell Biol* 116 (6):1487–96.

Yap, A. S., C. M. Niessen, and B. M. Gumbiner. 1998. The juxtamembrane region of the cadherin cytoplasmic tail supports lateral clustering, adhesive strengthening, and interaction with p120ctn. *J Cell Biol* 141 (3):779–89.

Yip, P. M., X. Zhao, A. M. Montgomery, and C. H. Siu. 1998. The Arg-Gly-Asp motif in the cell adhesion molecule L1 promotes neurite outgrowth via interaction with the alpha v beta 3 integrin. *Mol Biol Cell* 9 (2):277–90.

Yonemura, S., Y. Wada, T. Watanabe, A. Nagafuchi, and M. Shibata. 2010. alpha-Catenin as a tension transducer that induces adherens junction development. *Nat Cell Biol* 12 (6):533–42.

Yoshida, Y., N. Kojima, N. Kurosawa, T. Hamamoto, and S. Tsuji. 1995. Molecular cloning of Sia alpha 2,3Gal beta 1,4GlcNAc alpha 2,8-sialyltransferase from mouse brain. *J Biol Chem* 270 (24):14628–33.

Yoshida-Moriguchi, T., L. Yu, S. H. Stalnaker, S. Davis, S. Kunz, M. Madson, M. B. Oldstone, H. Schachter, L. Wells, and K. P. Campbell. 2010. O-mannosyl phosphorylation of alpha-dystroglycan is required for laminin binding. *Science* 327 (5961):88–92.

Yoshida-Noro, C., N. Suzuki, and M. Takeichi. 1984. Molecular nature of the calcium-dependent cell-cell adhesion system in mouse teratocarcinoma and embryonic cells studied with a monoclonal antibody. *Dev Biol* 101 (1):19–27.

Zaidel-Bar, R., S. Itzkovitz, A. Ma'ayan, R. Iyengar, and B. Geiger. 2007. Functional atlas of the integrin adhesome. *Nat Cell Biol* 9 (8):858–67.

Zervas, C. G., S. L. Gregory, and N. H. Brown. 2001. *Drosophila* integrin-linked kinase is required at sites of integrin adhesion to link the cytoskeleton to the plasma membrane. *J Cell Biol* 152 (5):1007–18.

Zhang, Y., X. Zhang, D. Wu, J. Verhaagen, P. M. Richardson, J. Yeh, and X. Bo. 2007a. Lentiviral-mediated expression of polysialic acid in spinal cord and conditioning lesion promote regeneration of sensory axons into spinal cord. *Mol Ther* 15 (10):1796–804.

Zhang, Y., X. Zhang, J. Yeh, P. Richardson, and X. Bo. 2007b. Engineered expression of poly-sialic acid enhances Purkinje cell axonal regeneration in L1/GAP-43 double transgenic mice. *Eur J Neurosci* 25 (2):351–61.

Zhao, X., and C. H. Siu. 1995. Colocalization of the homophilic binding site and the neuritogenic activity of the cell adhesion molecule L1 to its second Ig-like domain. *J Biol Chem* 270 (49):29413–21.

Zhou, R., and O. Skalli. 2000. Identification of cadherin-11 down-regulation as a common response of astrocytoma cells to transforming growth factor-alpha. *Differentiation* 66 (4–5):165–72.

Zhou, Y. W., D. B. Thomason, D. Gullberg, and H. W. Jarrett. 2006. Binding of laminin alpha1-chain LG4–5 domain to alpha-dystroglycan causes tyrosine phosphorylation of syntrophin to initiate Rac1 signaling. *Biochemistry* 45 (7):2042–52.

4 Contribution of Astrocytes to CNS Immunity
Roles of Pattern Recognition Receptors (PRRs)

Nilufer Esen
Department of Neurology, University of Michigan
Medical School, Ann Arbor, Michigan

Tammy Kielian
Department of Pathology and Microbiology, University
of Nebraska Medical Center, Omaha, Nebraska

CONTENTS

4.1 HISTORICAL OVERVIEW OF ASTROCYTES

In the first half of the nineteenth century, advances in microscopy enabled scientists to examine many different tissues at the cellular level, including nervous tissue, of which Purkinje neurons were the first cells described. In 1846, Virchow first used the term *neuroglia*, noting that the area surrounding neurons resembled connective tissue formed by a sort of cement (glia) in which the nervous system elements were embedded. Otto Deiters (1834–1863) is considered the first scientist who provided illustrations of a cell type that resembles our modern view of astrocytes. In the 1870s, Camillo Golgi described the diversity of glial cells in gray and white matter and characterized them as roundish, oval, or star-shaped cells, from which numerous long, fine, and never arborized prolongations originate. In addition, he was able to show the structural association of astrocytes with blood vessels; however, their involvement in the blood–brain barrier (BBB) would require several decades to become established. In 1893, Micheal von Lenbossek introduced the term *astrocyte* to refer to the star-shaped neuroglial cells. Once astrocytes and neurons were recognized as unique cell types in the early 1900s, this sparked the interest of scientists who pondered their possible link to brain diseases. During this time, improvements in electrophysiological techniques made neurons the primary cell of interest, resulting in neuroglia being largely overlooked. Therefore, until recent decades astrocytes were considered to be supportive cells only.

It is currently acknowledged that astrocytes influence neuronal activity and homeostasis in the central nervous system (CNS). Additionally, it is now clear that astrocytes affect neuronal function at several levels. Besides guiding neuronal development, astrocytes release neurotrophic factors, contribute to neurotransmitter metabolism, regulate extracellular pH and K^+ levels (Chen and Swanson 2003; Walz and Hertz 1982, 1983), provide nutritional molecules such as glucose (Brown et al. 2004), and influence BBB formation and maintenance (Wolburg and Risav 1995). It has also been shown that astrocytes increase the number of mature functional synapses on neurons and are required for synaptic maintenance, suggesting that astrocytes may actively participate in synaptic plasticity (Ullian et al. 2001; Perea et al. 2009). Furthermore, evidence has accumulated demonstrating that astrocytes participate in both innate and adaptive immune responses in the CNS (Farina et al. 2007). Therefore, it is not surprising that the number of published reports documenting the role and importance of astrocytes in several aspects of CNS homeostasis/pathology continue to expand at a rapid rate. This chapter will focus on the role of astrocytes in CNS innate immune responses, with an emphasis on pattern recognition receptors (PRRs), in particular Toll-like receptors (TLRs).

4.2   CNS: TO BE OR NOT TO BE IMMUNE PRIVILEGED?

Due to the existence of tight junctions between CNS endothelial cells, absence of constitutively primed antigen-presenting cells (APCs), and limited major histocompatibility complex (MHC) expression in normal CNS tissue, the CNS was once considered an immune-privileged organ. However, it is now appreciated that immune surveillance does occur in the CNS under normal conditions (Bailey et al. 2006; Ransohoff et al. 2003). If activated T cells penetrate the BBB and invade the CNS parenchyma and they do not encounter their cognate antigen in the context of MHC class I or II molecules, they will not be retained and exit the CNS compartment. As mentioned above, under physiological conditions CNS cells express low levels of MHC molecules, making it more difficult to establish lymphocyte responses in the context of this immunological quiescence. However, MHC molecules are induced during most inflammatory and degenerative diseases (Neumann and Wekerle 1998) by proinflammatory cyto-kines such as IFN-γ and TNF-α, which are released by either infiltrating cells or activated glia (Neumann 2001). This upregulation of MHC expression facilitates lymphocyte activation events in the context of other requisite activation signals. Astrocytes were the first CNS cell type documented to express MHC class II molecules upon IFN-γ stimulation in vitro (Wong et al. 1984); however, their lack of costimulatory molecules prevent naïve T cell activation, and therefore their ability to function as APCs remains controversial (Dong and Benveniste 2001). In addition, it is not clear whether activated astrocytes express MHC class II in vivo. In comparison, activated microglia are the main APC in the CNS parenchyma and are important in activating memory T cell populations (Aloisi et al. 1998, 1999; Carson et al. 1998).

4.3 PATTERN RECOGNITION RECEPTORS (PRRs)

Numerous CNS resident cell types express pattern recognition receptors (PRRs), which recognize pathogen-associated molecular patterns (PAMPs), revealing the ability of the CNS to initiate a local immune response to infectious insults (Kielian 2006). Besides PAMPs, some PRRs can also recognize endogenous stress-related or degradation molecules termed danger-associated molecular patterns (DAMPs), which are elaborated following brain insults other than infections, such as ischemia and neurodegenerative disorders (Bianchi 2007; Carta et al. 2009). Currently, PRRs can be classified into three groups: membrane bound, cytosolic, or secreted (Box 4.1). The variety of PRRs and their expression in both immune-competent as well as nonimmune CNS cell types support the concept that their relevance extends far beyond a role in antimicrobial defense. Among the PRRs, Toll-like receptors (TLRs) form the largest and most diverse group in terms of their ability to recognize a broad spectrum of PAMPs (Takeda et al. 2003; Medzhitov 2001; van Noort and Bsibsi 2009). Since their discovery, we continue to acquire new information about TLR ligands, their locations, mechanisms of activation, and signaling outcomes. For example, TLR2, which has the largest ligand repertoire of all TLR family members identified to date, has recently been documented to be important for the recognition of viral proteins (Barbalat et al. 2009; Martinez et al. 2010). In addition, TLR4, which is the only TLR capable of inducing both MyD88-dependent and -independent signaling pathways following activation by Gram-negative bacteria, has been recently shown to translocate into an endosomal compartment upon ligation and MyD88-dependent activation, where it induces MyD88-independent, i.e., TRIF-related adaptor molecule (TRAM)-TIR domain-containing adapter-inducing interferon-beta (TRIF) signaling (Kagan et al. 2008). Therefore, in addition to its traditional membrane-associated form, TLR4 can now be classified as an endosomal TLR similar to the other classical TLRs residing in this subcellular compartment (i.e., TLR3, TLR7/8, and TLR9). Although both microglia and astrocytes express various PRRs, the role of astrocytes in CNS immunity will be the main focus of this chapter, although microglia are consid-ered the main innate immune cell type in the CNS parenchyma and express the full repertoire of all TLRs currently identified (Kielian 2006).

BOX 4.1   CLASSIFICATION OF PRRs

1. Membrane-associated (both plasma and endosome) PRRs
   1a. Toll-like receptors: TLR1–11
   1b. Mannose receptors: MR-C1, MR-C2, SIGNR1
   1c. Scavenger receptors: SR-AI, SR-AII, LOX, MARCO
   1d. Glucan receptors: Dectin-1
   1e. Complement receptors: CR1–4, C3aR, C5aR
2. Cytosolic PRRs
   2a. NOD-like receptors:
       NOD1 and NOD2
       NLRPs: NLRP1–14
       IPAF and NAIP5/Birc1
   2b. RNA helicases:
       RIG-I
       MDA5
       dsRNA-dependent PKR
       LGP2
3. Secreted PRRs
   3a. Complement factors: C1–9
   3b. Pentraxin proteins: C-reactive protein, pentraxin, amyloid, PGR, and LRR
   3c. Collectins: Mannose-binding lectin (MBL)

4.3.1   TLR EXPRESSION IN ASTROCYTES

Under physiological conditions, basal TLR2 and TLR4 expression has been demonstrated in the meninges, choroid plexus, and circumventricular organs of the brain, namely, in CNS areas that possess an incomplete BBB and are more exposed to invading pathogens (Chakravarty and Herkenham 2005; Laflamme et al. 2001, 2003; Laflamme and Rivest 2001; Mishra et al. 2006). Additionally, the basal expression of TLR3 is detected on glial fibrillary acidic protein (GFAP)-positive astrocytes in the hippocampus and striatum (Park et al. 2006; Mishra et al. 2006). Cultured astrocytes have been shown to express TLR2, TLR3, TLR4, TLR5, TLR7, and TLR9 (Bowman et al. 2003; Bsibsi et al. 2002; Carpentier et al. 2005; Farina et al. 2007; Gorina et al. 2009; Esen et al. 2004), as well as downstream adaptor molecules, including MyD88, Toll-interleukin 1 receptor (TIR) domain-containing adaptor protein (TIRAP), and TRIF (Farina et al. 2005). There have been conflicting reports about the cadre of TLRs expressed by astrocytes; in particular, questions have been raised as to whether astrocytes express TLR4 or whether receptor expression may be the result of low-level microglial contamination of astrocyte cultures (Lehnardt et al. 2002, 2003; Saura 2007). However, there is consensus agreement for TLR3 expression in astrocytes, which has been demonstrated both in vivo and in vitro (Bsibsi et al. 2002; Carpentier et al. 2005; Farina et al. 2007; Park et al. 2006). Furthermore, it has been suggested that TLR3 is the only TLR constitutively expressed in human fetal astrocytes (Farina et al. 2005; Jack et al. 2005). TLR3 is commonly located in intracellular compartments, such as the endosome, and requires ligand internalization (i.e., dsRNA) to be activated. Interestingly, TLR3 has also been detected on the cell membrane of human astrocytes as well as intracellularly (Bsibsi et al. 2002; Jack et al. 2005). Injection of the synthetic dsRNA analog poly(I:C) into the brain triggered microglial and astrocytic activation in wild type (WT) but not TLR3 knockout (KO) mice, suggesting an important role for TLR3 in the CNS response to viral infection (Town et al. 2006). In cultured astrocytes, poly(I:C) induced the expression of several cytokines, including TNF, IL-6, IFN-β, granulocyte-macrophage colony-stimulating factor (GM-CSF), and transforming growth factor (TGF)-β, as well as chemokines such as CCL2, CCL5, CCL20, CXCL8, CXCL10, and CXCL13 (Bsibsi et al. 2006; Farina et al. 2005; Jack et al. 2005; Park et al. 2006; N. Esen, unpublished data). Moreover, dsRNA treatment elicits inducible nitric oxide synthase (iNOS) and nitric oxide production (Brahmachari et al. 2006; Carpentier et al. 2005; Scumpia et al. 2005), but at the same time decreases glutamate uptake by astrocytes (Scumpia et al. 2005) and downregulates connexin43 expression with a disturbance of gap-junction-mediated intercellular communication (Zhao et al. 2006). These effects are suggestive that TLR3-dependent astrocyte activation may elicit/perpetuate tissue damage and neurotoxicity during CNS viral infections. Similarly, in vitro infection of mouse astrocytes with Theiler's murine encephalomyelitis virus (TMEV) triggered an initial inflammatory response (CCL2 and CXCL10 secretion) mediated via TLR3 (So et al. 2006). However, viral replication was equivalent between WT and TLR3 KO astrocytes, suggesting that TLR3-mediated activation is insufficient to block TMEV replication. In adult human astrocytes, poly(I:C) treatment induced the expression of numerous genes, many of which were involved in antiviral mechanisms as well as cell growth, differentiation, and neuroprotection, including GM-CSF, vascular endothelial growth factor (VEGF)-C, neurotrophin (NT)-4, and ciliary neurotrophic factor (CNTF), suggesting a role for astrocytes in tissue repair pathways through TLR signaling (Bsibsi et al. 2006). Similarly, recent data suggest a nonimmune, regulatory function of TLR3 signaling in neurogenesis. Lathia and colleagues have reported that TLR3 expression was highest during the early period of cortical development, when neuron progenitor cells (NPCs) are highly proliferative, whereupon TLR3 levels declined as neurogenesis and gliogenesis ensued (i.e., mid-gestation through early postnatal periods) (Lathia et al. 2008a). In addition, the authors demonstrated that TLR3 activation inhibited the proliferation of embryonic mouse NPC, and NPC proliferation was increased in the developing cerebral cortex of TLR3-deficient mice (Lathia et al. 2008a). These data suggest that TLR3 may modulate neuronal development, and the predominant expression of TLR3 on embryonic astrocytes might be one of the mechanisms by which astrocytes influence neurogenesis.

Astrocytes express TLR2, which is augmented upon exposure to various PAMPs (Bowman et al. 2003; Carpentier et al. 2005; Esen et al. 2004). Strong evidence demonstrating that astrocytes express TLR2 in vivo was provided by Mishra et al. (2006) using immunofluorescence staining. In this study, robust TLR2 immunoreactivity was detected in astrocytes in both the normal and infected CNS (Mishra et al.

2006). Further support for astrocytic TLR2 expression was provided by a recent report by Kigerl et al. (2007) that utilized laser capture microdissection for astrocyte enrichment from control and injured spinal cord tissues and demonstrated TLR2 expression associated with astrocytes, although maximal expression was detected in microglia. However, studies examining TLR2 in other systems have produced some conflicting results with regard to astrocytic expression (Bsibsi et al. 2002; Farina et al. 2005; Owens 2005; Rivest 2003). It is likely that the context of PAMP exposure or the strength of the activation signal received following astrocyte activation may dictate whether TLR2 expression is induced. Alternative explanations may include the species from which astrocytes were procured, the route of PAMP administration during in vivo studies, and the length of time that astrocytes are cocultured with microglia prior to purification for in vitro studies. Since TLR2 expression levels fluctuate during development (Lathia et al. 2008a), the age of the species at the time of astrocyte procurement could be another factor for the discrepancies reported in the literature. Nevertheless, a recent study reports that unstimulated astrocytes express very low levels of TLR2, but receptor expression is dramatically induced upon stimulation (So et al. 2006); this is in agreement with the general consensus that astrocytes express lower levels of TLR2, both constitutively and upon activation, than microglia. Interestingly, in response to TMEV infection, the levels of TLR2 induced in TLR3 KO astrocytes were lower than for WT cells (So et al. 2006). These data raise two important issues: First, TLR2 and TLR3 exhibit complex interactions and regulatory pathways, which may not be unique to these TLRs, since other TLR ligands can also induce elevated TLR2 expression. Second, TLR2 may participate in the induction of a proinflammatory response subsequent to viral infection, a possibility that has not received much attention given the traditional role for TLR3 in recognition of viral dsRNA.

TLR4 is responsible for recognizing the Gram-negative cell wall component lipopolysaccharide (LPS) (Heine et al. 1999; Hirschfeld et al. 2000; Hoshino et al. 1999; Lien et al. 2000; Poltorak et al. 1998; Qureshi et al. 1999; Takeuchi et al. 1999; Tapping et al. 2000). LPS is delivered to CD14 by LPS-binding protein (LBP) and transferred to MD-2 to form a monomeric endotoxin: MD-2 complex that binds and activates TLR4 (Teghanemt et al. 2005; Visintin et al. 2005). TLR4 activation can occur in the absence of LPB and CD14; however, this requires several orders of magnitude more endotoxin to elicit TLR4 signaling (Lynn et al. 1993; Perera et al. 1997). In addition, membrane-bound CD14 directly binds to TLR2 ligands, inducing physical proximity of CD14 and lipopeptides with TLR2/TLR1, thus augmenting the signaling pathway (Manukyan et al. 2005) (Figure 4.1). As shown in macrophages and microglia, CD14 interacts with TLR4 to induce maximal responses to LPS (Dobrovolskaia and Vogel 2002; Esen and Kielian 2005; Palsson-McDermott and O'Neill 2004). As opposed to microglia, astrocytic TLR4 expression could not be demonstrated by several laboratories either in vitro (Farina et al. 2005) or in vivo (Laflamme and Rivest 2001; Lehnardt et al. 2002, 2003); however, other groups have detected low, constitutive expression of TLR4 in astrocytes that is increased upon cell activation (Bowman et al. 2003; Bsibsi et al. 2002; Carpentier et al. 2005). Similarly, astrocytes were considered non-CD14 expressing cells by some groups (Cauwels et al. 1999; Willis and Nisen 1996), while the others reported the expression of CD14 mRNA (Johann et al.

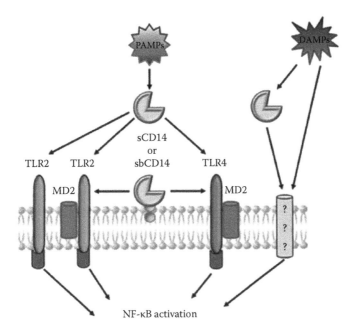

FIGURE 4.1 Both soluble (s) and surface-bound (sb) CD14 play a role in NF-κB activation following exposure to pathogen-associated (PAMPs) or danger-associated molecular patterns (DAMPs) through TLR-dependent or -independent mechanisms.

2008; Lehnardt et al. 2002) as well as another molecule of the LPS receptor complex, MD-2, in cultured murine astrocytes (Johann et al. 2008). It is important to acknowledge that great care must be taken when working with primary astrocytes to ensure that contamination with microglia is relatively low (Saura 2007). Since microglia express high levels of TLR4, a small number of residual microglia could introduce artifact signals that are not reflective of astrocytic receptor expression. Further studies using primary astrocyte cultures where microglia have been depleted by immunological means (i.e., magnetic bead purification or fluorescence-activated to cell sorting [FACS]) should help to resolve this lingering issue.

Despite the conflicting reports regarding astrocytic expression of membrane-bound CD14, astrocytes are responsive to soluble CD14. Soluble CD14 serves as a coreceptor for LPS recognition by TLR4 as well as coreceptor for peptidoglycan (PGN) recognition via TLR2, thus promoting cellular responses to these agonists by cells that do not express surface-bound CD14, including astrocytes (Rallabhandi et al. 2006; Remer et al. 2006; Orr and Tobias 2000; Vita et al. 1997). Recent data suggest that soluble CD14 can also directly trigger TLR2 in the absence of other ligands (Bsibsi et al. 2007), since soluble CD14 was capable of eliciting IL-6, CXCL8, and IL-12p40 production in a TLR2-dependent manner (Bsibsi et al. 2007). Recently, a new function for CD14 has been described in recognizing both DAMPs and PAMPs, as revealed by the fact that CD14, but not MD-2, played a significant role in NF-κB activation in response to necrotic cells in the presence or absence of TLR2 (Chun and Seong 2010) (Figure 4.1). Therefore, increased CD14 expression during inflammatory conditions suggests that a CD14-TLR2 combination may have particular relevance in controlling the nature and extent of CNS inflammation (Esen et al. 2004; Lin et al. 2000).

TLR9 mediates cellular responses to bacterial DNA, viral DNA, and synthetic oligodeoxynucleotides (ODNs), all of which contain unmethylated CpG motifs, whereas TLR7, which also resides in the endosomal compartment, recognizes ssRNA (Takeda et al. 2003). Recently, TLR9 became a focus of research in several autoimmune diseases, since its interaction with self nucleic acids can also induce an inflammatory response (Avalos et al. 2010; Krieg and Vollmer 2007). Under normal conditions, self-derived nucleic acids that are generated during apoptosis are degraded immediately and cannot stimulate TLR7 or TLR9 intracellularly (Barton et al. 2006). Another mechanism to prevent inappropriate activation of TLR7 and TLR9 by mammalian nucleic acids is afforded by the fact that these receptors are sequestered in the endoplasmic reticulum (ER) until cell activation, whereupon they traffic to endolysosomes by an Unc93B1- (a multitransmembrane ER-resident protein) controlled event (Figure 4.2). However, self nucleic acids may reach endolysosomes during inflammatory or autoimmune situations, where a variety of nucleic-acid-binding proteins such as autoantibodies, antimicrobial peptides, and HMGB-1 are complexed with host nucleic acids (Lande et al. 2007; Tian et al. 2007). These complexes are resistant to degradation, which allows them to reach endolysosomes and stimulate TLR7 and TLR9. It is also possible that defects in cell death processes could lead to the enrichment of certain sequences that can engage TLR9. Another function attributed to TLR7 and TLR9 activation is antibody class switching to IgG2a, either indirectly through the induction of type I IFNs (He

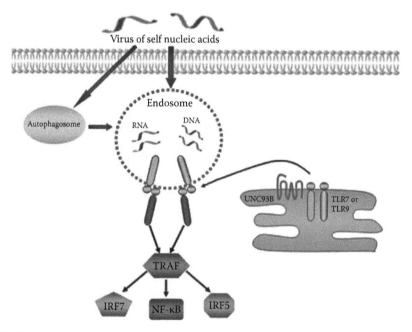

FIGURE 4.2 Pathways for nucleic acid recognition by Toll-like receptor 7 (TLR7) and TLR9.

et al. 2004) or via direct interaction with B cells (Jegerlehner et al. 2007). This direct activation is presumably because the B cell receptor (BCR) is the primary conduit into the late endosomal compartment where these TLRs reside (Siemasko and Clark 2001; Trinchieri and Sher 2007). This evidence implies that TLR7 or TLR9 signaling may exacerbate autoimmunity by stimulating autoreactive B cells to switch from an IgM to a more pathogenic IgG isotype. Both microglia (Dalpke et al. 2002; Iliev et al. 2004; Olson and Miller 2004; Takeshita et al. 2001; Zhang et al. 2005) and astrocytes (Bowman et al. 2003; Carpentier et al. 2005; Hosoi et al. 2004) express TLR9, and engagement of this PRR leads to a robust induction of proinflammatory mediators (Butchi et al. 2010). In addition, CpG ODN has been reported to induce the expression of adhesion molecules and matrix metalloproteinase-9 (MMP-9) in mouse astrocytes (Lee et al. 2004). CpG ODN stimulation in astrocytes induces the activation of IkappaB kinase (IKK) and c-Jun N-terminal kinase (JNK), whereas it inhibits constitutive ERK1/2 activation (Lee et al. 2004). The abrogation of JNK activity by pharmacological means showed that JNK activation is essential for the induction of cytokine and chemokine expression. This effect of JNK does not require c-Jun phosphorylation; rather, it works via the potentiation of NF-κB signaling (Lee et al. 2004). TLR7 and TLR9 signaling is controlled and balanced (Wang et al. 2006), which is considered another mechanism to avoid autoimmunity (Fukui et al. 2009). Indeed, astrocyte treatment with the TLR7 agonist, imiquimod, inhibited TLR9 agonist-induced innate immune responses in a concentration-dependent manner (Butchi et al. 2010). Surprisingly, this inhibition was not mediated by TLR7 directly, as deficiency in TLR7 activity did not alter TLR9 agonist-induced responses. This study suggested that the suppression of innate immune responses was not due to an inhibition of TLR9 agonist uptake; instead, it may be a direct effect, possibly by blocking CpG-ODN binding or signaling via TLR9, thus limiting cell activation (Butchi et al. 2010). Taken together, TLR9 engagement in astrocytes may contribute to CNS autoimmunity under certain circumstances, and the manipulation of TLR7 and TLR9 interactions might have potential therapeutic value for autoimmune conditions.

4.3.2 MANNOSE AND SCAVENGER RECEPTORS IN ASTROCYTES

Both mannose (MR) and scavenger receptors (SR) promote the attachment, uptake, and destruction of microorganisms by phagocytic cells. MR recognizes mannosylated ligands of endogenous or microbial origin, while SRs have high affinity for acetylated low-density lipoproteins and bind and internalize many unrelated ligands, such as fibrillar β-amyloid, lipids, glycated collagen, thrombospondin, and apoptotic cells (Husemann et al. 2002; Murphy et al. 2005). Both cultured astrocytes (Burudi et al. 1999) and microglia (Aloisi 2001; Kielian et al. 2005) express MR. They also express various SRs, including SR-B, macrophage receptor with collagenous structure (MARCO), receptor for advanced glycation endproducts (RAGE), and scavenger receptor with C-type lectin (SRCL) (Alarcon et al. 2005; Husemann et al. 2001; Nakamura et al. 2006; Sasaki et al. 2001). In addition, SR-A, LOX-1 (Esen and Kielian 2005; Husemann et al. 2002; Kielian et al. 2005), and CD36 (Coraci et al. 2002; Kielian, unpublished data) have been reported in microglia. Originally, it

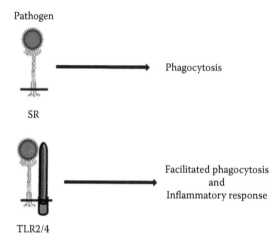

FIGURE 4.3   Cooperation between phagocytic and Toll-like receptors.

was suggested that MRs could not relay an intracellular signal. However, MR binding to human immunodeficiency virus-1 (HIV-1) not only induced CD4-independent
viral invasion into astrocytes, but also triggered intracellular signaling leading to
matrix metalloproteinase-2 production (Lopez-Herrera et al. 2005; Liu et al. 2004).
Similarly, a mutual cross talk between TLRs and SRs has been demonstrated
(Underhill and Gantner 2004; Underhill and Ozinsky 2002) (Figure 4.3). For example, LOX-1 and CD36 act as coreceptors for TLR2/4 ligands (Areschoug and Gordon
2009). In addition, it is suggested that TLRs may regulate phagosome formation
and maturation as well as modulate the transcription of some phagocytic receptors,
while activation of SR following ligand binding may also modulate TLR signaling
(Underhill and Gantner 2004). However, deficiency of TLR2/4 or MyD88 does not
affect the phagocytosis of pathogens in microglia (Esen and Kielian 2006; Kielian
et al. 2005) or macrophages (Henneke et al. 2002; Underhill et al. 1999), while
the inflammatory response is impaired. Recent data have offered a more mechanistic explanation on this issue by showing that when a heterozygous deficiency of
SR-A and TLR4 coexists (SR-A[+/-]TLR4[+/-]), phagocytic trafficking is reduced due to
impaired rates of bacterial internalization, but not bacterial recognition (Amiel et
al. 2009). This deficit was apparent at early time points following induction of the
phagocytic process (i.e., within 10 minutes). Since the same effect was observed in
SR-A[+/-]MyD88[+/-] DCs, this suggests that a rapid, TLR4-dependent process facilitates phagocytosis, and that alteration of this signaling process (i.e., through TLR4
heterozygosity) makes phagocytes sensitive to SR-A heterozygosity, leading to defective pathogen internalization.

Expression of MARCO in astrocytes (Alarcon et al. 2005; Brandenburg et al.
2010) has been associated with recognition and internalization of *N. meningitidis*
(Mukhopadhyay et al. 2006) and suggests a role for astrocytes during the control of
CNS bacterial infections. In addition, it is possible that MARCO, SR-B1, RAGE, and
SRCL, which bind and mediate the internalization of β-amyloid peptides and fibrils,
may play a major role in the ability of astrocytes to degrade β-amyloid deposits in

vitro and in situ (Wyss-Coray et al. 2003), along with formyl-peptide-receptor-like 1 (FPRL1) (Brandenburg et al. 2007, 2008). Therefore, astrocytes have the potential to influence the course of neurodegenerative and infectious disease by virtue of SR expression.

4.3.3 Astrocytes and Complement Receptors

The complement (C) system consists of approximately 30 fluid phase and cell-membrane-associated proteins. They are not only components of the CNS innate immune system, but also participate in neurogenesis and neuronal synaptic plasticity (Stevens et al. 2007; Rahpeymai et al. 2006; Rutkowski et al. 2010; Shinjyo et al. 2009). In addition to their primary function of participating in the complement cascade, complement proteins are linked to the adaptive immune system and several signaling pathways. Notably, brain cells, including astrocytes, microglia, oligodendrocytes, and neurons, synthesize the full range of C components, particularly after cytokine stimulation (Gasque et al. 1993, 1995a). IFN-γ is the most effective cytokine for inducing the expression of almost all C components by glia and neurons, whereas TNF-α and IL-1β are less potent (Barnum and Jones 1995; Barnum et al. 1992; Gasque et al. 1995a, 2000; Gasque and Morgan 1996). As a part of innate immunity, C proteins opsonize bacterial, viral, apoptotic cells, or even pathological protein deposits such as fibrillar Aβ plaques, for elimination especially by microglia and astrocytes in the CNS (Gasque et al. 2000; Griffiths et al. 2009; Matsuoka et al. 2001; Rogers et al. 2006). Astrocytes can also recognize apoptotic cells directly by binding phosphotidylserine (PS) via their PS receptors (Chang et al. 2000). Microglia and astrocytes protect themselves from complement-mediated lysis via the membrane attack complex by expressing several membrane-bound (CD59, DAF) and soluble (fH, FI, C1-INH) C inhibitors (Gasque et al. 2000). In addition to C proteins, astrocytes have been reported to express receptors for complement split products such as C5a and C3a receptors (Boos et al. 2005; Gasque et al. 1995b, 1996b; Woodruff et al. 2009), as well as CR1 (CD35) and CR2 (CD21) for C3b or C4b, and iC3b or C3d, respectively (Gasque et al. 1996b; van Beek et al. 2003). Astrocytic expression of CR2, which can also bind to Epstein-Barr virus (Yefenof and Klein 1976), along with CR1, emphasizes the importance of glia for removing C-opsonized viruses and apoptotic cells from the CNS and cerebrospinal fluid. Conversely, upon activation via C3a or C5a binding, the anaphylatoxin receptors can induce PLC and MAPK pathways (Sayah et al. 2003), which in turn may lead to cell activation and subsequent induction of cytokines, such as IL-6 and cytoskeleton-related proteins, to induce reactive astrocytes at sites of injury (Sayah et al. 1999). Evidence has shown that C3aR deletion or C5aR blockade improves disease outcome in animal models of experimental allergic encephalomyelitis (EAE) and amyotrophic lateral sclerosis (ALS), respectively (Woodruff et al. 2008). Conversely, C3a expression in the CNS protects against lipopolysaccharide (LPS)-induced septic shock (Boos et al. 2005), while C3 deficiency led to impaired neurogenesis (Rahpeymai et al. 2006) and increased Aβ-induced toxicity and neuronal loss (Maier et al. 2008; Wyss-Coray et al. 2002). Therefore, targeting C proteins or their receptors as a treatment modality shows great promise, although more specific and detailed studies are needed.

4.3.4 Cytoplasmic PRRs in Astrocytes

Cytosolic proteins that function as intracellular PRRs (Box 4.1) can be subgrouped into two categories, one for virus-related ligands (RIG-I, MDA5, PKR) and the second for bacteria-related molecules (nucleotide-binding oligomerization domain (NOD)-like receptors (NLRs); NOD1/2, NALP, NAIP). NOD2 recognizes a specific peptidoglycan motif from bacteria, namely, muramyl dipeptide (MDP) (Fournier and Philpott 2005; Girardin et al. 2003; Takada and Uehara 2006). Expression of NOD2 and its downstream effector molecule Rip2 kinase has been recently demonstrated in primary cultures of murine glial cells (Sterka and Marriott 2006; Sterka et al. 2006) and required for maximal inflammatory immune responses of primary glial cells to the Gram-negative bacteria *N. meningitidis* and *B. burgdorferi* (Chauhan et al. 2009), as well as the Gram-positive bacteria *Streptococcus pneumonia* and *Staphylococcus aureus* (Liu et al. 2010a, 2010b). In addition, NF-κB activation and subsequent production of key inflammatory cytokines in both astrocytes and microglia following the challenge with intact *S. pneumoniae* were significantly reduced in the absence of NOD2 expression (Liu et al. 2010a). On the other hand, recent studies have demonstrated the involvement of NLRP3 (also known as NALP3 or cryopyrin) in IL-1β secretion from adenovirus and vaccinia virus-infected cells (Delaloye et al. 2009; Muruve et al. 2008), which attributes new functions to NLR other than bacterial recognition. Taken together, these studies confirm the complementary role of NLRs to TLRs in recognition of, and subsequent inflammatory response to, bacteria- and virus-associated molecules by glia (Figure 4.4).

As mentioned previously, TLR3, which is strongly expressed in astrocytes, is activated by viral dsRNA inducing downstream signaling for expression of NF-κB- and IRF-associated inflammatory mediators. A second intracellular pathway detects viral nucleic acids via the cytosolic RNA helicases, which are grouped into RIG-I-like receptors (RLRs) and include melanoma differentiation-associated gene 5 (MDA-5), retinoic-acid-inducible gene I (RIG-I), and laboratory of genetics and physiology 2 (LGP2) (Kawai and Akira 2008; Takahasi et al. 2009). These cytosolic proteins are members of the DexD/H RNA box helicase family with an N-terminal caspase recruitment domain (CARD) and a C-terminal helicase domain (Kato et al. 2006; Takeuchi and Akira 2008). LGP2 serves as the upstream regulator of RIG-I and MDA-5 (Satoh et al. 2010; Vitour and Meurs 2007) (Figure 4.4). Following ligation, MDA-5/RIG-I associates with the adaptor protein, IFN-β promoter stimulator 1 (IPS-1; also known as mitochondrial antiviral-signaling protein (MAVS)) (Kawai et al. 2005). IPS-1 then recruits both IKKε/TBK1 and IKKα/β/γ complexes, resulting in the activation of several transcription factors, including interferon regulatory factor 3 (IRF3), NF-κB, and activating protein 1 (AP-1) (Honda et al. 2006). RLR cytosolic sensors are also functional in astrocytes as reported by De Miranda and coworkers, showing that naked and complexed poly(I:C) leads to differential astrocyte activation (De Miranda et al. 2009). Namely, while naked poly(I:C) triggered TLR3-dependent signaling responses culminating in the release of IFN-β, IL-6, TNF-α, IL-8, and CCL5, complexed poly(I:C) augmented IFN-β and IL-6 release in a MDA-5-dependent manner (De Miranda et al. 2009). These data further highlight the importance of astrocytes in innate immune pathogen recognition in the CNS.

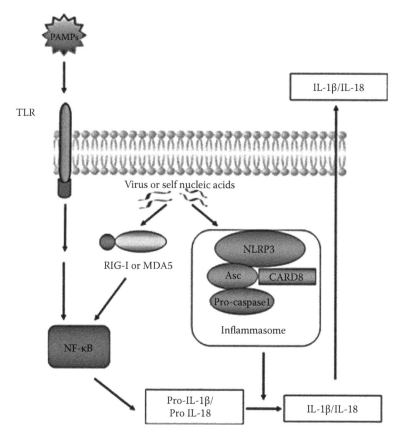

FIGURE 4.4 Inflammasome activation in response to bacterial or viral infections. After activation via its leucine rich-repeat domain, NLRP3/NALP3 interacts with ASC (apoptosis-associated speck-like protein containing a caspase recruitment domain) and CARD8 (caspase recruitment domain containing protein 8). Subsequently, ASC interacts with pro-caspase 1 to mediate its conversion to caspase 1. Caspase 1, in turn, activates the interleukin (IL) precursors pro-IL-1β and pro-IL-18 to their mature bioactive forms.

A third sensor, ds-RNA-dependent protein kinase (PKR), was originally characterized prior to TLR3 and RIG-1/MDA-5 for its activation by dsRNA, and its participation was acknowledged in the innate immune response to viruses (Levin et al. 1980; Williams 1999). PKR is a ubiquitously expressed serine/threonine kinase, which is activated in response to a variety of cellular stressors. In particular, PKR has major roles in the blockade of viral translation through the phosphorylation of the α subunit of eukaryotic initiation factor 2, as well as in both the induction and response to type I interferon (IFN) (Balachandran et al. 2000; Williams 1999). It may also be able to induce cytokines and chemokines through its ability to activate NF-κB (Zamanian-Daryoush et al. 2000). The availability of PKR-deficient mice led scientists to evaluate its role more clearly, which revealed that in astrocytes, PKR is critical for responses to Theiler's murine encephalitis virus (TMEV), inducing type I IFNs, cytokines, and chemokines (Carpentier et al. 2007), but not for responses to extracellular poly(I:C)

(Carpentier et al. 2007; De Miranda et al. 2009). In addition, cytokines released from HIV-infected macrophages in the brain trigger CXCL8 production by astrocytes via PKR activity (Zheng et al. 2008). Together these data, along with reports demonstrating the functional involvement of both TLR3 and RLR in astrocytes, imply that astrocytes are adequately equipped to rapidly respond to CNS viral infections.

4.3.5 ASTROCYTES AND SECRETED PRRs

As mentioned above, astrocytes possess an extensive arsenal of molecules necessary for the opsonization and eradication of pathogens and apoptotic cells, as well as complement inhibitory factors to protect themselves from the detrimental effects of complement activation.

Another secreted protein, long pentraxin 3 (PTX3), functions as a PRR by opsonizing and leading to the phagocytosis of apoptotic cells (Griffiths et al. 2009). PTX3 is expressed upon activation by various cells, including microglia (Esen and Kielian 2005) and astrocytes (Ravizza et al. 2001), and has important roles in innate resistance against selected bacteria and fungi (Garlanda et al. 2002; Mantovani et al. 2003). It binds to C1q, whereupon PTX3 can either inhibit or activate the classical pathway, depending on its context (Nauta et al. 2003).

4.4 CONSEQUENCES OF PRR SIGNALING IN ASTROCYTES

In the previous sections, we have outlined the types of PRRs that are expressed by astrocytes and their importance in cell activation. Following PRR engagement, several cytokines, chemokines, growth factors, and other soluble molecules are produced by the induction of various nuclear transcription factors, such as NF-κB, IRF3, IRF7, and AP-1. These mediators are summarized in Box 4.2 (Farina et al. 2007; van Noort and Bsibsi 2009) and typically exert their effects via either autocrine/paracrine or endocrine routes. For example, astrocyte release of IL-1β, and TNF-α can feed back through their respective receptors (also expressed on astrocytes) to induce the production of complement factors or PTX3 (Gasque et al. 2000; Ravizza et al. 2001), adhesion molecules such as IIntercellular adhesion molecule 1 (ICAM-1) and vascular cell adhesion molecule 1 (VCAM-1), as well as other cytokines and chemokines (John et al. 2003). TNF-α, in concert with IFN-γ, also induces the production of B cell activating factor (BAFF) in astrocytes, which is the main factor for the recruitment

---

BOX 4.2   MEDIATORS RELEASED FROM ASTROCYTES FOLLOWING PRR ACTIVATION

1. **Cytokines:** IL-1β, IL-6, IL-9, IL-10, IL-11, IL-12p40, IFNα/β TNF-α, TGF-β2, GM-CSF
2. **Chemokines:** CCL2, CCL3, CCL4, CCL5, CCL20, CXCL1, CXCL2, CxCL6, CXCL8, CXCL9, CXCL10, CXCL12, CXCL13, CX3CL1
3. **Growth factors:** NGF, CNTF, NT-4, VEGF, BDNF, GDNF, GGF-1/2, IGF-1

of antibody-secreting cells and sustained B cell responses in autoimmune and infectious diseases (Krumbholz et al. 2005). Binding of TNF-α to its receptor TNFR also activates the caspase pathway and leads to apoptosis. However, this effect is quite minor compared to CD95/CD95L (Fas/FasL)-induced apoptosis, for which astrocytes are particularly resistant (Song et al. 2006). While astrocytes express IFN-α/β, they do not produce IFN-γ. However, astrocytes respond to IFN-γ as evident by the induction of Major Histocompatibility Complex (MHC) class II, CD40, production of type I IFNs, and Suppressor of cytokine signaling (SOCS-1), which in turn negatively regulates the IFN-γ-induced effects (Wesemann and Benveniste 2004). In addition to SOCS-1, IL-4, IL-10, and TGF-β exert immunosuppressive effects on microglia and astrocytes by inhibiting the expression of MHC class II, TNF-α, and ICAM-1, subsequently suppressing T cell responses. The existence of these negative feedback mechanisms could explain the controversy surrounding MHC class II expression on astrocytes, which could be affected by the net balance of suppressors or activators at the moment of detection. In addition to its immunosuppressive effects, TGF-β along with IL-6 is an essential mediator for the differentiation of naïve Th0 T lymphocytes into Th17 cells, which play an important role in autoimmune diseases, including multiple sclerosis and psoriasis (Bettelli et al. 2007; Tesmer et al. 2008). Indeed, a recent report has demonstrated that specific deletion of Act1, a critical component of IL-17 signaling, was essential in resident neuroectoderm-derived CNS cells for peripheral immune cell recruitment and disease severity in EAE (Kang et al. 2010). In addition, Act1-deficient astrocytes were less responsive to IL-17 in terms of gene induction, collectively suggesting that astrocytes play a critical role in leukocyte recruitment during CNS inflammation (Kang et al. 2010). Moreover, IL-6 drives the proliferation and differentiation of B cells into plasma cells, and induces the synthesis of acute phase proteins such as α-antichymotrypsin in astrocytes (Kordula et al. 1998). Interestingly, astrocytes do not express membrane-bound IL-6 receptor, but respond to IL-6 in the presence of soluble IL-6 (sIL-6). The interaction of IL-6/sIL-6 complex with the membrane-bound gp130 receptor, which is ubiquitously expressed on astrocytes, is necessary for proper signaling (Wesemann and Benveniste 2005). Most of the growth factors listed in Box 4.2 are constitutively expressed by resting astrocytes, with the exception of brain-derived neurotrophic factor (BDNF) and Glial cell-derived neurotrophic factor (GDNF), which are induced upon stimulation. Induction of growth factors is usually an indirect effect, subsequent to astrocyte activation by inflammatory cytokines, but they play an important role in neuronal survival and plasticity.

As the name implies, chemokines recruit neutrophils, monocytes, dendritic cells, T cells, and B cells into the CNS via actions at the BBB. The effects of released cytokines on either endothelial cells or astrocytic endfeet are usually destructive to BBB integrity and lead to increased permeability. While cytokines affect BBB permeability and open the barrier for infiltrating cells, chemokines provoke the recruitment of those cells into the CNS along with the correct adhesion molecule repertoire, and are responsible for triggering immunity in the inflamed CNS (Ransohoff et al. 2003). In addition, several chemokines modulate microglial migration, survival, and activity; astrocyte proliferation and survival; as well myelination and migration of neural progenitors (Cardona et al. 2008; Engelhardt and Ransohoff 2005). Furthermore, chemokines such as CXCL12 and CCL5 induce glutamate release and cytokine and chemokine synthesis and, therefore, participate in glia-glia and glia-neuron communication (Ambrosini and Aloisi 2004; Farina et al. 2007). Although most of the chemokines produced by astrocytes are induced upon cell activation, chemokines such as CX3CL1, CXCL12, and CXCL1 are constitutively expressed during CNS development, regulating the migration of neuronal and oligodendrocyte precursors, microglial chemotaxis, as well as protecting microglia from Fas-induced death (Aloisi 2005).

The other consequence of PRR activation, particularly TLR signaling, is the induction of Notch ligands such as delta-like ligand 4 (dll4) and Jagged1. Interaction of Notch receptors with these ligands leads to a cascade of proteolytic cleavage events (Itoh et al. 2004; Lathia et al. 2008b; Yoon and Gaiano 2005; Amsen et al. 2009). Following two cleavages, the Notch intracellular domain (NICD) is released from the cell membrane. NICD then translocates to the nucleus, where it interacts with the DNA-binding protein recombination recognition sequence binding protein at the J kappa site (RBP-Jkappa) (also known as a highly conserved DNA-binding transcription factor (CBF)) and cooperates with a stabilizing protein Mastermind to displace corepressor proteins, thus activating the transcription of Notch target genes (Yoon and Gaiano 2005). The basic helix-loop-helix proteins' hairy/enhancer of split (such as Hes1, 5, and 7), Hes-related proteins (Hey1, 2, L), and EphrinB2 are the best characterized downstream targets. The Notch system was originally shown to be involved in neuronal differentiation and survival, as its signaling inhibits neuronal differentiation but maintains the pool of neural progenitors in the ventricular zone until the appropriate time (Lathia et al. 2008b). Notch also inhibits oligodendrocyte precursor differentiation, but does not interfere with astrocyte differentiation (Liu et al. 2006).

Notch ligands, especially dll4 and Jagged1, are induced upon stimulation with TLR ligands in DC and astrocytes (Morga et al. 2009; Mukherjee et al. 2009). The increases in glial fibrillary acidic protein (GFAP) and endothelin B (ETB), hallmarks of astrocyte activation or gliosis, have been attributed to Jagged1-Notch signaling since the suppression of Jagged1 repressed GFAP and ETB expression (Morga et al. 2009). In addition, inhibition of Jagged1 exerted anti-inflammatory effects on activated astrocytes by suppressing LPS-induced proinflammatory cytokine production through modulating IkBα phosphorylation (Morga et al. 2009). These data suggest that during CNS inflammation, Jagged/Notch signaling is an important event for astrocyte activation. On the other hand, the target proteins of Notch signaling, such as Hes5, have been shown to have inhibitory effects on remyelination due to interference with oligodendrocyte maturation (Liu et al. 2006), which could have a debilitating impact in demyelinating diseases like MS.

4.5 BLOOD–BRAIN BARRIER (BBB), ASTROCYTES, AND INFLAMMATION

The blood–brain barrier (BBB) is a complex structure that tightly regulates the entry of peripheral immune cells into the CNS under normal conditions. However, BBB disruption induced by infection, trauma, or autoimmunity is critical in initiating parenchymal inflammation. Although it is necessary for antimicrobial control, leukocyte access to CNS parenchyma can be detrimental, and is associated with chronic tissue damage and neuroinflammatory diseases. The structures forming the barrier are referred to as the neurovascular unit (NVU), and electron microscopy (EM) evaluation has determined that a healthy, intact NVU consists of cerebral endothelial cells (CECs), basal lamina, astrocytic endfoot processes, pericytes, and neurons (Hawkins and Davis 2005). Therefore, immune cells must breach several distinct physical barriers during their migration into the parenchyma (Owens et al. 2008; Ransohoff et al. 2003). The initial barrier is formed by the tight junctions between adjacent endothelial cells. Therefore, leukocytes at postcapillary venules must migrate across the vessel wall and enter the perivascular space. Once in the perivascular space, inflammatory cells must further penetrate the glia limitans to enter CNS parenchyma (Ransohoff et al. 2003). This barrier is composed of astrocyte foot processes associated with a distinct basement membrane (Janzer and Raff 1987). The process of leukocyte emigration into tissues involves a well-defined sequence of events, including tethering/rolling, chemokine-dependent activation, adhesion, and diapedesis, which is regulated by adhesion molecules, chemokines, and chemokine receptors (Engelhardt and Ransohoff 2005). It has been thought that activated lymphocytes migrate from the lumen to the perivascular area transcellularly without altering tight junctions (Engelhardt and Ransohoff 2005). However, tight junction disruption has been reported in some disease models where the BBB was compromised (Dimitrijevic et al. 2007). Tight junctions are composed of transmembranous proteins, including occludin, claudin-5, and the cytoplasmic proteins zona occludens 1, 2, and 3 (ZO-1, ZO-2, ZO-3) (Huber et al. 2001a, 2001b). Under certain conditions, CNS vascular permeability coincides with an alteration of CEC tight junctions and could be the main reason for mortality seen in rodent models of BBB breakdown (Johnson et al. 2005). It has been shown that antigen-specific $CD8^+$ T cells are responsible for the increased vascular permeability in TMEV encephalitis (Johnson et al. 2005). Removal of this antigen-specific population of $CD8^+$ T cells protected mice from becoming moribund, demonstrating the critical necessity of these cells in initiating this fatal condition (Johnson et al. 2001, 2005). Using specific KO mice and soluble receptor techniques, the authors determined that lethality from TEMV encephalitis was not mediated by obvious cytokine candidates, including TNF-α, lymphotoxin-β, or interleukin 1 (Johnson et al. 2005). Interestingly, perforin-deficient mice were resistant to CNS vascular permeability changes and perforin played a critical role in disruption of the NVU (Suidan et al. 2008).

Most studies have focused on the mechanisms by which the first barrier (i.e., endothelial cells) has been compromised during neuroinflammation. Our understanding of the factors governing migration through the glia limitans is not well established. The glia limitans is composed of astrocyte foot processes, which are in contact with 90% of abluminal capillary endothelial cells (CEC) (Abbott 2002). It has been widely accepted that the chemokine CCL2, produced by either astrocytes or endothelial cells, plays an important role in BBB integrity and is responsible for the disruption and increased infiltration of peripheral immune cells during CNS inflammation. CCL2 is associated with the decreased expression of tight junction proteins such as ZO-1, occludin, claudin-5, and increased adhesion molecules (Hoffman et al. 2009; Seguin et al. 2003; Song and Pachter 2004; Stamatovic et al. 2005; Van Der Voorn et al. 1999; Weiss et al. 1998). A recent study has shown that in the absence of CCL2, monocyte recruitment during acute viral encephalitis was disrupted, whereas neutrophil recruitment was not affected (Savarin et al. 2010). Interestingly, in CCL2-deficient mice, T cells had difficulty migrating deep into the parenchyma. Instead, T cells accumulated transiently in the perivascular space, which was associated with delayed disease onset and viral clearance (Savarin et al. 2010). Therefore, the authors suggested that monocytes, probably through the secretion of extracellular proteases distinct from matrix metalloproteinases (MMPs), have a direct effect on the glia limitans, creating a suitable environment for T cell migration. These studies further highlight the importance of astrocyte endfeet in the control of BBB integrity.

4.6 SUMMARY/PERSPECTIVES

A concept that emerges when considering the array of PRRs expressed by astrocytes is the marked degree of redundancy afforded by their receptor repertoire. For example, astrocytes possess key PRRs at the cell surface, endosomal compartments, and cytoplasm that are capable of sensing pathogens or endogenous danger signals. Therefore, although astrocytes are not classically considered as part of the CNS immune response, they are indeed capable of sensing and responding to pathogens. Since astrocytes represent the most numerous cell type in the CNS and are strategically located at the BBB, they are uniquely poised to provide an initial line of defense in sensing the presence of pathogenic motifs and initiating CNS inflammation. However, it is also important to note distinctions between astrocytes and microglia, the latter of which is well recognized as an innate immune effector cell in the CNS parenchyma. First, astrocytes express a more limited PRR repertoire than microglia, which may be explained by the fact that although astrocytes can participate in CNS immunity, this is not their primary role from an evolutionary perspective. Second, astrocytes are a major source of chemokines and, as such, play a critical role in peripheral immune cell recruitment into the CNS. In contrast, the ability of highly purified astrocytes to produce proinflammatory cytokines (i.e., TNF-α, IL-1β) is limited (Kielian and Liu, unpublished observations). Therefore, there are clear distinctions between the CNS immune potential of astrocytes from microglia, and because of the propensity for microglia to "contaminate" astrocyte cultures in vitro, great care should be taken when investigating the immune properties of astrocytes to avoid artifact signals (Saura 2007).

Traditionally, TLRs have been regarded as pathogen sensors, and as such, the early TLR literature in the CNS was focused on this topic. However, recent studies utilizing various TLR-deficient mouse strains have revealed that TLRs can also impact the course of distinct neurodegenerative diseases/pathologies (Babcock et al. 2006; Kigerl et al. 2007). Although the ligands responsible for triggering TLR involvement in the absence of infectious insults have not yet been elucidated, it is apparent that these PRRs play a role, at some level, in influencing the subsequent host immune response to injury/trauma and the subsequent regenerative response. Emerging evidence for the latter is provided by recent studies describing an important role for TLRs in neurogenesis and stem cell niches (Lathia et al. 2008a; Okun et al. 2010; Rolls et al. 2007). It is attractive to speculate that TLR engagement may direct stem cell populations toward a particular developmental pathway; in this instance there may be selective patterns leading to astrocyte development and formation of a glial scar during CNS injury, although this remains highly speculative.

There are several outstanding questions that remain with regard to astrocyte expression of PRRs and how they may impact CNS immune responses. First, a consensus is needed regarding the expression of TLRs whose expression in astrocytes is uncertain (i.e., TLR4). Second, additional insights into the astrocyte repertoire of TLRs expressed in vivo at the protein level are needed. Functional assessments regarding the role of astrocyte PRRs could be achieved with the use of Cre-Lox technology with floxed versions of PRRs targeted by either GFAP-Cre or nestin-Cre (the latter provides a more comprehensive, but not exclusive, deletion of target genes since not all astrocytes express GFAP). Collectively, based on their strategic location at the BBB and diffuse distribution throughout the CNS parenchyma, astrocytes have the potential to impact several facets of CNS inflammatory responses. A better understanding of the role of astrocytes during neuroinflammation may lead to the discovery of novel targets to thwart inappropriate inflammation and favor regenerative processes to restore CNS homeostasis.

ACKNOWLEDGMENTS

This work was supported by the NIH National Institute of Neurological Disorders and Stroke (R01s NS055385, NS40730, and NS053487) to T.K.

REFERENCES

Abbott, N. 2002. Astrocyte-endothelial interactions and blood-brain barrier permeability. *J Anat* 200 (5):527.

Alarcon, R., C. Fuenzalida, M. Santibanez, and R. von Bernhardi. 2005. Expression of scavenger receptors in glial cells. Comparing the adhesion of astrocytes and microglia from neonatal rats to surface-bound beta-amyloid. *J Biol Chem* 280 (34):30406–15.

Aloisi, F. 2001. Immune function of microglia. *Glia* 36 (2):165–79.

Aloisi, F. 2005. Cytokine production. In *Neuroglia*, ed. B. R. R. Helmut Kettenmann. New York: Oxford University Press.

Aloisi, F., F. Ria, S. Columba-Cabezas, H. Hess, G. Penna, and L. Adorini. 1999. Relative efficiency of microglia, astrocytes, dendritic cells and B cells in naive CD4+ T cell priming and Th1/Th2 cell restimulation. *Eur J Immunol* 29(9):2705–14.

Aloisi, F., F. Ria, G. Penna, and L. Adorini. 1998. Microglia are more efficient than astrocytes in antigen processing and in Th1 but not Th2 cell activation. *J Immunol* 160(10):4671–80.

Ambrosini, E., and F. Aloisi. 2004. Chemokines and glial cells: a complex network in the central nervous system. *Neurochem Res* 29 (5):1017–38.

Amiel, E., A. Alonso, S. Uematsu, S. Akira, M. E. Poynter, and B. Berwin. 2009. Pivotal advance: Toll-like receptor regulation of scavenger receptor-A-mediated phagocytosis. *J Leukoc Biol* 85 (4):595–605.

Amsen, D., A. Antov, and R. A. Flavell. 2009. The different faces of Notch in T-helper-cell differentiation. *Nat Rev Immunol* 9 (2):116–24.

Areschoug, T., and S. Gordon. 2009. Scavenger receptors: role in innate immunity and microbial pathogenesis. *Cell Microbiol* 11 (8):1160–69.

Avalos, A. M., L. Busconi, and A. Marshak-Rothstein. 2010. Regulation of autoreactive B cell responses to endogenous TLR ligands. *Autoimmunity* 43 (1):76–83.

Babcock, A. A., M. Wirenfeldt, T. Holm, H. H. Nielsen, L. Dissing-Olesen, H. Toft-Hansen, J. M. Millward, R. Landmann, S. Rivest, B. Finsen, and T. Owens. 2006. Toll-like receptor 2 signaling in response to brain injury: an innate bridge to neuroinflammation. *J Neurosci* 26 (49):12826–37.

Bailey, S. L., P. A. Carpentier, E. J. McMahon, W. S. Begolka, and S. D. Miller. 2006. Innate and adaptive immune responses of the central nervous system. *Crit Rev Immunol* 26 (2):149–88.

Balachandran, S., P. C. Roberts, L. E. Brown, H. Truong, A. K. Pattnaik, D. R. Archer, and G. N. Barber. 2000. Essential role for the dsRNA-dependent protein kinase PKR in innate immunity to viral infection. *Immunity* 13 (1):129–41.

Barbalat, R., L. Lau, R. M. Locksley, and G. M. Barton. 2009. Toll-like receptor 2 on inflammatory monocytes induces type I interferon in response to viral but not bacterial ligands. *Nat Immunol* 10 (11):1200–7.

Barnum, S. R., and J. L. Jones. 1995. Differential regulation of C3 gene expression in human astroglioma cells by interferon-gamma and interleukin-1 beta. *Neurosci Lett* 197 (2):121–24.

Barnum, S. R., J. L. Jones, and E. N. Benveniste. 1992. Interferon-gamma regulation of C3 gene expression in human astroglioma cells. *J Neuroimmunol* 38 (3):275–82.

Barton, G. M., J. C. Kagan, and R. Medzhitov. 2006. Intracellular localization of Toll-like receptor 9 prevents recognition of self DNA but facilitates access to viral DNA. *Nat Immunol* 7 (1):49–56.

Bettelli, E., M. Oukka, and V. K. Kuchroo. 2007. T(H)-17 cells in the circle of immunity and autoimmunity. *Nat Immunol* 8 (4):345–50.

Bianchi, M. E. 2007. DAMPs, PAMPs and alarmins: all we need to know about danger. *J Leukoc Biol* 81 (1):1–5.

Boos, L., A. J. Szalai, and S. R. Barnum. 2005. C3a expressed in the central nervous system protects against LPS-induced shock. *Neurosci Lett* 387 (2):68–71.

Bowman, C. C., A. Rasley, S. L. Tranguch, and I. Marriott. 2003. Cultured astrocytes express Toll-like receptors for bacterial products. *Glia* 43 (3):281–91.

Brahmachari, S., Y. K. Fung, and K. Pahan. 2006. Induction of glial fibrillary acidic protein expression in astrocytes by nitric oxide. *J Neurosci* 26 (18):4930–39.

Brandenburg, L. O., T. Koch, J. Sievers, and R. Lucius. 2007. Internalization of PrP106–126 by the formyl-peptide-receptor-like-1 in glial cells. *J Neurochem* 101 (3):718–28.

Brandenburg, L. O., M. Konrad, C. J. Wruck, T. Koch, R. Lucius, and T. Pufe. 2010. Functional and physical interactions between formyl-peptide-receptors and scavenger receptor MARCO and their involvement in amyloid beta 1–42-induced signal transduction in glial cells. *J Neurochem* 113(3):749–60.

Brandenburg, L. O., M. Konrad, C. Wruck, T. Koch, T. Pufe, and R. Lucius. 2008. Involvement of formyl-peptide-receptor-like-1 and phospholipase D in the internalization and signal transduction of amyloid beta 1–42 in glial cells. *Neuroscience* 156 (2):266–76.

Brown, A. M., S. Baltan Tekkok, and B. R. Ransom. 2004. Energy transfer from astrocytes to axons: the role of CNS glycogen. *Neurochem Int* 45 (4):529–36.

Bsibsi, M., J. J. Bajramovic, E. Van Duijvenvoorden, C. Persoon, R. Ravid, J. M. Van Noort, and M. H. Vogt. 2007. Identification of soluble CD14 as an endogenous agonist for Toll-like receptor 2 on human astrocytes by genome-scale functional screening of glial cell derived proteins. *Glia* 55 (5):473–82.

Bsibsi, M., C. Persoon-Deen, R. W. Verwer, S. Meeuwsen, R. Ravid, and J. M. Van Noort. 2006. Toll-like receptor 3 on adult human astrocytes triggers production of neuroprotective mediators. *Glia* 53 (7):688–95.

Bsibsi, M., R. Ravid, D. Gveric, and J. M. van Noort. 2002. Broad expression of Toll-like receptors in the human central nervous system. *J Neuropathol Exp Neurol* 61 (11):1013–21.

Burudi, E. M., S. Riese, P. D. Stahl, and A. Regnier-Vigouroux. 1999. Identification and functional characterization of the mannose receptor in astrocytes. *Glia* 25 (1):44–55.

Butchi, N. B., M. Du, and K. E. Peterson. 2010. Interactions between TLR7 and TLR9 agonists and receptors regulate innate immune responses by astrocytes and microglia. *Glia* 58 (6):650–64.

Cardona, A. E., M. Li, L. Liu, C. Savarin, and R. M. Ransohoff. 2008. Chemokines in and out of the central nervous system: much more than chemotaxis and inflammation. *J Leukoc Biol* 84 (3):587–94.

Carpentier, P. A., W. S. Begolka, J. K. Olson, A. Elhofy, W. J. Karpus, and S. D. Miller. 2005. Differential activation of astrocytes by innate and adaptive immune stimuli. *Glia* 49 (3):360–74.

Carpentier, P. A., B. R. Williams, and S. D. Miller. 2007. Distinct roles of protein kinase R and Toll-like receptor 3 in the activation of astrocytes by viral stimuli. *Glia* 55 (3):239–52.

Carson, M. J., C. R. Reilly, J. G. Sutcliffe, and D. Lo. 1998. Mature microglia resemble immature antigen-presenting cells. *Glia* 22 (1):72–85.

Carta, S., P. Castellani, L. Delfino, S. Tassi, R. Vene, and A. Rubartelli. 2009. DAMPs and inflammatory processes: the role of redox in the different outcomes. *J Leukoc Biol* 86 (3):549–55.

Cauwels, A., K. Frei, S. Sansano, C. Fearns, R. Ulevitch, W. Zimmerli, and R. Landmann. 1999. The origin and function of soluble CD14 in experimental bacterial meningitis. *J Immunol* 162 (8):4762–72.

Chakravarty, S., and M. Herkenham. 2005. Toll-like receptor 4 on nonhematopoietic cells sustains CNS inflammation during endotoxemia, independent of systemic cytokines. *J Neurosci* 25 (7):1788–96.

Chang, G. H., N. M. Barbaro, and R. O. Pieper. 2000. Phosphatidylserine-dependent phagocytosis of apoptotic glioma cells by normal human microglia, astrocytes, and glioma cells. *Neuro Oncol* 2 (3):174–83.

Chauhan, V. S., D. G. Sterka, Jr., S. R. Furr, A. B. Young, and I. Marriott. 2009. NOD2 plays an important role in the inflammatory responses of microglia and astrocytes to bacterial CNS pathogens. *Glia* 57 (4):414–23.

Chen, Y., and R. A. Swanson. 2003. Astrocytes and brain injury. *J Cereb Blood Flow Metab* 23 (2):137–49.

Chun, K. H., and S. Y. Seong. 2010. CD14 but not MD2 transmit signals from DAMP. *Int Immunopharmacol* 10 (1):98–106.

Coraci, I. S., J. Husemann, J. W. Berman, C. Hulette, J. H. Dufour, G. K. Campanella, A. D. Luster, S. C. Silverstein, and J. B. El-Khoury. 2002. CD36, a class B scavenger receptor, is expressed on microglia in Alzheimer's disease brains and can mediate production of reactive oxygen species in response to beta-amyloid fibrils. *Am J Pathol* 160 (1):101–12.

Dalpke, A. H., M. K. Schafer, M. Frey, S. Zimmermann, J. Tebbe, E. Weihe, and K. Heeg. 2002. Immunostimulatory CpG-DNA activates murine microglia. *J Immunol* 168 (10):4854–63.

Delaloye, J., T. Roger, Q. G. Steiner-Tardivel, D. Le Roy, M. Knaup Reymond, S. Akira, V. Petrilli, C. E. Gomez, B. Perdiguero, J. Tschopp, G. Pantaleo, M. Esteban, and T. Calandra. 2009. Innate immune sensing of modified vaccinia virus Ankara (MVA) is mediated by TLR2-TLR6, MDA-5 and the NALP3 inflammasome. *PLoS Pathog* 5 (6):e1000480.

De Miranda, J., K. Yaddanapudi, M. Hornig, and W. I. Lipkin. 2009. Astrocytes recognize intracellular polyinosinic-polycytidylic acid via MDA-5. *FASEB J* 23 (4):1064–71.

Dimitrijevic, O. B., S. M. Stamatovic, R. F. Keep, and A. V. Andjelkovic. 2007. Absence of the chemokine receptor CCR2 protects against cerebral ischemia/reperfusion injury in mice. *Stroke* 38 (4):1345–53.

Dobrovolskaia, M. A., and S. N. Vogel. 2002. Toll receptors, CD14, and macrophage activation and deactivation by LPS. *Microbes Infect* 4 (9):903–14.

Dong, Y., and E. N. Benveniste. 2001. Immune function of astrocytes. *Glia* 36 (2):180–90.

Engelhardt, B., and R. M. Ransohoff. 2005. The ins and outs of T-lymphocyte trafficking to the CNS: anatomical sites and molecular mechanisms. *Trends Immunol* 26 (9):485–95.

Esen, N., and T. Kielian. 2005. Recognition of *Staphylococcus aureus*-derived peptidoglycan (PGN) but not intact bacteria is mediated by CD14 in microglia. *J Neuroimmunol* 170 (1–2):93–104.

Esen, N., and T. Kielian. 2006. Central role for MyD88 in the responses of microglia to patho-gen-associated molecular patterns. *J Immunol* 176 (11):6802–11.

Esen, N., F. Y. Tanga, J. A. DeLeo, and T. Kielian. 2004. Toll-like receptor 2 (TLR2) mediates astrocyte activation in response to the Gram-positive bacterium *Staphylococcus aureus*. *J Neurochem* 88 (3):746–58.

Farina, C., F. Aloisi, and E. Meinl. 2007. Astrocytes are active players in cerebral innate immu-nity. *Trends Immunol* 28 (3):138–45.

Farina, C., M. Krumbholz, T. Giese, G. Hartmann, F. Aloisi, and E. Meinl. 2005. Preferential expression and function of Toll-like receptor 3 in human astrocytes. *J Neuroimmunol* 159 (1–2):12–19.

Fournier, B., and D. J. Philpott. 2005. Recognition of *Staphylococcus aureus* by the innate immune system. *Clin Microbiol Rev* 18 (3):521–40.

Fukui, R., S. Saitoh, F. Matsumoto, H. Kozuka-Hata, M. Oyama, K. Tabeta, B. Beutler, and K. Miyake. 2009. Unc93B1 biases Toll-like receptor responses to nucleic acid in dendritic cells toward DNA- but against RNA-sensing. *J Exp Med* 206 (6):1339–50.

Garlanda, C., E. Hirsch, S. Bozza, A. Salustri, M. De Acetis, R. Nota, A. Maccagno, F. Riva, B. Bottazzi, G. Peri, A. Doni, L. Vago, M. Botto, R. De Santis, P. Carminati, G. Siracusa, F. Altruda, A. Vecchi, L. Romani, and A. Mantovani. 2002. Non-redundant role of the long pentraxin PTX3 in anti-fungal innate immune response. *Nature* 420 (6912):182–86.

Gasque, P., P. Chan, M. Fontaine, A. Ischenko, M. Lamacz, O. Gotze, and B. P. Morgan. 1995b. Identification and characterization of the complement C5a anaphylatoxin recep-tor on human astrocytes. *J Immunol* 155 (10):4882–89.

Gasque, P., P. Chan, C. Mauger, M. T. Schouft, S. Singhrao, M. P. Dierich, B. P. Morgan, and M. Fontaine. 1996b. Identification and characterization of complement C3 receptors on human astrocytes. *J Immunol* 156 (6):2247–55.

Gasque, P., Y. D. Dean, E. P. McGreal, J. VanBeek, and B. P. Morgan. 2000. Complement components of the innate immune system in health and disease in the CNS. *Immunopharmacology* 49 (1–2):171–86.

Gasque, P., M. Fontaine, and B. P. Morgan. 1995a. Complement expression in human brain. Biosynthesis of terminal pathway components and regulators in human glial cells and cell lines. *J Immunol* 154 (9):4726–33.

Gasque, P., A. Ischenko, J. Legoedec, C. Mauger, M. T. Schouft, and M. Fontaine. 1993. Expression of the complement classical pathway by human glioma in culture. A model for complement expression by nerve cells. *J Biol Chem* 268 (33):25068–74.

Gasque, P., and B. P. Morgan. 1996. Complement regulatory protein expression by a human oli-godendrocyte cell line: cytokine regulation and comparison with astrocytes. *Immunology* 89 (3):338–47.

Girardin, S. E., I. G. Boneca, J. Viala, M. Chamaillard, A. Labigne, G. Thomas, D. J. Philpott, and P. J. Sansonetti. 2003. Nod2 is a general sensor of peptidoglycan through muramyl dipeptide (MDP) detection. *J Biol Chem* 278 (11):8869–72.

Gorina, R., T. Santalucia, V. Petegnief, A. Ejarque-Ortiz, J. Saura, and A. M. Planas. 2009. Astrocytes are very sensitive to develop innate immune responses to lipid-carried short interfering RNA. *Glia* 57 (1):93–107.

Griffiths, M. R., P. Gasque, and J. W. Neal. 2009. The multiple roles of the innate immune system in the regulation of apoptosis and inflammation in the brain. *J Neuropathol Exp Neurol* 68 (3):217–26.

Hawkins, B. T., and T. P. Davis. 2005. The blood-brain barrier/neurovascular unit in health and disease. *Pharmacol Rev* 57 (2):173–85.

He, B., X. Qiao, and A. Cerutti. 2004. CpG DNA induces IgG class switch DNA recombination by activating human B cells through an innate pathway that requires TLR9 and cooperates with IL-10. *J Immunol* 173 (7):4479–91.

Heine, H., C. J. Kirschning, E. Lien, B. G. Monks, M. Rothe, and D. T. Golenbock. 1999. Cutting edge: cells that carry A null allele for Toll-like receptor 2 are capable of responding to endotoxin. *J Immunol* 162 (12):6971–75.

Henneke, P., O. Takeuchi, R. Malley, E. Lien, R. R. Ingalls, M. W. Freeman, T. Mayadas, V. Nizet, S. Akira, D. L. Kasper, and D. T. Golenbock. 2002. Cellular activation, phagocytosis, and bactericidal activity against group B streptococcus involve parallel myeloid differentiation factor 88-dependent and -independent signaling pathways. *J Immunol* 169 (7):3970–77.

Hirschfeld, M., Y. Ma, J. H. Weis, S. N. Vogel, and J. J. Weis. 2000. Cutting edge: repurification of lipopolysaccharide eliminates signaling through both human and murine Toll-like receptor 2. *J Immunol* 165 (2):618–22.

Hoffman, W. H., S. M. Stamatovic, and A. V. Andjelkovic. 2009. Inflammatory mediators and blood brain barrier disruption in fatal brain edema of diabetic ketoacidosis. *Brain Res* 1254:138–48.

Honda, K., A. Takaoka, and T. Taniguchi. 2006. Type I interferon [corrected] gene induction by the interferon regulatory factor family of transcription factors. *Immunity* 25 (3):349–60.

Hoshino, K., O. Takeuchi, T. Kawai, H. Sanjo, T. Ogawa, Y. Takeda, K. Takeda, and S. Akira. 1999. Cutting edge: Toll-like receptor 4 (TLR4)-deficient mice are hyporesponsive to lipopolysaccharide: evidence for TLR4 as the Lps gene product. *J Immunol* 162 (7):3749–52.

Hosoi, T., S. Suzuki, J. Nomura, A. Ono, Y. Okuma, S. Akira, and Y. Nomura. 2004. Bacterial DNA induced iNOS expression through MyD88-p38 MAP kinase in mouse primary cultured glial cells. *Brain Res Mol Brain Res* 124 (2):159–64.

Huber, J. D., R. D. Egleton, and T. P. Davis. 2001a. Molecular physiology and pathophysiology of tight junctions in the blood-brain barrier. *Trends Neurosci* 24 (12):719–25.

Huber, J. D., K. A. Witt, S. Hom, R. D. Egleton, K. S. Mark, and T. P. Davis. 2001b. Inflammatory pain alters blood-brain barrier permeability and tight junctional protein expression. *Am J Physiol Heart Circ Physiol* 280 (3):H1241–48.

Husemann, J., J. D. Loike, R. Anankov, M. Febbraio, and S. C. Silverstein. 2002. Scavenger receptors in neurobiology and neuropathology: their role on microglia and other cells of the nervous system. *Glia* 40 (2):195–205.

Husemann, J., J. D. Loike, T. Kodama, and S. C. Silverstein. 2001. Scavenger receptor class B type I (SR-BI) mediates adhesion of neonatal murine microglia to fibrillar beta-amyloid. *J Neuroimmunol* 114 (1–2):142–50.

Iliev, A. I., A. K. Stringaris, R. Nau, and H. Neumann. 2004. Neuronal injury mediated via stimulation of microglial Toll-like receptor-9 (TLR9). *FASEB J* 18 (2):412–14.

Itoh, F., S. Itoh, M. J. Goumans, G. Valdimarsdottir, T. Iso, G. P. Dotto, Y. Hamamori, L. Kedes, M. Kato, and P. ten Dijke Pt. 2004. Synergy and antagonism between Notch and BMP receptor signaling pathways in endothelial cells. *EMBO J* 23 (3):541–51.

Jack, C. S., N. Arbour, J. Manusow, V. Montgrain, M. Blain, E. McCrea, A. Shapiro, and J. P. Antel. 2005. TLR signaling tailors innate immune responses in human microglia and astrocytes. *J Immunol* 175 (7):4320–30.

Janzer, R. C., and M. C. Raff. 1987. Astrocytes induce blood-brain barrier properties in endothelial cells. *Nature* 325 (6101):253–57.

Jegerlehner, A., P. Maurer, J. Bessa, H. J. Hinton, M. Kopf, and M. F. Bachmann. 2007. TLR9 signaling in B cells determines class switch recombination to IgG2a. *J Immunol* 178 (4):2415–20.

Johann, S., E. Kampmann, B. Denecke, S. Arnold, M. Kipp, J. Mey, and C. Beyer. 2008. Expression of enzymes involved in the prostanoid metabolism by cortical astrocytes after LPS-induced inflammation. *J Mol Neurosci* 34 (2):177–85.

John, G. R., S. C. Lee, and C. F. Brosnan. 2003. Cytokines: powerful regulators of glial cell activation. *Neuroscientist* 9 (1):10–22.

Johnson, A. J., Y. Mendez-Fernandez, A. M. Moyer, C. R. Sloma, I. Pirko, M. S. Block, M. Rodriguez, and L. R. Pease. 2005. Antigen-specific CD8+ T cells mediate a peptide-induced fatal syndrome. *J Immunol* 174 (11):6854–62.

Johnson, A. J., J. Upshaw, K. D. Pavelko, M. Rodriguez, and L. R. Pease. 2001. Preservation of motor function by inhibition of CD8+ virus peptide-specific T cells in Theiler's virus infection. *FASEB J* 15 (14):2760–62.

Kagan, J. C., T. Su, T. Horng, A. Chow, S. Akira, and R. Medzhitov. 2008. TRAM couples endocytosis of Toll-like receptor 4 to the induction of interferon-beta. *Nat Immunol* 9 (4):361–68.

Kang, Z., C. Z. Altuntas, M. F. Gulen, C. Liu, N. Giltiay, H. Qin, L. Liu, W. Qian, R. M. Ransohoff, C. Bergmann, S. Stohlman, V. K. Tuohy, and X. Li. 2010. Astrocyte-restricted ablation of interleukin-17-induced Act1-mediated signaling ameliorates autoimmune encephalomyelitis. *Immunity* 32 (3):414–25.

Kato, H., O. Takeuchi, S. Sato, M. Yoneyama, M. Yamamoto, K. Matsui, S. Uematsu, A. Jung, T. Kawai, K. J. Ishii, O. Yamaguchi, K. Otsu, T. Tsujimura, C. S. Koh, C. Reis e Sousa, Y. Matsuura, T. Fujita, and S. Akira. 2006. Differential roles of MDA5 and RIG-I helicases in the recognition of RNA viruses. *Nature* 441 (7089):101–5.

Kawai, T., and S. Akira. 2008. Toll-like receptor and RIG-I-like receptor signaling. *Ann NY Acad Sci* 1143:1–20.

Kawai, T., K. Takahashi, S. Sato, C. Coban, H. Kumar, H. Kato, K. J. Ishii, O. Takeuchi, and S. Akira. 2005. IPS-1, an adaptor triggering RIG-I- and Mda5-mediated type I interferon induction. *Nat Immunol* 6 (10):981–88.

Kielian, T. 2006. Toll-like receptors in central nervous system glial inflammation and homeostasis. *J Neurosci Res* 83 (5):711–30.

Kielian, T., N. Esen, and E. D. Bearden. 2005. Toll-like receptor 2 (TLR2) is pivotal for recognition of *S. aureus* peptidoglycan but not intact bacteria by microglia. *Glia* 49 (4):567–76.

Kigerl, K. A., W. Lai, S. Rivest, R. P. Hart, A. R. Satoskar, and P. G. Popovich. 2007. Toll-like receptor (TLR)-2 and TLR-4 regulate inflammation, gliosis, and myelin sparing after spinal cord injury. *J Neurochem* 102 (1):37–50.

Kordula, T., R. E. Rydel, E. F. Brigham, F. Horn, P. C. Heinrich, and J. Travis. 1998. Oncostatin M and the interleukin-6 and soluble interleukin-6 receptor complex regulate alpha1-antichymotrypsin expression in human cortical astrocytes. *J Biol Chem* 273 (7):4112–18.

Krieg, A. M., and J. Vollmer. 2007. Toll-like receptors 7, 8, and 9: linking innate immunity to autoimmunity. *Immunol Rev* 220:251–69.

Krumbholz, M., U. Specks, M. Wick, S. L. Kalled, D. Jenne, and E. Meinl. 2005. BAFF is elevated in serum of patients with Wegener's granulomatosis. *J Autoimmun* 25 (4):298–302.

Laflamme, N., and S. Rivest. 2001. Toll-like receptor 4: the missing link of the cerebral innate immune response triggered by circulating gram-negative bacterial cell wall components. *FASEB J* 15 (1):155–163.

Laflamme, N., G. Soucy, and S. Rivest. 2001. Circulating cell wall components derived from gram-negative, not gram-positive, bacteria cause a profound induction of the gene-encoding Toll-like receptor 2 in the CNS. *J Neurochem* 79 (3):648–57.

Lande, R., J. Gregorio, V. Facchinetti, B. Chatterjee, Y. H. Wang, B. Homey, W. Cao, B. Su, F. O. Nestle, T. Zal, I. Mellman, J. M. Schroder, Y. J. Liu, and M. Gilliet. 2007. Plasmacytoid dendritic cells sense self-DNA coupled with antimicrobial peptide. *Nature* 449 (7162):564–69.

Lathia, J. D., M. P. Mattson, and A. Cheng. 2008b. Notch: from neural development to neurological disorders. *J Neurochem* 107 (6):1471–81.

Lathia, J. D., E. Okun, S. C. Tang, K. Griffioen, A. Cheng, M. R. Mughal, G. Laryea, P. K. Selvaraj, C. ffrench-Constant, T. Magnus, T. V. Arumugam, and M. P. Mattson. 2008a. Toll-like receptor 3 is a negative regulator of embryonic neural progenitor cell proliferation. *J Neurosci* 28 (51):13978–84.

Lee, S., J. Hong, S. Y. Choi, S. B. Oh, K. Park, J. S. Kim, M. Karin, and S. J. Lee. 2004. CpG oligodeoxynucleotides induce expression of proinflammatory cytokines and chemokines in astrocytes: the role of c-Jun N-terminal kinase in CpG ODN-mediated NF-kappaB activation. *J Neuroimmunol* 153 (1–2):50–63.

Lehnardt, S., C. Lachance, S. Patrizi, S. Lefebvre, P. L. Follett, F. E. Jensen, P. A. Rosenberg, J. J. Volpe, and T. Vartanian. 2002. The Toll-like receptor TLR4 is necessary for lipopolysaccharide-induced oligodendrocyte injury in the CNS. *J Neurosci* 22 (7):2478–86.

Lehnardt, S., L. Massillon, P. Follett, F. E. Jensen, R. Ratan, P. A. Rosenberg, J. J. Volpe, and T. Vartanian. 2003. Activation of innate immunity in the CNS triggers neurodegeneration through a Toll-like receptor 4-dependent pathway. *Proc Natl Acad Sci USA* 100 (14):8514–19.

Levin, D. H., R. Petryshyn, and I. M. London. 1980. Characterization of double-stranded-RNA-activated kinase that phosphorylates alpha subunit of eukaryotic initiation factor 2 (eIF-2 alpha) in reticulocyte lysates. *Proc Natl Acad Sci USA* 77 (2):832–36.

Lien, E., T. K. Means, H. Heine, A. Yoshimura, S. Kusumoto, K. Fukase, M. J. Fenton, M. Oikawa, N. Qureshi, B. Monks, R. W. Finberg, R. R. Ingalls, and D. T. Golenbock. 2000. Toll-like receptor 4 imparts ligand-specific recognition of bacterial lipopolysaccharide. *J Clin Invest* 105 (4):497–504.

Lin, B., R. Noring, A. C. Steere, M. S. Klempner, and L. T. Hu. 2000. Soluble CD14 levels in the serum, synovial fluid, and cerebrospinal fluid of patients with various stages of Lyme disease. *J Infect Dis* 181 (3):1185–88.

Liu, A., J. Li, M. Marin-Husstege, R. Kageyama, Y. Fan, C. Gelinas, and P. Casaccia-Bonnefil. 2006. A molecular insight of Hes5-dependent inhibition of myelin gene expression: old partners and new players. *EMBO J* 25 (20):4833–42.

Liu, X., V. S. Chauhan, and I. Marriott. 2010b. NOD2 contributes to the inflammatory responses of primary murine microglia and astrocytes to *Staphylococcus aureus*. *Neurosci Lett* 474 (2):93–98.

Liu, X., V. S. Chauhan, A. B. Young, and I. Marriott. 2010a. NOD2 mediates inflammatory responses of primary murine glia to *Streptococcus pneumoniae*. *Glia* 58 (7):839–47.

Liu, Y., H. Liu, B. O. Kim, V. H. Gattone, J. Li, A. Nath, J. Blum, and J. J. He. 2004. CD4-independent infection of astrocytes by human immunodeficiency virus type 1: requirement for the human mannose receptor. *J Virol* 78 (8):4120–33.

Lopez-Herrera, A., Y. Liu, M. T. Rugeles, and J. J. He. 2005. HIV-1 interaction with human mannose receptor (hMR) induces production of matrix metalloproteinase 2 (MMP-2) through hMR-mediated intracellular signaling in astrocytes. *Biochim Biophys Acta* 1741 (1–2):55–64.

Lynn, W. A., Y. Liu, and D. T. Golenbock. 1993. Neither CD14 nor serum is absolutely necessary for activation of mononuclear phagocytes by bacterial lipopolysaccharide. *Infect Immun* 61 (10):4452–61.

Maier, M., Y. Peng, L. Jiang, T. J. Seabrook, M. C. Carroll, and C. A. Lemere. 2008. Complement C3 deficiency leads to accelerated amyloid beta plaque deposition and neurodegeneration and modulation of the microglia/macrophage phenotype in amyloid precursor protein transgenic mice. *J Neurosci* 28 (25):6333–41.

Mantovani, A., C. Garlanda, and B. Bottazzi. 2003. Pentraxin 3, a non-redundant soluble pattern recognition receptor involved in innate immunity. *Vaccine* 21 (Suppl 2):S43–47.

Manukyan, M., K. Triantafilou, M. Triantafilou, A. Mackie, N. Nilsen, T. Espevik, K. H. Wiesmuller, A. J. Ulmer, and H. Heine. 2005. Binding of lipopeptide to CD14 induces physical proximity of CD14, TLR2 and TLR1. *Eur J Immunol* 35 (3):911–21.

Martinez, J., X. Huang, and Y. Yang. 2010. Direct TLR2 signaling is critical for NK cell activation and function in response to vaccinia viral infection. *PLoS Pathog* 6 (3):e1000811.

Matsuoka, Y., M. Picciano, J. La Francois, and K. Duff. 2001. Fibrillar beta-amyloid evokes oxidative damage in a transgenic mouse model of Alzheimer's disease. *Neuroscience* 104 (3):609–13.

Medzhitov, R. 2001. Toll-like receptors and innate immunity. *Nat Rev Immunol* 1 (2):135–45.

Mishra, B. B., P. K. Mishra, and J. M. Teale. 2006. Expression and distribution of Toll-like receptors in the brain during murine neurocysticercosis. *J Neuroimmunol* 181 (1–2):46–56.

Morga, E., L. Mouad-Amazzal, P. Felten, T. Heurtaux, M. Moro, A. Michelucci, S. Gabel, L. Grandbarbe, and P. Heuschling. 2009. Jagged1 regulates the activation of astrocytes via modulation of NFkappaB and JAK/STAT/SOCS pathways. *Glia* 57 (16):1741–53.

Mukherjee, S., M. A. Schaller, R. Neupane, S. L. Kunkel, and N. W. Lukacs. 2009. Regulation of T cell activation by Notch ligand, DLL4, promotes IL-17 production and Rorc activation. *J Immunol* 182 (12):7381–88.

Mukhopadhyay, S., Y. Chen, M. Sankala, L. Peiser, T. Pikkarainen, G. Kraal, K. Tryggvason, and S. Gordon. 2006. MARCO, an innate activation marker of macrophages, is a class A scavenger receptor for *Neisseria meningitidis*. *Eur J Immunol* 36 (4):940–49.

Murphy, J. E., P. R. Tedbury, S. Homer-Vanniasinkam, J. H. Walker, and S. Ponnambalam. 2005. Biochemistry and cell biology of mammalian scavenger receptors. *Atherosclerosis* 182 (1):1–15.

Muruve, D. A., V. Petrilli, A. K. Zaiss, L. R. White, S. A. Clark, P. J. Ross, R. J. Parks, and J. Tschopp. 2008. The inflammasome recognizes cytosolic microbial and host DNA and triggers an innate immune response. *Nature* 452 (7183):103–7.

Nakamura, K., W. Ohya, H. Funakoshi, G. Sakaguchi, A. Kato, M. Takeda, T. Kudo, and T. Nakamura. 2006. Possible role of scavenger receptor SRCL in the clearance of amyloid-beta in Alzheimer's disease. *J Neurosci Res* 84 (4):874–90.

Nauta, A. J., B. Bottazzi, A. Mantovani, G. Salvatori, U. Kishore, W. J. Schwaeble, A. R. Gingras, S. Tzima, F. Vivanco, J. Egido, O. Tijsma, E. C. Hack, M. R. Daha, and A. Roos. 2003. Biochemical and functional characterization of the interaction between pentraxin 3 and C1q. *Eur J Immunol* 33 (2):465–73.

Neumann, H. 2001. Control of glial immune function by neurons. *Glia* 36 (2):191–99.

Neumann, H., and H. Wekerle. 1998. Neuronal control of the immune response in the central nervous system: linking brain immunity to neurodegeneration. *J Neuropathol Exp Neurol* 57 (1):1–9.

Okun, E., K. J. Griffioen, T. Gen Son, J. H. Lee, N. J. Roberts, M. R. Mughal, E. Hutchison, A. Cheng, T. V. Arumugam, J. D. Lathia, H. van Praag, and M. P. Mattson. 2010. TLR2 activation inhibits embryonic neural progenitor cell proliferation. *J Neurochem* 114(2):462–74.

Olson, J. K., and S. D. Miller. 2004. Microglia initiate central nervous system innate and adaptive immune responses through multiple TLRs. *J Immunol* 173 (6):3916–24.

Orr, S. L., and P. Tobias. 2000. LPS and LAM activation of the U373 astrocytoma cell line: differential requirement for CD14. *J Endotoxin Res* 6 (3):215–22.

Owens, T. 2005. Toll-like receptors on astrocytes: patterning for immunity. *J Neuroimmunol* 159 (1–2):1–2.

Owens, T., I. Bechmann, and B. Engelhardt. 2008. Perivascular spaces and the two steps to neuroinflammation. *J Neuropathol Exp Neurol* 67 (12):1113–21.

Palsson-McDermott, E. M., and L. A. O'Neill. 2004. Signal transduction by the lipopolysaccharide receptor, Toll-like receptor-4. *Immunology* 113 (2):153–62.

Park, C., S. Lee, I. H. Cho, H. K. Lee, D. Kim, S. Y. Choi, S. B. Oh, K. Park, J. S. Kim, and S. J. Lee. 2006. TLR3-mediated signal induces proinflammatory cytokine and chemokine gene expression in astrocytes: differential signaling mechanisms of TLR3-induced IP-10 and IL-8 gene expression. *Glia* 53 (3):248–56.

Perea, G., M. Navarrete, and A. Araque. 2009. Tripartite synapses: astrocytes process and control synaptic information. *Trends Neurosci* 32 (8):421–31.

Perera, P. Y., S. N. Vogel, G. R. Detore, A. Haziot, and S. M. Goyert. 1997. CD14-dependent and CD14-independent signaling pathways in murine macrophages from normal and CD14 knockout mice stimulated with lipopolysaccharide or taxol. *J Immunol* 158 (9):4422–29.

Poltorak, A., X. He, I. Smirnova, M. Y. Liu, C. Van Huffel, X. Du, D. Birdwell, E. Alejos, M. Silva, C. Galanos, M. Freudenberg, P. Ricciardi-Castagnoli, B. Layton, and B. Beutler. 1998. Defective LPS signaling in C3H/HeJ and C57BL/10ScCr mice: mutations in Tlr4 gene. *Science* 282 (5396):2085–88.

Qureshi, S. T., P. Gros, and D. Malo. 1999. The Lps locus: genetic regulation of host responses to bacterial lipopolysaccharide. *Inflamm Res* 48 (12):613–20.

Rahpeymai, Y., M. A. Hietala, U. Wilhelmsson, A. Fotheringham, I. Davies, A. K. Nilsson, J. Zwirner, R. A. Wetsel, C. Gerard, M. Pekny, and M. Pekna. 2006. Complement: a novel factor in basal and ischemia-induced neurogenesis. *EMBO J* 25 (6):1364–74.

Rallabhandi, P., J. Bell, M. S. Boukhvalova, A. Medvedev, M. Lorenz, M. Arditi, V. G. Hemming, J. C. Blanco, D. M. Segal, and S. N. Vogel. 2006. Analysis of TLR4 polymorphic variants: new insights into TLR4/MD-2/CD14 stoichiometry, structure, and signaling. *J Immunol* 177 (1):322–32.

Ransohoff, R. M., P. Kivisakk, and G. Kidd. 2003. Three or more routes for leukocyte migration into the central nervous system. *Nat Rev Immunol* 3 (7):569–81.

Ravizza, T., D. Moneta, B. Bottazzi, G. Peri, C. Garlanda, E. Hirsch, G. J. Richards, A. Mantovani, and A. Vezzani. 2001. Dynamic induction of the long pentraxin PTX3 in the CNS after limbic seizures: evidence for a protective role in seizure-induced neurodegeneration. *Neuroscience* 105 (1):43–53.

Remer, K. A., M. Brcic, K. S. Sauter, and T. W. Jungi. 2006. Human monocytoid cells as a model to study Toll-like receptor-mediated activation. *J Immunol Methods* 313 (1–2):1–10.

Rivest, S. 2003. Molecular insights on the cerebral innate immune system. *Brain Behav Immun* 17 (1):13–19.

Rogers, J., R. Li, D. Mastroeni, A. Grover, B. Leonard, G. Ahern, P. Cao, H. Kolody, L. Vedders, W. P. Kolb, and M. Sabbagh. 2006. Peripheral clearance of amyloid beta peptide by complement C3-dependent adherence to erythrocytes. *Neurobiol Aging* 27 (12):1733–39.

Rolls, A., R. Shechter, A. London, Y. Ziv, A. Ronen, R. Levy, and M. Schwartz. 2007. Toll-like receptors modulate adult hippocampal neurogenesis. *Nat Cell Biol* 9 (9):1081–88.

Rutkowski, M. J., M. E. Sughrue, A. J. Kane, S. A. Mills, S. Fang, and A. T. Parsa. 2010. Complement and the central nervous system: emerging roles in development, protection and regeneration. *Immunol Cell Biol* 88(8):781–6.

Sasaki, N., S. Toki, H. Chowei, T. Saito, N. Nakano, Y. Hayashi, M. Takeuchi, and Z. Makita. 2001. Immunohistochemical distribution of the receptor for advanced glycation end products in neurons and astrocytes in Alzheimer's disease. *Brain Res* 888 (2):256–62.

Satoh, T., H. Kato, Y. Kumagai, M. Yoneyama, S. Sato, K. Matsushita, T. Tsujimura, T. Fujita, S. Akira, and O. Takeuchi. 2010. LGP2 is a positive regulator of RIG-I- and MDA5-mediated antiviral responses. *Proc Natl Acad Sci USA* 107 (4):1512–17.

Saura, J. 2007. Microglial cells in astroglial cultures: a cautionary note. *J Neuroinflammation* 4:26.

Savarin, C., S. A. Stohlman, R. Atkinson, R. M. Ransohoff, and C. C. Bergmann. 2010. Monocytes regulate T cell migration through the glia limitans during acute viral encephalitis. *J Virol.* 84(10):4878–88.

Sayah, S., A. M. Ischenko, A. Zhakhov, A. S. Bonnard, and M. Fontaine. 1999. Expression of cytokines by human astrocytomas following stimulation by C3a and C5a anaphylatoxins: specific increase in interleukin-6 mRNA expression. *J Neurochem* 72 (6):2426–36.

Sayah, S., A. C. Jauneau, C. Patte, M. C. Tonon, H. Vaudry, and M. Fontaine. 2003. Two different transduction pathways are activated by C3a and C5a anaphylatoxins on astrocytes. *Brain Res Mol Brain Res* 112 (1–2):53–60.

Scumpia, P. O., K. M. Kelly, W. H. Reeves, and B. R. Stevens. 2005. Double-stranded RNA signals antiviral and inflammatory programs and dysfunctional glutamate transport in TLR3-expressing astrocytes. *Glia* 52 (2):153–62.

Seguin, R., K. Biernacki, R. L. Rotondo, A. Prat, and J. P. Antel. 2003. Regulation and functional effects of monocyte migration across human brain-derived endothelial cells. *J Neuropathol Exp Neurol* 62 (4):412–19.

Shinjyo, N., A. Stahlberg, M. Dragunow, M. Pekny, and M. Pekna. 2009. Complement-derived anaphylatoxin C3a regulates *in vitro* differentiation and migration of neural progenitor cells. *Stem Cells* 27 (11):2824–32.

Siemasko, K., and M. R. Clark. 2001. The control and facilitation of MHC class II antigen processing by the BCR. *Curr Opin Immunol* 13 (1):32–36.

So, E. Y., M. H. Kang, and B. S. Kim. 2006. Induction of chemokine and cytokine genes in astrocytes following infection with Theiler's murine encephalomyelitis virus is mediated by the Toll-like receptor 3. *Glia* 53 (8):858–67.

Song, J. H., A. Bellail, M. C. Tse, V. W. Yong, and C. Hao. 2006. Human astrocytes are resistant to Fas ligand and tumor necrosis factor-related apoptosis-inducing ligand-induced apoptosis. *J Neurosci* 26 (12):3299–308.

Song, L., and J. S. Pachter. 2004. Monocyte chemoattractant protein-1 alters expression of tight junction-associated proteins in brain microvascular endothelial cells. *Microvasc Res* 67 (1):78–89.

Stamatovic, S. M., P. Shakui, R. F. Keep, B. B. Moore, S. L. Kunkel, N. Van Rooijen, and A. V. Andjelkovic. 2005. Monocyte chemoattractant protein-1 regulation of blood-brain barrier permeability. *J Cereb Blood Flow Metab* 25 (5):593–606.

Sterka, D., Jr., and I. Marriott. 2006. Characterization of nucleotide-binding oligomerization domain (NOD) protein expression in primary murine microglia. *J Neuroimmunol* 179 (1–2):65–75.

Sterka, D., Jr., D. M. Rati, and I. Marriott. 2006. Functional expression of NOD2, a novel pattern recognition receptor for bacterial motifs, in primary murine astrocytes. *Glia* 53 (3):322–30.

Stevens, B., N. J. Allen, L. E. Vazquez, G. R. Howell, K. S. Christopherson, N. Nouri, K. D. Micheva, A. K. Mehalow, A. D. Huberman, B. Stafford, A. Sher, A. M. Litke, J. D. Lambris, S. J. Smith, S. W. John, and B. A. Barres. 2007. The classical complement cascade mediates CNS synapse elimination. *Cell* 131 (6):1164–78.

Suidan, G. L., J. R. McDole, Y. Chen, I. Pirko, and A. J. Johnson. 2008. Induction of blood brain barrier tight junction protein alterations by CD8 T cells. *PLoS One* 3 (8):e3037.

Takada, H., and A. Uehara. 2006. Enhancement of TLR-mediated innate immune responses by peptidoglycans through NOD signaling. *Curr Pharm Des* 12 (32):4163–72.

Takahasi, K., H. Kumeta, N. Tsuduki, R. Narita, T. Shigemoto, R. Hirai, M. Yoneyama, M. Horiuchi, K. Ogura, T. Fujita, and F. Inagaki. 2009. Solution structures of cytosolic RNA sensor MDA5 and LGP2 C-terminal domains: identification of the RNA recognition loop in RIG-I-like receptors. *J Biol Chem* 284 (26):17465–74.

Takeda, K., T. Kaisho, and S. Akira. 2003. Toll-like receptors. *Annu Rev Immunol* 21:335–76.

Takeshita, S., F. Takeshita, D. E. Haddad, N. Janabi, and D. M. Klinman. 2001. Activation of microglia and astrocytes by CpG oligodeoxynucleotides. *Neuroreport* 12 (14):3029–32.

Takeuchi, O., and S. Akira. 2008. MDA5/RIG-I and virus recognition. *Curr Opin Immunol* 20 (1):17–22.

Takeuchi, O., K. Hoshino, T. Kawai, H. Sanjo, H. Takada, T. Ogawa, K. Takeda, and S. Akira. 1999. Differential roles of TLR2 and TLR4 in recognition of gram-negative and gram-positive bacterial cell wall components. *Immunity* 11 (4):443–51.

Tapping, R. I., S. Akashi, K. Miyake, P. J. Godowski, and P. S. Tobias. 2000. Toll-like receptor 4, but not Toll-like receptor 2, is a signaling receptor for *Escherichia* and *Salmonella* lipopolysaccharides. *J Immunol* 165 (10):5780–87.

Teghanemt, A., D. Zhang, E. N. Levis, J. P. Weiss, and T. L. Gioannini. 2005. Molecular basis of reduced potency of underacylated endotoxins. *J Immunol* 175 (7):4669–76.

Tesmer, L. A., S. K. Lundy, S. Sarkar, and D. A. Fox. 2008. Th17 cells in human disease. *Immunol Rev* 223:87–113.

Tian, J., A. M. Avalos, S. Y. Mao, B. Chen, K. Senthil, H. Wu, P. Parroche, S. Drabic, D. Golenbock, C. Sirois, J. Hua, L. L. An, L. Audoly, G. La Rosa, A. Bierhaus, P. Naworth, A. Marshak-Rothstein, M. K. Crow, K. A. Fitzgerald, E. Latz, P. A. Kiener, and A. J. Coyle. 2007. Toll-like receptor 9-dependent activation by DNA-containing immune complexes is mediated by HMGB1 and RAGE. *Nat Immunol* 8 (5):487–96.

Town, T., D. Jeng, L. Alexopoulou, J. Tan, and R. A. Flavell. 2006. Microglia recognize double-stranded RNA via TLR3. *J Immunol* 176 (6):3804–12.

Trinchieri, G., and A. Sher. 2007. Cooperation of Toll-like receptor signals in innate immune defence. *Nature Rev Immunol* 7 (3):179–190.

Ullian, E. M., S. K. Sapperstein, K. S. Christopherson, and B. A. Barres. 2001. Control of synapse number by glia. *Science* 291 (5504):657–61.

Underhill, D. M., and B. Gantner. 2004. Integration of Toll-like receptor and phagocytic signaling for tailored immunity. *Microbes Infect* 6 (15):1368–73.

Underhill, D. M., and A. Ozinsky. 2002. Toll-like receptors: key mediators of microbe detection. *Curr Opin Immunol* 14 (1):103–10.

Underhill, D. M., A. Ozinsky, A. M. Hajjar, A. Stevens, C. B. Wilson, M. Bassetti, and A. Aderem. 1999. The Toll-like receptor 2 is recruited to macrophage phagosomes and discriminates between pathogens. *Nature* 401:811–15.

van Beek, J., K. Elward, and P. Gasque. 2003. Activation of complement in the central nervous system: roles in neurodegeneration and neuroprotection. *Ann NY Acad Sci* 992:56–71.

Van Der Voorn, P., J. Tekstra, R. H. Beelen, C. P. Tensen, P. Van Der Valk, and C. J. De Groot. 1999. Expression of MCP-1 by reactive astrocytes in demyelinating multiple sclerosis lesions. *Am J Pathol* 154 (1):45–51.

van Noort, J. M., and M. Bsibsi. 2009. Toll-like receptors in the CNS: implications for neurodegeneration and repair. *Prog Brain Res* 175:139–48.

Visintin, A., K. A. Halmen, E. Latz, B. G. Monks, and D. T. Golenbock. 2005. Pharmacological inhibition of endotoxin responses is achieved by targeting the TLR4 coreceptor, MD-2. *J Immunol* 175 (10):6465–72.

Vita, N., S. Lefort, P. Sozzani, R. Reeb, S. Richards, L. K. Borysiewicz, P. Ferrara, and M. O. Labeta. 1997. Detection and biochemical characteristics of the receptor for complexes of soluble CD14 and bacterial lipopolysaccharide. *J Immunol* 158 (7):3457–62.

Vitour, D., and E. F. Meurs. 2007. Regulation of interferon production by RIG-I and LGP2: a lesson in self-control. *Sci STKE* 2007 (384):pe20.

Walz, W., and L. Hertz. 1982. Ouabain-sensitive and ouabain-resistant net uptake of potassium into astrocytes and neurons in primary cultures. *J Neurochem* 39 (1):70–77.

Walz, W., and L. Hertz. 1983. Intracellular ion changes of astrocytes in response to extracellular potassium. *J Neurosci Res* 10 (4):411–23.

Wang, J., Y. Shao, T. A. Bennett, R. A. Shankar, P. D. Wightman, and L. G. Reddy. 2006. The functional effects of physical interactions among Toll-like receptors 7, 8, and 9. *J Biol Chem* 281 (49):37427–34.

Weiss, J. M., S. A. Downie, W. D. Lyman, and J. W. Berman. 1998. Astrocyte-derived monocyte-chemoattractant protein-1 directs the transmigration of leukocytes across a model of the human blood-brain barrier. *J Immunol* 161 (12):6896–903.

Wesemann, D. R., H. Qin, N. Kokorina, and E. N. Benveniste, 2004. TRADD interacts with STAT1-alpha and influences interferon-gamma signaling. *Nat. Immunol* 5(2):199–207.

Williams, B. R. 1999. PKR; a sentinel kinase for cellular stress. *Oncogene* 18 (45):6112–20.

Willis, S. A., and P. D. Nisen. 1996. Differential induction of the mitogen-activated protein kinase pathway by bacterial lipopolysaccharide in cultured monocytes and astrocytes. *Biochem J* 313 (Pt 2):519–24.

Wolburg, H., Risau W. 1995. Formation of the blood brain barrier. In *Neuroglia*, ed. K. H. Ransom BR. New York: Oxford University Press.

Wong, G. H., P. F. Bartlett, I. Clark-Lewis, F. Battye, and J. W. Schrader. 1984. Inducible expression of H-2 and Ia antigens on brain cells. *Nature* 310 (5979):688–91.

Woodruff, T. M., R. R. Ager, A. J. Tenner, P. G. Noakes, and S. M. Taylor. 2010. The role of the complement system and the activation fragment C5a in the central nervous system. *Neuromol Med.* 12(2):179–92.

Woodruff, T. M., K. J. Costantini, J. W. Crane, J. D. Atkin, P. N. Monk, S. M. Taylor, and P. G. Noakes. 2008. The complement factor C5a contributes to pathology in a rat model of amyotrophic lateral sclerosis. *J Immunol* 181 (12):8727–34.

Wyss-Coray, T., J. D. Loike, T. C. Brionne, E. Lu, R. Anankov, F. Yan, S. C. Silverstein, and J. Husemann. 2003. Adult mouse astrocytes degrade amyloid-beta *in vitro* and *in situ*. *Nat Med* 9 (4):453–57.

Wyss-Coray, T., F. Yan, A. H. Lin, J. D. Lambris, J. J. Alexander, R. J. Quigg, and E. Masliah. 2002. Prominent neurodegeneration and increased plaque formation in complement-inhibited Alzheimer's mice. *Proc Natl Acad Sci USA* 99 (16):10837–42.

Yefenof, E., and G. Klein. 1976. Difference in antibody induced redistribution of membrane IgM in EBV genome free and EBV positive human lymphoid cells. *Exp Cell Res* 99 (1):175–78.

Yoon, K., and N. Gaiano. 2005. Notch signaling in the mammalian central nervous system: insights from mouse mutants. *Nat Neurosci* 8 (6):709–15.

Zamanian-Daryoush, M., T. H. Mogensen, J. A. DiDonato, and B. R. Williams. 2000. NF-kappaB activation by double-stranded-RNA-activated protein kinase (PKR) is mediated through NF-kappaB-inducing kinase and IkappaB kinase. *Mol Cell Biol* 20 (4):1278–90.

Zhang, Z., K. Guo, and H. J. Schluesener. 2005. The immunostimulatory activity of CpG oligonucleotides on microglial N9 cells is affected by a polyguanosine motif. *J Neuroimmunol* 161 (1–2):68–77.

Zhao, Y., M. A. Rivieccio, S. Lutz, E. Scemes, and C. F. Brosnan. 2006. The TLR3 ligand polyI: C downregulates connexin 43 expression and function in astrocytes by a mechanism involving the NF-kappaB and PI3 kinase pathways. *Glia* 54 (8):775–85.

Zheng, J. C., Y. Huang, K. Tang, M. Cui, D. Niemann, A. Lopez, S. Morgello, and S. Chen. 2008. HIV-1-infected and/or immune-activated macrophages regulate astrocyte CXCL8 production through IL-1beta and TNF-alpha: involvement of mitogen-activated protein kinases and protein kinase R. *J Neuroimmunol* 200 (1–2):100–10.

Section II

Introduction to Cluster B: Establishment of Functional Units: Multimodality at the Tips

The chapters in this section cover the functional bases for connections of astrocytes to all other neural cells and how these connections may be altered under physiological and pathological conditions. In addition, chapters consider how manipulation of astrocyte interconnections affects networks of other neuronal cells.

The first chapter in this cluster (Chapter 5), by Christian Giaume, contrasts the concept of neuroglial networks with that of syncytia, including the rules of organization and control mechanisms, and then describes type of networks (metabolic, intercellular calcium wave spread, potassium dissipation, and the perivascular neurovascular unit). Conclusions are that the widespread gap junctions provide direct exchange of information among astrocytes that is integrated into the interaction between astroglial networks and neuronal circuits. Thus, there are discrete networks of astrocytes that can be modulated by neurons and environmental factors, and these networks have important functions in participation in synaptic units and modulation of vascular tone (two projections of astrocytes).

Chapter 6, by Daniela Rossi and Andrea Volterra, details what goes wrong in neuron-astrocyte communication, focusing primarily on glutamate release from astrocytes evoked by G-protein-coupled receptors modulating synaptic transmission. In response to injury, TNF-alpha and prostaglandins may cause Ca^{2+}-dependent glutamate release from astrocytes, likely affecting glutamatergic input to neurons, and excessive glutamate release from astrocytes, as in AIDS dementia, causes neural degeneration through excitotoxicity. In addition, astrocyte Ca^{2+} signaling is altered in Alzheimer's disease, so that increased waves could affect synaptic transmission, and in ALS pathogenesis glutamate uptake by astrocytes is reduced due to down-regulation of the glutamate transporter. Astrocyte-neuronal paracrine signaling due to a Ca^{2+}-dependent mechanism can control synaptic function; certain diseases may have as a major cause the dysregulation of astrocyte glutamate release due to morphological alterations or release of glutamate or impaired glutamate buffering. Thus, these diseases may be astrocytic rather than neuronal in origin.

Chapter 7, by Rolf Dermietzel and David Spray, deals with two aspects of astrocyte function at the blood–brain barrier: regulating blood flow through participation in the neurovascular unit and regulating tightness of the endothelial barrier. Overlapping sets of molecules are shown to participate in each function, and dysregulation in astrocytic control has severe consequences for brain function.

Chapter 8, by Kimberly Davidson and John Rash, provides an overview of oligodendrocyte homocellular networks as well as heterocellular networks (the panglial syncytium). The gap junction proteins that participate in both types of gap junctions (Cx32 and Cx47 in oligodendrocytes, Cx30 and Cx43 in astrocytes) are characterized and pathological consequences of deletion or mutation in these proteins are detailed. Other analogous diseases involve mutation or deletion of the channel involved in potassium shunting (Kir4.1), neuromyelitis optica resulting from mutations in the endfoot water channel (AQP4) and glial fibrillary acidic protein (GFAP) deletion causing Alexander's disease.

The final chapter in this section (Chapter 9), by Emma Wilson and Monica Carson, deals with neuroimmune circuits involving astrocytes in the central nervous system (CNS). It is proposed that astrocytes are active regulators of CNS immunity, responding to both CNS resident and infiltrating immune cells through the release of chemokines and activation of microglia and macrophages. The plasticity of immune choice is also stressed, with either the classical pro-inflammatory pathogen activation, alternative wound-healing activation states, or regulatory anti-inflammatory activation states.

5 Neuroglial Networks
Glial Wiring Also Matters

Christian Giaume
CIRB, UMR CNRS 7241/INSERM U1050,
Collège de France, Paris, France

CONTENTS

5.1 INTRODUCTION

Syncytium: A multinucleate mass of cytoplasm resulting from fusion of cells.*
Network: An interconnected or intercalated chain, group, or systems.*

When applied to glial cells, the difference between these two definitions illustrates the conceptual changes that have been operated during the last three decades to account for their property of intercellular communication. Indeed, the presence of a large number of gap junctions has led to the proposal that glia are organized as a syncytium (Mugnaini 1986; see also Theis et al. 2005), and so far many works relative

---

* From Merriam-Webster Online Dictionary.

to gap junctions in astrocytes refer to this syncytial concept. The purpose of this chapter is to update what is known about intercellular communication in astrocytes and their network organization in relation to dynamic interactions that astrocytes establish with neurons and the vascular system. These aspects have been recently presented in reviews (Giaume et al. 2010; Orellana et al. 2009); thus, in this chapter, I will place more emphasis on the potential roles of such astroglial networks and discuss future directions of research to better understand the role of astrocytes.

The molecular basis of gap junction channels (GJCs), typically made of two facing hexamers, each of them contributed by adjacent membranes, is a family of multigenic membrane proteins named connexins (Cxs). Such channels are poorly selective for ions and for small molecular weight signaling molecules, providing ionic and biochemical or metabolic coupling, respectively, between neighboring cells. In defined conditions (divalent free solution, metabolic inhibition, inflammation), Cxs can also work as hemichannels (HCs) providing an inside-out signaling pathway (see Spray et al. 2006). Finally, another three-member family of proteins, the pannexins (Panxs), homologous to innexins, the GJC-forming proteins in invertebrates, was shown to form HCs but not GJCs (see Scemes et al. 2007). Sequence analysis indicates a transmembrane topology of Panxs similar to that of Cxs, but a great divergence in primary sequence (Baranova et al. 2004; Panchin et al. 2000); also, Cxs and Panxs show distinct biophysics and pharmacological properties (Giaume and Theis 2009). So far at least 11 Cxs and 2 Panxs have been identified in various brain cell populations, including neurons (Sohl et al. 2005; MacVicar and Thompson 2010). In astrocytes, two major Cxs, Cx43 and Cx30, have been detected while the expression of functional Panx1 HCs is still a matter of debate (see Iglesias et al. 2009 and Retamal et al. 2007). Although HCs can contribute to intercellular signaling with neurons or endothelial cells through the release of active molecules in the extracellular space in this chapter, I will focus on GJC-mediated function of astroglial Cxs.

5.2 GLIAL NETWORKS RATHER THAN A PANGLIAL SYNCYTIUM

The notion of a glial syncytium, and even a panglial syncytium (Theis et al. 2005), was mainly based on two properties of glial cells: their high level of Cx expression compared to other brain cell types and the strong degree of intercellular communication, at least in astrocytes. Indeed, in the hippocampus (Konietzko and Muller 1994; D'Ambrosio et al. 1998; Blomstrand et al. 2004; Wallraff et al. 2004), the cortex (Houades et al. 2006, 2008), the striatum (Adermark and Lovinger 2008; Ball et al. 2007), the superior colliculus (Ball et al. 2007), and the cerebellum (Muller et al. 1996) the injection of a low molecular weight tracer (Lucifer yellow, sulforhodamine B, Alexa 350, biocytin, neurobiotin, etc.) into one astrocyte results in the staining of a number (up to hundreds) of other cells (Figure 5.1). Although there is evidence for heterotypic communication between astrocytes and either neurons (mainly in cocultures (Froes et al. 1999; Rozental et al. 2001) and in brain slices (Alvarez-Maubecin et al. 2000; Bittman et al. 2002; Pakhotin and Verkhratsky 2005)) or oligodendrocytes (in coculture (Venance et al. 1995) and in situ (Maglione et al. 2010)), when an astrocyte is initially injected, most of the coupled cells, if not all, are identified as astrocytes based on postfixation glial fibrillary acidic protein

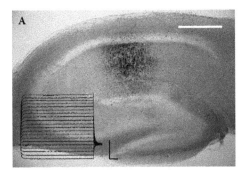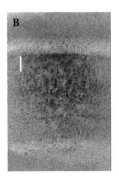

FIGURE 5.1 Example of an astroglial network in the CA1 region of an acute hippocampal slice of a young rat. Whole-cell recording of an astrocyte, identified by its I/V relationship (family of voltage traces in A, calibrations: 50 ms, 20 mV), was achieved with a patch pipette containing biocytin, an intercellular tracer passing through GJCs. After 20 minutes of recording, the slice was fixed and biocytin was revealed by an avidin-biotinylated horseradish peroxidase treatment that then reacted using 3,3-diaminobenzidine as a chromogen. Note that more than 200 cells are stained, indicating that there is a high level of gap-junction-mediated communication. This coupling can be blocked by carbenoxolone, and double staining with either neuronal, microglial, or astrocytic-specific markers indicates that most of the coupled cells (>80%) are identified as astrocytes (see Blomstrand et al. 2004 for complementary information). Calibration bars: 500 and 200 m in A and B, respectively.

(GFAP) or S100_ staining (Houades et al. 2008; Blomstrand et al. 2004). In comparison, coupling between neurons (Peinado et al. 1993; Roerig and Feller 2000) is weak, particularly after the first postnatal weeks in rodents. Similarly, coupling studied in oligodendrocytes is not large when investigated in the corpus callosum or the striatum (Von Blankenfeld et al. 1993). Interestingly, using eGFP-hGFAP transgenic mouse it was demonstrated that in the CA1 region of the hippocampus not all astrocytes present in the coupling area receive the dye, indicating that a certain proportion of the astrocytes is excluded from these networks (Houades et al. 2006), suggesting the prevalence of a certain degree of selectivity. Recently, a similar exclusion of glial cells in communicating networks has been observed in glial olfactory ensheathing cells from the olfactory bulb (Rela et al. 2010). However, as indicated below, astroglial networks are plastic; consequently, the number of connected astrocytes may change with time and may depend on neuronal activity or the reactive status of astrocytes and microglia.

5.3 ASTROGLIAL NETWORKS HAVE THEIR OWN RULES OF ORGANIZATION

Although the spatial distribution of astroglial networks never reaches the complexity and specificity of neuronal circuits based on synaptic connection, their shape varies according to the age, the brain area considered, and the physiopathological context. Of course, their extent depends directly on the level of expression of Cx43 and Cx30 that varies between and within brain regions; for instance, astrocytes are more extensively dye coupled in the CA1 region than in CA3 of the hippocampus (D'Ambrosio et al. 1998). Also, their shape varies within the cerebral cortex, depending on the deepness of the studied layers, or in the hippocampus, depending on the proximity to the pyramidal layer (Houades et al. 2006). More precisely, when studied in brain structures characterized by anatomofunctional neuronal units, astroglial networks also show a specific spatial organization. This is the case in the somatosensory cortex, where neurons in layer IV are organized as "barrels" that respond specifically to the stimulation of a single whisker with a perfect topology (Woolsey and Van der Loos 1970). There, immunofluorescence analysis indicates that Cx43 and Cx30 expression is enriched in the barrels, although the distribution of astrocytes is homogenous all along layer IV. Interestingly, dye injections reveal that gap junctional communication is favored within a single barrel rather than between adjacent barrels (Houades et al. 2008). Such oriented communication toward the barrel center indicates that astroglial networking parallels the columnar compartmentalization of neurons, and that communicating astrocytes may contribute to the confinement of neuroglial interaction within a defined barrel as demonstrated for excitatory neuronal networks (Petersen and Sakmann 2000). A similar situation is encountered in the olfactory glomeruli, as coupling is also oriented toward the center of these neuronal units (Roux and Giaume 2009).

5.4 REFLEXIVE GAP JUNCTIONS PROVIDE A SUBCELLULAR NETWORKING

It is now well established that astrocytes, instead of having intricate processes as initially thought (see Rohlmann and Wolff 1998), occupy individual domains with a very restricted contact area (Bushong et al. 2002; Ogata and Kosaka 2002). Accordingly, expected gap junctions should be only found at these zones of contact; however, this not the case when analyzing immunofluorescence stainings. Indeed, while typical junctional plaques are observed at the interface between astrocytic domains and at contacts between endfeet enwrapping blood vessels (Nagy and Rash 2000; Rouach et al. 2008), immunohistochemical staining revealed that Cx43 or Cx30 expression is not restricted to these sole locations. This suggests that, in addition to intercellular communication, gap junctions could also operate between processes originating from a single astrocyte, i.e., within its own domain. In fact, such reflexive gap junctions have already been described at the ultrastructural level (Wolff et al. 1998). In situ, the functional study of these reflexive gap junctions is limited due to the presence of intercellular communication that prevents the possibility to isolate and observe the pattern of dye coupling within the processes of a single astrocyte. However, it is obvious that several of the roles attributed to astroglial networks (see below) are expected to operate there too. Consequently, reflexive GJCs constitute an element that should be taken into account when considering the kinetics and spatial distribution of subcellular events taking place in astroglial microdomains (Grosche et al. 1999). Indeed, this property may interfere with or modulate local calcium responses, amino acid or ion uptake as well as dissipation, and second-messenger mobilization in response to metabotropic receptor stimulation. Thus, through reflexive gap junctions, Cxs channels may contribute actively to integrative responses of a single astrocyte.

5.5   ASTROGLIAL NETWORKS ARE UNDER CONTROL

The permeability of astrocyte GJCs is regulated by signaling molecules that stimulate membrane receptors, including neurotransmitters, pro-inflammatory cytokines, bioactive peptides, and lipids (see Giaume and McCarthy 1996; Giaume et al. 2010). However, does neuronal activity per se regulate communication within astroglial networks? Up to now, only a few studies have addressed this question. For instance, dye coupling is increased in cerebellar and striatal astrocytes that were cocultured with neurons (Fischer and Kettenmann 1985; Rouach et al. 2000) and in glial cells of the frog optic nerve when neurons are stimulated (Marrero and Orkand 1996). Also, acoustic stimulation was reported to increase astroglial coupling in the inferior colliculus (Cruz et al. 2007), and seasonal changes affect GJC-mediated communication in the song control area HVc of the canary (Kafitz et al. 1999). However, the mechanisms underlying these short- and long-term activity-dependent regulations have not been elucidated yet. Elevated levels of extracellular potassium depolarize astrocytes and increase dye coupling in cultured cortical astrocytes (Enkvist and McCarthy 1994), an effect mediated by phosphorylation of Cx43 by CaM kinase II (De Pina-Benabou et al. 2001). The effects of glutamatergic activity are more complex and depend on the brain area studied, the type of preparation used, as well as the type of glutamate receptor stimulated in the cerebellum (Muller et al. 1996) and in the hippocampus (Serrano et al. 2008). Only a few studies have been carried out to assess the regulation of GJCs with electrophysiological techniques, mainly because this approach requires simultaneous recordings of two neighboring astrocytes with double voltage-clamp technique. In addition, the 3D connection of astrocytes in situ and the short length constant of their processes do not favor accurate monitoring of truly junctional currents. However, this electrophysiological approach applied in pairs of astrocytes in situ provides reliable information about the kinetic changes in GJC's conductance. Up to now, this has been performed in cerebellar Bergmann glial cells for glutamate (Muller et al. 1996) and in the hippocampus for endothelin-1 (Meme et al. 2009). Both studies indicate that agonist-induced changes in junctional currents can be triggered very rapidly (within less than 1 minute), illustrating the occurrence of a short-term regulation of GJCs. Finally, the extent of these networks is also reported to be affected in pathological situations, such as in a transgenic mouse model of Alzheimer's disease (Peters et al. 2009) and in a tuberous sclerosis complex mouse (Xu et al. 2009), for which an increase in dye coupling is correlated with enhanced Cx43 expression in astrocytes (Nagy et al. 1996; Xu et al. 2009).

5.6   A NETWORK OR SEVERAL NETWORKS?

As dye coupling experiments are based on the injection of a low molecular weight tracer in a cell and the analysis of its diffusion toward coupled cells, it is expected that the number of coupled cells is directly related to the duration of the injection and the time allowed for diffusion (see Muller et al. 1996; Rouach et al. 2008). This means that when a network is visualized, its extent could be different if the time of injection and the delay before analysis is either shorter or longer. As GJC's permeability for a molecule depends on its size, charge, and conformation, the kinetic of intercellular diffusion depends on the selected tracer: for instance, in the barrel cortex or the hippocampus for the same time of injection the networks observed with biocytin are much larger (two to three times more) than those with sulforhodamine B (Houades et al. 2008; Rouach et al. 2008). On the other hand, if the time allowed for diffusion is increased, dye coupling can be observed even when only few GJCs are functional. Based on these properties, it is difficult to state that astroglial networks are independent and distinct. However, when considering the passage of signaling molecules with a biological significance and known to permeate GJCs such as second messengers, amino acids, or metabolites (see Harris 2007), the definition of what is an astroglial network becomes different. Indeed, the initial concentration increase of a signaling molecule in a stimulated astrocyte is limited. In addition, these molecules have a defined radius of action and time life. Consequently, the size of the observed networks resulting from these physiological responses is likely more restricted than the size of those visualized by passive dyes. Accordingly, I propose that astroglial networks involved in physiological responses to neuronal activity are independent from each other provided that the sites of activation are distant (Figure 5.2A). Their size increases with the strength of the stimuli, and they become more extensive as the number of activation sites is enhanced (Figure 5.2B,C), and finally, they can overlap when widespread, large, and synchronous stimuli are generated (Figure 5.2D).

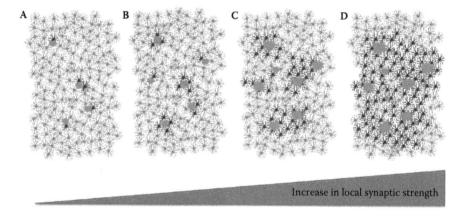

Increase in local synaptic strength

FIGURE 5.2 Schematic illustration of changes in astroglial networks triggered by neuronal activity. (A) When weak and sparse neuronal discharges (light symbols) are generated in a brain area, only weak and rare responses are triggered in astrocytes (dark symbols). (B,C) As the number and intensity of neuronal foci increase, the size and number of responding astrocytes are enhanced and clusters of active astrocytes start to be observed. (D) Finally, for a more activated situation there is an overlapping of the groups of responding astrocytes that form a single network. Of course, this scheme is valid for different properties of astrocytes. For instance, the release of neurotransmitters by neurons could account for this sequence of increasing responses. Another way to interpret this feature is to consider that initial responses are triggered in astrocytes directly in contact with the active neuronal site (A). Then when the neuronal signal is increased, the initial astroglial responses are stronger and the generated second messengers (calcium and IP3, for instance) pass through GJCs and evoke a response in more distant astrocytes (B,C) until all astrocytes are affected for stronger and numerous stimuli (D).

5.7 METABOLIC NETWORKS CONTRIBUTE TO DYNAMIC NEUROGLIAL INTERACTION

The diffusion of energy metabolites through GJCs was recently studied in hippocampal slices using fluorescent glucose derivatives (Rouach et al. 2008), and for lactate using a real-time enzymatic fluorescent assay in inferior colliculus slices (Gandhi et al. 2009). In the hippocampus, glutamate released by spontaneous, epileptic-like, or evoked activities increases glucose trafficking in astroglial networks through a mechanism that involves the activation of i-amino-3 hydroxy-5-methyl-4-isoxalate-propionic acid (AMPA) receptors. As hippocampal astrocytes connected by gap junctions lack AMPA receptors (Matthias et al. 2003), it remains unclear whether glutamate itself or a downstream effector of postsynaptic AMPA receptor activation regulates glucose trafficking through astroglial GJCs. Surprisingly, these activity-dependent regulations of glucose trafficking are not observed with passive dyes or tracers. This suggests that, at least in the hippocampus, glutamatergic activity does not regulate GJCs, but rather triggers a local energetic sink generating a diffusion gradient for glucose to the sites of high neuronal demand.

Astrocytes can supply glucose to neurons, but of course glucose can also be directly taken up by neurons from the extracellular space. By exposing hippocampal slices to glucose-free solution, it was possible to investigate the metabolic contribution of astroglial networks by using wild-type and double knockout ($Cx43^{-/-}/Cx30^{-/-}$) mice. Indeed, while in control slices the introduction of glucose or lactate in the patch-clamp pipette allows maintaining evoked or epileptic-like activity induced by pharmacological treatments, when gap junctional communication is absent, this rescue is not observed. This indicates that glucose as well as lactate diffuses through GJCs and provides sufficient energy substrates to fuel active neurons (Rouach et al. 2008). In the inferior colliculus, it was reported that astrocytes have a faster and higher capacity for lactate uptake and lactate dissipation through astroglial networks than neuronal lactate uptake from extracellular fluid or shuttling of lactate to neurons from neighboring astrocytes (Gandhi et al. 2009). Astroglial networks can provide neuronal fuel by metabolic intercellular diffusion and rapid removal of lactate from activated glycolytic domains. In addition, lactate can be dispersed widely throughout astrocyte GJCs to endfeet along the vasculature for release to blood or other brain regions via perivascular fluid flow. Thus, as astrocytes occupy a strategic place between the vasculature, the source of energy compounds, and the neurons, the main energy consumers in the brain, their network organization provides intercellular pathways contributing to establish a metabolic homeostasis in the brain through the balance between delivery and dissipation of energy compounds.

5.8 WHAT IS THE CONTRIBUTION FOR CONNEXIN-MEDIATED ASTROGLIAL NETWORKS TO THE PROPAGATION OF INTERCELLULAR WAVES?

The first observation of intercellular calcium waves in astrocytes (Cornell-Bell et al. 1990) and the subsequent description of calcium-dependent long-range signaling from astrocytes to neurons (Nedergaard 1994) have opened the door for a role of Cxs in astrocytes. Initially, their contribution to this process was only considered on the basis of GJCs. Then, the demonstration that Cxs can form functional HCs and the discovery of the Panx family have provided other modalities of contribution of these proteins to calcium wave propagation. Several reviews have listed and discussed the pros and cons of a contribution of these different actors (Giaume and Venance 1998; Charles and Giaume 2002; Nedergaard et al. 2003; Fiacco and McCarthy 2006; Scemes and Giaume 2006). Although the final mechanisms have not been fully understood, the present agreement is that these alternative mechanisms and molecular elements are not exclusive and could all participate in the propagation process with different weights, depending on the triggering stimulus, the brain area considered, and the pathological status of the preparation. In this domain, the most exciting finding that has recently emerged is that the occurrence of these astroglial calcium waves has been observed in vivo in the cerebellum from anesthetized mice (Hoogland et al. 2009) and in a transgenic model of Alzheimer's disease in the cortex (Kuchibhotla et al. 2009). Interestingly, an increase in Cx43 expression has been reported to occur in reactive astrocytes at amyloid plaques in brain from Alzheimer's disease patients (Nagy et al. 1996), and a similar observation has been recently made for Cx43 and Cx30 in a double transgenic mouse model (APP/PS1) making amyloid plaques (Mei et al. 2010). Although there is currently a debate about the involvement of astrocyte calcium-dependent processes in dynamic neuroglial interactions (see Kirchhoff 2010), these in vivo observations should certainly stimulate future research to fully understand the how and why of calcium waves in astrocytes.

5.9 OTHER ROLES FROM ASTROGLIAL NETWORKS

5.9.1 POTASSIUM DISSIPATION

The earliest proposed role for glial gap-junction-mediated communication is related to potassium dissipation secondary to neuronal activity and firing (Orkand et al. 1966). The starting point of this hypothesis was the high potassium permeability of glial cells and their highly coupling properties that led to the spatial potassium buffering hypothesis. This question has then been addressed specifically in astrocytes. However, one major obstacle to really address this question is the difficulty to have a specific pharmacological control of GJC activity without affecting neuronal properties. So far this is not possible due to the poor selectivity of GJC blockers (see Giaume and Theis 2009). An alternative is the use of transgenic animals with deletion of the genes coding for the two astroglial Cxs. As the total Cx43 knockout (KO) is lethal (Reaume et al. 1995), a cell type restricted deletion was achieved by the *cre lox P* system, resulting in an astrocyte-directed deletion of Cx43 in a viable mouse (Theis et al. 2003). In addition, the total Cx30 KO can be used to study coupling since in the brain this Cx is only expressed by astrocytes and the meninges (Dere et al. 2003). From the functional point of view, when the contribution of the two Cxs is studied using these KO mice, each Cx seems to participate to about half of the coupling strength in the hippocampus (Wallraff et al. 2006; Rouach et al. 2008). Finally, the crossbreeding of these two KO mice results in a double KO without evidence of dye coupling

(Wallraff et al. 2006; Rouach et al. 2008). Using these animals it was reported that in hippocampus astroglial gap junctions accelerate potassium clearance, limit potassium accumulation during synchronized neuronal firing, and aid in radial potassium relocation in the stratum lacunosum moleculare. Furthermore, slices of mice with coupling-deficient astrocytes displayed a reduced threshold for the generation of epileptiform events. However, in this work it was evident that radial relocation of potassium in the stratum radiatum was not dependent on gap junctional coupling. Finally, a large capacity for potassium clearance was conserved in mice with coupling-deficient astrocytes, indicating that gap-junction-dependent processes only partially account for potassium buffering in the hippocampus (Wallraff et al. 2006).

5.9.2 Masking of Pharmacological and Biochemical Heterogeneity

There is increasing evidence for morphological and functional heterogeneity of astrocytes, including the expression of membrane receptors that respond to neurotransmitters (see Matyash and Kettenmann 2009). This heterogeneity is observed when considering different brain areas, but is also true within a defined structure, implicating that neighboring astrocytes can differentially integrate neuronal activity. Considering that the stimulation of metabotropic receptors generated changes in second-messenger concentration, and that GJCs are permeable to a number of second messengers, the existence of a high level of gap-junction-mediated communication is expected to attenuate or mask the consequences of a pharmacological heterogeneity in a group of coupled astrocytes. Such a question has been addressed using primary cultures of astrocytes and analyzing their pattern of calcium responses to agonist stimulation (Venance et al. 1998). While the expression of endothelin receptors is detected in all astrocytes, α-adrenergic and muscarinic receptors are only observed in a subpopulation of astrocytes (70 and 60%, respectively). When the occurrence (i.e., the number of responding cells) of agonist-induced calcium responses is compared in a situation of open or closed GJCs, differences appear concerning the adrenergic and muscarinic stimulations. Indeed, a reduction of about 29 and 22%, respectively, in the number of responding cells is observed in noncommunicating astrocytes. In addition, the pattern of the calcium responses is also affected. In controls two distinct types of responses are monitored, either a rapid calcium peak or a delayed and slowly rising increase, but in the absence of gap junctional communication, only the first type of responses are observed. In noncommunicating astrocytes there is no difference in the number of responding cells and in the pattern of endothelin-1-induced calcium responses; in contrast, those generated by the α-adrenergic or muscarinic agonists are decreased in number and exhibit a sole peak response pattern. Altogether, these observations indicate that functional GJCs contribute to mask a pharmacological heterogeneity (Figure 5.3). A rather similar observation has been made when monitoring the level of sodium in cultured astrocytes (Rose and Ransom 1997). While the basal sodium intracellular sodium concentration is homogenous in a population of communicating astrocytes, when GJCs are blocked differences in sodium level appear between neighboring cells, demonstrating that gap junctions equalize internal ionic concentrations. Of course, the next step would be to study such role in vivo and to determine whether and how gap junctional communication

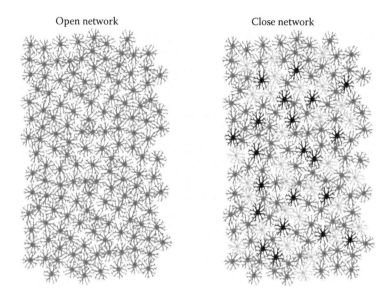

FIGURE 5.3 Astroglial gap junctions mask pharmacological heterogeneity. There is increasing evidence for morphological and functional heterogeneity of astrocytes. This includes also a pharmacological heterogeneity, i.e., the fact that not all astrocytes in a defined brain area express the same amount or the same catalog of membrane receptors. This implies that only subpopulations of astrocytes initially respond to a defined neuronal signal. However, the expression of a large number of GJCs allows communication with nonresponding astrocytes, but when they communicate, almost astrocytes can sense the signal with time (left diagram). In contrast, when communication is absent, the pharmacological heterogeneity of the astroglial population is unmasked and the concentration of the second messenger resulting from receptor stimulation is expected to vary among the cells from weak (light symbols) to high (dark symbols) responses, with the possibility of intermediary situations (medium light symbols). This hypothesis is based on experimental data obtained from calcium imaging experiments performed in cultured astrocytes exposed to endothelin-1, methoxamine, and carbachol (see Venance et al. 1998). A similar model should be encountered when considering the intracellular sodium concentration that is not homogenous within a population of culture astrocytes, but equalized when GJCs are open (see Rose and Ransom 1998).

in astrocytes affects the occurrence and shape of pharmacological or biochemical responses and ionic homeostasis.

5.9.3 SYNAPSE AND NEUROGLIAL NETWORKING

Up to now, the involvement of astrocytes in central functions has mostly been considered to result from the activity of individual bystander cells (Araque et al. 1999; Haydon and Carmignoto 2006; Perea et al. 2009) rather than from an active partnership between neuroglial networks. Indeed, the tripartite synapse puts together the pre- and postsynaptic elements with processes from a single astrocyte that contact this synapse (Araque et al. 1999). A more integrated view of neuroglial dynamic interactions is to take into account the network organization of astrocytes and their permeability for signaling molecules (Giaume et al. 2010). Indeed, due to the proximity of astrocyte gap junctions and neuronal synapses (Rohlmann and Wolff 1998; Giaume et al. 2010), astroglial networks could be actively associated with a group of synapses. Hence, specific astroglial circuits, thanks to their permeability to glutamate and glutamine (Giaume et al. 1997) and possibly to D-serine, could coordinate local groups of synapses. Moreover, through the release of gliotransmitters over a wide area, astroglial networks could affect neuronal network activity during synaptic plasticity or epileptic discharges. Alternatively, instead of involving a group of connected astrocytes, a similar role of GJCs in interfering with the tripartite synapse can be hypothesized at a more local level if reflexive gap junctions are considered. Such a working hypothesis could be tested by increasing glutamate (glutamine or D-serine) concentration within an astrocyte for instance by whole-cell loading, let it diffuse through GJCs within the astroglial network, and determine whether this infusion affects neuronal activity. In fact, this has recently been demonstrated for the purinergic pathway. Indeed, due to their calcium signaling properties, astroglial networks in the hippocampus are involved in heterosynaptic depression: ATP is released by astrocytes beyond the area of activated synapses and converted into adenosine that causes presynaptic inhibition of transmitter release in the surrounding synapses (Serrano et al. 2006). Finally, astroglial networks have also been implicated in hippocampal neuronal epileptic discharges; indeed infusion of astrocytes with IP_3 triggers glutamate release that in turn induces neuronal depolarization and epileptic discharges (Cotrina et al. 1998).

5.9.4 Perivascular Astroglial Networks as Part of the Gliovascular Unit

Recently, time-dependent analysis of dye coupling initiated from perivascular astrocytes demonstrated that a preferential coupling occurs along the vasculature (Rouach et al. 2008). This property is a direct consequence of the high level of Cx43 and Cx30 expression at the contact area between astrocytic endfeet that surround blood vessels (Yamamoto et al. 1992; Nedergaard et al. 2003; Rouach et al. 2008). Moreover, emerging evidence implicates astrocytes as one of the key players in coordinating neurovascular coupling. Indeed, local $[Ca^{2+}]_i$ increases in a single perivascular digit spread throughout its processes, including endfeet, and then propagates to adjacent endfeet (Mulligan and MacVicar 2004). As Cx channels (HCs and GJCs) contribute to the intercellular propagation of calcium waves (see Scemes and Giaume 2006), it is suggested that they participate in the regulation of blood flow by increasing the number of endfeet involved in calcium responses. Indeed, $[Ca^{2+}]_i$ increases in astrocytes generate signals that, depending on their nature, trigger either a vasodilation or a vasoconstriction (Gordon et al. 2007; Koehler et al. 2009). As the production of these molecules is calcium dependent, the efficacy of the control of blood flow is expected to be related to the extent of the perivascular propagation of calcium signaling through endfeet of the same or a group of astrocytes connected by either reflexive or intercellular gap junctions, respectively. Alternatively, HCs could be involved in this process, as they have also been proposed to contribute to the propagation of intercellular calcium waves.

5.10   CONCLUSIONS AND PERSPECTIVES

Interestingly, the demonstration of a multicellular network organization in the brain, with identified regulatory mechanisms and functions, might allow reconciling the supporters of the reticularists with the defenders of the "neuron doctrine." Indeed, more than one century ago these two views of the brain architecture opposed Golgi and Cajal when they tried to understand how neuronal circuitry is built and works. Although neurons also express gap junctions at electrical synapses, it is now clear that they are casual and that chemical synapses provide the structural basis for information processing between nerve cells. Nevertheless, the widespread gap-junction-mediated communication in astrocytes and the demonstration of astroglial networking are certainly the best illustration that a direct cell-to-cell communication between cells also exists in the nervous system. Now that the concept of dynamic interaction between astrocytes and neurons has gained interest, it is necessary to approach this property at a more integrated level by considering interaction between neuronal circuits and astroglial networks. Thus, glial wiring also matters when considering information processing and brain functions.

However, to fully address this question much progress has to be made. As a first step, we need to be able to act specifically on Cx and Panx channels. To achieve this goal, two strategies are offered by either pharmacology or molecular biology and genetics. So far, both of them have been used and have shown their limits concerning Cx and Panxs channels (see Giaume and Theis 2009). Indeed, as in the central nervous system, 11 Cxs and 2 Panxs are expressed in glia and in neurons; several conditions have to be fulfilled to make correct interpretations: (1) the selected approach neither must have side effects in glia nor affect per se neuronal excitability, development, or survival; (2) as more than one Cx is expressed in a defined cell type, several Cxs have to be targeted to fully prevent gap junctions and HC communication; (3) as Cxs support two channel functions, it is important to be able to discriminate between GJC and HC activity; and finally, (4) since Cxs but also Panxs can form HCs, their respective contribution should be distinguished. The second step is to identify what passes through GJCs and which message is transmitted between these nonexcitable cells. So far only few works have been oriented in this direction due to the lack of appropriate tools. To this purpose we need to be able to follow in a group of astrocytes the diffusion of small molecules that have a biological significance in vitro, but also in vivo. This has been possible quite rapidly for calcium in the 1990s and more recently for glucose, with Z-[N-(7-nitrobenz-2-oxa-1, 3-dizol-4yl)amino]-2-deoxyglucose compounds (Rouach et al. 2008). However, efforts are necessary to make available other probes (fluorescence will be the best) to follow ionic coupling (sodium ions, for instance) and biochemical coupling in astroglial networks with a focus on amino acids or second messengers (IP3, cAMP, cGMP, etc.). In fact, some of them are already available and need to be adapted to acute slices or in vivo approaches in the brain. This approach could also be undertaken to understand the role of glial HCs, as they are also permeable to glutamate, ATP, glutathione, and glucose. Third, the permeability properties of astroglial GJCs can help to act on the physiological status of a population of coupled astrocytes. This is the case for infusion of BAPTA in hippocampal astroglial networks that results in the silencing of calcium signaling in a group of coupled astrocytes (Serrano et al. 2006). It could also be applied for channel, enzyme, or G-protein inhibitors, provided that the molecular weight of the used agents is sufficiently low to pass through GJCs. Interestingly, Cx43-specific cell-to-cell transfer of short interfering RNA has been reported (Valiunas et al. 2005). This work supports the novel hypothesis that nonhybridized and possible hybridized forms of siRNA can move between mammalian cells through Cx-specific GJCs and could be used as an alternative pharmacological approach to target a group of astrocytes.

A number of studies have reported that remodeling of Cx expression in astrocytes occurs in many brain pathologies and lesions (see Giaume et al. 2010). In most of the cases, these observations are limited to immunocytochemical and ultrastructural description, but in some of them functional changes have been also reported. For instance, the expression of Cx43 is increased in reactive astrocytes associated to withdraw amyloid plaques in brain of patients suffering from Alzheimer's disease (Nagy et al. 1996), and dye coupling is increased in a transgenic mouse model that exhibits amyloid plaques (Peters et al. 2009). In contrast, in Parkinson's disease animal models, Cx43 is increased but coupling is not affected (Rufer et al. 1996). In the tuberous sclerosis complex mouse, Cx43 expression is reduced in astrocytes and coupling is also reduced (Xu et al. 2009). Moreover, similar observations are made in the traumatic cortex after mechanical lesion (Dermietzel et al. 1995). Finally, in an oxygen-glucose deprivation model coupling is reduced and associated with a dephosphorylation of Cx43 (Cotrina et al. 1998), while in an epilepsy model both properties were increased in parallel (Samoilova et al. 2003). Altogether these observations indicate that astroglial networks are affected in brain diseases, suggesting that they should impact on neuroglial and gliovascular interactions. However, it is not clear yet whether these changes are the cause or the consequence of such pathological situations. In addition, in vitro and in vivo results indicate that pharmacological blockade or genetic ablation of Cxs affect neuronal survival in pathological models, strengthening the notion that Cxs likely play a role during the progression of central diseases (Giaume et al. 2007; Kielian 2008; Orellana et al. 2009; Bargiotas et al. 2009). Based on these remarks, it is possible that a better understanding of the role of astroglial Cxs (and Panxs) may contribute to the development of new therapeutic approaches to treat diseases and injuries of the central nervous system.

Dynamic interactions between neurons and astrocytes are not any more considered an emerging field in neuroscience. Indeed, the increasing number of laboratories working in this domain illustrates the importance of this partnership in major questions relative to information processing, synaptic plasticity, brain functions such as sleep or pain, and brain pathologies. Now, I think that the demonstration of the existence of astroglial networks provides a basis to propose that neuroglial interactions should also be considered at a more integrated level. Indeed, the level of Cx expression in astrocytes is so unique in the nervous system that these proteins certainly play a critical role in the glial contribution to brain processing and metabolism, as well as to their dysfunction.

ACKNOWLEDGMENTS

I thank Dr. A. Koulakoff for her critical comments on this chapter and express my gratitude to all the collaborators and colleagues who along the years have contributed to improving my understanding of gap-junction-mediated communication in the brain. Their work, ideas, and shared discussions constitute the basement of this chapter.

REFERENCES

Adermark, L., and D. M. Lovinger. 2008. Electrophysiological properties and gap junction coupling of striatal astrocytes. *Neurochem Int* 52 (7):1365–72.

Alvarez-Maubecin, V., F. Garcia-Hernandez, J. T. Williams, and E. J. Van Bockstaele. 2000. Functional coupling between neurons and glia. *J Neurosci* 20 (11):4091–98.

Araque, A., V. Parpura, R. P. Sanzgiri, and P. G. Haydon. 1999. Tripartite synapses: glia, the unacknowledged partner. *Trends Neurosci* 22 (5):208–15.

Ball, K. K., G. K. Gandhi, J. Thrash, N. F. Cruz, and G. A. Dienel. 2007. Astrocytic connexin distributions and rapid, extensive dye transfer via gap junctions in the inferior colliculus: implications for [(14)C]glucose metabolite trafficking. *J Neurosci Res* 85 (15):3267–83.

Baranova, A., D. Ivanov, N. Petrash, A. Pestova, M. Skoblov, I. Kelmanson, D. Shagin, S. Nazarenko, E. Geraymovych, O. Litvin, A. Tiunova, T. L. Born, N. Usman, D. Staroverov, S. Lukyanov, and Y. Panchin. 2004. The mammalian pannexin family is homologous to the invertebrate innexin gap junction proteins. *Genomics* 83 (4):706–16.

Bargiotas, P., H. Monyer, and M. Schwaninger. 2009. Hemichannels in cerebral ischemia. *Curr Mol Med* 9 (2):186–94.

Bittman, K., D. L. Becker, F. Cicirata, and J. G. Parnavelas. 2002. Connexin expression in homotypic and heterotypic cell coupling in the developing cerebral cortex. *J Comp Neurol* 443 (3):201–12.

Blomstrand, F., L. Venance, A. L. Siren, P. Ezan, E. Hanse, J. Glowinski, H. Ehrenreich, and C. Giaume. 2004. Endothelins regulate astrocyte gap junctions in rat hippocampal slices. *Eur J Neurosci* 19 (4):1005–15.

Bushong, E. A., M. E. Martone, Y. Z. Jones, and M. H. Ellisman. 2002. Protoplasmic astrocytes in CA1 stratum radiatum occupy separate anatomical domains. *J Neurosci* 22 (1):183–92.

Charles, A. C., and Giaume, C. 2002. Intercellular calcium waves in astrocytes: underlying mechanisms and functional significance. In *The tripartite synapse: glia in synaptic transmission*, ed. A. Volterra, P. Magistretti, and P. G. Haydon. Oxford: Oxford University Press. 100–126.

Cornell-Bell, A. H., S. M. Finkbeiner, M. S. Cooper, and S. J. Smith. 1990. Glutamate induces calcium waves in cultured astrocytes: long-range glial signaling. *Science* 247 (4941):470–73.

Cotrina, M. L., J. Kang, J. H. Lin, E. Bueno, T. W. Hansen, L. He, Y. Liu, and M. Nedergaard. 1998. Astrocytic gap junctions remain open during ischemic conditions. *J Neurosci* 18 (7):2520–37.

Cotrina, M. L., J. H. Lin, and M. Nedergaard. 1998. Cytoskeletal assembly and ATP release regulate astrocytic calcium signaling. *J Neurosci* 18 (21):8794–804.

Cruz, N. F., K. K. Ball, and G. A. Dienel. 2007. Functional imaging of focal brain activation in conscious rats: impact of [(14)C]glucose metabolite spreading and release. *J Neurosci Res* 85 (15):3254–66.

D'Ambrosio, R., J. Wenzel, P. A. Schwartzkroin, G. M. McKhann 2nd, and D. Janigro. 1998. Functional specialization and topographic segregation of hippocampal astrocytes. *J Neurosci* 18 (12):4425–38.

De Pina-Benabou, M. H., M. Srinivas, D. C. Spray, and E. Scemes. 2001. Calmodulin kinase pathway mediates the K+-induced increase in Gap junctional communication between mouse spinal cord astrocytes. *J Neurosci* 21 (17):6635–43.

Dere, E., M. A. De Souza-Silva, C. Frisch, B. Teubner, G. Sohl, K. Willecke, and J. P. Huston. 2003. Connexin30-deficient mice show increased emotionality and decreased rearing activity in the open-field along with neurochemical changes. *Eur J Neurosci* 18 (3):629–38.

Dermietzel, R. A. Hofer, A. Roll Mann, C. M. Müller, and J. E. Trosko, 1995. Functional plasticity and cell specific expression of connexins in manual and pathological glia tissues. In *Intercellular communication through gap functions,* ed. Y. Kanno, K. Kataoka, Y. Shiba, Y. Shibata, and T. Shimazu. Amsterdam and New York: Elsevier, 235–237.

Enkvist, M. O., and K. D. McCarthy. 1994. Astroglial gap junction communication is increased by treatment with either glutamate or high K+ concentration. *J Neurochem* 62 (2):489–95.

Fiacco, T. A., and K. D. McCarthy. 2006. Astrocyte calcium elevations: properties, propagation, and effects on brain signaling. *Glia* 54 (7):676–90.

Fischer, G., and H. Kettenmann. 1985. Cultured astrocytes form a syncytium after maturation. *Exp Cell Res* 159 (2):273–79.

Froes, M. M., A. H. Correia, J. Garcia-Abreu, D. C. Spray, A. C. Campos de Carvalho, and M. V. Neto. 1999. Gap-junctional coupling between neurons and astrocytes in primary central nervous system cultures. *Proc Natl Acad Sci USA* 96 (13):7541–46.

Gandhi, G. K., N. F. Cruz, K. K. Ball, S. A. Theus, and G. A. Dienel. 2009. Selective astrocytic gap junctional trafficking of molecules involved in the glycolytic pathway: impact on cellular brain imaging. *J Neurochem* 110 (3):857–69.

Giaume, C., F. Kirchhoff, C. Matute, A. Reichenbach, and A. Verkhratsky. 2007. Glia: the fulcrum of brain diseases. *Cell Death Differ* 14 (7):1324–35.

Giaume, C., A. Koulakoff, L. Roux, D. Holcman, and N. Rouach. 2010. Astroglial networks: a step further in neuroglial and gliovascular interactions. *Nat Rev Neurosci* 11 (2):87–99.

Giaume, C., and K. D. McCarthy. 1996. Control of gap-junctional communication in astrocytic networks. *Trends Neurosci* 19 (8):319–25.

Giaume, C., A. Tabernero, and J. M. Medina. 1997. Metabolic trafficking through astrocytic gap junctions. *Glia* 21 (1):114–23.

Giaume, C., and M. Theis. 2010. Pharmacological and genetic approaches to study connexin-mediated channels in glial cells of the central nervous system. *Brain Res Rev.* 63:160–176.

Giaume, C., and L. Venance. 1998. Intercellular calcium signaling and gap junctional communication in astrocytes. *Glia* 24 (1):50–64.

Gordon, G. R., S. J. Mulligan, and B. A. MacVicar. 2007. Astrocyte control of the cerebrovasculature. *Glia* 55 (12):1214–21.

Grosche, J., V. Matyash, T. Moller, A. Verkhratsky, A. Reichenbach, and H. Kettenmann. 1999. Microdomains for neuron-glia interaction: parallel fiber signaling to Bergmann glial cells. *Nat Neurosci* 2 (2):139–43.

Harris, A. L. 2007. Connexin channel permeability to cytoplasmic molecules. *Prog Biophys Mol Biol* 94 (1–2):120–43.

Haydon, P. G., and G. Carmignoto. 2006. Astrocyte control of synaptic transmission and neurovascular coupling. *Physiol Rev* 86 (3):1009–31.

Hoogland, T. M., B. Kuhn, W. Gobel, W. Huang, J. Nakai, F. Helmchen, J. Flint, and S. S. Wang. 2009. Radially expanding transglial calcium waves in the intact cerebellum. *Proc Natl Acad Sci USA* 106 (9):3496–501.

Houades, V., A. Koulakoff, P. Ezan, I. Seif, and C. Giaume. 2008. Gap junction-mediated astrocytic networks in the mouse barrel cortex. *J Neurosci* 28 (20):5207–17.

Houades, V., N. Rouach, P. Ezan, F. Kirchhoff, A. Koulakoff, and C. Giaume. 2006. Shapes of astrocyte networks in the juvenile brain. *Neuron Glia Biol* 2 (1):3–14.

Iglesias, R., G. Dahl, F. Qiu, D. C. Spray, and E. Scemes. 2009. Pannexin 1: the molecular substrate of astrocyte "hemichannels." *J Neurosci* 29 (21):7092–97.

Kafitz, K. W., H. R. Guttinger, and C. M. Muller. 1999. Seasonal changes in astrocytes parallel neuronal plasticity in the song control area HVc of the canary. *Glia* 27 (1):88–100.

Kielian, T. 2008. Glial connexins and gap junctions in CNS inflammation and disease. *J Neurochem* 106 (3):1000–16.

Kirchhoff, F. 2010. Neuroscience. Questionable calcium. *Science* 327 (5970):1212–13.

Koehler, R. C., R. J. Roman, and D. R. Harder. 2009. Astrocytes and the regulation of cerebral blood flow. *Trends Neurosci* 32 (3):160–69.

Konietzko, U., and C. M. Muller. 1994. Astrocytic dye coupling in rat hippocampus: topography, developmental onset, and modulation by protein kinase C. *Hippocampus* 4 (3):297–306.

Kuchibhotla, K. V., C. R. Lattarulo, B. T. Hyman, and B. J. Bacskai. 2009. Synchronous hyperactivity and intercellular calcium waves in astrocytes in Alzheimer mice. *Science* 323 (5918):1211–15.

MacVicar, B. A., and R. J. Thompson. 2010. Non-junction functions of pannexin-1 channels. *Trends Neurosci* 33 (2):93–102.

Marrero, H., and R. K. Orkand. 1996. Nerve impulses increase glial intercellular permeability. *Glia* 16 (3):285–89.

Matthias, K., F. Kirchhoff, G. Seifert, K. Huttmann, M. Matyash, H. Kettenmann, and C. Steinhauser. 2003. Segregated expression of AMPA-type glutamate receptors and glutamate transporters defines distinct astrocyte populations in the mouse hippocampus. *J Neurosci* 23 (5):1750–58.

Matyash, V., and H. Kettenmann. 2010. Heterogeneity in astrocyte morphology and physiology. *Brain Res Rev.* 63:2–10

Meme, W., M. Vandecasteele, C. Giaume, and L. Venance. 2009. Electrical coupling between hippocampal astrocytes in rat brain slices. *Neurosci Res* 63 (4):236–43.

Mugnaini, E. 1986. Cell junctions of astrocytes, ependymal and related cells in the mammal central nervous system, with emphasis on the hypothesis of a generalized syncytium of supporting cells. In *Astrocytes*, ed. F. S. a. V. A. New York: Academic Press. 329–371.

Muller, T., T. Moller, J. Neuhaus, and H. Kettenmann. 1996. Electrical coupling among Bergmann glial cells and its modulation by glutamate receptor activation. *Glia* 17 (4):274–84.

Mulligan, S. J., and B. A. MacVicar. 2004. Calcium transients in astrocyte endfeet cause cerebrovascular constrictions. *Nature* 431 (7005):195–99.

Nagy, J. I., W. Li, E. L. Hertzberg, and C. A. Marotta. 1996. Elevated connexin43 immunoreactivity at sites of amyloid plaques in Alzheimer's disease. *Brain Res* 717 (1–2):173–78.

Nagy, J. I., and J. E. Rash. 2000. Connexins and gap junctions of astrocytes and oligodendrocytes in the CNS. *Brain Res Brain Res Rev* 32 (1):29–44.

Nedergaard, M. 1994. Direct signaling from astrocytes to neurons in cultures of mammalian brain cells. *Science* 263 (5154):1768–71.

Nedergaard, M., B. Ransom, and S. A. Goldman. 2003. New roles for astrocytes: redefining the functional architecture of the brain. *Trends Neurosci* 26 (10):523–30.

Ogata, K., and T. Kosaka. 2002. Structural and quantitative analysis of astrocytes in the mouse hippocampus. *Neuroscience* 113 (1):221–33.

Orellana, J. A., D. E. Hernandez, P. Ezan, V. Velarde, M. V. Bennett, C. Giaume, and J. C. Saez. 2009. Hypoxia in high glucose followed by reoxygenation in normal glucose reduces the viability of cortical astrocytes through increased permeability of connexin 43 hemichannels. *Glia* 58 (3):329–43.

Orellana, J. A., P. J. Saez, K. F. Shoji, K. A. Schalper, N. Palacios-Prado, V. Velarde, C. Giaume, M. V. Bennett, and J. C. Saez. 2009. Modulation of brain hemichannels and gap junction channels by pro-inflammatory agents and their possible role in neurodegeneration. *Antioxid Redox Signal* 11 (2):369–99.

Orkand, R. K., J. G. Nicholls, and S. W. Kuffler. 1966. Effect of nerve impulses on the membrane potential of glial cells in the central nervous system of amphibia. *J Neurophysiol* 29 (4):788–806.

Pakhotin, P., and A. Verkhratsky. 2005. Electrical synapses between Bergmann glia cells and Purkinje neurones in rat cerebellar slices. *Molecular Cellular Neuroscience* 28:79–84.

Panchin, Y., I. Kelmanson, M. Matz, K. Lukyanov, N. Usman, and S. Lukyanov. 2000. A ubiquitous family of putative gap junction molecules. *Curr Biol* 10 (13):R473–74.

Peinado, A., R. Yuste, and L. C. Katz. 1993. Extensive dye coupling between rat neocortical neurons during the period of circuit formation. *Neuron* 10 (1):103–14.

Perea, G., M. Navarrete, and A. Araque. 2009. Tripartite synapses: astrocytes process and control synaptic information. *Trends Neurosci* 32 (8):421–31.

Peters, O., C. G. Schipke, A. Philipps, B. Haas, U. Pannasch, L. P. Wang, B. Benedetti, A. E. Kingston, and H. Kettenmann. 2009. Astrocyte function is modified by Alzheimer's disease-like pathology in aged mice. *J Alzheimers Dis* 18 (1):177–89.

Petersen, C. C., and B. Sakmann. 2000. The excitatory neuronal network of rat layer 4 barrel cortex. *J Neurosci* 20 (20):7579–86.

Reaume, A. G., P. A. de Sousa, S. Kulkarni, B. L. Langille, D. Zhu, T. C. Davies, S. C. Juneja, G. M. Kidder, and J. Rossant. 1995. Cardiac malformation in neonatal mice lacking connexin43. *Science* 267 (5205):1831–34.

Rela, L., A. Bordey, and C. A. Greer. 2010. Olfactory ensheathing cell membrane properties are shaped by connectivity. *Glia* 58 (6):665–78.

Retamal, M. A., N. Froger, N. Palacios-Prado, P. Ezan, P. J. Saez, J. C. Saez, and C. Giaume. 2007. Cx43 hemichannels and gap junction channels in astrocytes are regulated oppositely by proinflammatory cytokines released from activated microglia. *J Neurosci* 27 (50):13781–92.

Roerig, B., and M. B. Feller. 2000. Neurotransmitters and gap junctions in developing neural circuits. *Brain Res Brain Res Rev* 32 (1):86–114.

Rohlmann, A., and J.R. Wolff. 1998. Subcellular topography and plasticity of gap junction distribution in astrocytes. *Gap Junctions Nervous Syst* 154:175–92.

Rose, C. R., and B. R. Ransom. 1997. Gap junctions equalize intracellular Na+ concentration in astrocytes. *Glia* 20 (4):299–307.

Rouach, N., J. Glowinski, and C. Giaume. 2000. Activity-dependent neuronal control of gap-junctional communication in astrocytes. *J Cell Biol* 149 (7):1513–26.

Rouach, N., A. Koulakoff, V. Abudara, K. Willecke, and C. Giaume. 2008. Astroglial metabolic networks sustain hippocampal synaptic transmission. *Science* 322 (5907):1551–55.

Roux, L. and C. Giaume. 2009. Two astroglial networks are differentially regulated in the olfactory glomerular layer. *Glia* 57(suppl. 13) 57 (Abstract)

Rozental, R., A. F. Andrade-Rozental, X. Zheng, M. Urban, D. C. Spray, and F. C. Chiu. 2001. Gap junction-mediated bidirectional signaling between human fetal hippocampal neurons and astrocytes. *Dev Neurosci* 23 (6):420–31.

Rufer, M., S. B. Wirth, A. Hofer, R. Dermietzel, A. Pastor, H. Kettenmann, and K. Unsicker. 1996. Regulation of connexin-43, GFAP, and FGF-2 is not accompanied by changes in astroglial coupling in MPTP-lesioned, FGF-2-treated parkinsonian mice. *J Neurosci Res* 46 (5):606–17.

Samoilova, M., J. Li, M. R. Pelletier, K. Wentlandt, Y. Adamchik, C. C. Naus, and P. L. Carlen. 2003. Epileptiform activity in hippocampal slice cultures exposed chronically to bicuculline: increased gap junctional function and expression. *J Neurochem* 86 (3):687–99.

Scemes, E., and C. Giaume. 2006. Astrocyte calcium waves: what they are and what they do. *Glia* 54 (7):716–25.

Scemes, E., S. O. Suadicani, G. Dahl, and D. C. Spray. 2007. Connexin and pannexin mediated cell-cell communication. *Neuron Glia Biol* 3 (3):199–208.

Serrano, A., N. Haddjeri, J. C. Lacaille, and R. Robitaille. 2006. GABAergic network activation of glial cells underlies hippocampal heterosynaptic depression. *J Neurosci* 26 (20):5370–82.

Serrano, A., R. Robitaille, and J. C. Lacaille. 2008. Differential NMDA-dependent activation of glial cells in mouse hippocampus. *Glia* 56 (15):1648–63.

Sohl, G., S. Maxeiner, and K. Willecke. 2005. Expression and functions of neuronal gap junctions. *Nat Rev Neurosci* 6 (3):191–200.

Spray, D. C., Z. C. Ye, and B. R. Ransom. 2006. Functional connexin "hemichannels": a critical appraisal. *Glia* 54 (7):758–73.

Theis, M., R. Jauch, L. Zhuo, D. Speidel, A. Wallraff, B. Doring, C. Frisch, G. Sohl, B. Teubner, C. Euwens, J. Huston, C. Steinhauser, A. Messing, U. Heinemann, and K. Willecke. 2003. Accelerated hippocampal spreading depression and enhanced locomotory activity in mice with astrocyte-directed inactivation of connexin43. *J Neurosci* 23 (3):766–76.

Theis, M., G. Sohl, J. Eiberger, and K. Willecke. 2005. Emerging complexities in identity and function of glial connexins. *Trends Neurosci* 28 (4):188–95.

Valiunas, V., Y. Y. Polosina, H. Miller, I. A. Potapova, L. Valiuniene, S. Doronin, R. T. Mathias, R. B. Robinson, M. R. Rosen, I. S. Cohen, and P. R. Brink. 2005. Connexin-specific cell-to-cell transfer of short interfering RNA by gap junctions. *J Physiol* 568 (Pt 2):459–68.

Venance, L., J. Cordier, M. Monge, B. Zalc, J. Glowinski, and C. Giaume. 1995. Homotypic and heterotypic coupling mediated by gap junctions during glial cell differentiation *in vitro*. *Eur J Neurosci* 7 (3):451–61.

Venance, L., J. Premont, J. Glowinski, and C. Giaume. 1998. Gap junctional communication and pharmacological heterogeneity in astrocytes cultured from the rat striatum. *J Physiol (Lond)* 510 (Pt 1):429–40.

Von Blankenfeld, G., B. R. Ransom, and H. Kettenmann. 1993. Development of cell-cell coupling among cells of the oligodendrocyte lineage. *Glia* 7 (4):322–28.

Wallraff, A., R. Kohling, U. Heinemann, M. Theis, K. Willecke, and C. Steinhauser. 2006. The impact of astrocytic gap junctional coupling on potassium buffering in the hippocampus. *J Neurosci* 26 (20):5438–47.

Wallraff, A., B. Odermatt, K. Willecke, and C. Steinhauser. 2004. Distinct types of astroglial cells in the hippocampus differ in gap junction coupling. *Glia* 48 (1):36–43.

Wolff, J. R., K. Stuke, M. Missler, H. Tytko, P. Schwarz, A. Rohlmann, and T. I. Chao. 1998. Autocellular coupling by gap junctions in cultured astrocytes: a new view on cellular autoregulation during process formation. *Glia* 24 (1):121–40.

Woolsey, T. A., and H. Van der Loos. 1970. The structural organization of layer IV in the somatosensory region (SI) of mouse cerebral cortex. The description of a cortical field composed of discrete cytoarchitectonic units. *Brain Res* 17 (2):205–42.

Xu, L., L. H. Zeng, and M. Wong. 2009. Impaired astrocytic gap junction coupling and potassium buffering in a mouse model of tuberous sclerosis complex. *Neurobiol Dis* 34 (2):291–99.

Yamamoto, T., J. Vukelic, E. L. Hertzberg, and J. I. Nagy. 1992. Differential anatomical and cellular patterns of connexin43 expression during postnatal development of rat brain. *Brain Res Dev Brain Res* 66 (2):165–80.

6 Astrocyte-Neuron Communication

What Goes Wrong in Pathology?

Daniela Rossi
Laboratory for Research on Neurodegenerative
Disorders, Salvatore Maugeri Foundation, Pavia, Italy

Andrea Volterra
Department of Cell Biology and Morphology,
University of Lausanne, Lausanne, Switzerland

CONTENTS

6.1   INTRODUCTION: A HISTORICAL PERSPECTIVE

The initial identification of glial cells goes back to the mid-nineteenth century, when the German pathologist Rudolph Virchow indicated the existence in the brain tissue of a nonneuronal component that he described as an interstitial, cement-like substance interposed between the actual nervous elements and holding them together. He called this unusual substance neuroglia and thought it was connective tissue

(Virchow 1856). Basic illustrations of what we currently consider astrocytes were made a few years later by Otto Deiters, who depicted stellate cells within the connective tissue reported by Virchow, and represented them with morphological features already distinguishing the fibrous and protoplasmic cells of white and grey matter (Dieters 1865). It took another 4 years until Jakob Henle and Friedrich Merkel represented a network of star-shaped cells in the white matter of the spinal cord (Henle and Merkel 1869). The next, impressive contribution toward comprehension of the functional roles of astrocytes came, a few years later, from Camillo Golgi and Santiago Ramon y Cajal, thanks to the development of new staining techniques, notably the famous "black reaction." With this approach, Golgi noted that the processes of astrocytes contact blood vessels and proposed the innovative concept that these cellular elements play a central role in the distribution of energy substrates from capillaries to neurons (Golgi 1871). This hypothesis was, however, questioned by Cajal, who believed that the function of glial cells was related to their specific location in the microenvironment of neurons and proposed the *isolation* theory, according to which astrocytes act as neuronal insulators to avoid incorrect interactions between neurons and to preserve neuronal activity (Cajal 1995). Furthermore, together with his collaborators, Cajal conceived the additional theory that glial cells secrete trophic factors to support neuronal growth (Tello 1911). Although the above concepts have limitations, one needs to acknowledge that the ideas on glial function put forward by Golgi and Cajal are still largely valid nowadays (for a thorough overview see Kettenmann and Ransom 2005). A new impulse to the comprehension of astrocyte physiology came only several decades later, in the 1980s, when electrophysiological recordings with the patch-clamp technique (Neher et al. 1978; Hamill et al. 1981) revealed the presence of voltage- and ligand-gated ion channels in cultured astrocytes (MacVicar 1984; Bevan et al. 1985; Barres et al. 1988, 1990; MacVicar and Tse 1988; Sontheimer et al. 1988) and molecular studies identified expression of a wide array of neurotransmitter receptors in both cultured and tissue astrocytes, often mirroring the one of neighboring synapses (for review see Verkhratsky et al. 1998). The latter discovery represented an important conceptual breakthrough, as it introduced the idea that astrocytes can sense neuronal activity and be activated by it. In the last two decades this view was extraordinarily expanded, and several new studies have shown that astrocytes can respond to neuronal inputs by releasing a variety of chemical messengers that produce modulatory effects on both neuronal and vascular functions (a process defined as gliotransmission).

6.2   ASTROCYTES, INTERFACES, AND PHYSIOLOGY

6.2.1   TERRITORIAL ORGANIZATION AND FUNCTIONS OF PROTOPLASMIC ASTROCYTES

Astrocytes represent the major glial cell population in the mammalian brain and constitute up to 50% of its volume (Tower and Young 1973). Morphologically, they can be distinguished in two main classes, the fibrous astrocytes, mainly located in the central nervous system (CNS) white matter, and the protoplasmic astrocytes, found in the grey matter. In this chapter, we will specifically focus on the features of protoplasmic astrocytes (Volterra and Meldolesi 2005). This class of astrocytes displays a peculiar anatomical organization, occupying discrete, nonoverlapping parenchymal territories. Within one such territory, an individual astrocyte sends out from the cell body numerous and highly ramified processes that can establish contacts and local interactions with other cells (neurons, glia, and blood vessel cells) present in the territory (Bushong et al. 2002, 2004; Halassa et al. 2007). At the vascular interface, astrocytes exert an active control on the cerebral blood flow by releasing vasoactive mediators inducing local vasodilation or contraction (Kacem et al. 1998; Simard et al. 2003; Zonta et al. 2003; Mulligan and MacVicar 2004; Takano et al. 2006; Gordon et al. 2008). Moreover, they take up from the circulation glucose, the main energy source for neurons, and favor its entrance into the brain parenchyma (Figure 6.1). Within the astrocytes, glucose is either stored in the form of glycogen or transformed glycolytically into lactate, which can be then exported to neurons and utilized as metabolic substrate (Tsacopoulos and Magistretti 1996). At the neuronal interface, astrocytes exert a number of homeostatic functions that collectively contribute to maintain the microenvironment in conditions ensuring optimal neuronal function. It has been estimated that the territorial volume that a rodent astrocyte could oversee individually in the CA1 region of the hippocampus contains more than 100,000 synapses (Bushong et al. 2002) and a few hundred dendrites (Halassa et al. 2007). By expressing a high number of inward rectifier potassium channels on their plasma membrane, astrocytes contribute to reduce the extracellular potassium concentration during intense synaptic activity (Karwoski et al. 1989) and to avoid excessive neuronal depolarization and hyperexcitability (Figure 6.1). Likewise, by means of specific plasma membrane uptake proteins, astrocytes control the extracellular concentration of synaptically released neurotransmitters (Danbolt 2001). At glutamatergic synapses, astrocyte glutamate uptake is the main mechanism for removing the amino acid from the extracellular compartment and preventing excitotoxicity (Rothstein et al. 1996) (Figure 6.1).

Astrocyte-neuron interactions are not, however, limited to these activities. There is compelling evidence that receptors present on the astrocytic plasma membrane are activated by neurotransmitters released during synaptic activity and induce intracellular calcium concentration ($[Ca^{2+}]_i$) elevation in astrocytes (for review see Volterra and Meldolesi 2005; Jourdain et al. 2007; Navarrete and Araque 2008). On the other hand, $[Ca^{2+}]_i$ elevation in astrocytes was found to trigger release of chemical mediators (dubbed gliotransmitters, i.e., transmitters of glial origin, as opposed to neurotransmitters), such as glutamate, D-serine, Adenosine Triphosphate (ATP), and its metabolite adenosine (Parpura et al. 1994; Pasti et al. 1997; Bezzi et al. 1998) (Figure 6.1). Although not all studies agree (see Agulhon et al. 2010; Fiacco et al. 2007), many reports from several laboratories show that gliotransmitters can activate neuronal receptors and play potentially physiological roles in the control of neuronal excitability, synaptic transmission, and plasticity in brain slice preparations (for reviews see Volterra and Meldolesi 2005; Perea et al. 2009; Hamilton and Attwell 2010; for original studies see Henneberger et al. 2010; Pasti et al. 1997; Angulo et al. 2004; Fellin et al. 2004; Kozlov et al. 2006; Jourdain et al. 2007; Perea and Araque 2007; Navarrete and Araque 2008; Shigetomi et al. 2008).

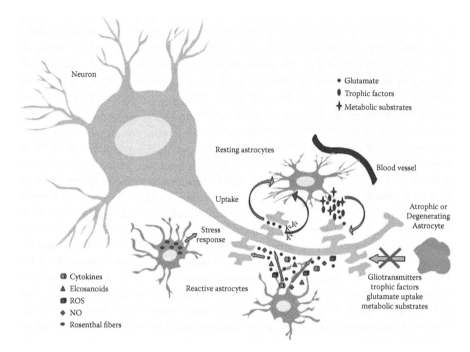

FIGURE 6.1 **(See color insert.)** Schematic representation of astrocyte-neuron interactions in physiopathological conditions. In physiological conditions, astrocytes exhibit relatively long and highly branched cell processes that can perform many functions: (1) Owing to their morphological connections with capillaries at the endfeet, they can take up nutrients from the blood and transform and transfer them to neurons. (2) They are highly permeable to potassium ions and can rapidly clear their excess from the peri-synaptic space. (3) They express plasma membrane transporters for several neurotransmitters, including glutamate, and clear them from the extracellular space upon synaptic release. (4) They provide neurons with trophic factors and release gliotrasmitters. In pathological conditions, astrocytes become activated, i.e., show thicker cell processes, and start to secrete factors, such as cytokines, reactive oxygen species (ROS), nitric oxide (NO), eicosanoids, and excitotoxins, by which they can damage nearby neurons. Furthermore, in specific pathological instances, they show fibrous, eosinophilic deposits known as Rosenthal fibers. Astrocyte degeneration deprives neurons of trophic factors, metabolic substrates, and glutamate uptake, and interrupts astrocyte-neuron communication.

In addition to being released in response to $[Ca^{2+}]_i$ elevations, glutamate can be discharged from astrocytes through Ca^{2+}-independent pathways, possibly activated under conditions and at loci different from those involved in Ca^{2+}-dependent release. Among the Ca^{2+}-independent mechanisms, we can mention the transport via reversed operation of reuptake carriers, probably taking place under ischemic conditions (Rossi et al. 2000); the exchange with cystine, mediated by specific cystine-glutamate antiporters (Warr et al. 1999); and the molecular permeation of large pore channels, including $P2X_7$ receptors (Duan et al. 2003), gap junction hemichannels (Ye et al. 2003; Kang et al. 2008), and volume-sensitive organic anion channels (Kimelberg et al. 1990).

In this chapter, we will focus on the mechanism(s) underlying astrocytic Ca^{2+}-dependent glutamate release and we will discuss the implications that alterations of this mechanism may have in specific neuroinflammatory and degenerative processes. Furthermore, we will report current evidence that not only neurons, but also astrocytes, in pathological conditions can become vulnerable to physiological concentrations of glutamate and degenerate.

6.2.2 C_{A}^{2+}-DEPENDENT GLUTAMATE RELEASE FROM ASTROCYTES

The first evidence that activation of endogenous receptors in astrocytes triggers Ca^{2+}-dependent glutamate release and signaling to neurons dates back to the early 1990s. In a pioneering study in mixed neuron-glia cultures, Haydon, Parpura, and colleagues demonstrated that stimulation with the peptide bradykinin triggers $[Ca^{2+}]_i$ rises and glutamate release from astrocytes, followed by N-Methyl-D-Aspartate (NMDA) receptor activation in neurons (Parpura et al. 1994). A few years later, two related studies (Pasti et al. 1997; Bezzi et al. 1998) showed that the events initially observed in cell cultures occur also in the intact brain tissue. The first study showed that stimulation of neuronal afferents in acute cortical or hippocampal slices induces $[Ca^{2+}]_i$ oscillations in the surrounding astrocytes, most likely activated by synaptically released glutamate, as they were blocked by metabotropic receptor (mGluR) antagonists. Astrocytic $[Ca^{2+}]_i$ oscillations were followed by GluR-dependent $[Ca^{2+}]_i$ rises in the surrounding neurons, most likely resulting from glutamate release from astrocytes. The second study provided direct demonstration that activation of astrocyte glutamate receptors (mGluRs, either alone or in combination with alpha-amino-3-hydroxy-5-methyl-4- isoxazole-propionic acid (AMPA) receptors) triggers glutamate release from astrocytes via a Ca^{2+}-dependent mechanism and indicated a role for prostaglandins (PGs) in this process. In the hippocampal stratum radiatum, such glutamate- and prostaglandin-dependent glutamate release from astrocytes induced glutamate receptor-dependent $[Ca^{2+}]_i$ responses in neighboring pyramidal neurons, revealing the existence of a glutamatergic bidirectional communication between synapses and astrocytes in situ. The role of prostaglandins in the process was not fully clarified, although application of micromolar exogenous PGE_2 induced astrocytic $[Ca^{2+}]i$ elevation and glutamate release, suggesting that PGs could act as autocrine or paracrine amplifiers of the signaling cascades activated by glutamate receptor stimulation. Interestingly, the same study by Bezzi et al. (1998) showed that glutamate-induced glutamate release from astrocytes is occlusive with the Ca^{2+}-dependent release triggered by bradykinin, suggesting that distinct astrocyte receptors are coupled to a common Ca^{2+}-dependent transduction pathway ultimately leading to glutamate release. Glutamate release via this Ca^{2+}-dependent pathway was, in contrast, additive to Ca^{2+}-independent release mediated by glutamate transporters, indicating that Ca^{2+}-dependent and Ca^{2+}-independent release mechanisms use different glial glutamate pools. Later studies provided evidence that the Ca^{2+}-dependent pathway is generally triggered by activation of G-protein-coupled receptors (GPCRs) linked to inositol 1,4,5-triphosphate (IP_3)-induced Ca^{2+} release from stores of the endoplasmic reticulum (Sanzgiri et al. 1999; Bezzi et al. 2001; Jeremic et al. 2001; Kang et al. 2005; Takano et al. 2005). The release process would involve

Ca^{2+}-regulated exocytosis of glutamate-containing vesicles, for which there is over-whelming evidence in cultured astrocytes and good evidence in tissue astrocytes (see below). However, a report exists that GPCR- and Ca^{2+}-dependent glutamate release may also occur via permeation of large anion channels by cytosolic glutamate (Takano et al. 2005), so the issue of the mechanism controlling Ca^{2+}-dependent glutamate release is not conclusively resolved (Hamilton and Attwell 2010). The existence of a regulated exocytotic pathway in cultured astrocytes has been documented by a variety of experimental approaches, including optical detection methods (Bezzi et al. 2004; Crippa et al. 2006; Bowser and Khakh 2007; Marchaland et al. 2008), membrane capacitance measurements (Kreft et al. 2004; Zhang et al. 2004b), electrochemical amperometry (Chen et al. 2005), and selective interference with proteins of the exocytotic machinery (Montana et al. 2004; Zhang et al. 2004a). Several GPCRs, including group I mGluRs, $P2Y_1$ purinergic receptors, CXCR4 chemokine receptors, and thrombin PAR-1 receptors, were shown to respond to ligand stimulation with Ca^{2+}-dependent glutamate exocytosis in cultured astrocytes (Domercq et al. 2006; Bowser and Khakh 2007; Cali et al. 2008). An initial indication of glutamate exocytosis from astrocytes in situ was provided by the ultrastructural identification of synaptic-like microvesicles (SLMVs) in thin astrocytic processes surrounding synapses in the hippocampus (Bezzi et al. 2004; Jourdain et al. 2007). SLMVs were found to express on their membranes' vesicular glutamate transporters (VGLUTs), responsible for glutamate uptake into vesicles, and the vesicle-associated membrane protein (VAMP), a protein essential for the formation of the **SNAP RE**ceptor (SNARE) fusion pore that drives vesicle exocytosis. The critical role of VAMP in Ca^{2+}-dependent glutamate release from astrocytes in situ was subsequently demonstrated in functional experiments where astrocyte-evoked glutamate-dependent synaptic modulation was abolished by infusing tetanus neurotoxin (TeNT) selectively in the astrocytes (Jourdain et al. 2007; Perea and Araque 2007). This toxin is known to block exocytosis by specifically cleaving and inactivating VAMP (either VAMP2, the isoform present in neurons, or VAMP3, the isoform present in astrocytes). It has been correctly pointed out that TeNT sensitivity of the release process is not final proof of exocytosis of glutamate-containing vesicles in situ, as the toxin could block vesicle-mediated insertion of ion channels, ultimately mediating glutamate release (Hamilton and Attwell 2010). However, this latter hypothesis is incompatible with other observations, notably that Ca^{2+}-dependent glutamate release in situ is also sensitive to bafilomycin A1 (Domercq et al. 2006), a toxin that prevents refilling of glutamatergic vesicles of their transmitter content, but not membrane insertion of the transmitter-emptied vesicles (Ikeda and Bekkers 2009).

Interestingly, glutamate exocytosis from cultured astrocytes in response to stimulation of purinergic $P2Y_1$ receptors ($P2Y_1R$) is regulated not only by prostaglandins, but also by the cytokine tumor necrosis factor α (TNFα), another pro-inflammatory mediator (Domercq et al. 2006). The precise mechanism of action of TNFα remains to be established, but it is conceivable that the cytokine exerts a tonic control on key steps of the secretory process initiated by $P2Y_1R$ stimulation. The same TNFα and PG dependence of the glutamatergic response was observed upon stimulation of astrocytes with the chemotactic cytokine (chemokine) stromal-derived factor-1α (SDF-1α)/CXCL12, acting on its GPCR, CXCR4 (Bezzi et al. 2001). The fact that factors typically implicated in inflammatory processes control the Ca^{2+}-dependent glutamate release cascade of astrocytes suggests that this signaling pathway could be importantly altered in pathological conditions and participate in the neurodamaging processes.

6.3 REACTION OF ASTROCYTES TO INJURY

The nervous system (NS) responds to many forms of injury and to diverse pathological conditions with an "atypical" inflammatory process that occurs locally in the brain parenchyma and does not involve the peripheral immune system with infiltration of leukocytes from the blood. Such a response, defined *neuroinflammation*, typically involves a modification of glial cells, particularly microglia and astrocytes. It is commonly claimed that these cells become active (or reactive), i.e., undergo morphological and functional changes and can start to proliferate.

Glial morphological alterations have been traditionally described in terms of cellular hypertrophy (Duchen 1992; Ridet et al. 1997). However, this concept has been recently revised with regard to astrocytes, as it has been reported that reactive astrocytes display hypertrophic cell processes with a correspondingly reduced extracellular space, but the territorial volume that each astrocyte occupies in the tissue remains unchanged (Wilhelmsson et al. 2006). Furthermore, it is now clear that astrocytes do not respond in a stereotypic fashion to all forms of injury, with the mode and extent of astrocytic activation varying in dependence on both the severity of the insult and the context of the injury site. An additional factor that may condition the characteristics of the astrocytic reaction is represented by the intrinsic nature of the cells, considering that reactive astrocytes are a biochemically heterogeneous population (for review see Ridet et al. 1997). Whether this heterogeneity is due to original differences between subsets of astrocytes or to the interaction of homogeneous astrocytes with different mediators secreted in their microenvironment (by damaged neurons, astrocytes, microglia, and endothelial cells) remains to be clarified (Hewett 2009). The signals exchanged between different neural cell types during injury or in the course of neuropathologies are largely unknown, yet the transition of glial cells from the resting to the activated state appears to be associated with marked upregulation of several genes and secretion of a number of factors, like cytokines, eicosanoids, reactive oxygen species, nitric oxide, and excitatory amino acids (Perry et al. 1995) (Figure 6.1). Little is known about the impact that these functional changes in astrocytes have on the function of the synaptic circuitry and on neuronal survival. The inflammatory reaction was originally thought to be mostly beneficial for tissue-repairing processes. However, in some cases, reactive astrocytes may be also deleterious. In view of the control exerted by TNFα and prostaglandins on the Ca^{2+}-dependent glutamate release process of astrocytes, it is reasonable to expect that overproduction of these mediators during neuroinflammation affects the properties of the process and, consequently, the glutamatergic input to neurons, resulting in an unbalance of the synaptic circuitry. Furthermore, it was recently reported that selective induction of reactive astrocytosis produces a circuital unbalance by reducing inhibitory synaptic transmission in hippocampal slices (Ortinski et al. 2010).

Another observation, still not widely considered, is that in specific pathological circumstances, the reaction of astrocytes to injury or disease may take the form of atrophy or degeneration (Figure 6.1). There is little information on the morphological changes accompanying astrocytic degeneration. Duchen identified three distinctive features: disintegration of the astrocytic processes (*clasmatodendrosis*), cytoplasmic swelling, and appearance of lipid droplets in the cytoplasm (Duchen 1992). Another abnormality, identified in a variety of chronic pathologic conditions, is the appearance of Rosenthal fibers, i.e., homogeneous, hyaline, and eosinophilic oval bodies located in the astrocytic cell bodies and processes (Janzer and Friede 1981; Duchen 1992). Such structures typically characterize disorders like Alexander's disease, which is caused by missense mutations in the gene encoding glial fibrillary acidic protein (GFAP) (Brenner et al. 2001), and also other conditions involving chronic GFAP upregulation, including slow-growth astrocytoma (Gessaga and Anzil 1975), multiple sclerosis (Herndon et al. 1970), and tuberous sclerosis (Jay et al. 1998). Lowe and colleagues reported that Rosenthal fibers are immunopositive for ubiquitin, a protein associated with intracellular protein degradation, and suggested that the formation of ubiquitin-GFAP conjugates may cause the degradation of the astrocytic cytoskeleton with the consequent formation of Rosenthal fibers (Lowe et al. 1989). Noteworthy, induction of GFAP overexpression in transgenic mice with formation of Rosenthal fibers in astrocytes produced a stress response that compromised neuronal function (Hagemann et al. 2005) (Figure 6.1). More recent work reconstructing the three-dimensional morphology of GFAP-positive astrocytic profiles during neurodegenerative processes has added further detail to the description of atrophic or degenerating astrocytes, reporting changes in the cell body size and reduction in the number of main processes and in the complexity of their arborization (Rossi et al. 2008; Rodriguez et al. 2009). In addition, astrocyte degeneration was associated to the activation of cell death effector enzymes (e.g., caspases) and to DNA damage (Su et al. 2000; Giffard and Swanson 2005; Danilov and Fiskum 2008; Rossi et al. 2008; for review see Takuma et al. 2004). Although the mechanisms of astrocyte degeneration are still only imprecisely defined, it is becoming more and more clear that, in pathological conditions, these cells can cause neuronal damage not only directly, by secreting neurotoxic factors, but also indirectly, by undergoing themselves degeneration and, consequently, losing functions that are critically important for neurons.

6.4 EXCESS GLUTAMATE RELEASE FROM ASTROCYTES AND NEURONAL CELL DEATH: THE EXAMPLE OF AIDS DEMENTIA COMPLEX

The neurological complications resulting from Human Immunodeficiency Virus (HIV) infection of the brain are associated to loss of neuronal cells (Kaul et al. 2005; Mattson et al. 2005). Cells die by apoptosis (Bagetta et al. 1996; Shi et al. 1996), apparently via NMDA receptor-mediated excitotoxicity. This can eventually result in cognitive impairment and progressively evolve to Acquired Immune Deficiency Syndrome (AIDS) dementia complex (ADC). Importantly, several observations indicate that neurotropic strains of HIV usually gain access to the brain by targeting microglial cells, rather than neurons, suggesting that neuronal cell death is non-cell autonomous and occurs via interactions with microglia and astrocytes (Meucci and Miller 1996; Toggas et al. 1996; Kaul et al. 2001). In this context of damaging cell-cell interactions, an important role might be played by viral particles shed from the infected cells, including the HIV-1 coat protein gp120. This was initially suggested by the observation that expression of gp120 in transgenic mice reproduced several features of ADC neuropathology (Toggas et al. 1996). In addition, a gp120 isoform, $gp120_{IIIB}$, derived from the T-tropic HIV-1 IIIB strain, was found to act as CXCR4 agonist on both astrocytes and microglia. Based on the observation that astrocytes and microglia form local foci of reactive cells in the brains of subjects with HIV neuropathology, we established in our laboratory a model of cocultures of reactive microglia and astrocytes so as to mimic the pathological situation in vivo (Bezzi et al. 2001). Stimulation of the cultures with $gp120_{IIIB}$ activated CXCR4 signaling in both astrocytes and microglia, leading to the production of unusually high concentrations of TNFα (>500 pM), which in turn led to a dramatically increased Ca^{2+}-dependent glutamate release from astrocytes compared to the response observed in cultures devoid of reactive microglia. Since production and shedding of viral proteins in HIV-1-infected brains is an uncontrolled process, we postulated that one of its consequences could be overstimulation of glial CXCR4 with resultant excess glutamate release from astrocytes eventually triggering excitotoxic neuronal cell death. This hypothesis was corroborated by experimental evidence. Thus, challenging hippocampal cultures containing neurons, astrocytes, and microglia with $gp120_{IIIB}$ led to slow apoptotic death of a subpopulation of neurons via a glutamate-dependent mechanism mediated by NMDA receptors (Bezzi et al. 2001). This provided a novel mechanistic hypothesis for HIV-associated neuropathology, potentially relevant for therapeutic intervention. Current antiretroviral agents, although significantly prolonging the life expectancy of AIDS-affected subjects, do not fully control AIDS-associated neuropathology. Therefore, drugs limiting the development of ADC and ensuring a better quality of life are highly needed. Clinical trials have been conducted with memantine, an inhibitor of the NMDA receptor channels, and proved helpful in limiting excitotoxicity (Zhao et al. 2010). A series of second-generation memantine derivatives are currently in development and may have even greater neuroprotective properties than the original drug (Lipton 2007). However, several additional potential targets have been identified in our study in relation to the glutamatergic toxic pathway of astrocytes, including PG- and TNFα-forming enzymes and receptors. A number of drugs acting at the level of such targets are already available and should be tested for beneficial effects against ADC.

6.5 ASTROCYTE ALTERATIONS IN ALZHEIMER'S DISEASE

Alzheimer's disease (AD) is a neurodegenerative disorder characterized by progressive loss of cognitive functions. One of its biochemical hallmarks is accumulation of the amyloid-β (Aβ) protein into amyloid plaques that can be found in the extracellular space of forebrain regions (Glenner and Wong 1984). Early morphological and biochemical studies established a correlation between amyloid deposition and cognitive deficits (Cummings et al. 1996; Naslund et al. 2000; Bussiere et al. 2002). More specifically, results obtained in various transgenic AD models suggest that cognitive deficits are the consequence of synaptic failure induced by Aβ (Selkoe 2002). Recent observations with two-photon calcium (Ca^{2+}) imaging in vivo provided further insight into the nature of synaptic alterations revealing that many individual neurons, located near Aβ plaques, are functionally hyperactive, showing an increased frequency of spontaneous Ca^{2+} transients, likely because of a relative decrease in synaptic inhibition. In addition, a number of cells throughout the cortex of AD mice were found to be silent, showing no Ca^{2+} transients. This suggested that the altered functionality of the cortical circuitry in Alzheimer's disease is not due to an overall decrease in synaptic activity, but rather to a redistribution of synaptic drive leading to a global unbalance (Busche et al. 2008). Concerning astrocytes, it has been proposed that they play an homeostatic function against amyloid plaque formation by actively clearing up Aβ peptide when it forms, and that this function is lost in AD (Wyss-Coray et al. 2003). A possible cause is a defect in apolipoprotein E (Koistinaho et al. 2004). This protein, whose allele ε4 represents a recognized genetic risk factor for AD, is mainly produced by astrocytes in the mammalian brain and seems to play a central role in the degradation and clearance of deposited Aβ. Once Aβ starts to accumulate, not only neurons but also astrocytes are perturbed by its presence. For example, incubation of cultured astrocytes with Aβ peptides led to alterations of astrocyte metabolism that affected neuronal viability in astrocyte-neuron cocultures (Allaman et al. 2010). Furthermore, administration of Aβ peptides to mixed cultures of hippocampal neurons and astrocytes caused abnormal $[Ca^{2+}]_i$ transients and mitochondrial depolarization in astrocytes long before any impairment was visible in neurons. Therefore, the subsequent neuronal death might be the result of oxidative stress generated by the astrocytic dysfunction (Abramov et al. 2003, 2004). Finally, important alterations of astrocytic calcium signaling, leading to the spread of large intercellular waves in the astrocytic network, were recently described by two-photon Ca^{2+} imaging in vivo in the cerebral cortex of a mouse model of AD (Kuchibhotla et al. 2009). While the specific implications of astrocytic Ca^{2+} signaling alterations are not yet known, the existence of this phenomenon suggests that in AD the circuital dysfunction may be generalized, involving not only synaptic transmission but also neuron-glia communications, and be collectively responsible for cognitive decline.

The presence of chronically inflamed glial cells around Aβ plaques is a well-documented feature of AD (for review see Wyss-Coray 2006). While it is not clear whether inflammation plays a causative role in the disease, recent data from animal models suggest that inflammatory processes might contribute to at least accelerate its progression. High levels of pro-inflammatory cytokines such as interleukin (IL) 1β, IL 6, and TNFα, produced by reactive microglia and astrocytes, are detected in the brain of AD subjects (McGeer and McGeer 2002b). These agents might have a toxic action in AD because inhibition of their production results in restoration of hippocampal synaptic dysfunction and attenuation of the behavioral deficits induced by Aβ in mice (Ralay et al. 2006). Given the reported control of TNFα on the Ca^{2+}-dependent glutamate release process of astrocytes, it is reasonable that this mechanism constitutes one of the targets of the deleterious cytokine action. This hypothesis was tested by our group using the PDAPP mice, a transgenic line that overexpresses a mutant human amyloid precursor protein (APP) implicated in familial AD under control of platelet-derived growth factor promoter (Rossi et al. 2005). In particular, we utilized aged animals (12 months old), presenting abundant amyloid plaque deposition and reactive gliosis in the forebrain, and presymptomatic animals (4 months old) with little or no amyloid deposits and no apparent glial alteration. Ca^{2+}-dependent glutamate release from astrocytes was stimulated in brain slices from PDAPP animals and controls by direct application of high concentrations of exogenous $TNF\alpha$. The amount of glutamate detected in hippocampal slices from aged PDAPP animals was lower than that found in presymptomatic mice and age-matched controls (Rossi et al. 2005). The defect was region selective, as the glutamate release response from cerebellar slices of aged PDAPP mice was identical to that of controls. These observations were unexpected, particularly considering previous data in cultures where, in an acute experiment, production of high $TNF\alpha$ levels by reactive microglia strongly potentiated the astrocytic release (Bezzi et al. 2001). However, in AD, glial inflammation is a chronic phenomenon, particularly around amyloid plaques, where endogenous $TNF\alpha$ levels are presumably constantly high and may overstimulate receptors causing functional uncoupling. Indeed, there is evidence that $TNF\alpha$ signal transduction is altered in AD, in particular the protein DENN (**d**ifferentially **e**xpressed in **n**ormal vs. **n**eoplastic)/ MADD (**M**APK-**a**ctivating **d**eath **d**omain), which transduces activation of the TNF receptor 1 (TNFR1) to multiple downstream signaling pathways, and is downregulated in the hippocampi of both Alzheimer patients and Tg2576 mice (Zhao et al. 2003; Del Villar and Miller 2004). In agreement, we found that this protein is defective also in the hippocampus (not cerebellum) of aged PDAPP mice (Rossi et al. 2005), suggesting that alteration of $TNF\alpha$ signaling downstream of TNFR1 may account for the loss of the normal control exerted by the cytokine on Ca^{2+}-dependent glutamate release. At present, the functional significance of this defect is not established, nor is its relation with the reported increase in astrocyte Ca^{2+} signaling in AD mice (Kuchibotla et al. 2009). However, one can speculate that a reduced astrocytic glutamate input may result in weakened connectivity of excitatory hippocampal synapses (Jourdain et al. 2007). Moreover, astrocytic $TNF\alpha$-dependent signaling was shown to favor synaptic strength (Beattie et al. 2002) and induce homeostatic synaptic upscaling (Beattie et al. 2002; Stellwagen and Malenka 2006; Kaneko et al. 2008) by promoting insertion of post-synaptic AMPA receptor subunits. Therefore, disruption of $TNF\alpha$ signal transduction in astrocytes could lead to several alterations converging in a progressive reduction of synaptic efficacy and underlie cognitive decline in AD.

6.6 ASTROCYTES AS PLAYERS IN AMYOTROPHIC LATERAL SCLEROSIS PATHOGENESIS

Amyotrophic lateral sclerosis (ALS) is a neurodegenerative disease characterized by the progressive loss of corticospinal and spinal motor neurons. Most forms of ALS are sporadic (sALS), but about 10% of patients have an inherited familial form of the disease (fALS). While various genes and genetic loci have been recently linked to this disorder (Kwiatkowski et al. 2009; Vance et al. 2009; for review see Wijesekera and Leigh 2009), most of the current knowledge on ALS pathogenesis is based on the discovery of mutations of the ubiquitously expressed enzyme Cu-Zn superoxide dismutase (SOD1). Thus, the seminal finding that ~20% of fALS and ~1% of sALS are linked to dominant mutations in the gene coding for SOD1 (Rosen et al. 1993) prompted the generation of transgenic mice (Gurney et al. 1994; Wong et al. 1995; Bruijn et al.

1997; for review see Turner and Talbot 2008) and rats (Howland et al. 2002; Nagai et al. 2007) carrying single-amino-acid mutations of human SOD1, which were found to recapitulate several features of the human pathology, confirming the existence of a cause-effect relationship between mutant SOD1 expression and ALS.

Because SOD1 activity was found to be reduced in the blood and brain of fALS patients (Deng et al. 1993), it was initially proposed that mutations in the enzyme could be toxic because they diminish free radical scavenging capability. However, this hypothesis was later challenged by studies in transgenic mice showing that motor neuron disease developed even in the presence of elevated or unchanged SOD1 activity levels (Gurney et al. 1994; Wong et al. 1995; Bruijn et al. 1997; for review see Turner and Talbot 2008). It was therefore suggested that mutant SOD1s become toxic to motor neurons because of a gain of function, the nature of which remains obscure despite intense investigations. While the primary toxic property of mutant SOD1s is unresolved, a hint of the cascade of events implicated in motor neuron degeneration came by the landmark observation that death of the motor cells is not a cell-autonomous process, but instead involves interaction with neighboring non-neuronal cells, particularly microglia and astrocytes (Clement et al. 2003). Massive activation of microglia and astrocytes in the areas of motor neuron loss was reported in both sporadic and familial human cases, as well as in transgenic animal models (McGeer and McGeer 2002a; Boillee et al. 2006; Yamanaka et al. 2008). Microglial alterations were shown to be directly implicated in favoring disease progression in vivo (Beers et al. 2006; Boillee et al. 2006). Silencing mutant SOD1 expression in astrocytes also affected disease progression in transgenic mice, and the deleterious role of astrocytes was ascribed to their ability to amplify the inflammatory response of microglia (Yamanaka et al. 2008). Noteworthy, however, the impact of glial cells on the course of the disease does not depend on their proliferation, as ablation of both dividing microglia and astrocytes had no effects on disease outcome in transgenic mice (Gowing et al. 2008; Lepore et al. 2008). New insight on the possible mechanistic contribution of astrocytes to motor neuron degeneration was provided by a number of recent in vitro studies, which found that mutant SOD1-expressing astrocytes in culture release factors selectively toxic for motor neurons, such as reactive oxygen species (ROS) and prostaglandin D_2 (Di Giorgio et al. 2007, 2008; Nagai et al. 2007; Marchetto et al. 2008). In mutant SOD1-bearing astrocytes, oxygen consumption and Adenosine Diphosphate (ADP)-dependent respiratory control in mitochondria are impaired, suggesting that motor neuron survival may be affected by ROS generated in the astrocytic mitochondria (Cassina et al. 2008). Expression of SOD1 mutants would also disrupt the protective functions of astrocytes against excitotoxic neuronal damage. For example, wild-type astrocytes exert a positive control on the expression of the GluR2 AMPAR subunit in motor neurons, thereby limiting AMPAR permeability to calcium ions, a potential source of excitotoxicity, but this property is lost in astrocytes expressing mutant SOD1s (Van Damme et al. 2007). Moreover, astrocyte glutamate uptake is downregulated due to loss of the Excitatory Amino Acid Transporter 2 (EAAT2, also known as Glutamate Transporter 1 (GLT1)), a landmark defect in ALS, observed in both human patients (Rothstein et al. 1995) and transgenic animals (Bruijn et al. 1997; Bendotti et al. 2001; Howland et al. 2002; Yang et al. 2009). EAAT2 crucially controls extracellular glutamate clearance in the spinal cord, and its loss may enhance extracellular glutamate up to excitotoxic levels (Rothstein et al. 1996). EAAT2 transcription is controlled by functional interactions between axons and astrocytes, which are compromised in ALS in view of the progressive damage and eventually loss of corticospinal motor neuron terminals (Yang et al. 2009). Partial restoration of EAAT2 expression in ALS transgenics via genetic (Guo et al. 2003; Pardo et al. 2006) or pharmacological approaches (Rothstein et al. 2005) was reported to have a positive impact on disease progression.

Another peculiar abnormality in SOD1 mutant mice in vivo is the presence of astrocytes containing protein inclusions made of SOD1, ubiquitin, or activated caspase-3 (Bruijn et al. 1997; Pasinelli et al. 2000) (Figure 6.2A). To gain insight into the significance of such astrocytic inclusions in the context of ALS pathology, our group performed a thorough histopathological analysis of the lumbar tract of the spinal cord from transgenic mice carrying the Gly93 → Ala substitution in the human

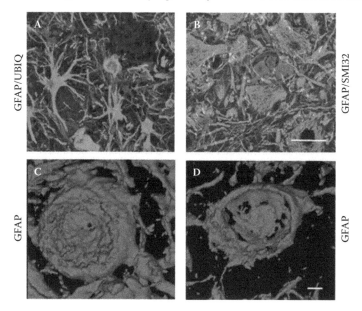

FIGURE 6.2 **(See color insert.)** A subpopulation of astrocytes undergoes degeneration in ALS transgenic mice during disease progression. (A) All spheroid-shaped astrocytes are positive for ubiquitin and often are located near vacuolated motor neurons. Scale bar, 20 μm (A and B). (C, D) 3D reconstruction of spheroid-shaped astrocytes. 3D reconstruction of z-stack confocal microscopy images rendered by ImageJ software (National Institute of Health) and VolumeJ plug in. (B) Lumbar spinal cord sections from ALS mice were simultaneously labeled for nonphosphorylated neurofilaments (SMI32 antibody, green) and GFAP (red). The image is representative of reactive and spheroid-shaped astrocytes, directly contiguous to motor neurons, at the age of disease onset. (C) The 3D section of this spheroid astrocyte shows a compact GFAP cytoskeleton around a central core and the presence of a few GFAP-positive processes, which appear short and abnormally thick. This cell is immunonegative for the active caspase-3, indicative of an early stage of degeneration. (D) The 3D section of this spheroid astrocyte displays a much thinner and less compact GFAP cytoskeleton around the central core compared to the example in (C), with rare GFAP-positive processes. This cell is positive for the active caspase-3 (not shown), indicative of a late stage of degeneration. Scale bar, 2 μm (C and D).

SOD1 amino acid sequence (hSOD1^{G93A}). We localized a subpopulation of astrocytes harboring protein inclusions specifically in the neighborhood of motor cells (Figure 6.2B). These astrocytes displayed morphological and biochemical features reminiscent of degenerating cells (Rossi et al. 2008), including a spheroid cell body with increased diameter and a reduced number or even absence of GFAP-positive cell processes that, when present, appeared short and abnormally thick compared to those of normal astrocytes (Figure 6.2C,D). Degenerating astrocytes were first observed at a presymptomatic stage, when motor neurons show axonal damage but are still alive (Pun et al. 2006), and significantly increased in number concomitant with the onset of neuronal degeneration and the appearance of ALS symptoms. We then studied the mechanism underlying astrocyte degeneration in cultured spinal cord astrocytes and surprisingly found that mutant SOD1s, either G93A or G85R, do not exert a direct pro-apoptotic action, but rather make the astrocytes vulnerable to glutamate, even at low concentrations. The glutamatergic mechanism responsible for the deleterious effect involves metabotropic glutamate receptor 5 (mGluR5) signaling. Thus, blockage of this receptor reduced apoptosis of mutant SOD1-expressing astrocytes in response to glutamate challenges. Moreover, administration of an mGluR5 antagonist in vivo reduced astrocyte degeneration in the lumbar spinal cord, delayed the appearance of ALS symptoms, and extended survival in hSOD1^{G93A} transgenic mice (Rossi et al. 2008). The involvement of mGluR5 is particularly intriguing because this receptor was found to be selectively overexpressed in astrocytes in the ventral spinal cord of ALS patients (Aronica et al. 2001; Anneser et al. 2004). Astrocytes are intimately associated with synapses and can sense neurotransmitter released during synaptic activity (Volterra and Meldolesi 2005). Therefore, it is possible that spinal cord astrocytes, endangered by the expression of ALS-linked mutant SOD1s, become vulnerable to physiological glutamate concentrations released at neighboring synapses and start to degenerate. This in turn may deprive the neighboring motor neurons of the optimal microenvironment and accelerate their degeneration in an interactive process of reciprocal damage (Figure 6.1).

6.7   CONCLUDING REMARKS

The increasing awareness that astrocytes are active partners of neurons in brain communication and possess both neurotransmitter receptors and regulated forms of transmitter release necessarily changes the classical view of neurodegenerative disorders as cell-autonomous processes, largely or totally consisting in an intrinsic dysfunction of neuronal cells. The evidence that a number of microenvironmental stimuli trigger $[Ca^{2+}]_i$ elevations in astrocytes, and that this may lead to release of gliotransmitters, which in turn affect the activity of neighboring neurons, transmits the crucial message that the contribution of astrocytes, and of glia in general, to synaptic physiology and pathology must be thoroughly considered in order to fully comprehend brain function and disease. Paradoxically, in some cases, like for ADC, neurodegeneration could result mainly from glial rather than neuronal alterations. In other cases, like in ALS and perhaps AD, neurodegeneration could instead depend on multiple alterations and unbalances occurring in parallel and interactively in neurons and glia. In this context, we have shown that astrocytes undergo two distinct types of transformation during injury and disease, i.e., activation and atrophy/degeneration. We predict that a better understanding of these phenomena, as well as of the signaling alterations underlying astrocyte-neuron communication in pathology, will lead to significant progress toward the cure of neurodegenerative disorders.

ACKNOWLEDGMENTS

We thank Drs. Liliana Brambilla and Francesca Martorana for figure drawings and artwork. The original studies on ALS were supported by grants from Fondazione Telethon (GGP02052 and GGP05244), EMBO (ALTF 279–2002), Fondazione Cariplo, and Ministero Italiano della Università e Ricerca. Work in AV Lab is supported by grants from the Swiss National Foundation (3100A0–120398) and from the Synapsis Foundation (Zurich, CH) and the MBF Foundation (Triesen, LI).

REFERENCES

Abramov, A. Y., L. Canevari, et al. (2003). Changes in intracellular calcium and glutathione in astrocytes as the primary mechanism of amyloid neurotoxicity. *J Neurosci* 23(12): 5088–95.

Abramov, A. Y., L. Canevari, et al. (2004). Beta-amyloid peptides induce mitochondrial dysfunction and oxidative stress in astrocytes and death of neurons through activation of NADPH oxidase. *J Neurosci* 24(2): 565–75.

Agulhon, C., T. A. Fiacco, et al. (2010). Hippocampal short- and long-term plasticity are not modulated by astrocyte Ca2+ signaling. *Science* 327(5970): 1250–54.

Allaman, I., M. Gavillet, et al. (2010). Amyloid-beta aggregates cause alterations of astrocytic metabolic phenotype: impact on neuronal viability. *J Neurosci* 30(9): 3326–38.

Angulo, M. C., A. S. Kozlov, et al. (2004). Glutamate released from glial cells synchronizes neuronal activity in the hippocampus. *J Neurosci* 24(31): 6920–27.

Anneser, J. M., C. Chahli, et al. (2004). Glial proliferation and metabotropic glutamate receptor expression in amyotrophic lateral sclerosis. *J Neuropathol Exp Neurol* 63(8): 831–40.

Aronica, E., M. V. Catania, et al. (2001). Immunohistochemical localization of group I and II metabotropic glutamate receptors in control and amyotrophic lateral sclerosis human spinal cord: upregulation in reactive astrocytes. *Neuroscience* 105(2): 509–20.

Bagetta, G., M. T. Corasaniti, et al. (1996). The HIV-1 gp120 causes ultrastructural changes typical of apoptosis in the rat cerebral cortex. *Neuroreport* 7(11): 1722–24.

Barres, B. A., L. L. Chun, et al. (1988). Ion channel expression by white matter glia. I. Type 2 astrocytes and oligodendrocytes. *Glia* 1(1): 10–30.

Barres, B. A., W. J. Koroshetz, et al. (1990). Ion channel expression by white matter glia: the type-1 astrocyte. *Neuron* 5(4): 527–44.

Beattie, E. C., D. Stellwagen, et al. (2002). Control of synaptic strength by glial TNFalpha. *Science* 295(5563): 2282–85.

Beers, D. R., J. S. Henkel, et al. (2006). Wild-type microglia extend survival in PU.1 knockout mice with familial amyotrophic lateral sclerosis. *Proc Natl Acad Sci USA* 103(43): 16021–26.

Bendotti, C., M. Tortarolo, et al. (2001). Transgenic SOD1 G93A mice develop reduced GLT-1 in spinal cord without alterations in cerebrospinal fluid glutamate levels. *J Neurochem* 79(4): 737–46.

Bevan, S., S. Y. Chiu, et al. (1985). The presence of voltage-gated sodium, potassium and chloride channels in rat cultured astrocytes. *Proc R Soc Lond B Biol Sci* 225(1240): 299–313.

Bezzi, P., G. Carmignoto, et al. (1998). Prostaglandins stimulate calcium-dependent glutamate release in astrocytes. *Nature* 391(6664): 281–85.

Bezzi, P., M. Domercq, et al. (2001). CXCR4-activated astrocyte glutamate release via TNFalpha: amplification by microglia triggers neurotoxicity. *Nat Neurosci* 4(7): 702–10.

Bezzi, P., V. Gundersen, et al. (2004). Astrocytes contain a vesicular compartment that is competent for regulated exocytosis of glutamate. *Nat Neurosci* 7(6): 613–20.

Boillee, S., K. Yamanaka, et al. (2006). Onset and progression in inherited ALS determined by motor neurons and microglia. *Science* 312(5778): 1389–92.

Bowser, D. N., and B. S. Khakh (2007). Two forms of single-vesicle astrocyte exocytosis imaged with total internal reflection fluorescence microscopy. *Proc Natl Acad Sci USA* 104(10): 4212–17.

Brenner, M., A. B. Johnson, et al. (2001). Mutations in GFAP, encoding glial fibrillary acidic protein, are associated with Alexander disease. *Nat Genet* 27(1): 117–20.

Bruijn, L. I., M. W. Becher, et al. (1997). ALS-linked SOD1 mutant G85R mediates damage to astrocytes and promotes rapidly progressive disease with SOD1-containing inclusions. *Neuron* 18(2): 327–38.

Busche, M. A., G. Eichhoff, et al. (2008). Clusters of hyperactive neurons near amyloid plaques in a mouse model of Alzheimer's disease. *Science* 321(5896): 1686–89.

Bushong, E. A., M. E. Martone, et al. (2002). Protoplasmic astrocytes in CA1 stratum radiatum occupy separate anatomical domains. *J Neurosci* 22(1): 183–92.

Bushong, E. A., M. E. Martone, et al. (2004). Maturation of astrocyte morphology and the establishment of astrocyte domains during postnatal hippocampal development. *Int J Dev Neurosci* 22(2): 73–86.

Bussiere, T., P. D. Friend, et al. (2002). Stereologic assessment of the total cortical volume occupied by amyloid deposits and its relationship with cognitive status in aging and Alzheimer's disease. *Neuroscience* 112(1): 75–91.

Cajal, S. R. (1995). *Histology of the nervous system*. New York: Oxford University Press.

Cali, C., J. Marchaland, et al. (2008). SDF 1-alpha (CXCL12) triggers glutamate exocytosis from astrocytes on a millisecond time scale: imaging analysis at the single-vesicle level with TIRF microscopy. *J Neuroimmunol* 198(1–2): 82–91.

Cassina, P., A. Cassina, et al. (2008). Mitochondrial dysfunction in SOD1G93A-bearing astrocytes promotes motor neuron degeneration: prevention by mitochondrial-targeted antioxidants. *J Neurosci* 28(16): 4115–22.

Chen, X., L. Wang, et al. (2005). "Kiss-and-run" glutamate secretion in cultured and freshly isolated rat hippocampal astrocytes. *J Neurosci* 25(40): 9236–43.

Clement, A. M., M. D. Nguyen, et al. (2003). Wild-type nonneuronal cells extend survival of SOD1 mutant motor neurons in ALS mice. *Science* 302(5642): 113–17.

Crippa, D., U. Schenk, et al. (2006). Synaptobrevin2-expressing vesicles in rat astrocytes: insights into molecular characterization, dynamics and exocytosis. *J Physiol* 570(Pt 3): 567–82.

Cummings, B. J., C. J. Pike, et al. (1996). Beta-amyloid deposition and other measures of neuropathology predict cognitive status in Alzheimer's disease. *Neurobiol Aging* 17(6): 921–33.

Danbolt, N. C. (2001). Glutamate uptake. *Prog Neurobiol* 65(1): 1–105.

Danilov, C. A., and G. Fiskum (2008). Hyperoxia promotes astrocyte cell death after oxygen and glucose deprivation. *Glia* 56(7): 801–8.

Del Villar, K., and C. A. Miller (2004). Down-regulation of DENN/MADD, a TNF receptor binding protein, correlates with neuronal cell death in Alzheimer's disease brain and hippocampal neurons. *Proc Natl Acad Sci USA* 101(12): 4210–15.

Deng, H. X., A. Hentati, et al. (1993). Amyotrophic lateral sclerosis and structural defects in Cu,Zn superoxide dismutase. *Science* 261(5124): 1047–51.

Dieters, O. (1865). *Untersuchungen uber Gehirn und Ruckenmark des Menschen und der Saugethiere*. Braunschweig: Vieweg.

Di Giorgio, F. P., G. L. Boulting, et al. (2008). Human embryonic stem cell-derived motor neurons are sensitive to the toxic effect of glial cells carrying an ALS-causing mutation. *Cell Stem Cell* 3(6): 637–48.

Di Giorgio, F. P., M. A. Carrasco, et al. (2007). Non-cell autonomous effect of glia on motor neurons in an embryonic stem cell-based ALS model. *Nat Neurosci* 10(5): 608–14.

Domercq, M., L. Brambilla, et al. (2006). P2Y1 receptor-evoked glutamate exocytosis from astrocytes: control by tumor necrosis factor-alpha and prostaglandins. *J Biol Chem* 281(41): 30684–96.

Duan, S., C. M. Anderson, et al. (2003). P2X7 receptor-mediated release of excitatory amino acids from astrocytes. *J Neurosci* 23(4): 1320–28.

Duchen, L. W. (1992). *General pathology of neurons and neuroglia.* New York: Oxford.

Fellin, T., O. Pascual, et al. (2004). Neuronal synchrony mediated by astrocytic glutamate through activation of extrasynaptic NMDA receptors. *Neuron* 43(5): 729–43.

Fiacco, T. A., C. Agulhon, et al. (2007). Selective stimulation of astrocyte calcium *in situ* does not affect neuronal excitatory synaptic activity. *Neuron* 54(4): 611–26.

Gessaga, E. C., and A. P. Anzil (1975). Rod-shaped filamentous inclusions and other ultrastructural features in a cerebellar astrocytoma. *Acta Neuropathol* 33(2): 119–27.

Giffard, R. G., and R. A. Swanson (2005). Ischemia-induced programmed cell death in astrocytes. *Glia* 50(4): 299–306.

Glenner, G. G., and C. W. Wong (1984). Alzheimer's disease: initial report of the purification and characterization of a novel cerebrovascular amyloid protein. *Biochem Biophys Res Commun* 120(3): 885–90.

Golgi, C. (1871). *Contribuzione alla Fina Anatomia degli organi centrali del sistema nervoso.* Bologna: Rivista Clinica di Bologna.

Gordon, G. R., H. B. Choi, et al. (2008). Brain metabolism dictates the polarity of astrocyte control over arterioles. *Nature* 456(7223): 745–49.

Gowing, G., T. Philips, et al. (2008). Ablation of proliferating microglia does not affect motor neuron degeneration in amyotrophic lateral sclerosis caused by mutant superoxide dismutase. *J Neurosci* 28(41): 10234–44.

Guo, H., L. Lai, et al. (2003). Increased expression of the glial glutamate transporter EAAT2 modulates excitotoxicity and delays the onset but not the outcome of ALS in mice. *Hum Mol Genet* 12(19): 2519–32.

Gurney, M. E., H. Pu, et al. (1994). Motor neuron degeneration in mice that express a human Cu,Zn superoxide dismutase mutation. *Science* 264(5166): 1772–75.

Hagemann, T. L., S. A. Gaeta, et al. (2005). Gene expression analysis in mice with elevated glial fibrillary acidic protein and Rosenthal fibers reveals a stress response followed by glial activation and neuronal dysfunction. *Hum Mol Genet* 14(16): 2443–58.

Halassa, M. M., T. Fellin, et al. (2007). Synaptic islands defined by the territory of a single astrocyte. *J Neurosci* 27(24): 6473–77.

Hamill, O. P., A. Marty, et al. (1981). Improved patch-clamp techniques for high-resolution current recording from cells and cell-free membrane patches. *Pflugers Arch* 391(2): 85–100.

Hamilton, N. B., and D. Attwell. (2010). Do astrocytes really exocytose neurotransmitters? *Nat Rev Neurosci* 11(4): 227–38.

Henle, J., and F. Merkel (1869). *Ueber die sogenannte Bindesubstanz der Centralorgane des Nervensystems. Z. Ration. Med.* 34: 49–82.

Henneberger, C., T. Papouin, et al. (2010). Long-term potentiation depends on release of D-serine from astrocytes. *Nature* 463(7278): 232–36.

Herndon, R. M., L. J. Rubinstein, et al. (1970). Light and electron microscopic observations on Rosenthal fibers in Alexander's disease and in multiple sclerosis. *J Neuropathol Exp Neurol* 29(4): 524–51.

Hewett, J. A. (2009). Determinants of regional and local diversity within the astroglial lineage of the normal central nervous system. *J Neurochem* 110(6): 1717–36.

Howland, D. S., J. Liu, et al. (2002). Focal loss of the glutamate transporter EAAT2 in a transgenic rat model of SOD1 mutant-mediated amyotrophic lateral sclerosis (ALS). *Proc Natl Acad Sci USA* 99(3): 1604–9.

Ikeda, K., and Bekkers, J. M. (2009). Counting the number of releasable synaptic vesicles in a presynaptic terminal. *Proc Natl Acad Sci USA* 106: 2945–59.

Janzer, R. C., and R. L. Friede (1981). Do Rosenthal fibers contain glial fibrillary acid protein? *Acta Neuropathol* 55(1): 75–76.

Jay, V., V. Edwards, et al. (1998). Cerebellar pathology in tuberous sclerosis. *Ultrastruct Pathol* 22(4): 331–39.

Jeremic, A., K. Jeftinija, et al. (2001). ATP stimulates calcium-dependent glutamate release from cultured astrocytes. *J Neurochem* 77(2): 664–75.

Jourdain, P., L. H. Bergersen, et al. (2007). Glutamate exocytosis from astrocytes controls synaptic strength. *Nat Neurosci* 10(3): 331–39.

Kacem, K., P. Lacombe, et al. (1998). Structural organization of the perivascular astrocyte endfeet and their relationship with the endothelial glucose transporter: a confocal microscopy study. *Glia* 23(1): 1–10.

Kaneko, M., D. Stellwagen, et al. (2008). Tumor necrosis factor-alpha mediates one component of competitive, experience-dependent plasticity in developing visual cortex. *Neuron* 58(5): 673–80.

Kang, J., N. Kang, et al. (2008). Connexin 43 hemichannels are permeable to ATP. *J Neurosci* 28(18): 4702–11.

Kang, N., J. Xu, et al. (2005). Astrocytic glutamate release-induced transient depolarization and epileptiform discharges in hippocampal CA1 pyramidal neurons. *J Neurophysiol* 94(6): 4121–30.

Karwoski, C. J., H. K. Lu, et al. (1989). Spatial buffering of light-evoked potassium increases by retinal Muller (glial) cells. *Science* 244(4904): 578–80.

Kaul, M., G. A. Garden, et al. (2001). Pathways to neuronal injury and apoptosis in HIV-associated dementia. *Nature* 410(6831): 988–94.

Kaul, M., J. Zheng, et al. (2005). HIV-1 infection and AIDS: consequences for the central nervous system. *Cell Death Differ* 12(Suppl 1): 878–92.

Kettenmann, H., and B. R. Ransom, Eds. (2005). *Neuroglia.* Oxford: Oxford University Press.

Kimelberg, H. K., S. K. Goderie, et al. (1990). Swelling-induced release of glutamate, aspartate, and taurine from astrocyte cultures. *J Neurosci* 10(5): 1583–91.

Koistinaho, M., S. Lin, et al. (2004). Apolipoprotein E promotes astrocyte colocalization and degradation of deposited amyloid-beta peptides. *Nat Med* 10(7): 719–26.

Kozlov, A. S., M. C. Angulo, et al. (2006). Target cell-specific modulation of neuronal activity by astrocytes. *Proc Natl Acad Sci USA* 103(26): 10058–63.

Kreft, M., M. Stenovec, et al. (2004). Properties of Ca(2+)-dependent exocytosis in cultured astrocytes. *Glia* 46(4): 437–45.

Kuchibhotla, K. V., C. R. Lattarulo, et al. (2009). Synchronous hyperactivity and intercellular calcium waves in astrocytes in Alzheimer mice. *Science* 323(5918): 1211–15.

Kwiatkowski, T. J., Jr., D. A. Bosco, et al. (2009). Mutations in the FUS/TLS gene on chromosome 16 cause familial amyotrophic lateral sclerosis. *Science* 323(5918): 1205–8.

Lepore, A. C., C. Dejea, et al. (2008). Selective ablation of proliferating astrocytes does not affect disease outcome in either acute or chronic models of motor neuron degeneration. *Exp Neurol* 211(2): 423–32.

Lipton, S. A. (2007). Pathologically-activated therapeutics for neuroprotection: mechanism of NMDA receptor block by memantine and S-nitrosylation. *Curr Drug Targets* 8(5): 621–32.

Lowe, J., K. Morrell, et al. (1989). Rosenthal fibres are based on the ubiquitination of glial filaments. *Neuropathol Appl Neurobiol* 15(1): 45–53.

MacVicar, B. A. (1984). Voltage-dependent calcium channels in glial cells. *Science* 226(4680): 1345–47.

MacVicar, B. A., and F. W. Tse (1988). Norepinephrine and cyclic adenosine 3':5'-cyclic monophosphate enhance a nifedipine-sensitive calcium current in cultured rat astrocytes. *Glia* 1(6): 359–65.

Marchaland, J., C. Cali, et al. (2008). Fast subplasma membrane Ca2+ transients control exo-endocytosis of synaptic-like microvesicles in astrocytes. *J Neurosci* 28(37): 9122–32.

Marchetto, M. C., A. R. Muotri, et al. (2008). Non-cell-autonomous effect of human SOD1 G37R astrocytes on motor neurons derived from human embryonic stem cells. *Cell Stem Cell* 3(6): 649–57.

Mattson, M. P., N. J. Haughey, et al. (2005). Cell death in HIV dementia. *Cell Death Differ* 12(Suppl 1): 893–904.

McGeer, P. L., and E. G. McGeer (2002a). Inflammatory processes in amyotrophic lateral sclerosis. *Muscle Nerve* 26(4): 459–70.

McGeer, P. L., and E. G. McGeer (2002b). Local neuroinflammation and the progression of Alzheimer's disease. *J Neurovirol* 8(6): 529–38.

Meucci, O., and R. J. Miller (1996). gp120-induced neurotoxicity in hippocampal pyramidal neuron cultures: protective action of TGF-beta1. *J Neurosci* 16(13): 4080–88.

Montana, V., Y. Ni, et al. (2004). Vesicular glutamate transporter-dependent glutamate release from astrocytes. *J Neurosci* 24(11): 2633–42.

Mulligan, S. J., and B. A. MacVicar. (2004). Calcium transients in astrocyte endfeet cause cerebrovascular constrictions. *Nature* 431(7005): 195–99.

Nagai, M., D. B. Re, et al. (2007). Astrocytes expressing ALS-linked mutated SOD1 release factors selectively toxic to motor neurons. *Nat Neurosci* 10(5): 615–22.

Naslund, J., V. Haroutunian, et al. (2000). Correlation between elevated levels of amyloid beta-peptide in the brain and cognitive decline. *JAMA* 283(12): 1571–77.

Navarrete, M., and A. Araque (2008). Endocannabinoids mediate neuron-astrocyte communication. *Neuron* 57(6): 883–93.

Neher, E., B. Sakmann, et al. (1978). The extracellular patch clamp: a method for resolving currents through individual open channels in biological membranes. *Pflugers Arch* 375(2): 219–28.

Ortinski, P. I., J. Dong, et al. (2010). Selective induction of astrocytic gliosis generates deficits in neuronal inhibition. *Nat Neurosci* 13(5): 584–91.

Pardo, A. C., V. Wong, et al. (2006). Loss of the astrocyte glutamate transporter GLT1 modifies disease in SOD1(G93A) mice. *Exp Neurol* 201(1): 120–30.

Parpura, V., T. A. Basarsky, et al. (1994). Glutamate-mediated astrocyte-neuron signalling. *Nature* 369(6483): 744–47.

Pasinelli, P., M. K. Houseweart, et al. (2000). Caspase-1 and -3 are sequentially activated in motor neuron death in Cu,Zn superoxide dismutase-mediated familial amyotrophic lateral sclerosis. *Proc Natl Acad Sci USA* 97(25): 13901–6.

Pasti, L., A. Volterra, et al. (1997). Intracellular calcium oscillations in astrocytes: a highly plastic, bidirectional form of communication between neurons and astrocytes *in situ*. *J Neurosci* 17(20): 7817–30.

Perea, G., and A. Araque (2007). Astrocytes potentiate transmitter release at single hippocampal synapses. *Science* 317(5841): 1083–86.

Perea, G., Navarrete, M. et al. (2009). Tripartite synapses: astrocytes process and control synaptic information. *Trends Neurosci* 32: 421–31.

Perry, V. H., M. D. Bell, et al. (1995). Inflammation in the nervous system. *Curr Opin Neurobiol* 5(5): 636–41.

Pun, S., A. F. Santos, et al. (2006). Selective vulnerability and pruning of phasic motoneuron axons in motoneuron disease alleviated by CNTF. *Nat Neurosci* 9(3): 408–19.

Ralay Ranaivo, H., J. M. Craft, et al. (2006). Glia as a therapeutic target: selective suppression of human amyloid-beta-induced upregulation of brain proinflammatory cytokine production attenuates neurodegeneration. *J Neurosci* 26(2): 662–70.

Ridet, J. L., S. K. Malhotra, et al. (1997). Reactive astrocytes: cellular and molecular cues to biological function. *Trends Neurosci* 20(12): 570–77.

Rodriguez, J. J., M. Olabarria, et al. (2009). Astroglia in dementia and Alzheimer's disease. *Cell Death Differ* 16(3): 378–85.

Rosen, D. R., T. Siddique, et al. (1993). Mutations in Cu/Zn superoxide dismutase gene are associated with familial amyotrophic lateral sclerosis. *Nature* 362(6415): 59–62.

Rossi, D., L. Brambilla, et al. (2005). Defective tumor necrosis factor-alpha-dependent control of astrocyte glutamate release in a transgenic mouse model of Alzheimer disease. *J Biol Chem* 280(51): 42088–96.

Rossi, D., L. Brambilla, et al. (2008). Focal degeneration of astrocytes in amyotrophic lateral sclerosis. *Cell Death Differ* 15(11): 1691–700.

Rossi, D. J., T. Oshima, et al. (2000). Glutamate release in severe brain ischaemia is mainly by reversed uptake. *Nature* 403(6767): 316–21.

Rothstein, J. D., M. Dykes-Hoberg, et al. (1996). Knockout of glutamate transporters reveals a major role for astroglial transport in excitotoxicity and clearance of glutamate. *Neuron* 16(3): 675–86.

Rothstein, J. D., S. Patel, et al. (2005). Beta-lactam antibiotics offer neuroprotection by increasing glutamate transporter expression. *Nature* 433(7021): 73–77.

Rothstein, J. D., M. Van Kammen, et al. (1995). Selective loss of glial glutamate transporter GLT-1 in amyotrophic lateral sclerosis. *Ann Neurol* 38(1): 73–84.

Sanzgiri, R. P., A. Araque, et al. (1999). Prostaglandin E(2) stimulates glutamate receptor-dependent astrocyte neuromodulation in cultured hippocampal cells. *J Neurobiol* 41(2): 221–29.

Selkoe, D. J. (2002). Alzheimer's disease is a synaptic failure. *Science* 298(5594): 789–91.

Shi, B., U. De Girolami, et al. (1996). Apoptosis induced by HIV-1 infection of the central nervous system. *J Clin Invest* 98(9): 1979–90.

Shigetomi, E., D. N. Bowser, et al. (2008). Two forms of astrocyte calcium excitability have distinct effects on NMDA receptor-mediated slow inward currents in pyramidal neurons. *J Neurosci* 28(26): 6659–63.

Simard, M., G. Arcuino, et al. (2003). Signaling at the gliovascular interface. *J Neurosci* 23(27): 9254–62.

Sontheimer, H., H. Kettenmann, et al. (1988). Glutamate opens Na+/K+ channels in cultured astrocytes. *Glia* 1(5): 328–36.

Stellwagen, D., and R. C. Malenka. (2006). Synaptic scaling mediated by glial TNF-alpha. *Nature* 440(7087): 1054–59.

Su, J. H., K. E. Nichol, et al. (2000). DNA damage and activated caspase-3 expression in neurons and astrocytes: evidence for apoptosis in frontotemporal dementia. *Exp Neurol* 163(1): 9–19.

Takano, T., J. Kang, et al. (2005). Receptor-mediated glutamate release from volume sensitive channels in astrocytes. *Proc Natl Acad Sci USA* 102(45): 16466–71.

Takano, T., G. F. Tian, et al. (2006). Astrocyte-mediated control of cerebral blood flow. *Nat Neurosci* 9(2): 260–67.

Takuma, K., A. Baba, et al. (2004). Astrocyte apoptosis: implications for neuroprotection. *Prog Neurobiol* 72(2): 111–27.

Tello, F. (1911). La influencia del neurotrofismo en la regeneracion de los centros nerviosos. *Trab Lab Inv Biol Univ Madrid* 9: 123–59.

Toggas, S. M., E. Masliah, et al. (1996). Prevention of HIV-1 gp120-induced neuronal damage in the central nervous system of transgenic mice by the NMDA receptor antagonist memantine. *Brain Res* 706(2): 303–7.

Tower, D. B., and O. M. Young. (1973). The activities of butyrylcholinesterase and carbonic anhydrase, the rate of anaerobic glycolysis, and the question of a constant density of glial cells in cerebral cortices of various mammalian species from mouse to whale. *J Neurochem* 20(2): 269–78.

Tsacopoulos, M., and P. J. Magistretti. (1996). Metabolic coupling between glia and neurons. *J Neurosci* 16(3): 877–85.

Turner, B. J., and K. Talbot. (2008). Transgenics, toxicity and therapeutics in rodent models of mutant SOD1-mediated familial ALS. *Prog Neurobiol* 85(1): 94–134.

Vance, C., B. Rogelj, et al. (2009). Mutations in FUS, an RNA processing protein, cause familial amyotrophic lateral sclerosis type 6. *Science* 323(5918): 1208–11.

Van Damme, P., E. Bogaert, et al. (2007). Astrocytes regulate GluR2 expression in motor neurons and their vulnerability to excitotoxicity. *Proc Natl Acad Sci USA* 104(37): 14825–30.

Verkhratsky, A., R. K. Orkand, et al. (1998). Glial calcium: homeostasis and signaling function. *Physiol Rev* 78(1): 99–141.

Virchow, R. (1856). *Gesammelte Abhandlungen zur wissenschaftlichen Medizin*. Frankfurt: Meidinger Sohn.

Volterra, A., and J. Meldolesi. (2005). Astrocytes, from brain glue to communication elements: the revolution continues. *Nat Rev Neurosci* 6(8): 626–40.

Warr, O., M. Takahashi, et al. (1999). Modulation of extracellular glutamate concentration in rat brain slices by cystine-glutamate exchange. *J Physiol* 514(Pt 3): 783–93.

Wijesekera, L. C., and P. N. Leigh. (2009). Amyotrophic lateral sclerosis. *Orphanet J Rare Dis* 4: 3.

Wilhelmsson, U., E. A. Bushong, et al. (2006). Redefining the concept of reactive astrocytes as cells that remain within their unique domains upon reaction to injury. *Proc Natl Acad Sci USA* 103(46): 17513–18.

Wong, P. C., C. A. Pardo, et al. (1995). An adverse property of a familial ALS-linked SOD1 mutation causes motor neuron disease characterized by vacuolar degeneration of mitochondria. *Neuron* 14(6): 1105–16.

Wyss-Coray, T. (2006). Inflammation in Alzheimer disease: driving force, bystander or beneficial response? *Nat Med* 12(9): 1005–15.

Wyss-Coray, T., J. D. Loike, et al. (2003). Adult mouse astrocytes degrade amyloid-beta *in vitro* and *in situ*. *Nat Med* 9(4): 453–57.

Yamanaka, K., S. J. Chun, et al. (2008). Astrocytes as determinants of disease progression in inherited amyotrophic lateral sclerosis. *Nat Neurosci* 11(3): 251–53.

Yang, Y., O. Gozen, et al. (2009). Presynaptic regulation of astroglial excitatory neurotransmitter transporter GLT1. *Neuron* 61(6): 880–94.

Ye, Z. C., M. S. Wyeth, et al. (2003). Functional hemichannels in astrocytes: a novel mechanism of glutamate release. *J Neurosci* 23(9): 3588–96.

Zhang, Q., M. Fukuda, et al. (2004a). Synaptotagmin IV regulates glial glutamate release. *Proc Natl Acad Sci USA* 101(25): 9441–46.

Zhang, Q., T. Pangrsic, et al. (2004b). Fusion-related release of glutamate from astrocytes. *J Biol Chem* 279(13): 12724–33.

Zhao, M., D. H. Cribbs, et al. (2003). The induction of the TNF alpha death domain signaling pathway in Alzheimer's disease brain. *Neurochem Res* 28(2): 307–18.

Zhao, Y., B. A. Navia, et al. (2010). Memantine for AIDS dementia complex: open-label report of ACTG 301. *HIV Clin Trials* 11(1): 59–67.

Zonta, M., M. C. Angulo, et al. (2003). Neuron-to-astrocyte signaling is central to the dynamic control of brain microcirculation. *Nat Neurosci* 6(1): 43–50.

7 Blood–Brain Barrier and the Neural Vascular Unit

Rolf Dermietzel
Department of Neuroanatomy and Molecular Brain
Research, University of Bochum, Bochum, Germany

David C. Spray
Dominick P. Purpura Department of Neuroscience
Department of Neuroscience, Albert Einstein
College of Medicine, Bronx, New York

CONTENTS

7.1   INTRODUCTION

This chapter focuses on some recent discoveries regarding molecular driving forces for cerebral angio- and barriergenesis and the formation of the gliovascular complex, which comprises the structural and functional unit of the blood–brain barrier (BBB). Progress made in our understanding of the ontogeny of the BBB, its regulation, and interaction with the brain parenchyma in the adult brain is amazing and will certainly be helpful in approaching new strategies to treat a variety of human diseases of the brain in the future. Since numerous reviews, handbooks, and monographs dealing with the blood–brain barrier have been published (e.g., Dermietzel et al. 2006) within recent years, we primarily emphasize discoveries that are not of common knowledge and which in some way mark turnarounds in our former understanding of the BBB establishment and function. We start with a brief appraisal on the evolution of the BBB and continue with respect to molecular clues that drive the maturation of the BBB during ontogenesis. A focus is placed on the driving force of oxygen tension for angiogenesis and barriergenesis, which features the most characteristic aspect of the brain vasculature. This is followed by discussion of the impact of two cellular entities, neuroglial cells and pericytes, on the development and maintenance of the BBB. Two features of the BBB are of particular interest: the tight junction, which enables the cerebral endothelium to perform its barrier function, and the intimate association of the vasculature with the polarized astroglia. The formation of the glial endfeet has long been a descriptive term in histology. The discovery that these endfeet constitute highly differentiated plasmalemmal domains furnished with a complement of channels and regulatory, membrane-associated proteins that provide key features attributing to mechanisms responsible for brain homeostasis is a hallmark in BBB research. Finally, we give a brief account of some pathological aspects of the BBB and its underlying molecular mechanisms before this chapter finishes, with some recent, in part controversial, aspects of the dynamics of vascular perfusion of brain tissue.

7.1.1   THE BLOOD–BRAIN BARRIER: AN EVOLUTIONARY PERSPECTIVE

The achievement of the blood–brain barrier (BBB) during evolution parallels the centralization and concentration of neuronal functions. Increase and remodeling of neural tissues as a consequence of adaptive mechanisms of organisms required efficient ways of delivering metabolites to and from the nervous system and constitutive clearing of waste products (Niewenhuis et al. 1998, 2002, Abbott et al. 1986, 2006, 2010; Bundgart and Abbott 2008). The elaboration of a potent blood supply was one of the major advances that allowed fueling of the emerging brain tissue. Obviously, optimal nourishing of the brain was not only a logistic problem, but necessitated an intricate surveillance of the delivery processes that allowed the creation of a specialized *micromilieu* necessary to maintain optimal neuronal functioning. This selective advantage was made possible by the development of the BBB. The barrier is essential for controlling the influx and efflux of metabolites and ions into the brain parenchyma and keeps out toxic substrates, which may potentially be harmful for brain functions. The identification of the structural locus of the blood–brain barrier was provided by Reese, Brightman, and Karnovsky (Brightman and Reese 1969, Reese and Karnovsky 1967), who determined that the electron opaque tracer's horseradish peroxidase (HRP) and lanthanum penetrated only part way between endothelial cells lining the central nervous system (CNS) capillaries, its diffusion being blocked by the presence of sealing tight junctions, together with complex tortuosity contributed by the presence of other junctional types.

Endothelial cells line all vessels, but in the brain they are specialized for optimization of impermeability, reduced transcytotic activity, or fenestrae that characterize other endothelia; as a consequence, the barrier formed by brain capillaries is 50–100 times tighter than that of the peripheral vasculature (see Cardoso et al. 2010). Moreover, CNS endothelial cells are rich in mitochondria and the capillaries are embedded in the parenchyma, with a continuous basement membrane surrounding the endothelium and separating it from the endfeet of astrocytes.

As considered in more detail below, breakdown of the blood–brain barrier is a hallmark feature of numerous neuropathological conditions. In response to such insults, penetration of endothelial junctions is an initial event, with compensatory changes in glial cells, which Abbott and colleagues (2006) have termed a "second line of defense." Phylogenetically, there has been a switch in the cell type that provides the barrier from peripheral blood to brain. In those invertebrates that have such a barrier, it is formed by perivascular or boundary glial cells, whereas in most vertebrates, the barrier is at the endothelium (for review see Bundgaard and Abbott 2008). As shown in electron micrographs of brain vasculature following horseradish peroxidase injection into the circulation, a transition occurred in primitive fishes. In elasmobranchs (sharks, skates, and rays) the restrictive boundary is at the level of the glial cells, and in sturgeon, even though the endothelium is continuous in the vessel wall, perivascular glia form the barrier. By contrast, in lungfish, hagfish, lamprey, and teleosts, the barrier is endothelial, with no transendothelial HRP diffusion. The authors of these interesting comparative studies point out that during mammalian development there is a glial barrier formed beneath the pia and in the ependyma, demonstrating the capability of glia to provide a barrier function. Moreover, treatment of human astrocytes with inflammatory cytokines induces expression of tight junction proteins while connexin expression is reduced, again providing evidence that such sealing junctions may be made by glia under some conditions (Duffy et al. 2000).

Two structural components have become paramount for the development of the BBB: the endothelium of blood vessels and its glial and pericytic linings. Perivascular cells, primarily astrocytes, do not only account for the structural barrier in invertebrates (Abbott et al. 1986), but help to establish barrier properties of the endothelium during embryogenesis in companion with pericytes, and presumably serve for maintenance of barrier properties in the adult brain.

7.1.2 BARRIERGENESIS AND THE CONTRIBUTION OF THE NEUROGLIAL SURROUNDING

While the historical definition of the BBB consisted primarily of the exclusion capacity of the endothelium toward basic dyes like Trypan blue (Ehrlich 1885, Goldmann 1913), a plethora of data became available during the second half of the last century that painted the picture of a complex dynamic machinery responsible for selective tightening of the endothelial lining. It became clear that neural progenitor cells, pericytes, and astrocytes play a decisive role in the establishment and maintenance of the barrier. First evidence that the microenvironmental milieu of the brain vesicle, which surrounds the endothelial cells when invading from the perineural plexus, may be responsible for some inductive mechanism that triggers the development of the cerebral endothelial phenotype came from transplantation studies of Stewart and Wiley (1981), who used a classical experimental model of chick/quail microtransplantation. This model provides a useful means to study morphogenetic interactions of heterologous transplants since it allows tracing of the fate of transplanted tissue in the host, because of striking differences in cell nuclei morphology of these avian species. While quails have an ellipsoidal shape of their nuclei, chickens reveal a rounded structure. When avascular quail brain was transplanted into the chick coelomic cavity the visceral chick blood vessels acquired a BBB phenotype upon invasion of the quail transplant. The neuronal tissue of the quail seems to provide instructive cues to the advancing chicken endothelial cells, and triggered the achievement of a tight endothelium. Conversely, when avascular quail coelomic tissue was transplanted into chick brain, the invading chick vessels did not differentiate to a BBB phenotype. Follow-up studies by Janzer and Raff (1987) demonstrated the contribution of astrocytes to barriergenesis. When astrocytes (type I) were injected into the anterior eye chamber of adult syngeneic rats, they formed aggregates on the surface of the iris that became vascularized over time. Those microvessels proved tight toward intravascularly injected dye (Evans blue), indicating an inductive mechanism initiated by the astrocytic surrounding. Conversely, cells obtained from meningeal layers (nonglial cells) did not induce tightness of blood vessels to the injected dye in the meningeal aggregates. Although the contribution of astrocytes to BBB induction in this particular model was questioned in part (see Holash et al. 1993), data in support of neural progenitor cell contribution to barriergenesis have accumulated, allowing a picture of the part that is played by neuroglia to be drawn.

Apparently, an intimate heterologous cross talk between endothelial cells and surrounding neuroglial cells in concert with additional cellular and noncellular components is necessary to promote the maturation of the BBB. The following scenario, at least from in vitro studies, has emerged (for reviews see Engelhardt 2003, Engelhardt and Sorokin 2009, Park et al. 2006, Greenberg et al. 2008, Abbott et al. 2006) that places endothelial-neuroglia interplay at the very center of barriergenesis.

7.1.3 Molecular Cues for Angio- and Barriergenesis

Angio- and barriergenesis in the CNS have long been considered to follow a two-step model, described as angiogenesis followed by barriergenesis (Englehardt and Sorokin 2009, Engelhardt 2003). Recent evidence, however, suggests that both processes are tightly coupled such that distinct molecular mechanisms may coactivate angiogenetic events and BBB formation (Daneman et al. 2009). Insofar as separate discussions of both mechanisms does not seem appropriate, we will have combined these topics in the following.

A decisive component for angiogenesis in the embryonic brain is the vascular-derived endothelial growth factor (VEGF) (Ferrara 2000, Engelhardt 2003), which is secreted by neural progenitor cells. VEGF secretion is regulated by the transcription factor HIF-1α (hypoxia-induced factor), which is responsive to oxygen tension. Reduction of oxygen availability (hypoxia) activates neuroglial HIF-1α, which further increases VEGF production. Since VEGF is the most potent angiogenic component, angiogenesis is effectively promoted in brain domains with low oxygen tension, i.e., in the developing brain the subependymal zone, which is most distant from the pial perineural vascular plexus. As long as vascular density and consequently oxygen tension is low, angiogenesis is sustained. With further maturation of the brain vesicles and multiplication of intraparenchymal blood vessels, which concomitantly increase the oxygen supply, VEGF is downregulated as a consequence of HIF-1α decay. At this event a further oxygen-related factor SSeCKS (src-suppressed C kinase substrate, rat homolog of human AKAP12) comes into play (see Park et al. 2006). SSeCKS is a potential tumor suppressor and constitutes a multivalent scaffolding protein for protein kinases A and C (PKA and PKC), calmodulin, and cyclines. In the context of BBB genesis SSeCKS has been shown to reduce VEGF in cultured astrocytes (Lee et al. 2003). This observation correlates with immunohistochemical findings that revealed low levels of SSeCKS in early embryonic stages, when VEGF expression is high, and a progressive increase in SSeCKS in late embryonic stages and postnatally, when VEGF levels decline. However, the morphogenetic driving force at this stage is not a simple monofactorial event. At the same time when SSeCKS increases, angiotensin 1 (Ang-1), a known pro-differentiation factor for endothelial cells, is expressed in astrocytes with the advent of an increased localization of the tight junction proteins ZO1, claudin1, claudin3, and occludin in interendothelial contact areas (Lee et al. 2003). SSeCKS secretion is also induced by the production of endothelial PDGF-β, which provides the signal for the recruitment of pericytes to blood vessels (Coats et al. 2002). Pericytes constitute a major source for Ang-1 production. It seems that the impact of SSeCKS on barriergenesis is mainly due to enhanced Ang-1 signaling activating endothelial Tie-2 (the receptor for Ang-1) and thereby ceasing angiogenesis and finalizing BBB differentiation. Besides the endothelial PDGF-β effect on pericyte angiogenic potency, endothelial cells are a reliable source of leukemia inhibitory factor (LIF), which specifically elicits differentiation of astrocytes (Mi et al. 2001). Likewise, Wnt/β-catenin signaling seems to be essentially involved in early angiogenesis and postnatal development of the CNS-specific signature of endothelial gene expression. As has been shown by Liebner et al. (2008) and Daneman et al. (2009), the canonical Wnt/β-catenin signaling is specifically activated in CNS blood vessels during development with a differential expression of Wnt ligands, including Wnt1, Wnt3, and Wnt3a in ventral regions and Wnt4a in dorsal regions (Daneman et al. 2009). Inhibition of Wnt/β-catenin leads to severe CNS-specific angiogenesis defects without affecting non-CNS angiogenesis. Wnt/β-catenin signaling also regulates the expression of cerebral endothelial specific transporters such as Glut1, supporting the view that brain angiogenesis and the acquisition of a BBB phenotype are in part linked. Similar robust changes of angiogenic invasion of the telencephalon have been described when the orphan G protein-coupled receptor (GPCR) GPR124 was deleted. Beginning at E11 GPR$^{-/-}$ mice embryos exhibited completely penetrant, progressive hemorrhage starting in the forebrain telencephalon and ventral neural tube, with resulting lethality. Interestingly, GPR$^{-/-}$ embryos revealed no deficits in angiogenic sprouting into the diencephalon, mid-, and hindbrain (Kuhnert et al. 2010). As for Wnt/β-catenin signaling, GPR124 deletion also compromised the expression of Glut1 in brain endothelial cells, reenforcing the notion that modeling of the BBB signature and angiogenesis seems to be closely linked. Interestingly, the maturation of the cerebral vasculature and its specification is also accompanied by a shift in the expression pattern for Glut1. While the neuroepithelium of the brain vesicle at embryonic days 12 and 13 shows a high expression level of Glut1, subsequent development of vascular tightness is followed by a dramatic reduction of the transporter in the brain parenchyma and upregulation of its expression in the cerebral epithelium (Dermietzel et al. 1992).

Several additional growth and differentiation factors besides those mentioned above have been described. A major focus was given to the angiogenic potency of Erythropoietin (EPO), which is under control of widespread oxygen-sensing mechanisms (Maxwell et al. 1993, Fan et al. 2009). In addition to its well-known effects on erythropoiesis, EPO seems to provide a mobilization effect on endothelial progenitor cells during brain development and displays stimulation of postnatal neovascularization (Buemi et al. 2009) after brain and spinal cord injury (Goldman and Needergard 2002, Grasso et al. 2007). Fibroblast growth factor and its receptor (FGF/FGFR) system comprises a further candidate for implications in angiogenesis and barrier function. It was shown that moderate FGF-2 increases vessel density and preserves the composition of cerebral endothelial tight junction proteins (occludin, claudin-5, and claudin-3) in organotypic cultures. FGFs are likely to promote the formation of new vessels and exert a stabilizing influence on BBB tight junction proteins (Park et al. 2006).

A further fascinating issue of angiogenesis in brain tissue is the phenomenon that similar factors are employed for neuronal patterning and angiogenesis. Shared developmental principles have evolved to orchestrate the formation of both the vascular and the nervous system (see Tam and Watts 2010). Specific guidance cues have been identified that drive both systems to their target tissues, including the ephrin (EPH) ligands and their cognate EPH receptors, and the neuropilin-semaphoring-plexin complex. Since both systems share an intricate interface that in the developing brain is evident by the radial orientation of both components during growth and migration, and in the adult by neuroglial/vascular columnar organization, it is evident that these modular patterning elements in the form of neurovascular units (NVUs) seem to play key roles not only in the developing brain, but also in adult brain organization.

7.1.4 PERICYTES AND THEIR SPECIFIC ROLE IN BBB DEVELOPMENT

The particular contribution of distinct cellular species to CNS angio- and barriergenesis is still an open issue. While the original experiments by Janzer and Raff (1987) focused on the contribution of astrocytes to BBB development, more recent data indicate that pericytes play a remarkable role in the induction of BBB properties. First evidence was contributed by in vitro models employing triculture systems, which consisted of purified cerebral endothelial cells, pericytes, and astrocytes. Under these triculture conditions (Minakawa et al. 1991), cerebral endothelial cells assembled into a capillary-like structure (CLS). The presence of pericytes in the cultures enhanced the formation and arborization of CLSs and rescued endothelial cells from apoptosis (Ramsauer et al. 2002). Similar results were reported by Al Ahmad et al. (2011).

Since in vitro data cannot be extrapolated in a simple one-to-one fashion to the in vivo situation, two recent papers using transgenic mouse models with pericyte deficiency (PDGF null or mutant animals) are extremely supportive to appreciate the role of pericytes for the blood–brain barrier generation. Daneman et al. (2010b) reported that pericytes are recruited to immature vessels before astrocytes appear. In the mutant mice, tight junction morphology was altered. Surprisingly, gene expression analysis revealed that pericytes did not induce expression of blood–brain-barrier-specific genes, but rather inhibited expression of genes encoding proteins causing vascular permeability and immune cell infiltration.

Similarly, Armulik et al. (2010) found that pericyte depletion caused a dose-dependent increased vascular permeability to water and various molecular weight tracers that was attributable to transcytosis. Astrocyte polarization was also affected in the mutant animals, as evidenced by decreased endfeet expression of aquaporin4 (Aqp4), Lana2, and α-syntrophin, although there was no alteration in expression of the endothelial junction proteins ZO1, claudin-5, and VE cadherin.

Based on their data, Daneman et al. (2010b) suggest a new model of barriergenesis, which emphasizes a sequential, cell-driven development of the BBB with an induction of BBB-specific genes by interaction of endothelial cells and neural progenitor cells and subsequent organization of functional integrity of the BBB through pericytes during brain development. The role of astrocytes was considered mainly to form a safeguard for maintenance of BBB integrity in the adult brain.

In summary, angio- and barriergenesis do not constitute a one-way road in terms of differential cues at the endothelial-neuroglial interface, but consist of a multidimensional network with mutual interdependence of different cellular components. A comprehensive summary of the paracrine/autocrine interplay and its cellular components involved in the development of the BBB is indicated in Figure 7.1.

7.1.5 MOLECULAR ANATOMY OF THE BLOOD–BRAIN BARRIER

Several recent studies have attempted to define the molecular components of the blood–brain barrier by applying proteomic or gene expression approaches to determine differences between brain and nonneural microvasculature (e.g., Enerson and Drewes 2006, Agarwal and Shusta 2009, Pottiez et al. 2009, Murugesan et al. 2011). Although these studies have identified transporters, receptors, junctional and signaling proteins, or their genes as being differentially expressed in brain vasculature, it was not clear in which cell type the differential expression was manifest. A recent study has addressed this issue through the use of mice, in which GFP expression is driven by the Tie1 promoter, so that it is expressed in all endothelial cells but not elsewhere; endothelial cells in various vascular beds were harvested and purified twice with fluorescence- activated cell sorting (FACS) to maximize purity (Daneman et al. 2010b). This study reported blood–brain barrier enrichment of the tight junction proteins Cldn5, marveld2, cingulin-like-1, and pard3, signaling pathways Wnt/β-catenin (see above) and RXR-α, as well as integrins and semaphorins and a number of ion channels and transporters (see Table 7.1). This study thus identifies a number of genes that might be targeted to modify the tightness of the endothelial barrier and confirms previous data obtained from "non-omic" studies.

Possible answers to the questions concerning the molecular complement that serves for maintenance of the BBB in the adult may be given by comparative microarray analyses of certain brain regions that lack a BBB, i.e., the circumventricular organs

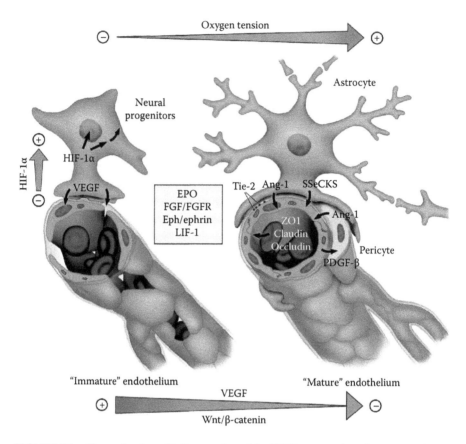

FIGURE 7.1 **(See color insert.)** Proposed model of BBB maturation. Oxygen tension is suggested to comprise the driving force for angiogenesis in the early brain anlage. Invading endothelial cells from the perineural plexus settle within the brain vesicle where neural progenitor cells produce VEGF according to activation of the oxygen-responsive hypoxia-induced factor (HIF-1α). Increasing levels of oxygen elicit reduction of VEGF secretion, and maturation of the BBB is induced by differentiation factors like SSeCKS (src-suppressed C kinase substrate) and Ang-1 (Angiopoetin-1 (Tie-2 is the cognate Ang-1 receptor)). Endothelial PDGF-β (platelet-derived growth factor) is a strong attractor for pericytes, which also produce PDGF-β. The center box indicates some further growth and differentiation factors responsible for angiogenesis and barriergenesis in brain tissues. Likewise, Wnt/β-catenin signaling controls the development of the BBB indicated by the gradient marker in the lower part of the graph.

(CVOs) with the choroid plexuses, with their tight parenchymal counterparts. One striking structural difference between these leaky segments of the brain microvasculature and the BBB vessels is that the microvasculature of the CVOs is substituted with an expanded perivascular space, including common extracellular components such as collagens, glycosaminoglycans, and proteoglycans.

With respect to molecular candidates responsible for the sustained leakiness of the BBB in these areas is the observation that VEGF continues to be expressed, at least in the choroid plexuses (Marti and Risau 1998). Noteworthy, a pan-endothelial antigen

TABLE 7.1
The BBB Transcriptome

UniGene ID	Gene Title	Gene Symbol
Mm.207354	ATP-binding cassette, subfamily B (MDR/TAP), member 1A	Abcb1a
Mm.217161	Actin-binding LIM protein 1	Ablim1
Mm.267478	Acyl-CoA synthetase long-chain family member 6	Acsl6
Mm.42040	Adenylate kinase 3 α-like 1	Ak3l1
Mm.391102	Adenomatosis polyposis coli downregulated 1	Apcdd1
Mm.29756	Aquaporin 11	Aqp11
Mm.135129	ATPase, class V, type 10A	Atp10a
Mm.250719	Butyrylcholinesterase	Bche
Mm.726	Basigin	Bsg
Mm.332961	Cysteine-rich BMP regulator 2 (chordin-like)	Crim2
Mm.12906	Dopa decarboxylase	Ddc
Mm.82598	DIX domain containing 1	Dixdc1
Mm.215971	Early B cell factor 1	Ebf1
Mm.286127	ELOVL family member 7, elongation of long-chain fatty acids (yeast)	Elovl7
Mm.12949	Forkhead box C1	Foxc1
Mm.390554	Forkhead box F2	Foxf2
Mm.44235	Forkhead box Q1	Foxq1
Mm.272019	G protein-coupled receptor 160	Gpr160
Mm.28597	G protein-coupled receptor 172B	Gpr172b
Mm.167625	G protein-coupled receptor 85	Gpr85
Mm.289131	3-Hydroxy-3-methylglutaryl-coenzyme A synthase 2	Hmgcs2
Mm.313876	Inter-α (globulin) inhibitor H5	Itih5
Mm.193	Integral membrane protein 2A	Itm2a
Mm.46675	Kelch repeat and BTB (POZ) domain containing 11	Kbtbd11
Mm.245119	LanC lantibiotic synthetase component C-like 3 (bacterial)	Lancl3
Mm.331760	Low-density lipoprotein receptor-related protein 8, apolipoprotein e receptor	Lrp8
Mm.331842	Major facilitator superfamily domain containing 2	Mfsd2
Mm.29357	Myo-inositol 1-phosphate synthase A1	Lsyna1
Mm.4067	Liver-specific bHLH-Zip transcription factor	Lsr
Mm.259470	Oxysterol binding protein-like 1A	Osbpl1a
Mm.34650	Paralemmin	Palm
Mm.273267	Progestin and adipoQ receptor family member V	Paqr5
Mm.39738	Protocadherin 19	Pcdh19
Mm.1845	Pyruvate carboxylase	Pcx
Mm.206159	Pyridoxal (pyridoxine, vitamin B6) kinase	Pdxk
Mm.21855	Peptidoglycan recognition protein 1	Pglyrp1
Mm.6250	Prominin 1	Prom1
Mm.279690	Pleiotrophin	Ptn
Mm.336316	Protein tyrosine phosphatase, receptor type, R	Ptprr
Mm.249437	RAD54 homolog B (*S. cerevisiae*)	Rad54b

(*Continued*)

TABLE 7.1 (CONTINUED)
The BBB Transcriptome

UniGene ID	Gene Title	Gene Symbol
Mm.38384	Solute carrier family 16 (monocarboxylic acid transporters), member 4	Slc16a4
Mm.246670	Solute carrier family 1 (neuronal/epithelial high-affinity glutamate transporter, member 1)	Slc1a1
Mm.285294	Solute carrier family 22 (organic anion transporter), member 8	Slc22a8
Mm.21002	Solute carrier family 2 (facilitated glucose transporter), member 1	Slc2a1
Mm.26159	Solute carrier family 35, member F2 (Slc35f2), mRNA	Slc35f2
Mm.296560	Solute carrier family 38, member 3	Slc38a3
Mm.6055	Solute carrier family 38, member 5	Slc38a5
Mm.233889	Solute carrier family 39 (zinc transporter), member 10	Slc39a10
Mm.27943	Solute carrier family 7 (cationic amino acid transporter, y+ system), member 5	Slc7a5
Mm.255586	Solute carrier organic anion transporter family, member 1a4	Slco1a4
Mm.284495	Solute carrier organic anion transporter family, member 1c1	Slco1c1
Mm.11249	Solute carrier organic anion transporter family, member 2b1	Slco2b1
Mm.211096	Sorbin and SH3 domain containing 2	Sorbs2
Mm.153429	Sparc/osteonectin, cwcv and kazal-like domains proteoglycan 2	Spock2
Mm.330004	ST8 α-N-acetyl-neuraminide α-2,8-sialyltransferase 6	St8sia6
Mm.295194	T-box 1	Tbx1
Mm.28683	Transferrin receptor	Tfrc
Mm.69380	Transmembrane channel-like gene family 7	Tmc7
Mm.281356	Tumor necrosis factor receptor superfamily, member 19	Tnfrsf19
Mm.7821	Tumor protein D52-like 1	Tpd52l1
Mm.26515	von Willebrand factor A domain containing 1	Vwa1
Mm.87599	WAP four-disulfide core domain 1	Wfdc1
Mm.255890	Zinc finger protein of the cerebellum 3	Zic3
Mm.380672	Opr	Zic5

Note: This table shows genes for which expression in brain endothelial cells was found to be at least 10 times as high as in liver and lung endothelial cells. Data are taken from Supplemental Figure 8 of Daneman et al. (2010a), with redundancies removed and ambiguous annotations deleted.

(plasmalemmal vesicle associated protein 1, Plvap1) remains present in the leaky epithelium of the CVOs, while it becomes downregulated in the mature cerebral endothelium (Hallman et al. 1995).

7.2 TYPES OF CELL JUNCTIONS AT THE BLOOD–BRAIN BARRIER

The tight sealing at the endothelial margins in higher vertebrates is provided by a combination of basal lamina and the interactions among the cells of the NVU: pericytes, astrocytes, microglia, and neurons. The endothelial basal lamina induces astrocytic cell polarity, with the astroglial endfeet covered in turn by a glial basal lamina. The proteins of the basal lamina include the laminins, fibonectins, collagens, and heparan sulfate proteoglycans, including agrin (Hallmann et al. 2005). Agrin, in particular, is associated with blood–brain barrier integrity and is widely believed to play a major role in establishing and maintaining astrocyte polarity (Wolburg et al. 2009a).

From the standpoint of junctions between cells of the blood–brain barrier, adhesive and sealing (tight) junctions play major roles in holding cells together and preventing permeation and separating apical from basal domains of endothelial cells. In addition, gap junctions formed between astrocytes, between endothelial cells, and between pericytes and endothelial cells allow direct intercellular signal transfer, and thus play major roles in coordinating cell functions (for reviews see Giaume et al. 2010, Bennett et al. 1991, Scemes et al. 2007, Meier and Dermietzel 2006). Each junction type is made up of several molecular components, including the membrane-embedded molecules, which provide the function of adhesion, sealing, or communication, and associated adaptor molecules link them to one another and to the cytoskeleton.

7.2.1 Adhesive Junctions

Endothelial cells of brain capillaries have numerous adhesion molecules that are used for intercellular adhesion of endothelial cells and are exploited as receptors by leukocytes for invasion (Faustmann and Dermietzel 1985, Wolburg et al. 2009b). Adherens junctions mediate adhesion between neighboring cells by linking the actin cytoskeleton of one cell to that of the next cell via transmembrane adhesion molecules and their associated protein complexes (for review see Derangeon et al. 2009). The core of these junctions consists of two basic adhesive units, the interactions among transmembrane glycoproteins of the classical cadherin superfamily and the catenin family members (including p120-catenin, β-catenin, and α-catenin) and the nectin/afadin complexes (for review see Niessen and Gottardi 2008). These molecules include cadherins and integrins (such as ICAM-1 or CD54, VCAM-1 or CD106, PECAM-1 or CD31, which bind blood cells) and selectins and components of associated signal cascades.

The cadherins are calcium-dependent adhesive molecules, of which VE cadherin is the best known in endothelial cells. These bind to scaffolding molecules, α- and β-catenin, and also γ-catenin (plakoglobin), thereby forming a linkage of the surface membrane to the cytoskeleton. β-Catenin controls Wnt signaling, in which Wnt binds to its receptor (frizzled), thereby stabilizing β-catenin, which activates transcription factors altering genes such as tight junction molecules (for review see Paolinelli et al. 2011); moreover, Wnt signaling through β-catenin regulates the size of Cx43 gap junction plaques on cardiac myocytes (Ai et al. 2000). Integrins are formed by α- and β-heterodimers whose role may be primarily in angiogenesis.

Endothelial cells are tethered to the basement membrane and to one another by focal adhesions (formed of selectins, the immunoglobulin superfamily, and integrins). The adhesion complex also contains signaling molecules (such as focal adhesion kinase and src tyrosine kinase). Contractile and adhesive components interact with each other, so that the endothelial cytoskeleton establishes junctional integrity with other endothelial cells through actin filaments, intermediate filaments (primarily vimentin), and microtubules. *Cadherins* comprise an important family of trans-membrane glycoproteins that mediate calcium-dependent cell-cell adhesion and are linked to the actin cytoskeleton via catenins. In NIH3T3 cells, Cx43, N-cadherin, and multiple N-cadherin-associated proteins were found colocalized and coimmu-noprecipitated, suggesting that Cx43 and N-cadherin are coassembled in form of a multiprotein complex containing various N-cadherin-associated proteins (Wei et al. 2005). However, no evidence was found for direct binding between N-cadherin and the C-terminal part of Cx43, suggesting that weak protein-protein interactions might exist between them, or that interactions occur with the short N-terminus or intracellular loop domains of Cx43 or via a protein partner acting as an anchoring bridge.

Extracellular calcium has long been known to affect BBB permeability, as was first shown in epithelial cells. The original interpretation of the finding that adhesive junctions fell apart in the absence of calcium was that calcium was itself linking adhesive junctions to substrate or other adhesive junctions. However, calcium could also act to change the conformation of the cadherin extracellular domains or could stimulate signaling cascades, leading to the disassembly of ZO-1 and occludin away from the apical-lateral borders.

7.2.2 TIGHT JUNCTIONS

Tight junctions, the most apical junctional type in epithelium, are responsible for polarization of cells and the separation of components of apical from those of basolateral membranes (the so-called fence function) and also the blockade of paracellular permeability (the barrier function). In addition, tight junction pro-teins are involved in coordinating signals impinging on or emanating from the plasma membrane (see Forster et al. 2008). Tight junction molecules include the integral membrane proteins occludin and claudins, junction adhesion molecules (JAMs) and the coxsackie, and adenovirus receptor (CAR). Adapter proteins are those that bind directly to the integral membrane proteins (ZO-1,2,3) and second-order adapters that create a multilevel complex (cingulin-related junction-asso-ciated coil protein (JACOP)). The peripherally associated scaffolding molecules ZO-1,2,3, AF6, and cingulin couple the tight junction proteins to the cytoskeleton. Additional signaling proteins such as ZO-1 associated nucleic acid binding pro-tein (ZONAB) play roles in junciton assembly as well as transcription (for review see Bazzoni, 2003).

Occludin was the first tight junction transmembrane protein discovered (Furuse et al. 1993) and probably modulates tight junctions through signaling that is depen-dent on phosphorylation state. It is facilitory but not required for tight junction formation. Occludin has a short cytoplasmic amino terminus and a very long car-boxyl terminal tail. Overexpression increases transendothelial resistance, but this protein is not sufficient to form tight junctions by itself. Nevertheless, it is recruited to junctions formed by claudins 1 and 2 (Furuse et al. 1998).

Claudins are the true tight junction molecules (Tsukita and Furuse 2000, Krause et al. 2008); although they are also four transmembrane domain proteins with two extracellular loops, they show no sequence homology with occludin. There are more than 20 members forming both homotypic and heterotypic paired strands of particles in the membrane under freeze-fracture conditions. The peripheral membrane protein ZO-1 has a molecular weight of 220 Kd.

In freeze-fracture preparations tight junctions reveal anastomosing strands of intramembrane particles in grooves. The degree of complexity (number of rows and overlaps) of the endothelial tight junctions is positively correlated with the transepithelial resistance, and it has been proposed that the association of the freeze-fracture particles with either the P or E face is an essential feature of the tightness of the permeability barrier in brain, which displays a high number of P face particles (Wolburg et al. 2009a). The latter authors argue that of the three claudins in the blood–brain barrier (Cldn3, 5, 12), the particles in fibroblasts transfected with Cldn3, are on the P face, whereas in Cldn5 transfectants, particles are E face associated. Therefore, modulation of expression stoichiometry of *Cldn* genes may vary the percentage of particles in each configuration, reflecting the tightness of the endothelial barrier. With respect to the permeability barrier that claudins provide, it is of particular interest that the extracellular loops of the individual claudins differ with respect to the number of charged amino acids; presumably repulsive charges result in leaky junctions and attractive charges tight ones (see van Italie and Anderson 2003, Krause et al. 2008).

Interleukins influence the barrier function of tight junctions and the blood–brain barrier is severely compromised in multiple sclerosis (van Horssen et al. 2005, McQuaid et al. 2009, Bennett et al. 2010). The increased migration of leukocytes in multiple sclerosis leads to the reorganization of the actin cytoskeleton and the breakdown of the permeability barrier, and reorganization of the tight junction proteins occludin and ZO1. Figure 7.2 depicts a diagrammatic representation of the different tight junction components as discussed above, emphasizing residence of claudins and occludin within the membranes of adjacent cells and associated proteins linking these molecules to the cytoskeleton.

7.2.3 GAP JUNCTIONS

Gap junctions interconnect endothelial cells and astrocytes, and the presence of gap junctions and intercellular dye transfer between endothelial and pericytes has been demonstrated in vitro (Larson et al. 1987, Lai and Kuo 2005). Astrocytic gap junctions are formed primarily of Cx43 (Dermietzel et al. 1991, Giaume et al. 1991), although Cx30 is also present (Kunzelmann et al. 1999), as are minor amounts of other connexins; endothelial cells express Cx37 as well as Cx43, and pericytes express Cx43 (see Ramsauer et al. 2002). Interestingly, astrocytic Cx30 shows a delayed expression with a late onset around 3–4 weeks postpartum in rats. Moreover, it shows a distinct regional distribution with a higher expression in perivascular astrocytic processes in adult rodent brains (Kunzelmann et al. 1999, Nagy et al. 2001). Both the late onset of Cx30 expression, which parallels postpartum maturation of the BBB, and its region-specific localization have been correlated with BBB specific activities of this connexin.

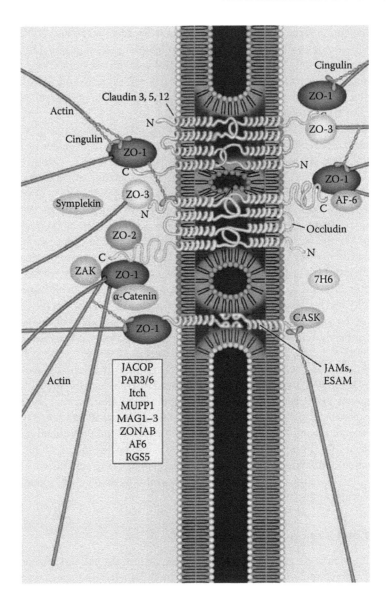

FIGURE 7.2 **(See color insert.)** Cumulative sketch of proteins involved in tight junction formation. CASK, Ca^{2+}-dependent serine protein kinase; JAM, junctional adhesion molecule; ZAK, leucine-zipper and sterile-α-motif kinase; 7H6/barmontin, a tight junction-associated protein recognized by the 7H6 monoclonal antibody. The following tight junction-associated proteins are listed in the box: JACOB, junction-associated coiled-coiled protein; PAR3/6, the partitioning defective protein 3 and 6; MUPP1, multi-PDZ protein1; Itch, E3 ubiquitin-protein ligase; MAG1–3, membrane-associated guanylate kinase with inverted orientation of protein-protein interaction domains; ZONAB, ZO-1-associated nucleic-acid-binding protein; AF6, afadin, a large PDZ-containing family of actin-binding proteins concentrated at the adherens junction; RGS5, regulator of G protein signaling. For clarity, not all of the adherens junction-associated proteins are listed.

7.3   POLARIZATION AT THE BBB-GLIAL INTERFACE

The interface between the neuroglial and vascular compartments is established by glial endfeet, which are characterized by large numbers of orthogonal arrays of particles (OAPs) (Dermietzel 1973, Landis and Reese 1974, for review see Wolburg et al. 2009b), which contain the water channel protein aquaporin4 (Aqp4) (Rash et al. 2004). This protein forms functional complexes with other membrane proteins, including the inwardly rectifying potassium channel protein Kir4.1 and the dystrophin-dystroglycan complex (Fort et al. 2008), which have overlapping distributions in the astrocytic endfeet. Dystrophin is a scaffold protein, and mice in which this protein is deleted show a leaky blood–brain barrier. Dystroglycan consists of two isoforms, one of which is selectively cleaved by metalloproteases (specifically Mmp2 and Mmp9) providing a mechanism by which macrophages transgressing the endothelium can enter the CNS. The dystroglycan knockout shows disorganization of cortical layering with aberrant migration of granular cells and reduced leukocyte infiltration.

Complexing of the OAPs is accomplished by proteins like dystrobrevin and syntrophins, which connect to the carboxyl terminus of dystrophin and interact with the channel molecules. α-Syntrophin connects to Kir4.1 and Aqp4, and laminin may induce aggregation of Kir4.1 and Aqp4 in the astrocytic membrane. The Kir potassium channel and the aquaporin water channel likely couple potassium and water flux, although Kir4.1 is still present in glial endfeet in Aqp4 knockout mice (Noell et al. 2009). Aqp4 knockdown reduces volume-sensitive ion channel expression, perhaps through altering lipid rafts.

7.3.1   Disruption of the Gliovascular Interface in Pathologies

It is now clear that the orthogonal particle arrays (OAPs) are images of Aqp4 proteins, either alone or in complex with other molecules (Furman et al. 2003). It is also clear that the number of OAPs is related to the quality of the seal of the BBB. In brain tumors, Cldn3 is decreased and the BBB is leaky. It is therefore possible that Aqp4 dissociates from the OAPs when agrin is missing. Agrin is responsible for Aqp4 restriction to astrocytic endfeet, and agrin is provided by both astrocytes and endothelial cells. In tumors and in agrin-deficient mice OAPs but not aqp4 are lost (Fallier-Becker et al. 2011, Noell et al. 2009). Notably, Mmp3 inhibition decreases BBB degradation, and knockout of this enzyme is protective of the BBB.

The accomplishment of the polarized status of astrocytes has also been studied by coculture experiments of brain endothelial cells with astrocytes, which corroborate the importance of the endothelial glial interaction (See Nicchia et al. 2004).

Neuroinflammatory conditions disrupt the BBB and open tight junctions (Avison et al. 2004). In HIV encephalitis, multiple sclerosis, and Alzheimer's disease (AD), similar sequences of events occur. For example, the infection of macrophages and microglia by HIV leads to release of cytokines, chemokines, reactive oxygen species (ROS), glutamate, and matric metalloproteinsase (MMPs), any of which may alter the expression and function of tight junction proteins and viral proteins secreted by infected cells alter tight junction expression and function (Kanmogne et al. 2005).

In AD, activated microglia and astrocytes stimulated by β-amyloid produced proinflammatory factors (e.g., IL1-β, TNF-α, complement, ROS, and β-amyloid) around vessels, and thus may directly affect endothelial permeability. In multiple sclerosis, the reactive T lymphocytes interact with brain microglia and macrophages, which secrete proinflammatory cytokines, MMPs, and ROS.

7.3.2 PATHOGENS BREACH OF BLOOD–BRAIN BARRIER

Infectious agents and pathogenic organisms use multiple strategies to breach the blood–brain barrier. At the blood–brain interface itself, organisms, diverse proteins, and toxins can bind to surface receptors and be internalized through endocytosis and subsequently released on the abluminal site of the cell, thereby using a transcytotic pathway to cross the endothelial barrier. The receptors exploited for this purpose include the family of tight junction proteins, the claudins, cholinergic receptors, as well as glycosylated membrane-bound proteins whose internalization is triggered by binding.

One example of such an invasion where barrier proteins serve a role for invasion is that of clostridium leading to diarrhea as a consequence of food poisoning. The toxins secreted by this bacteria form pores in the membrane and then interact with the carboxyl terminal portion of claudins, in particular claudin4 (Van Itallie and Anderson 2006, Van Itallie et al. 2008).

The blood–brain interface is an extraordinarily vulnerable target for parasitic infection, with amoeboid, cestode, nematodes, protozoans, and the schistosome family of trematodes all affecting this interface with varying degrees of pathogenicity (dealt with in detail in Desreuisseuax et al. 2006). In recognition of this profound effect of parasitic infections on the endothelial cell in the brain, recent publications have stressed the etiology of cerebral malaria as being a vasculopathy, where the plasmodium packs and engorges cerebral vessels in end-stage human disease. African trypanosomiasis causes meningitis and reactive gliosis, whereas American trypanosomiasis affects both microglia and astrocytes.

Endothelial cells serve as the entry point of inflammatory cells where the resulting inflammation contributes to increased vascular leakage. There likely are both direct effects of parasites on the endothelial cells and later changes that are consequent to immune responses. Icam1 is one example of a surface receptor to which lymphocytes adhere that is thought to alter phosphorylation of cytoskeletal proteins in endothelial cells resulting in increased permeability (Adams et al. 2002).

In addition to the direct transcytotic mechanism of entry, there is internalization but sequestration, where the host cell continues to harbor the inflammatory cells or invading pathogenic organism (Faustmann et al. 1991, Wolburg et al. 2005), and paracellular entry, where either the fence function is weakened by the pathogen or the molecular binding site is the tight or adhesive junction component. Cerebral malaria is a prime example of a disease in which BBB damage correlates well with disease severity (Adams et al. 2002).

The endothelial cells with their sealing tight junctions provide an initial barrier to diffusion of all but the smallest nonpolar molecules, and this vascular barrier is reinforced by the neural elements behind it, forming a second line of defense. As reviewed extensively by Salinas et al. (2010), the naked axon extending far beyond the CNS vasculature provides an opportunity for viruses and bacterial toxins to invade the nerve terminal and to "hitchhike" along the vesicular transport system to the cell body. Such pathogens with major impact on human health include polio, rabies, and herpes viruses, as well as tetanus toxin.

The route of entry and pathology of the common neurotropic viruses and toxins were summarized as the "tool kit" in Salinas et al. (2010). For example, neural cell adhesion molecules (NCAMs) and nicotinic receptors form receptors for rabies virus (RABV), as well as other pathogens. An additional receptor is CAR (the coxsackie virus and adenovirus receptor). Salinas et al. termed the process of pathogen uptake "hitchhiking," to emphasize that noted proteins can be recruited by the pathogen once inside the cell. Examples include the retrograde motor dynein and the conventional kinesin molecule. However, because the pathogens do not move passively, it seems more appropriate to term their action hijacking, rather than hitchhiking.

7.4 STRUCTURE AND PLASTICITY OF NEUROVASCULAR BLOOD FLOW

The second function of the neurovascular unit is to control blood flow and thereby ensure appropriate oxygenation under conditions of high utilization. This function is achieved through release of vasoactive compounds from astrocytes and endothelial cells and to some extent, and in some regions, from neurons. This function is the basis for the detection of brain activity by magnetic resonance imaging (MRI) and the disruption that underlies ischemic events in stroke.

Functional hyperemia, defined as the active process of enlarging vessel diameter in response to rising metabolic demands of neuronal activity, is fundamental to normal brain function. Moreover, changes in blood flow and hemoglobin oxygenation status are the basis for imaging strategies that detect local differences, thereby generating maps of functional activity within the brain.

The brain is supplied with carotid and vertebral arteries that form the circle of Willis at the base of the brain, from which pial arteries emanate to run along the brain surface. Smaller vessels (parenchymal arterioles) sprout from pial arteries and penetrate the brain, ultimately branching to form a capillary network throughout the brain. The components of the walls of each type of vessel are distinct. Arteries have multiple smooth muscle layers, with an elastic lamina separating the smooth muscle cells from the endothelial layer; arterioles have astrocyte endfeet separated from smooth muscle and endothelial layers by a basal lamina; and capillaries have contractile pericytes covering the endothelium and its basal lamina.

7.4.1 GLIOVASCULAR INTERACTION AND CEREBRAL HEMODYNAMICS

Until very recently, it was believed that specific neuronal networks within the brain were responsible for constriction or dilation of arterial and arteriolar vessels (for review see Attwell and Iadecola 2002). However, astrocytic endfeet so completely enwrap arterioles that the majority of nerve endings are on astrocytes rather than on smooth muscle (see Paspalas and Papadopoulos 1996). Even though Eric Newman's group had proposed that glial K channels might play a role in controlling vessel diameter in retina more than 20 years ago (a concept revisited by work from Mark Nelson's laboratory on maxi-K (BK) channels described below), a major recent discovery regarding the role of glial cells in vascular perfusion was the publication of experiments claiming that glutamate acts through astrocytes to dilate vessels (Zonta et al. 2003). In this scheme, glutamate elevates calcium through activation of metabotropic astrocyte receptors, thereby activating phospholipase A2 and producing arachidonic acid, which is then converted by cylcooxygenase into highly effective vasodilating prostaglandin derivatives. The polarized nature of the astrocyte, with extensive neuronal contact and endfeet lining the arteriolar wall, seems ideally positioned for such activity.

Mulligan and MacVicar (2004) contradicted these results through calcium uncaging studies on brain slices. Rather than dilating vessels, they found arteriolar constriction, through the generation and release of arachidonic acid by astrocytes, which was then processed to 20-Hydroxyeicosatetraenoic Acid (20-HETE) by the enzyme Cytochrome P450 4A (CYP4A) in the smooth muscle. Reconciliation with Zonta et al. (2003) is the observation that increased Ca^{2+} in astrocytes can lead to production of both vasoconstricting and vasodilating derivatives of arachidonic acid, depending on the metabolic pathway enabled. Mulligan and MacVicar (2004) also reported that inhibiting nitric oxide (NO) synthase converted the constriction to dilation, which could have resulted from a more constricted baseline. Metea and Newman (2006), working with retinal slices, observed arteriolar constriction in response to Ca^{2+} uncaging that was due to arachidonic acid conversion to 20-HETE, but also in some cases dilation due to arachidonic acid conversion by a cytochrome P450 enzyme to epoxyeicosatrienoic acid (EET). They hypothesized that the retinal levels of nitric oxide were responsible for the switch and found that the NO donor S-Nitroso-N-acetyl-D,L-penicillamine (SNAP) converted the relaxations to constrictions.

An additional mechanism hypothesized to contribute to vascular reactivity is the Ca^{2+}-activated K^+ channel (BK) (Girouard et al. 2010). In this scheme, calcium elevation in astrocytic endfeet leads to vessel dilation, if elevation is small, and contraction, if Ca^{2+} levels increase above a certain threshold. The mechanism of vasodilation is through release of K^+ through the large conductance BK channel, which acts on the inward rectifying Kir channels to hyperpolarize smooth muscle cells; at higher endfeet Ca^{2+} levels, more K^+ is released, which depolarizes the smooth muscle cell.

Current knowledge of the biochemical processes responsible for vasoactivity in brain arterioles is summarized in Figure 7.3. A unifying scheme for activation of these pathways is the elevation of intracellular Ca^{2+} in astrocyte endfeet, which can result from activation of numerous receptors, such as both $\alpha 1$ and β adrenergic receptors by norepinephrine, activation of metabotropic glutamate receptors by glutamate, etc. As indicated in Figure 7.3, it is now clear that arachidonic acid production by Ca^{2+}-sensitive PLA2 is a common underpinning of both arteriolar constriction and dilation in the brain. Constriction results from conversion of arachidonic acid into 20-HETE by a cytochrome P450 enzyme, whereas dilation results from arachidonic acid conversion into PGE2 through cyclooxygenase or to EET through another P450

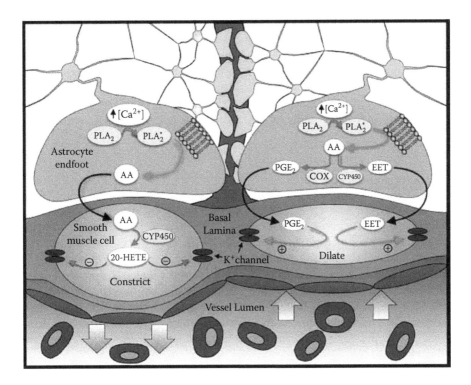

FIGURE 7.3 **(See color insert.)** Pathways of vasomotor control in brain arterioles. Two separate astroglial endfeet (green) are illustrated, the left causing vasoconstriction and the right one causing vasodilation of smooth muscle (yellow). (From Gordon et al., *Glia* 55, 1214–21, 2007. With permission of John W. Wiley Publishers.)

enzyme. Finally, as pointed out by Iadecola and Nedergaard (2007), Adenosine triphosphate (ATP) release from stressed blood cells through pannexin1 channels can cause vasodilation by activation of P2Y receptors on the endothelial wall and NO generation (Locovei et al. 2006).

It should be noted that the role of astrocytes in blood flow control is limited to the arterioles, due to the extensive coverage of their surfaces with astrocytic endfeet. In upstream regions, flow is controlled by neuronal assemblies specialized for this function. The lack of smooth muscle in capillaries has led to the belief that blood flow regulation could only occur at precapillary arterioles and upstream vessels. However, Peppiatt et al. (2006) have shown that electrical or pharmacological stimulation of pericytes caused local constriction that could spread to adjacent regions with a velocity of about 2 μm/a. Likewise, ATP release in the retina and norepinephrine in the cerebellum cause constriction, and glutamate reversed the norepinephrine (NE)-evoked constriction.

In a recent high-profile paper (Schummers et al. 2008), local influence of astrocytes on cerebral hemodynamics was studied by calcium imaging in the ferret visual cortex. There were several surprises in this report, in which both neurons and astrocytes were imaged using specific targeting of calcium indicators to each cell type. The first was the high degree of tuning of astrocytes with respect to orientation and frequency compared to neurons, and the second surprise was that the signals generated in individual astrocytes did not spread to their presumably coupled neighbors. The conclusion was that individual astrocytes have independent control over neuronal subsets. Unfortunately, the anesthetic used in these studies (isofluorane) is known to block gap-junction-mediated electrical signaling (Burt and Spray 1989), so that the sharp individual tuning of single astrocytes might well have resulted from such blockade.

7.5 CONCLUSIONS

The neurovascular unit (NVU) is comprised of neurons, astrocytes, endothelia, and pericytes (with supporting roles provided by microglia and oligodendrocytes) and performs two functionally distinct roles in the nervous system. The first role is accomplished by coinduction of proteins in individual cell types by proximity of partner cells (e.g., aquaporin4 induction in astrocytes cocultured with endothelial cells) and creates polarized distribution of cell adhesive and impermeant junctional complexes that functionally seal the brain from the blood, creating the blood–brain interface. This relatively impermeable barrier is a target both for therapeutic delivery of neuroactive compounds and for pathological attack by infectious agents and inflammation.

The second function of the neurovascular unit is to control blood flow and thereby ensure appropriate oxygenation under conditions of high utilization. This function is achieved through release of vasoactive compounds from astrocytes and endothelial cells and to some extent, and in some regions, from neurons. This function is the basis for the detection of brain activity by MRI, and disruption underlies ischemic events in stroke.

In this chapter, we have summarized recent progress in understanding of the ontogeny of the BBB, its regulation, and interaction with the brain parenchyma in the adult brain. One of these areas involves roles of oxygen tension and pericytes in angiogenesis and barriergenesis. Another is the consideration of the astrocyte endfeet as a highly specialized signaling domain regulating both vasculature perfusion and vessel leakiness. Thus, the components of the blood–brain barrier both mediate separation of these compartments and occupy a control position where they determine the degree of metabolic exchange with the blood supply.

REFERENCES

Abbott, N. J., N. J. Lane, and M. Bundgaard. The blood-brain interface in invertebrates. *Ann NY Acad Sci* 481 (1986): 20–42.

Abbott, N. J., A. A. Patabendige, D. E. Dolman, S. R. Yusof, and D. J. Begley. Structure and function of the blood-brain barrier. *Neurobiol Dis* 37, no. 1 (2010): 13–25.

Abbott, N. J., L. Ronnback, and E. Hansson. Astrocyte-endothelial interactions at the blood-brain barrier. *Nat Rev Neurosci* 7, no. 1 (2006): 41–53.

Adams, S., H. Brown, and G. Turner. Breaking down the blood-brain barrier: signaling a path to cerebral malaria? *Trends Parasitol* 18 (2002): 360–6.

Agarwal, N., and E. V. Shusta. Multiplex expression cloning of blood-brain barrier membrane proteins. *Proteomics* 9, no. 4 (2009): 1099–108.

Ai, Z., A. Fischer, D. C. Spray, A. M. Brown, and G. I. Fishman. Wnt-1 regulation of connexin43 in cardiac myocytes. *J Clin Invest* 105 (2000): 161–71.

Al Ahmad, A., C. B. Taboada, M. Gassmann, and O. O. Ogunshola. Astrocytes and pericytes differentially modulate blood-brain barrier characteristics during development and hypoxic insult. *J Cereb Blood Flow Metab* 31 (2011): 693–705.

Armulik, A., G. Genove, M. Mae, M. H. Nisancioglu, E. Wallgard, C. Niaudet, L. He, J. Norlin, P. Per Lindblom, K. Strittmatter, B. R. Johansson, and C. Betsholtz C. Pericytes regulate the blood-brain barrier. *Nature* 468 (2010): 557–61.

Attwell, D., and C. Iadecola. The neural basis of functional brain imaging signals. *Trends Neurosci.* 25 (2002): 621–25.

Avison, M. J., A. Nath, R. Greene-Avison, F. A. Schmitt, R. A. Bales, A. Ethisham, R. N. Greenberg, and J. R. Berger. Inflammatory changes and breakdown of microvascular integrity in early human immunodeficiency virus dementia. *J Neurovirol* 10 (2004): 23–32.

Bazzoni, G. The Jam family of junctional adhesion molecules. *Curr Opin Cell Biol* 15, no. 5 (2003): 525–30.

Bennett, J., J. Basivireddy, A. Kollar, K. E. Biron, P. Reickmann, W. A. Jefferies, and S. McQuaid. Blood-brain barrier disruption and enhanced vascular permeability in the multiple sclerosis model Eae. *J Neuroimmunol* 229, no. 1–2 (2010): 180–91.

Bennett, M. V., L. C. Barrio, T. A. Bargiello, D. C. Spray, E. Hertzberg, and J. C. Sáez. Gap junctions: new tools, new answers, new questions. *Neuron* 6 (1991): 305–20 (review).

Brightman, M. W., and T. S. Reese. Junctions between intimately apposed cell membranes in the vertebrate Brain. *J Cell Biol* 40, no. 3 (1969): 648–77.

Buemi, M., A. Lacquaniti, D. Bolignano, V. Cernaro, S. Campo, G. Grasso, A. Buemi, V. Donato, and A. Sturiale. Down with the erythropoietin. Long live the erythropoietin! *Curr Drug Targets* 10 (2009): 1028–32.

Bundgaard, M., and N. J. Abbott. All vertebrates started out with a glial blood-brain barrier 4–500 million years ago. *Glia* 56, no. 7 (2008): 699–708.

Burt, J. M., and D. C. Spray. Volatile anesthetics block intercellular communication between neonatal rat myocardial cells. *Circ Res* 65 (1989): 829–37.

Cardoso, F. L., D. Brites, and M. A. Brito. Looking at the blood-brain barrier: molecular anatomy and possible investigation approaches. *Brain Res Rev* 64, no. 2 (2010): 328–63.

Coats, S. R., L. M. Pabon-Pena, J. W. Covington, and D. E. Vaughan. Ligand-specific control of Src-suppressed C kinase substrate gene expression. *Biochem Biophys Res Commun* 297, no. 5 (2002): 1112–20.

Daneman, R., D. Agalliu, L. Zhou, F. Kuhnert, C. J. Kuo, and B. A. Barres. Wnt/beta-catenin signaling is required for CNS, but not non-CNS, angiogenesis. *Proc Natl Acad Sci USA* 106 (2009): 641–46.

Daneman, R., L. Zhou, D. Agalliu, J. D. Cahoy, A. Kaushal, and B. A. Barres. The mouse blood-brain barrier transcriptome: a new resource for understanding the development and function of brain endothelial cells. *PLoS One* 29, no. 5 (2010a): e13741.

Daneman, R., L. Zhou, A. A. Kebede, and B. A. Barres. Pericytes are required for blood-brain barrier integrity during embryogenesis. *Nature* 468, no. 7323 (2010b): 562–66.

Derangeon M., D. C. Spray, N. Bourmeyster, D. Sarrouilhe, and J. C. Hervé. Reciprocal influence of connexins and apical junction proteins on their expressions and functions. *Biochim Biophys Acta* (2009) 1788:768–78.

Dermietzel, R. Visualization by freeze-fracturing of regular structures in glial cell membranes. *Naturwissenschaften* 60, no. 4 (1973): 208.

Dermietzel R., E. L. Hertberg, J. A. Kessler, and D. C. Spray. Gap junctions between cultured astrocytes: immunocytochemical, molecular, and electrophysiological analysis. *J Neurosci.* 11 (1991): 1421–32.

Dermietzel, R., D. Krause, M. Kremer, C. Wang, and B. Stevenson. Pattern of glucose transporter (Glut 1) expression in embryonic brains is related to maturation of blood-brain barrier tightness. *Dev Dyn* 193, no. 2 (1992): 152–63.

Dermietzel, R., D. C. Spray, and M. Nedergaard. *Blood brain interfaces: from ontogeny to artificial interfaces*, ed. R. Dermietzel, D. C. Spray, and M. Nedergaard, 671. New York: Wiley, 2006.

Desruisseuax, M. S., L. M. Weiss, H. B. Tanowitz, A. Mott, and D. Milner. Human parasitic disease in the context of the blood-brain barrier—effects, interactions and transgressions. In *Blood brain interfaces: From ontogeny to artificial interfaces*, ed. R. Dermietzel, D. C. Spray, and M. Nedergaard, 671. New York: Wiley, 2006.

Duffy, H. S., G. R. John, S. C. Lee, C. F. Brosnan, and D. C. Spray. Reciprocal regulation of the junctional proteins claudin-1 and connexin43 by interleukin-1beta in primary human fetal astrocytes. *J Neurosci* 20, no. 23 (2000): RC114.

Enerson, B. E., and L. R. Drewes. The rat blood-brain barrier transcriptome. *J Cereb Blood Flow Metab* 26, no. 7 (2006): 959–73.

Engelhardt, B. Development of the blood-brain barrier. *Cell Tissue Res* 314, no. 1 (2003): 119–29.

Engelhardt, B., and L. Sorokin. The blood-brain and the blood-cerebrospinal fluid barriers: function and dysfunction. *Semin Immunopathol* 31, no. 4 (2009): 497–511.

Ehrlich P. Das Sauerstoff-Bedurfnis des Organismus: eine farbenanalytische Studie. Berlin: *Hirschward*, 1885.

Fallier-Becker, P., J. Sperveslage, H. Wolburg, and S. Noell. The impact of agrin on the formation of orthogonal arrays of particles in cultured astrocytes from wild-type and agrin-null mice. *Brain Res* 1367 (2011): 2–12.

Fan, X., C. J. Heijnen, M. A. van der Kooij, F. Groenendaal, and F. van Bel. The role and regulation of hypoxia-inducible factor-1alpha expression in brain development and neonatal hypoxic-ischemic brain injury. *Brain Res Rev* 62, no. 1 (2009): 99–108.

Faustmann, P. M., and R. Dermietzel. Extravasation of polymorphonuclear leukocytes from the cerebral microvasculature. Inflammatory response induced by alpha-bungarotoxin. *Cell Tissue Res* 242, no. 2 (1985): 399–407.

Faustmann, P. M., S. Teutrine, D. Krause and R. Dermietzel. Subarachnoidal macrophates share a common epitope with resident non-cerebral macrophages and show receptor-mediated endocytosis of albumin-gold and IgG-gold complexes. *J. Neuroimmunol.* 35 (1991): 79–88.

Ferrara, N. Vascular endothelial growth factor and the regulation of angiogenesis. *Recent Prog Horm Res* 55 (2000): 15–35.

Fort, P. E., A. Sene, T. Pannicke, M. J. Roux, V. Forster, D. Mornet, U. Nudel, D. Yaffe, A. Reichenbach, J. A. Sahel, and A. Rendon. Kir4.1 and Aqp4 associate with Dp71- and utrophin-daps complexes in specific and defined microdomains of Muller retinal glial cell membrane. *Glia* 56, no. 6 (2008): 597–610.

Foster, C. Tight Junctions and the modulation of barrier function in disease. *Histochem Cell Biol* 130 (2008): 55–70.

Furman C.S., Gorelick-Feldman D.A., Davidson K.G., Yasumura T., Neely J.D., Agre P., and Rash J.E. Aquaporin-4 square array assembly: opposing actions of M1 and M23 isoforms. *Proc Nati Acad Sci USA* 100 (2003): 13609–14.

Furuse, M., K. Fujita, T. Hiiragi, K. Fujimoto, and S. Tsukita. Claudin-1 and -2: novel integral membrane proteins localizing at tight junctions with no sequence similarity to occludin. *J Cell Biol* 141, no. 7 (1998): 1539–50.

Furuse M, T. Hirase, M. Itoh, A. Nagafuchi, S. Yonemura, S. Tsukita, and S. Tsukita. Occludin: a novel integral membrane protein localizing at tight junctions. *J Cell Biol* 123 (1993): 1777–88.

Giaume, C., C. Fromaget, A. el Aoumari, J. Cordier, J. Glowinski, and D. Gros. Gap junctions in cultured astrocytes: single-channel currents and characterization of channel-forming protein. *Neuron*. 6 (1991): 133–43.

Giaume, C., A. Koulakoff, L. Roux, D. Holcman, and N. Rouach. Astroglial networks: a step further in neuroglial and gliovascular interactions. *Nat Rev Neurosci* 11, no. 2 (2010): 87–99.

Girouard, H., A. D. Bonev, R. M. Hannah, A. Meredith, R. W. Aldrich, and M. T. Nelson. Astrocytic endfoot Ca2+ and BK channels determine both arteriolar dilation and constriction. *Proc Natl Acad Sci USA* 107 (2010): 3811–16.

Goldmann E. Vitalfarbung am Zentralnervensystem: beitrag zur Physiopathologie des plexus chorioideus der Hirnhaute. *Abh Preuss Akad Wiss Physik-Math* 1 (1913): 1–60.

Goldman, S. A., and M. Nedergaard. Erythropoietin strikes a new cord. *Nat Med* 8, no. 8 (2002): 785–87.

Gordon, G. R., S. J. Mulligan, and B. A. MacVicar. Astrocyte control of the cerebrovasculature. *Glia* 55 (2007): 1214–21.

Grasso, G., A. Sfacteria, F. Meli, M. Passalacqua, V. Fodale, M. Buemi, F. Giambartino, D. G. Iacopino, and F. Tomasello. The role of erythropoietin in neuroprotection: therapeutic perspectives. *Drug News Perspect* 20 (2007): 315–20.

Greenberg, J. I., D. J. Shields, S. G. Barillas, L. M. Acevedo, E. Murphy, J. Huang, L. Scheppke, C. Stockmann, R. S. Johnson, N. Angle, and D. A. Cheresh. A role for Vegf as a negative regulator of pericyte function and vessel maturation. *Nature* 456 (2008): 809–13.

Hallmann, R., D. N. Mayer, E. L. Berg, R. Broermann, and E. C. Butcher. Novel mouse endothelial cell surface marker is suppressed during differentiation of the blood brain barrier. *Dev Dyn* 202 (1995): 325–32.

Hallmann, R., N. Horn, M. Selg, O. Wendler, F. Pausch, and L. M. Sorokin. Expression and function of laminins in the embryonic and mature vasculature. *Physiol Rev* 85 (2005): 979–1000.

Holash J. A., D. M. Noden, and P. A. Stewart. Re-evaluating the role of astrocytes in blood-brain barrier iduction. *Dev Dyn* 197 (1993): 14–25.

Iadecola, C., and M. Nedergaard. Glial regulation of the cerebral microvasculature. *Nat Neurosci* 10, no. 11 (2007): 1369–76.

Janzer, R. C., and M. C. Raff. Astrocytes induce blood-brain barrier properties in endothelial cells. *Nature* 325, no. 6101 (1987): 253–57.

Kanmogne, G. D., C. Primeaux, and P. Grammas. HIV-1 gp120 proteins alter tight junction protein expression and brain endothelial cell permeability: implications for the pathogenesis of HIV-associated dementia. *J Neuropathol Exp Neurol.* 64 (2005): 498–505.

Krause, G., L. Winkler, S. L. Mueller, R. F. Haseloff, J. Piontek, and I. E. Blasig. Structure and function of claudins. *Biochim Biophys Acta* 1778 (2008): 631–45.

Kuhnert, F., M. R. Mancuso, A. Shamloo, H. T. Wang, V. Choksi, M. Florek, H. Su, M. Fruttiger, W. L. Young, S. C. Heilshorn, and C. J. Kuo. Essential regulation of Cns angiogenesis by the orphan G protein-coupled receptor Gpr124. *Science* 330 (2010): 985–89.

Kunzelmann, P., W. Schroder, O. Traub, C. Steinhauser, R. Dermietzel, and K. Willecke. Late onset and increasing expression of the gap junction protein connexin30 in adult murine brain and long-term cultured astrocytes. *Glia* 25 (1999): 111–19.

Lai, C. H., and K. H. Kuo. The critical component to establish in vitro BBB model: pericyte. *Brain Res Brain Res Rev* 50 (2005): 258–65.

Landis, D. M., and T. S. Reese. Arrays of particles in freeze-fractured astrocytic membranes. *J Cell Biol* 60 (1974): 316–20.

Larson, D. M., M. P. Carson, and C. C. Haudenschild. Junctional transfer of small molecules in cultured bovine brain microvascular endothelial cells and pericytes. *Microvasc Res* 34 (1987): 184–99.

Lee, S. W., W. J. Kim, Y. K. Choi, H. S. Song, M. J. Son, I. H. Gelman, Y. J. Kim, and K. W. Kim. SSeCKS regulates angiogenesis and tight junction formation in blood-brain barrier. *Nat Med* 9 (2003): 900–6.

Liebner, S., M. Corada, T. Bangsow, J. Babbage, A. Taddei, C. J. Czupalla, M. Reis, A. Felici, H. Wolburg, M. Fruttiger, M. M. Taketo, H. von Melchner, K. H. Plate, H. Gerhardt, and E. Dejana. Wnt/beta-catenin signaling controls development of the blood-brain barrier. *J Cell Biol* 183 (2008): 409–17.

Locovei, S., L. Bao, and G. Dahl. Pannexin 1 in erythrocytes: function without a gap. *Proc Natl Acad Sci USA* 103 (2006): 7655–59.

Marti, H. H., and W. Risau. Systemic hypoxia changes the organ-specific distribution of vascular endothelial growth factor and its receptors. *Proc Natl Acad Sci USA* 95, no. 26 (1998): 15809–14.

Maxwell, P. H., C. W. Pugh, and P. J. Ratcliffe. Inducible operation of the erythropoietin 3' enhancer in multiple cell lines: evidence for a widespread oxygen-sensing mechanism. *Proc Natl Acad Sci USA* 90 (1993): 2423–27.

McQuaid, S., P. Cunnea, J. McMahon, and U. Fitzgerald. The effects of blood-brain barrier disruption on glial cell function in multiple sclerosis. *Biochem Soc Trans* 37, pt. 1 (2009): 329–31.

Meier, C., and R. Dermietzel. Electrical synapses—gap junctions in the brain. *Results Probl Cell Differ* 43 (2006): 99–128.

Metea, M. R., and E. A. Newman. Glial cells dilate and constrict blood vessels: a mechanism of neurovascular coupling. *J Neurosci.* 26 (2006): 2862–70.

Mi, H., H. Haeberle, and B. A. Barres. Induction of astrocyte differentiation by endothelial cells. *J Neurosci* 21 (2001): 1538–47.

Minakawa, T., J. Bready, J. Berliner, M. Fisher, and P. A. Cancilla. In vitro interaction of astrocytes and pericytes with capillary-like structures of brain microvessel endothelium. *Lab Invest* 65 (1991): 32–40

Mulligan, S. J., and B. A. MacVicar. Calcium transients in astrocyte endfeet cause cerebrovascular constrictions. *Nature* 431 (2004): 195–99.

Murugesan, N., J. A. Macdonald, Q. Lu, S. L. Wu, W. S. Hancock, and J. S. Pachter. Analysis of mouse brain microvascular endothelium using laser capture microdissection coupled with proteomics. *Methods Mol Biol* 686 (2011): 297–311.

Nagy, J. I., X. Li, J. Rempel, G. Stelmack, D. Patel, W. A. Staines, T. Yasumura, and J. E. Rash. Connexin26 in adult rodent central nervous system: demonstration at astrocytic gap junctions and colocalization with connexin30 and connexin43. *J Comp Neurol* 441 (2001): 302–23.

Nicchia, G. P., B. Nico, L. M. Camassa, M. G. Mols, N. Loh, R. Dermietzel, D. C. Spray, M. Svetto, and A. Frigeri. The role of aquaporin-4 in the blood-brain-barrier development and integrity: studies in animal and cell culture models. *Neuroscience* 129 (2004): 935–45.

Niessen, C. M., and C. J. Gottardi. Molecular compoenets of the adherens junctions. *Biochem Biophys Acta* 1778 (2008): 562–71.

Noell, S., P. Fallier-Becker, U. Deutsch, A. F. Mack, and H. Wolburg. Agrin defines polarized distribution of orthogonal arrays of particles in astrocytes. *Cell Tissue Res* 337, no. 2 (2009): 185–95.

Paolinelli, R., M. Corada, F. Orsenigo, and E. Dejana. The molecular basis of the blood brain barrier differentiation and maintenance. Is it still a mystery? *Pharmacol Res* 63, no. 3 (2011): 165–71.

Park, J. A., Y. K. Choi, S. W. Lee, and K-W. Kim. Brain angiogenesis and barriergenesis. In *Blood-brain-barriers*, ed. D. C. Spray, R. Dermietzel, and M. Nedergaard, 41–59. Weinheim: Wiley-VCH, 2006.

Paspalas, C. D., and G. C. Papadopoulos. Ultrastructural relationships between noradrenergic nerve fibers and non-neuronal elements in the rat cerebral cortex. *Glia* 17, no. 2 (1996): 133–46.

Peppiatt, C. M., C. Howarth, P. Mobbs, and D. Attwell. Bidirectional control of CNS capillary diameter by pericytes. *Nature* 443 (2006): 700–4.

Pottiez, G., C. Flahaut, R. Cecchelli, and Y. Karamanos. Understanding the blood-brain barrier using gene and protein expression profiling technologies. *Brain Res Rev* 62 (2009): 83–98.

Ramsauer, M., D. Krause, and R. Dermietzel. Angiogenesis of the blood-brain barrier in vitro and the function of cerebral pericytes. *FASEB J* 16 (2002): 1274–76.

Rash, J. E., K. G. Davidson, T. Yasumura, and C. S. Furman. Freeze-fracture and immunogold analysis of aquaporin-4 (AQP4) square arrays, with models of AQP4 lattice assembly. *Neuroscience* 129 (2004): 915–34.

Reese, T. S., and M. J. Karnovsky. Fine structural localization of a blood-brain barrier to exogenous peroxidase. *J Cell Biol* 34 (1967): 207–17.

Salinas, S., S. Giampietro, and E. J. Kremer. A hitchhiker's guide to the nervous system: the complex journey of viruses and toxins. *Nature Rev Microbiol* 8 (2010): 645–55.

Scemes, E., S. O. Suadicani, G. Dahl, and D. C. Spray. Connexin and pannexin mediated cell-cell communication. *Neuron Glia Biol 3* (2007): 199–208.

Schummers, J., H. Yu, and M. Sur. Tuned responses of astrocytes and their influence on hemodynamic signals in the visual cortex. *Science* 320 (2008): 1638–43.

Stewart, P. A., and M. J. Wiley. developing nervous tissue induces formation of blood-brain barrier characteristics in invading endothelial cells: a study using quail-chick transplantation chimeras. *Dev Biol* 84, no. 1 (1981): 183–92.

Tam, S. J., and R. J. Watts. Connecting vascular and nervous system development: angiogenesis and the blood-brain barrier. *Annu Rev Neurosci* 33 (2010): 379–408.

Tsukita, S., and M. Furuse. The structure and function of claudins, cell adhesion molecules at tight junctions. *Ann NY Acad Sci* 915 (2000): 129–35.

van Horssen, J., L. Bo, C. M. Vos, I. Virtanen, and H. E. de Vries. Basement membrane proteins in multiple sclerosis-associated inflammatory cuffs: potential role in influx and transport of leukocytes. *J Neuropathol Exp Neurol* 64 (2005): 722–29.

Van Itallie, C. M., and J. M. Anderson. Claudins and epithelial paracellular transport. *Annu Rev Physiol* 68 (2006): 403–29.

Van Itallie, C. M., L. Betts, J. G. Smedley 3rd, B. A. McClane, and J. M. Anderson. Structure of the claudin-binding domain of clostridium perfringens enterotoxin. *J Biol Chem* 283 (2008): 268–74.

Wei, C. J., R. Francis, X. Xu, and C. W. Lo. Conexin43 associated with an N-cadherin-containing multiprotein complex is required for gap junction formation in Nih3t3 cells. *J Biol Chem* 2005 (2005): 19925–36.

Wolburg, H., S. Noell, A. Mack, K. Wolburg-Buchholz, and P. Fallier-Becker. Brain endothelial cells and the glio-vascular complex. *Cell Tissue Res* 335 (2009a): 75–96.

Wolburg, H., S. Noell, K. Wolburg-Buchholz, A. Mack, and P. Fallier-Becker. Agrin, aquaporin-4, and astrocyte polarity as an important feature of the blood-brain barrier. *Neuroscientist* 15 (2009b): 180–93.

Wolburg, H., K. Wolburg-Buchholz, and B. Engelhardt. Diapedesis of mononuclear cells across cerebral venules during experimental autoimmune encephalomyelitis leaves tight junctions intact. *Acta Neuropathol* 109 (2005): 181–90.

Zonta, M., M. C. Angulo, S. Gobbo, B. Rosengarten, K. A. Hossmann, T. Pozzan, and G. Carmignoto. Neuron-to-astrocyte signaling is central to the dynamic control of brain microcirculation. *Nat Neurosci* 6 (2003): 43–50.

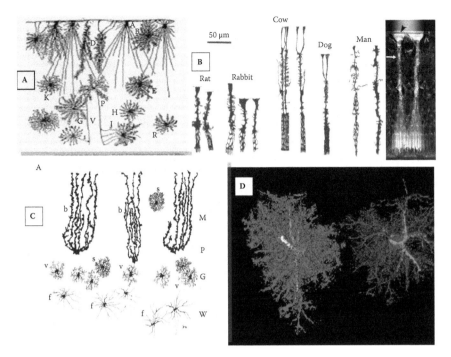

FIGURE 1.1 (A) Golgi-stained astrocytes from a 2-month-old human infant in the molecular layer and layers II and III of the cerebral cortex. A–D are cells in the first cortical lamina and E–H are cells in the second and third lamina. I and J are cells with endfeet contacting blood vessels. V, blood vessel. (From Ramón y Cajal, *Trab. Lab. Invest. Biol. Univ. Madrid*, 11, 255–315, 1913.) (B) Muller cells in different species as indicated. Furthest right, Muller cells as light guides. (Courtesy of S. Skatchov, from Franze et al., *Proc. Natl. Acad. Sci. USA*, 104, 8287–8292, 2007, cover page of that issue.) (C) Different astrocyte cells in mature mammalian cerebellum by Golgi staining (Golgi, 1885). Working down from the cerebellar surface A at the top, represented by dotted line: M, molecular; P, Purkinje; G, granule cell; and W, white matter layers. b, Bergmann glia; s, protoplasmic astrocytes; v, velate astrocytes; and f, fibrous astrocytes (From Kimelberg, 2010, Figure 1 on p. 83.) (D) Dye-filled astrocytes of rat hippocampus. (From Eric Bushong, personal communication.)

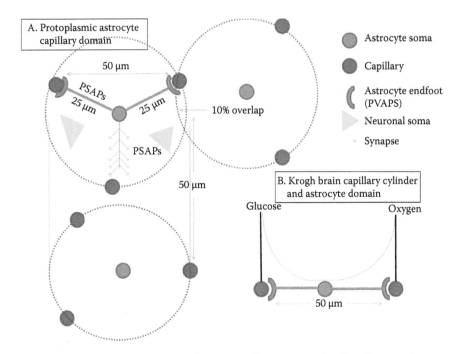

FIGURE 1.3 (A) Domain concept of protoplasmic astrocyte related to distances between capillaries in the grey matter. (B) The Krogh cylinder referring to the exponential decline of oxygen and glucose with increasing distance from the capillary creating a low mid-distance level (see Lubbers, 1977), which would correspond approximately to the position of the astrocyte soma.

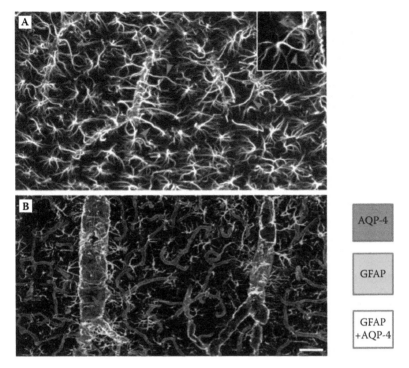

A

B

AQP-4

GFAP

GFAP
+AQP-4

FIGURE 1.4 AQP-4 and GFAP labeling of astrocytes in rat cortex. (A) GFAP immunola-beling of astrocytes. Individual astrocytes appear somewhat star shaped and are uniformly distributed, with little overlap with neighboring astrocytes. Astrocytic processes to blood vessels differ from other processes by being straight, unbranched, and of wider diameter (red arrowheads). The surfaces of large- to medium-size vessels were densely covered by GFAP+ astrocytic endfeet. Inset, an astrocyte with two vascular processes. (B) Double immunolabel-ing for AQP-4 (red) and GFAP (green). The larger and even smaller arterioles (precapillary sphincters; see text) are surrounded by endfeet that are positive for both aquaporin-4 and GFAP, possibly because the larger size of the processes leading to the arterioles allows GFAP filaments to localize there. AQP-4 is in the endfeet membranes. The capillaries are only AQP-4 positive, being covered by astrocytic processes, but these are GFAP negative, possibly because they are too small for the GFAP filament bundles to penetrate. Scale bar = 60 μm. (From Simard et al., *J. Neurosci.*, 23, 9254–9262, 2003. With Permission.)

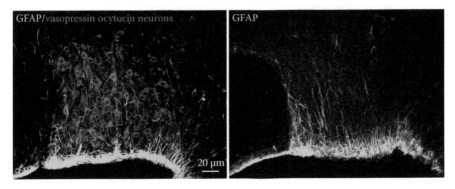

FIGURE 3.1 Polarized astrocytes in the hypothalamus. A coronal section of the ventral part of the adult rat hypothalamus was immunocytochemically labeled with antibodies to glial fibrillary acidic protein (green) to identify astrocytes and oxytocin- and vasopressin-associated neurophysins to identify neurons. Astrocytes (green) send out processes toward the neurons and are able to modify the responsiveness of these neurons (red). The optic tract is at left. A population of astrocytes with cell bodies sitting on the glial limitans send long processes dorsally and ventrally to the supraoptic nucleus, where they surround cell bodies and dendrites of magnocellular neurons. Bar = 20 μm. (Courtesy of Dr. Charles Bourque, McGill University.)

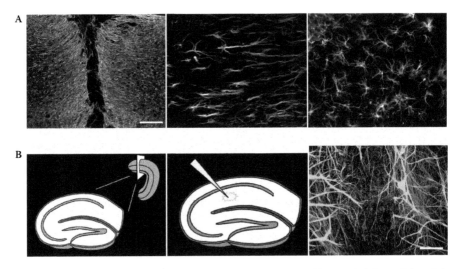

FIGURE 3.2 Polarized astrocytes orient toward a wound. (A) Traumatic injury of the adult mouse spinal cord was induced using iridectomy scissors. After 8 days, the spinal cord was fixed, removed, and serially sliced. Glial fibrillary acidic protein (GFAP) immunostaining shows that reactive astrocytes of the gray matter present two main morphologies. Astrocytes that directly surround the lesion site (dark area, left panel) most often have extended processes toward the wound (middle panel), whereas the ones that are more distant keep their stellate morphology (right panel). Scale bar = 150 μm in left panel and 40 μm in middle and right panels. (B) Rat hippocampal slices in culture were either wounded (left panel) or α-amino-adipic acid was injected to create a chemical injury in rat hippocampal slices in culture. After 5 days, the slice was fixed and labeled immunocytochemically with antibodies to GFAP that revealed astrocytic processes extending toward the injury (green). The injury site delineated by expression of SPARC (red), a protein secreted by astrocytes as well as the absence of any neurofilament positive axons (blue). Bar = 100 μm. (Courtesy of Drs. Emma V. Jones and Keith Mural, McGill University.)

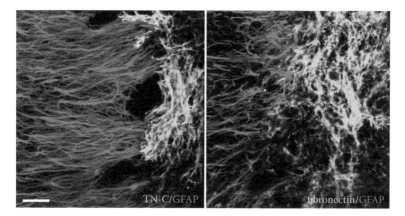

FIGURE 3.3 Astrocytes interact with tenascin-C and fibronectin in the glial scar. Eight days after traumatic injury of the adult mouse spinal cord, the lesion site is invaded with different cell types from nonnervous origin, such as inflammatory cells and fibroblasts. These cells express and secrete extracellular matrix molecules, such as tenascin-C (green, left panel) or fibronectin (green, right panel), organized in fibrillar structures that are in contact with reactive astrocytes (GFAP, red in both pictures) of the glial scar. Tenascin-C in particular is believed to inhibit axon regeneration through the scar and to participate in the formation of a new glia limitans. Scale bar = 80 μm.

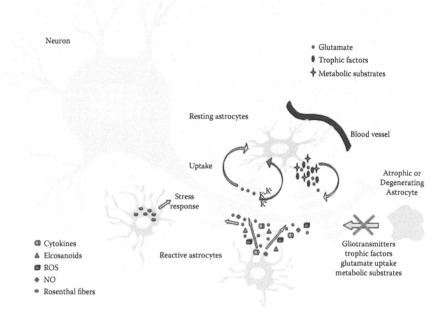

FIGURE 6.1 Schematic representation of astrocyte-neuron interactions in physiopathological conditions. In physiological conditions, astrocytes exhibit relatively long and highly branched cell processes that can perform many functions: (1) Owing to their morphological connections with capillaries at the endfeet, they can take up nutrients from the blood and transform and transfer them to neurons. (2) They are highly permeable to potassium ions and can rapidly clear their excess from the peri-synaptic space. (3) They express plasma membrane transporters for several neurotransmitters, including glutamate, and clear them from the extracellular space upon synaptic release. (4) They provide neurons with trophic factors and release gliotrasmitters. In pathological conditions, astrocytes become activated, i.e., show thicker cell processes, and start to secrete factors, such as cytokines, reactive oxygen species (ROS), nitric oxide (NO), eicosanoids, and excitotoxins, by which they can damage nearby neurons. Furthermore, in specific pathological instances, they show fibrous, eosinophilic deposits known as Rosenthal fibers. Astrocyte degeneration deprives neurons of trophic factors, metabolic substrates, and glutamate uptake, and interrupts astrocyte-neuron communication.

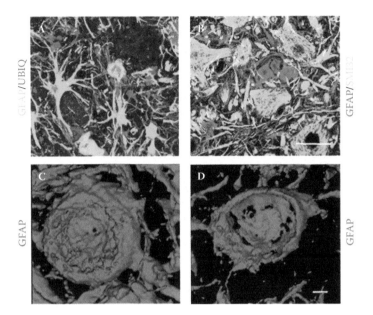

FIGURE 6.2 A subpopulation of astrocytes undergoes degeneration in ALS transgenic mice during disease progression. (A) All spheroid-shaped astrocytes are positive for ubiquitin and often are located near vacuolated motor neurons. Scale bar, 20 µm (A and B). (C, D) 3D reconstruction of spheroid-shaped astrocytes. 3D reconstruction of z-stack confocal microscopy images rendered by ImageJ software (National Institute of Health) and VolumeJ plug in. (B) Lumbar spinal cord sections from ALS mice were simultaneously labeled for nonphosphorylated neurofilaments (SMI32 antibody, green) and GFAP (red). The image is representative of reactive and spheroid-shaped astrocytes, directly contiguous to motor neurons, at the age of disease onset. (C) The 3D section of this spheroid astrocyte shows a compact GFAP cytoskeleton around a central core and the presence of a few GFAP-positive processes, which appear short and abnormally thick. This cell is immunonegative for the active caspase-3, indicative of an early stage of degeneration. (D) The 3D section of this spheroid astrocyte displays a much thinner and less compact GFAP cytoskeleton around the central core compared to the example in (C), with rare GFAP-positive processes. This cell is positive for the active caspase-3 (not shown), indicative of a late stage of degeneration. Scale bar, 2 µm (C and D).

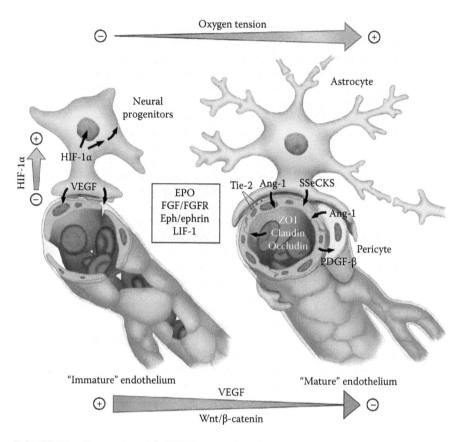

FIGURE 7.1 Proposed model of BBB maturation. Oxygen tension is suggested to comprise the driving force for angiogenesis in the early brain anlage. Invading endothelial cells from the perineural plexus settle within the brain vesicle where neural progenitor cells produce VEGF according to activation of the oxygen-responsive hypoxia-induced factor (HIF-1α). Increasing levels of oxygen elicit reduction of VEGF secretion, and maturation of the BBB is induced by differentiation factors like SSeCKS (src-suppressed C kinase substrate) and Ang-1 (Angiopoetin-1 (Tie-2 is the cognate Ang-1 receptor)). Endothelial PDGF-β (platelet-derived growth factor) is a strong attractor for pericytes, which also produce PDGF-β. The center box indicates some further growth and differentiation factors responsible for angiogenesis and barriergenesis in brain tissues. Likewise, Wnt/β-catenin signaling controls the development of the BBB indicated by the gradient marker in the lower part of the graph.

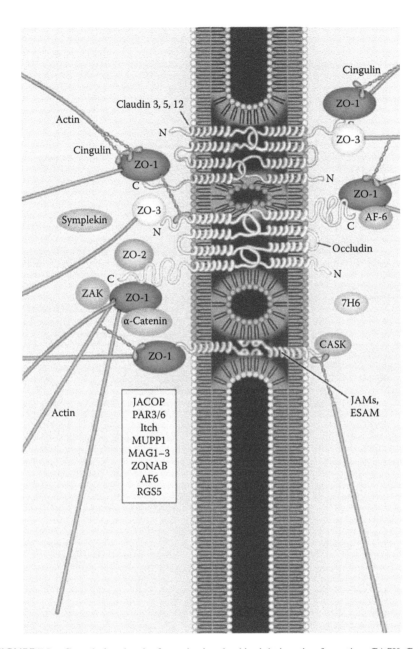

FIGURE 7.2 Cumulative sketch of proteins involved in tight junction formation. CASK, Ca²⁺-dependent serine protein kinase; JAM, junctional adhesion molecule; ZAK, leucine-zipper and sterile-α-motif kinase; 7H6/barmontin, a tight junction-associated protein recognized by the 7H6 monoclonal antibody. The following tight junction-associated proteins are listed in the box: JACOB, junction-associated coiled-coiled protein; PAR3/6, the partitioning defective protein 3 and 6; MUPP1, multi-PDZ protein1; Itch, E3 ubiquitin-protein ligase; MAG1–3, membrane-associated guanylate kinase with inverted orientation of protein-protein interaction domains; ZONAB, ZO-1-associated nucleic-acid-binding protein; AF6, afadin, a large PDZ-containing family of actin-binding proteins concentrated at the adherens junction; RGS5, regulator of G protein signaling. For clarity, not all of the adherens junction-associated proteins are listed.

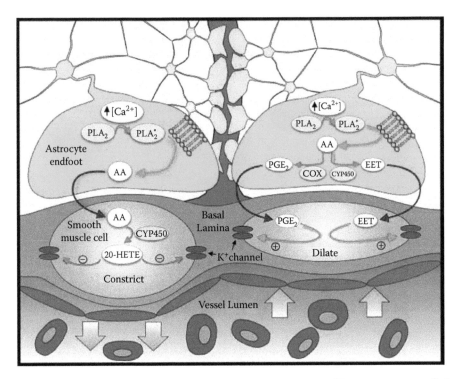

FIGURE 7.3 Pathways of vasomotor control in brain arterioles. Two separate astroglial endfeet (green) are illustrated, the left causing vasoconstriction and the right one causing vasodilation of smooth muscle (yellow). (From Gordon et al., *Glia* 55, 1214–21, 2007. With permission of John W. Wiley Publishers.)

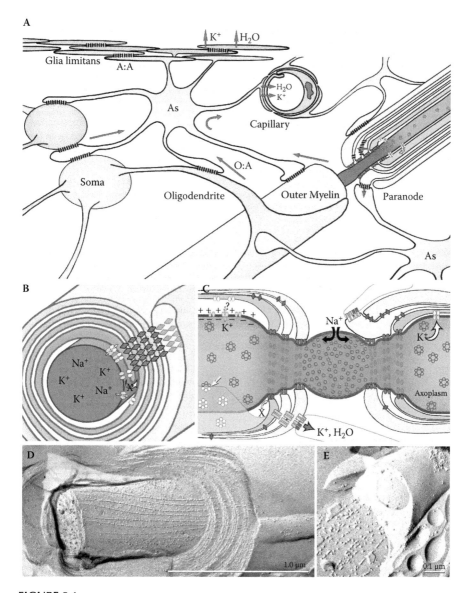

FIGURE 8.1

FIGURE 8.1 Gateway cells of the panglial syncytium (As, blue, astrocytes; green, "soma," oligodendrocyte) and their relationships to myelinated axons (red), capillaries (maroon), and the glia limitans (dark blue stack). Astrocytes, which are interlinked by abundant gap junctions, act as intermediaries that couple oligodendrocyte somata, their oligodendrite processes, and their myelin segments to the astrocyte endfeet that surround capillaries and form the glia limitans. At these locations, excess K^+ (K^+, red arrows) and osmotic water (H_2O, blue arrows) are released into the circulatory system or into the bathing cerebrospinal fluid. (Modified from Kamasawa, N., et al., *Neuroscience*, 136, 65–86, 2005.) (B) Diagram of cross-sectioned myelin, showing multiple tight junction strands (brown diamonds) that are stacked from inner mesaxon to outer mesaxon, isolating each layer of internodal myelin extracellular space into ionically and electrically distinct compartments, one per layer of myelin wrapping. K_V1 channels (blue hexagons) are drawn linked directly to innermost cytoplasmic myelin via Cx29 hemichannels (yellow hexagons), and conversely as spatially separated from Cx29 hemichannels. In the peri-internodal extracellular space, black X with white arrow indicates that K^+ cannot enter the next myelin extracellular space because of multiple tight junction barriers. Na^+ and K^+, excess positive ions after axonal depolarization. (Modified from Rash, J.E., *Neuroscience*, 168, 982–1008, 2010.) (C) Locations and interrelationships of principal proteins identified to date at nodes of Ranvier, paranodes, and juxtaparanodes. Diagram depicts entry of Na^+ through Na_V1 sodium channels (dark green circles) at nodes of Ranvier, and exit of K^+ at juxtaparanodal axonal plasma membrane via K_V1 channels (blue hexagons in hexagonal rosettes), either directly into the internodal extracellular space (designated by ?) or from axoplasm via K_V1 channels linked to innermost cytoplasmic myelin via Cx29 hemichannels (yellow hexagons in hexagonal array). K^+ is blocked at contactin/caspr-containing septate junctions (X with white curved arrow). Brown diamonds, tight junctions between myelin layers. Autologous gap junctions composed of Cx32 (green barrels) couple successive layers of paranodal myelin. The outermost layer of myelin is coupled to the astrocyte syncytium via biheterotypic gap junctions consisting of Cx47 (light green barrels) and Cx32 (dark green barrels) in oligodendrocyte hemiplaques coupled to Cx43 (blue barrels) and Cx30 (dark blue barrels) in astrocyte hemiplaques. Relative K^+ concentrations at successive paranodal layers grade from orange to light yellow to green (oligodendrocyte) to blue (astrocyte). (Modified from Rash, J.E., *Neuroscience*, 168, 982–1008, 2010.) Red ovals, caspr/contactin septate junctions in paranodal membrane; yellow rosettes, Cx29 (as immunogold labeled in PNS myelin; Li et al., 2002). Fractures that stepped from innermost myelin to the axon plasma membrane (larger straight white arrow) revealed that the myelin rosettes are structurally coupled to rosettes in the axon plasma membrane (Stolinski et al., 1981, 1985). K^+, H_2O with blue arrow depicts movement of K^+ and water from oligodendrocyte to astrocyte via gap junctions. (D) Freeze-fracture image of multiple tight junction strands stacked from inner mesaxon to outer mesaxon, separating each layer of myelin internodal extracellular space into ionically and electrically distinct compartments, one per layer of myelin wrapping. (Modified from Rash, J.E., *Neuroscience*, 168, 982–1008, 2010.) (E) Outer tongue of myelin linked to first compact layer via 12 parallel tight junction strands. The particle partitioning coefficient (Satir and Satir, 1979) of 50% for tight junctions in formaldehyde-fixed tissues used in FRIL resulted in rows of both IMPs and pits in this myelin P-face. Arrow, 10 nm immunogold bead labeling Cx32 in small gap junction on outer tongue.

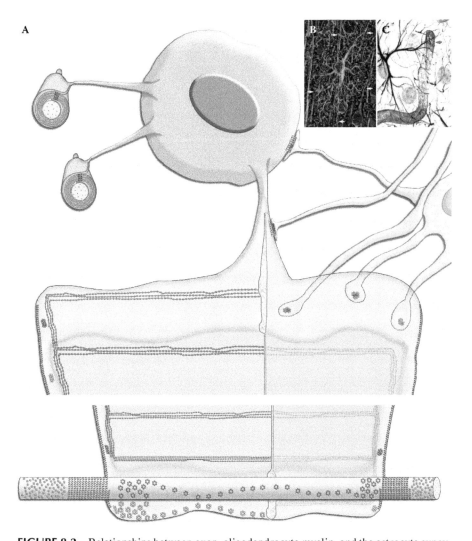

FIGURE 8.2 Relationships between axon, oligodendrocyte myelin, and the astrocyte syncytium. (A) Diagram of one segment of oligodendrocyte myelin, unrolled to show molecular specializations of both inner (left side) and outer (right side) surfaces. The oligodendrocyte soma (green) is connected by oligodendrites to three different myelin segments, all from different axons. Astrocyte (blue) processes form gap junctions with the outer surface of myelin; blue cylinder, axon; blue hexagonal rosettes, K_V1 channels in the axon internodal plasma membrane, concentrated at the juxtaparanodal surface and along the inner mesaxon; brown diamonds in lines, tight junction strands link successive layers of myelin (light vs. dark lines); green IMPs in hexagonal arrays, Cx32-containing gap junctions between paranodal loops of myelin; orange hexagonal arrays, Cx32/Cx47 to Cx30/Cx43 heterotypic heterologous gap junctions; triplet red ovals as dotted lines, septate junctions binding the paranodal loops to the axon plasma membrane, forming a spiral barrier separating the nodal extracellular space from the peri-internodal extracellular space; yellow rosettes, Cx29 in innermost myelin. (B) Image of astrocyte (pink staining for GFAP), with processes approaching myelin (bottom center) and capillaries (top center). Neurons are stained for MAP2 (blue) and all nuclei (including oligodendrocyte somata) are stained with DAPI (purple). (Image courtesy of Maiken Nedergard, Nancy Ann Oberheim, and Takahiro Takano.) (C) Staining for AQP4 in astrocyte endfeet (pink) precisely follows and delineates the capillary bed. Large GFAP filament bundles (black) in the cell body diverge and spread to fill the astrocyte fingers and extend to each endfoot process (adult rat cortex). Nuclei (blue) stained with DAPI. (From Xaiohai Wang, with permission.)

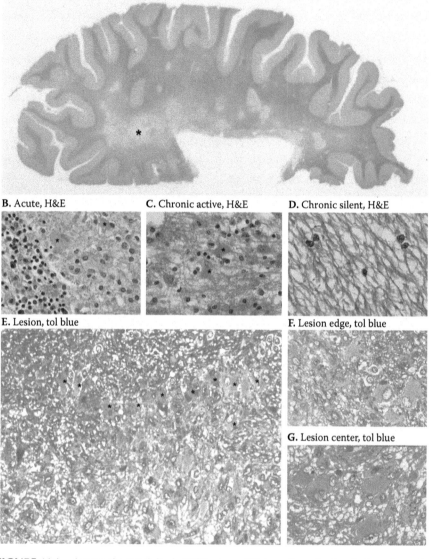

A. MS cerebrum, Luxol fast blue

B. Acute, H&E

C. Chronic active, H&E

D. Chronic silent, H&E

E. Lesion, tol blue

F. Lesion edge, tol blue

G. Lesion center, tol blue

FIGURE 11.1 Astrocytic reactivity in MS lesions. (A) A section of a hemisphere from the brain of a patient with chronic MS stained with Luxol-fast-blue for myelin. Note the many large periventricular lesions lacking myelin, some of which present with diffuse edges (*) suggestive of ongoing activity, whereas others possess sharper borders, indicative of chronic myelin loss. (B) An H&E preparation from an acute MS lesion shows hypertrophic astrocytes (*) in the CNS parenchyma adjacent to a perivascular accumulation of plasma cells and lymphocytes (at left). (C) An H&E-stained section from the center of a chronic active MS lesion reveals many hypertrophic astrocytes in the demyelinated parenchyma. (D) The center of a chronic silent lesion shows a network of fibrous astrocytes. (E) A one-micron epoxy section from the margin of an acute MS lesion shows a broad zone of hypertrophic astrocytes (asterisks) between myelinated white matter (upper left) and lesion center (below). Note how the same cells extend into the lesion proper. (F, G) Detail of lesion edge and center, respectively, to show hypertrophic astrocytes. Note how the cells are more globoid at the edge and more process bearing within the lesion.

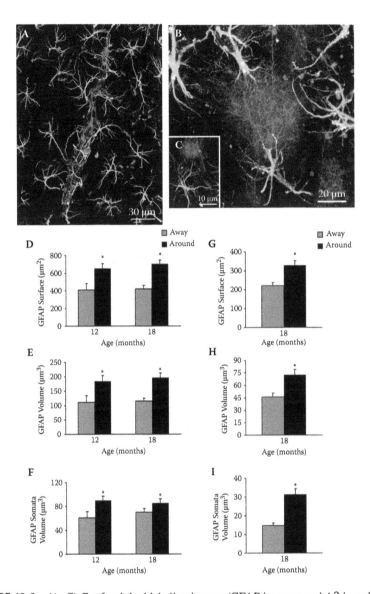

FIGURE 12.3 (A–C) Confocal dual labeling images (GFAP in green and Aβ in red) showing the accumulation of astrocytes around the Aβ plaques and vascular Aβ deposits in the hippocampus of 3xTg-AD mice. Astrocytes surrounding Aβ plaques (B, C) and Aβ-loaded blood vessel (A) undergo astrogliosis. (D–F) Bar graphs showing GFAP-positive astrocytic surface (D), volume (E), and soma volume (F) differences between astrocytes located around the amyloid plaques (Aβ) and those distant to the plaques in the CA1 of 3xTg-AD animals. (G–I) Similar astrocytic surface (G), volume (H), and soma volume (I) differences are observed in the DG at 18 months of age. Bars represent mean ± SEM ($p < 0.05$). (Modified from Olabarria, M., et al., *Glia*, 58, 831–38, 2010.)

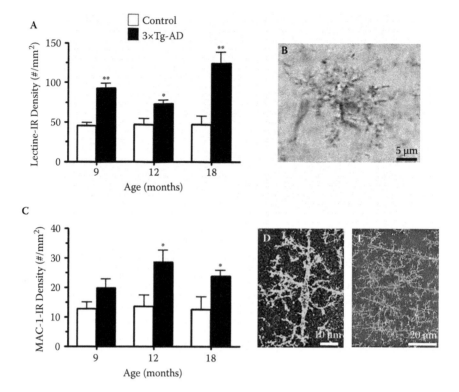

FIGURE 12.5 (A, B) Visualization and quantification of resting and activated microglia in the hippocampi of 3xTg-AD animals. (A) Bar graph showing the area density of resting microglia (Sv; number/mm^2) in the hippocampal CA1 of 3xTg-AD mice and non-Tg control mice at 9, 12, and 18 months. (B) Brightfield micrographs of typical resting TL-IR microglia with small cell body showing thin to medium ramified processes extending to the surrounding neuropil in the CA1 subfield, which is not modified either by age or by Aβ-amyloid plaques. (CE) Microglial activation in plaque-infested hippocampal tissue. (C) Bar graph showing the effect of age-related changes on MAC-1-IR reactive microglia within the CA1 subfield of the hippocampus at 9, 12, and 18 months between non-Tg control and 3xTg-AD mice. (D) Fluorescence micrograph illustrating the characteristic morphology of reactive microglia within the CA1 subfield of the hippocampus of an 18-month-old 3xTg-AD mouse. Reactive microglia appear with most enlarged cell bodies from which a greater number of numerous processes emanated, but with an enlarged and thicker appearance. (E). Confocal image showing recruitment of MAC-1- IR microglia (green) in the vicinity of Aβ-amyloid aggregates (red) in the CA1 subfield of the hippocampus of an 18-month-old 3xTg-AD mouse. (Modified from Rodríguez, J. J., et al., *Cell Death Dis*, 1, e1, 2010.)

8 Oligodendrocytes
Gateway to the Panglial Syncytium

Kimberly G. V. Davidson
Department of Biomedical Sciences, Colorado State
University, Fort Collins, Colorado

John E. Rash
Department of Biomedical Sciences and Program in
Molecular, Cellular, and Integrative Neurosciences,
Colorado State University, Fort Collins, Colorado

CONTENTS

8.1 INTRODUCTION

The nearly simultaneous evolution of myelin in the central and peripheral nervous systems (CNS and PNS) was a transformative event in early vertebrate evolution (Zalc et al., 2008). The much higher conduction velocity of myelinated axons than of the same-diameter unmyelinated axons allowed the development of much larger body plans, with meter-long peripheral nerves, larger and more complexly integrated brains, and meter-long spinal cords. Because myelin is not present in the brain or spinal cords of primitive, slow-moving, jawless fishes (i.e., hagfish and lampreys) (Bullock et al., 1984), it has been proposed that CNS myelin first appeared in the jawed placoderm fishes (Zalc et al., 2008). However, genomic analysis has revealed that some essential myelin molecules appeared much earlier, in primitive protochordates, including tunicate larvae (Gould et al., 2005). These ancient molecules include myelin basic protein (MBP), which is also expressed in some nonmyelinated cells (Li and Richardson, 2008). Regardless of which early vertebrate animals developed the first myelin, and whether myelin appeared first in the CNS or PNS, the resulting much faster response times afforded by saltatory conduction not only allowed the development of much larger free-swimming predator fish, but also allowed vertebrates more rapid escape responses from predators, which during the early Devonian period included giant arthropods (e.g., 2 m long sea scorpions) and large cephalopods (e.g., nautilus). In this neuronal arms race, invertebrate predators had already increased their neuronal conduction velocities based primarily on larger-diameter axons rather than on myelination (Hartline, 2008), whereas in vertebrate species, the smaller axons allowed by myelin led to the development of larger brains, more complex behaviors, and the evolution of animals with "massively-parallel processing" (Hartline, 2008), allowing the emergence of long-term memory, cunning, and intelligence.

8.2   EVOLUTIONARY DIVERGENCE OF OLIGODENDROCYTES VERSUS ASTROCYTES

The evolutionary development of saltatory conduction in the vertebrate PNS was based on each axonal segment being myelinated by a different Schwann cell. In contrast, the development of saltatory conduction in the CNS required simultaneous evolution and divergence of two distinct glial cell types—oligodendrocytes and astrocytes—to facilitate both (1) saltatory conduction and (2) removal of the large, potentially fatal amounts of K^+ released during high-frequency saltatory conduction. (This problem of K^+ removal had already been solved in primitive vertebrates based on widespread permeation of astrocyte processes within bundles of nonmyelinated axons.) The subsequent specialization of oligodendrocyte myelin for initial sequestration of axonally derived K^+ (detailed below), and the further specialization of astrocytes for long-distance K^+ siphoning and disposal of excess K^+ outside of the CNS parenchyma (also detailed below), allowed the development of the densely packed bundles of myelinated fibers that characterize the much larger brains and spinal cords of higher vertebrates. In addition, with the evolution of the skull as a rigid encasement for the brain, an efficient mechanism was required to reduce the ionic and osmotic stresses created in those confined spaces during high-frequency saltatory conduction within densely packed axon bundles. Potentially lethal levels of K^+ would otherwise have been released directly into the tightly confined CNS, with the resulting high external potassium ion concentration ($[K^+]_O$) causing prolonged axonal depolarization, seizures, and eventually complete failure of impulse propagation (Hodgkin, 1951; Hodgkin and Huxley, 1952).

In the CNS, oligodendrocytes typically form single segments of myelin on 20–60 different axons (Matthews and Duncan, 1971). As detailed below, this branching and distribution of myelin segments to many different axons in the CNS has important ramifications regarding the averaging of ionic and osmotic loads on individual oligodendrocytes. As an alternative method for minimizing increases of $[K^+]_O$ within PNS nerve bundles, the myelinated axons of both sensory and motor neurons seldom fire simultaneously, greatly reducing peak ionic and osmotic loads. Equally important, the comparatively vast extracellular space within PNS axon bundles allowed much greater dilution of the K^+ released during high-frequency saltatory conduction. Thus, any axonal swelling in the PNS would occur within a large and expandable volume, allowing for any residual osmotic swelling to occur without compressional damage.

Equally relevant to the evolution of CNS myelin, it was widely believed until very recently that K^+ efflux during axon repolarization occurred strictly at the nodes of Ranvier, directly into the surrounding extracellular space. According to that model, if the buildup of $[K]_O$ were from the normal 2–3 mM to only 5–6 mM (a value that, by calculation, could easily occur after only a few action potentials), this small change would nevertheless be sufficient to depolarize the node region and inactivate its voltage-gated sodium channels (i.e., Na_v1 channels), effectively blocking any subsequent axonal action potentials (Hodgkin, 1951; Hodgkin and Huxley, 1952). Thus, K^+ was thought either (1) to be transported rapidly and directly back into the axon via abundant $Na^+/K^+ATPase$ molecules that were postulated to be located within the plasma membranes of the nodes of Ranvier, or (2) to directly enter astrocyte fingers that occasionally were found abutting the axonal plasma membrane at nodes of Ranvier (Waxman and Black, 1984; Black and Waxman, 1988). However, recent immunocytochemical evidence revealed that neither of those previous models of K^+ movement were accurate or relevant to the mechanisms of saltatory conduction.

8.3   EVOLUTION OF THE CONCEPT OF POTASSIUM SIPHONING

Efforts to understand why there was no toxic buildup of K^+ near nodes of Ranvier during prolonged axonal activity began in the 1960s. Orkand et al. (1966) used a dissected optic nerve preparation from the salamander *Necturus*, which because it is cold-blooded and has a low metabolic rate, allowed long-term recordings to measure K^+ efflux and possible transport following direct electrical stimulation. In addition, the preparation is small, compact, and devoid of CNS parenchyma, allowing local ionic changes induced by axonal depolarization to be measured without contamination by other types of neuronal activity or by buffering by neuronal somata. Following single stimulations, as well as trains of stimulation up to 5 Hz, Orkand and coworkers detected no increase in perinodal $[K^+]_O$, but instead measured substantial K^+ release at astrocyte endfeet at the glia limitans surrounding the optic nerve, several millimeters from the site of stimulation. They also found that those distant astrocytes were depolarized from –85 mV to about –58 mV (i.e., to the reversal potential for K^+), suggesting that K^+ released from axons immediately entered closely associated astrocyte processes and was transported away ("siphoned" away), passing from astrocyte to astrocyte before being released at the glia limitans that forms the surface sheath of the optic nerve. (Note: These measurements were made at the blood–brain barrier (BBB) (here known as the blood optic nerve barrier), providing early evidence for high K^+ permeability of the BBB.) From these observations, they proposed a model of "potassium siphoning" or "potassium spatial buffering" (Orkand et al., 1966), which postulated that excess $[K^+]_O$ entered astrocyte processes within the bundles of axons and exited from astrocyte endfeet at the distant glia limitans and at endfeet that surround capillaries.

Twenty years were to elapse before Newman (1986) was able to show that K^+ permeability was high at the ends of astrocyte processes (i.e., endfeet) in primary cell culture but low at astrocyte somata and proximal processes. He suggested that astrocyte endfoot processes had high concentrations of molecules specialized for both K^+ influx and efflux, and that these K^+ fluxes occurred via passive diffusion processes at hypothetical K^+ "leak channels" (subsequently called inward rectifier channels, now designated Kir4 channels; see below). This localization of K^+ leak conductance to astrocyte endfoot processes seemed to validate the original observations of Orkand et al. (1966). However, those original measurements had been made in *Necturus*, which have few or no myelinated axons in their optic nerves. Thus, the original model of potassium siphoning near unmyelinated fibers seemed to many investigators to be irrelevant to the functioning of myelinated axons; so the concept of potassium siphoning slowly fell from favor (but see Kofuji and Newman, 2004).

8.4 EMERGING CONCEPTS OF THE PRIMARY FUNCTIONS OF THE PANGLIAL SYNCYTIUM

Over the past decade, overwhelming evidence has revealed that oligodendrocytes and astrocytes provide separate but essential functions for generating and maintaining saltatory conduction. As an essential corollary, it has also been found that mutations, pharmacological agents, and bacterial infections that directly damage components of the panglial syncytium that are specialized for transport of K^+ and water lead not only to the failure of saltatory conduction, but also to sclerotic damage to myelin (dysmyelination), including several forms of multiple sclerosis (MS). Therefore, in this chapter, we address the rapidly emerging concepts that oligodendrocytes do not simply form insulating layers of myelin, and that CNS myelin does not serve primarily to reduce axonal membrane capacitance, but instead that myelin provides structural and molecular components required for both generation of action potentials in myelinated fibers and maintenance of high-frequency saltatory conduction.

8.5 THE PANGLIAL SYNCYTIUM

It is now well established that virtually all macroglial cells (i.e., astrocytes, oligodendrocytes, and ependymocytes) are linked into a vast glial syncytium (Mugnaini, 1986; see also Morales and Duncan, 1975), with abundant gap junctions providing for strong ionic, metabolic, and osmotic coupling within this network (Figure 8.1). Not only are astrocytes strongly coupled to other astrocytes by abundant gap junctions (Brightman and Reese, 1969), but they are also strongly coupled to oligodendrocytes (Figure 8.1A,C) and to ependymocytes (Morales and Duncan, 1975; Mugnaini, 1986; Rash et al., 1997; Rash, 2010), with astrocytes acting as intermediaries for transport of ions and water away from oligodendrocytes, through successive astrocytes, to the pericapillary endfeet and to the glia limitans. As a consequence, it has become increasingly clear that damage to oligodendrocyte or astrocyte gap junctions (Altevogt et al., 2002; Menichella et al., 2003, 2006; Lutz et al., 2009), or to any molecules of the K^+ or water transport or exit pathways within the panglial syncytium (Menichella et al., 2003, 2006; Kleopa et al., 2004; Lennon et al., 2004; Roemer et al., 2007), leads to a cascade of effects that quickly disrupt ionic and osmotic regulation across broad expanses of the CNS (Menichella et al., 2003, 2006; Kleopa et al., 2004; Lennon et al., 2004; Roemer et al., 2007; reviewed in Rash

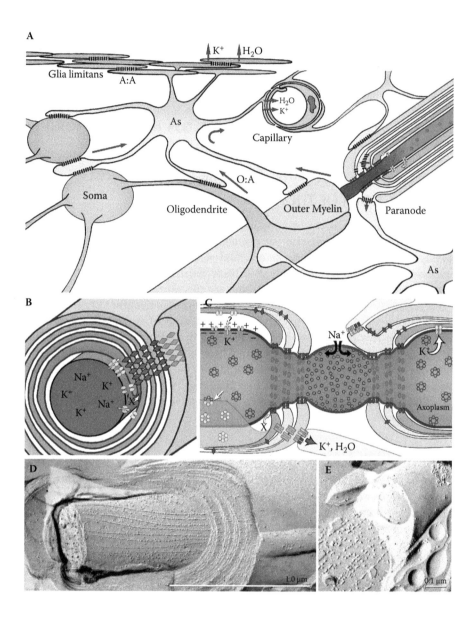

(2010). According to this new "potassium gateway" model, K^+ that is released from axons specifically during high-frequency saltatory conduction is initially confined within the internodal myelin compartment, where K^+ enters the innermost layer of myelin, then is transported via gap junctions to surrounding astrocytes, for release at the glia limitans that forms the outer surface of the brain and spinal cord, as well as at astrocyte endfeet that surround all CNS capillaries. As a consequence, any disruptions of the K^+ transport pathway, from innermost myelin to astrocyte endfeet at the

FIGURE 8.1 **(Opposite) (See color insert.)** Gateway cells of the panglial syncytium (As, blue, astrocytes; green, "soma," oligodendrocyte) and their relationships to myelinated axons (red), capillaries (maroon), and the glia limitans (dark blue stack). Astrocytes, which are interlinked by abundant gap junctions, act as intermediaries that couple oligodendrocyte somata, their oligodendrite processes, and their myelin segments to the astrocyte endfeet that surround capillaries and form the glia limitans. At these locations, excess K^+ (K^+, red arrows) and osmotic water (H_2O, blue arrows) are released into the circulatory system or into the bathing cerebrospinal fluid. (Modified from Kamasawa, N., et al., *Neuroscience*, 136, 65–86, 2005.) (B) Diagram of cross-sectioned myelin, showing multiple tight junction strands (brown diamonds) that are stacked from inner mesaxon to outer mesaxon, isolating each layer of internodal myelin extracellular space into ionically and electrically distinct compartments, one per layer of myelin wrapping. K_V1 channels (blue hexagons) are drawn linked directly to innermost cytoplasmic myelin via Cx29 hemichannels (yellow hexagons), and conversely as spatially separated from Cx29 hemichannels. In the peri-internodal extracellular space, black X with white arrow indicates that K^+ cannot enter the next myelin extracellular space because of multiple tight junction barriers. Na^+ and K^+, excess positive ions after axonal depolarization. (Modified from Rash, J.E., *Neuroscience*, 168, 982–1008, 2010.) (C) Locations and interrelationships of principal proteins identified to date at nodes of Ranvier, paranodes, and juxtaparanodes. Diagram depicts entry of Na^+ through Na_V1 sodium channels (dark green circles) at nodes of Ranvier, and exit of K^+ at juxtaparanodal axonal plasma membrane via K_V1 channels (blue hexagons in hexagonal rosettes), either directly into the internodal extracellular space (designated by ?) or from axoplasm via K_V1 channels linked to innermost cytoplasmic myelin via Cx29 hemichannels (yellow hexagons in hexagonal array). K^+ is blocked at contactin/caspr-containing septate junctions (X with white curved arrow). Brown diamonds, tight junctions between myelin layers. Autologous gap junctions composed of Cx32 (green barrels) couple successive layers of paranodal myelin. The outermost layer of myelin is coupled to the astrocyte syncytium via biheterotypic gap junctions consisting of Cx47 (light green barrels) and Cx32 (dark green barrels) in oligodendrocyte hemiplaques coupled to Cx43 (blue barrels) and Cx30 (dark blue barrels) in astrocyte hemiplaques. Relative K^+ concentrations at successive paranodal layers grade from orange to light yellow to green (oligodendrocyte) to blue (astrocyte). (Modified from Rash, J.E., *Neuroscience*, 168, 982–1008, 2010.) Red ovals, caspr/contactin septate junctions in paranodal membrane; yellow rosettes, Cx29 (as immunogold labeled in PNS myelin; Li et al., 2002). Fractures that stepped from innermost myelin to the axon plasma membrane (larger straight white arrow) revealed that the myelin rosettes are structurally coupled to rosettes in the axon plasma membrane (Stolinski et al., 1981, 1985). K^+, H_2O with blue arrow depicts movement of K^+ and water from oligodendrocyte to astrocyte via gap junctions. (D) Freeze-fracture image of multiple tight junction strands stacked from inner mesaxon to outer mesaxon, separating each layer of myelin internodal extracellular space into ionically and electrically distinct compartments, one per layer of myelin wrapping. (Modified from Rash, J.E., *Neuroscience*, 168, 982–1008, 2010.) (E) Outer tongue of myelin linked to first compact layer via 12 parallel tight junction strands. The particle partitioning coefficient (Satir and Satir, 1979) of 50% for tight junctions in formaldehyde-fixed tissues used in FRIL resulted in rows of both IMPs and pits in this myelin P-face. Arrow, 10 nm immunogold bead labeling Cx32 in small gap junction on outer tongue.

glia limitans, result in catastrophic, potentially fatal failure of saltatory conduction. Thus, for the remainder of this review, we describe:

1. The ultrastructural features involved in K^+ siphoning from myelinated axons, beginning with the recently discovered spatial separation of voltage-gated sodium channels from voltage-gated potassium channels (i.e., Na_V1 channels cleanly separated from K_V1 channels)
2. The ultrastructure, biochemistry, and function of the paranodal loops of oligodendrocyte myelin, including their interlamellar, homotypic gap junctions
3. The heterotypic gap junctional coupling relationships of oligodendrocytes with astrocytes, as well as between successive astrocytes, both of which facilitate long-distance K^+ siphoning
4. The mechanisms for directed/directional release of K^+ and water at astrocyte endfeet
5. Evidence that astrocyte glial fibrillary acidic protein (GFAP) filaments are not simply structural, but instead represent essential elements for K^+ siphoning and water transport
6. Evidence that when any of these vital components are disrupted, saltatory conduction ceases, and axons are progressively demyelinated, ultimately leading to several debilitating or fatal MS-like diseases.

These concepts regarding the role of oligodendrocyte myelin form the foundation for the potassium gateway hypothesis for maintenance of saltatory conduction and potassium siphoning in the CNS. This gateway hypothesis also provides a new paradigm for investigating MS, X-linked Charcot-Marie-Tooth disease (CMTX), Pelizaeus-Merzbacher-like disease (PMLD), oculodentodigital dysplasia (ODDD), neuromyelitis optica (NMO), and Alexander disease, as well as several experimentally induced white matter "leukodystrophies" that are created by genetic knockout of Cx32+Cx47, Cx30+Cx43, and Kir4.1.

8.6 NA_V1 CHANNELS AT NODES OF RANVIER; K_V1 CHANNELS ONLY AT JUXTAPARANODES AND INTERNODES

For more than 50 years, it has been known that in myelinated fibers, Na^+ influx occurs across the axonal plasma membrane only at nodes of Ranvier (Tasaki, 1939; Hodgkin, 1951; Hodgkin and Huxley, 1952), mediated by members of the Na_V1 family of sodium channels (reviewed in Catterall et al., 2005). Previously, it was also believed that voltage-gated potassium channels (K_V1 family; reviewed in Gutman et al., 2005) were colocalized and intermixed with voltage-gated Na_V1 at nodes of Ranvier, with sodium conductance/axonal depolarization preceding potassium conductance/axonal repolarization by about 1 mS (i.e., channel functions were temporally but not spatially separated). (Please refer to any recent physiology textbook.) However, K_V1 channels (and therefore K^+ efflux during saltatory conduction) were recently found to be localized beneath internodal myelin, concentrated in the "juxtaparanodal" axonal plasma membrane, with no K_V1 channels at the nodes of Ranvier (Rasband et al., 1999; Rasband and Shrager,

2000; Brophy, 2001; Bhat et al., 2001; Rios et al., 2003; Brophy, 2003; Rasband, 2004; Altevogt and Paul, 2004).

Almost immediately, it was determined that both physical and, more importantly, electrical isolation of Na^+ and K^+ conductances were essential for the process of saltatory conduction to occur. When paranodal septate junctions were disrupted/disconnected from the axolemma, either by genetic, enzymatic, or disease processes (Chiu and Ritchie, 1980; Sherratt et al., 1980; Chiu, 1991; Bhat et al., 2001), saltatory conduction ceased immediately. With these disruptions of the paranodal septate junction "fence" (Hirano and Dembitzer, 1982), K_V1 channels moved immediately adjacent to the Na_V1-containing nodes but did not intermix with the Na_V1 channels (Bhat et al., 2001; Rios et al., 2003). Although both Na_V1 and K_V1 channels remained at normal densities, saltatory conduction could no longer be elicited. Thus, it is now clear that it is not solely the number of Na_V1 channels or their localization at widely spaced nodes of Ranvier that is critical for saltatory conduction, nor is the physical separation of Na_V1 and K_V1 channels within nodal vs. juxtaparanodal membranes sufficient for saltatory conduction to occur. Rather, both the patency of the septate junctions and the electrical isolation of Na^+ and K^+ conductance into separate compartments appear to be required, either (1) for saltatory conduction to occur at all or (2) to maintain the conditions necessary for multiple action potentials to be propagated by high-frequency saltatory conduction.

Based on the pioneering studies of Bhat, Rios, Rosenbluth, Rasband, Brophy, Ellisman, Levinson, Trimmer, and others, it is now clear that previous models for saltatory conduction were based on several false assumptions (detailed below). Although K^+ efflux from stimulated unmyelinated axons is detected as the release of K^+ at the far distant astrocyte endfeet of the glia limitans (Orkand et al., 1966), there are no astrocyte processes at the sites of K^+ efflux from myelinated axons, i.e., deep within myelin at the juxtaparanodal and internodal axolemma, so K^+ cannot directly enter astrocytes. Therefore, the myelin sheath itself must be intimately involved in initial sequestration and removal of axonally released K^+ from its areas of highest concentration in the peri-internodal compartment within each myelinated segment (David et al., 1992, 1993). In the following sections, we discuss probable pathways for the excess K^+ that is released during high-frequency axonal activity.

8.7 OLIGODENDROCYTES FORM MYELIN IN THE CNS

Early electron microscopic studies revealed that each segment of CNS myelin is formed during postnatal development from a planar extension of oligodendrocyte plasma membrane (plus included cytoplasm) that wraps repeatedly around a ¼–½ mm segment of axon (i.e., 100–200 times the 1–2 μm axon diameter; Peters et al., 1991). During subsequent myelin maturation and compaction, the aqueous component of the cytoplasm is removed, leaving all except the outermost and innermost cytoplasmic layers as compact myelin (reviewed by Bunge, 1968; also see the proposed role of Cx32-containing gap junctions in water and ion transport during myelin compaction in the PNS (Meier et al., 2004)). In addition, a thin rim of cytoplasm remains within the edge of the spirally wrapped myelin (Figur 8.2A, seen as unrolled myelin), including the leading edge of the

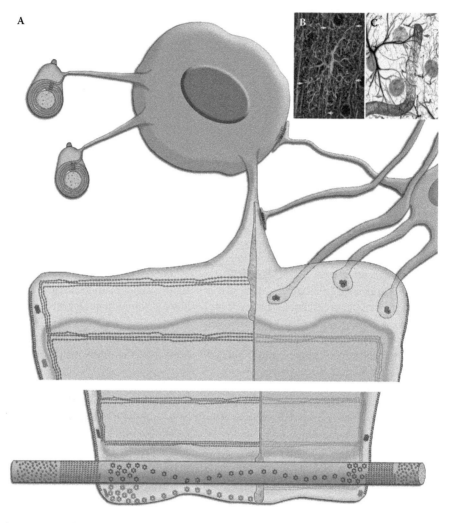

innermost oligodendrite "tongue" (Figure 8.2A, bottom). The lateral cytoplasmic rims remain noncompacted, becoming the paranodal loops. The innermost inter-nodal margin of the tongue is overlain by the next succeeding layer of myelin, which also contains a longitudinal rivulet of noncompact myelin, with these two overlapping cytoplasmic expansions forming the inner mesaxon (Figures 8.1B and 8.2A). In PNS myelin and occasionally in CNS myelin, additional periodic cytoplasmic expansions extend from outermost to innermost myelin, forming stair-step circumferential cytoplasmic pathways that are faintly resolvable by light microscopy as "Schmidt-Lanterman incisures" (not illustrated here, but see Mugnaini et al., 1977 and Schnapp and Mugnaini, 1978). In the PNS, at least, each cytoplasmic layer of each Schmidt-Lanterman incisure is linked to the next by Cx32-containing gap junctions (Meier et al., 2004), providing additional radial pathways for transport of K^+ and water at ca. 100 μm intervals within the hun-dreds-of-micrometers- to millimeter-long internodal segments.

FIGURE 8.2 **(Opposite)** **(See color insert.)** Relationships between axon, oligodendrocyte myelin, and the astrocyte syncytium. (A) Diagram of one segment of oligodendrocyte myelin, unrolled to show molecular specializations of both inner (left side) and outer (right side) surfaces. The oligodendrocyte soma (green) is connected by oligodendrites to three different myelin segments, all from different axons. Astrocyte (blue) processes form gap junctions with the outer surface of myelin; blue cylinder, axon; blue hexagonal rosettes, K_V1 channels in the axon internodal plasma membrane, concentrated at the juxtaparanodal surface and along the inner mesaxon; brown diamonds in lines, tight junction strands link successive layers of myelin (light vs. dark lines); green IMPs in hexagonal arrays, Cx32-containing gap junctions between paranodal loops of myelin; orange hexagonal arrays, Cx32/Cx47 to Cx30/Cx43 heterotypic heterologous gap junctions; triplet red ovals as dotted lines, septate junctions binding the paranodal loops to the axon plasma membrane, forming a spiral barrier separating the nodal extracellular space from the peri-internodal extracellular space; yellow rosettes, Cx29 in innermost myelin. (B) Image of astrocyte (pink staining for GFAP), with processes approaching myelin (bottom center) and capillaries (top center). Neurons are stained for MAP2 (blue) and all nuclei (including oligodendrocyte somata) are stained with DAPI (purple). (Image courtesy of Maiken Nedergard, Nancy Ann Oberheim, and Takahiro Takano.) (C) Staining for AQP4 in astrocyte endfeet (pink) precisely follows and delineates the capillary bed. Large GFAP filament bundles (black) in the cell body diverge and spread to fill the astrocyte fingers and extend to each endfoot process (adult rat cortex). Nuclei (blue) stained with DAPI. (From Xaiohai Wang, with permission.)

The edges of the paranodal cytoplasmic loops are bonded, spirally, to the axon via extracellular septate junctions (Figures 8.1C and 8.2A). In the myelin plasma membrane, these paracellular fences (Hirano and Dembitzer, 1982) are composed of contactin/paranodin/neurexin (which are evolutionarily homologous proteins found in both vertebrate and invertebrate septate junctions (Bhat et al., 2001)), binding to axonal membrane proteins that include Caspr (contactin-associated protein (Einheber et al., 1997)). These septate junctions, which are functionally analogous to tight junctions, form a strong but gated paracellular diffusion barrier to ions and water (Aschenbrenner and Walz, 1998), thereby ionically and osmotically isolating the internodal periaxonal extracellular space from the nodal extracellular space (Banerjee et al., 2006). Along with longitudinally oriented bands of closely spaced tight junctions that run from paranode to paranode (Figures 8.1D,E and 8.2A), the septate junctions and tight junctions link each successive layer of myelin into a series of osmotically and electrically isolated, concentric cylindrical compartments of extracellular space (Figure 8.1B). This combination of spirally circumferential septate junctions and tight junctions between paranodal loops (Kamasawa et al., 2005), plus multiple longitudinal tight junctions that run from paranode to paranode within a segment of myelin, electrically isolate nodal Na^+ conductance from internodal K^+ conductance (Rosenbluth, 2009). The tight junctions between successive layers of myelin preclude circumferential movement of K^+ from the periaxonal extracellular space, even precluding circumferential passage into the first interlamellar space (Figure 8.1B; arrow with X indicating the blockade of ion movements at tight junctions). Combined with evidence that axonally derived K^+ exits from the outermost layer of oligodendrocyte myelin directly into surrounding astrocyte endfeet, at the minimum, this means that K^+ released during saltatory conduction must enter cytoplasmic myelin, somewhere in the innermost turn of myelin, and perhaps from axon cytoplasm directly into myelin cytoplasm.

The above observations invoke several important questions:

1. If depolarization of the next successive node requires a "return electrical circuit" (aka shunt circuit; Hartline, 2008), does that same principle (shunt circuit) also apply to internodal repolarization? That is, does repolarization current also require an electrical return circuit?
2. If so, how do those hypothetical electrical circuits occur in the presence of electrically resistant paranodal septate junctions? More specifically, what intracellular or extracellular pathway provides the conduit for the repolarization circuit?
3. Alternatively, is the mere physical movement of ions into separate isolated electrical compartments, and possible capacitance discharging of those compartments, sufficient to account for the electrical events recorded during saltatory conduction?

In this regard, if one of the functions of myelin is truly to reduce capacitance (Huxley and Stämpfli, 1949; Hartline, 2008) by separating positive charges outside of myelin from axonal negative charges (Figure 8.1C, apposed + and − symbols), it should be noted that, by calculation, the extracellular space between the axon and the innermost layer of myelin within each peri-internodal compartment (2 µm diameter axon, surrounded by a 15 nm extracellular space within a 300 µm long internodal segment) contains ca. 10^{12} Na$^+$ and 5×10^{10} K$^+$ ions, which therefore provide ample ionic charges to create high internodal capacitance. Stated differently, the multiple compact myelin layers cannot significantly reduce axonal capacitance by widely separating negative charges within the axon from those positive ions outside of the myelin sheath, so long as there are abundant positive ions in the immediate periaxonal space. Moreover, any K$^+$ efflux from the axon into the peri-internodal extracellular space would *increase* capacitance for immediately succeeding action potentials. On the other hand, efflux of K$^+$ directly into the negatively charged innermost layer of myelin cytoplasm (possibly via Cx29 hemichannels, as described below) may account for the reduced capacitance detected in myelinated fibers (Huxley and Stämpfli, 1949).

8.8   HOW DOES AXONALLY-RELEASED K$^+$ ENTER MYELIN?

As a consequence of K_V1 channels localized beneath and within each myelin segment, K$^+$ efflux must occur either

1. Directly into the spatially isolated internodal periaxonal space (David et al., 1992, 1993) that is delimited by the paranodal septate junctions (Figures 8.1A–C)
2. Directly from the axon into the innermost layer of cytoplasmic myelin (Figure 8.1B, top arrow), possibly via K_V1 potassium channels in the axon plasma membrane coupled directly to Cx29 hemichannels in the oligodendrocyte plasma membrane (Figure 8.1C, right curved arrow), as detailed below

By freeze-fracture, distinctive hexagonal "rosettes" of intramembrane protein particles (IMPs) populate the extraplasmic leaflet (E-face) of the juxtaparanodal membrane in PNS axons (Miller and Pinto da Silva, 1977; Stolinski et al., 1981, 1985; also see the image from 1977 generously provided by Dr. Nancy Shinowara in Rash (2010)), precisely where K_V1 K^+ channels were subsequently found to be localized (Rasband et al., 1999; Rasband and Shrager, 2000; Brophy, 2001, 2003; Bhat et al., 2001; Rios et al., 2003; Rasband, 2004; Altevogt and Paul, 2004). Similarly distinctive rosettes of IMPs were observed in the protoplasmic leaflet (P-face) of the innermost layer of myelin, concentrated directly opposite the axonal E-face IMP rosettes. Where the fracture plane stepped from the E-face IMP rosettes in axonal plasma membranes to the P-face of the innermost layer of myelin (Stolinski et al., 1985), the axonal and myelin IMPs were precisely aligned—as if physically linked from axonal to myelin plasma membranes. Moreover, the IMP rosettes in PNS myelin were positively identified as Cx29 by freeze-fracture replica immunogold labeling (FRIL) (Li et al., 2002). In CNS myelin, Cx29 and K_V1 channels also occur at the same subcellular locations (Altevogt et al., 2002), suggesting similar molecular interactions in both PNS and CNS myelin. Significantly, Cx29 does not form gap junctions—either with itself or with any other connexin (Altevogt et al., 2002; Altevogt and Paul, 2004; Ahn et al., 2008). On the other hand, if the Cx29 hemichannels in the innermost layer of juxtaparanodal myelin are functionally linked to the E-face rosettes/presumptive K_V1 channels in the juxtaparanodal and internodal axonal plasma membrane, this would provide an efficient composite gap-junction-like ion channel for K^+ efflux through axonal K_V1 channels, directly into innermost myelin. Such composite channels might also couple and uncouple (via gating of either or both components) according to relative neuronal activity, transient local electrical potential, internal K^+ concentration, or changing osmotic conditions, just as K_V1 is voltage gated in all locations, and Cx32 connexons may also be gated by several of those same factors (Orthmann-Murphy et al., 2007).

8.9   RETHINKING THE PROPOSED ROLES OF MYELIN IN SALTATORY CONDUCTION

Previous models to account for increased axonal conduction velocity of myelinated axons were based on

1. Much higher axonal electrical resistance of myelinated vs. unmyelinated axons, which was widely interpreted as indicating few or no ion channels in the internodal axon membrane, thereby reducing internodal current leakage, presumably resulting in increased outward flow of "virtual current" at the next successive one or two nodes of Ranvier, thereby facilitating activation of those distant Na_V1 channels for virtually instantaneous long-distance saltation.
2. Presumptions that the many layers of internodal myelin widely separated the ionic charges in the axoplasm from the ionic charges outside myelin, thereby greatly reducing axonal membrane capacitance.

Both of these explanations must now be reconsidered in light of evidence that

1. The high densities of K_V1 ion channels in the juxtaparanodal and internodal axonal plasma membrane (Figure 8.2A) greatly reduce internodal

axolemmal resistance—at least during normal K$^+$ conductance/axon repolarization.

2. The septate junctions that link the paranodal loops to the axon plasma membrane prevent current leakage from internodal to nodal compartments, thereby precluding the generation of a repolarizing return circuit (Figure 8.1C).

3. Abundant Na$^+$ and K$^+$ in the peri-internodal extracellular space (Figure 8.1C) would likely produce high axonal membrane capacitance rather than the lowered capacitance attributed to wide charge separation by myelination. This suggests that the reduction of electrical capacitance by myelin may be due to its internal ultrastructure, including gap junctions in electrical series between paranodal loops of myelin, as detailed below. Specifically, the internodal capacitance effects might instead be reduced if during high-frequency saltatory conduction, excess K$^+$ directly entered the innermost layer of myelin cytoplasm, as proposed below.

In addition to abundant K$_V$1 and Cx29 in the juxtaparanodal membrane, K$_V$1 and Cx29 also form very narrow but continuous immunofluorescent ribbons, in precise coalignment, extending from one juxtaparanode to the next (Bhat et al., 2001; Altevogt et al., 2002; Altevogt and Paul, 2004), suggesting that substantial K$^+$ conductance occurs along the entire internodal segment, from juxtaparanode to juxtaparanode (Figure 8.2B). As a consequence, indirectly measured axonal resistance may be due in large part to resistance to return current flow created by the paranodal septate junction barrier (Rosenbluth, 2009). Likewise, axolemmal capacitance may appear to be reduced, not by the thickness of the myelin sheath, but instead by the paranodal gap junctions that ionically and electrically connect the innermost layer of cytoplasmic myelin directly to the outermost cytoplasmic layer of myelin, and thence via gap junctions linking to the astrocytes of the panglial syncytium, as summarized below.

8.10   K$^+$ FLUXES DURING DAILY TURNOVER OF CEREBROSPINAL FLUID (CSF) COMPARED WITH K$^+$ EFFLUXES DURING HIGH-FREQUENCY AXONAL ACTIVITY, WITH RELEVANCE TO SEIZURE ACTIVITY

During generalized epileptic seizures, the amount of K$^+$ released is sufficient to raise extracellular [K$^+$] to dangerous levels (Dudek et al., 2010), i.e., from 2.8 mM to 4.5mM, and in stimulated adult hippocampal slices, to 8.85 mM (Hablitz and Heinemann, 1989). To put this amount of neuronally derived excess [K$^+$] into perspective, it should be recalled that each day, the choroid plexus secretes ca. ½ L of CSF, most of which enters the parenchyma of the brain and spinal cord (Segal, 1993; Keep et al., 1999). Ions and small solutes freely diffuse between parenchymal extracellular spaces and the CSF, whose volume turns over three or four times per day. Bulk flow (including ions and even large macromolecules) exit freely at the arachnoid granulations (Kandel et al., 2000), so exit of excess neuronally-derived K$^+$

is not at issue. At 138 mM $[Na^+]_O$ and 2.8 mM $[K^+]_O$ in CSF, this means that ca. 10^{23} Na^+ and 10^{21} K^+ ions exit the CNS per day (or 10^{17} Na^+ and 10^{15} K^+ ions per second). This is to be contrasted with ca. 10^7 K^+ ions per action potential per juxtaparanode. Even at 100 Hz, this amounts to 10^9 K^+ per second, or six orders of magnitude less than normally exits across the BBB each second, a value not easily discernible from other sources of measurement "noise" (Keep et al., 1999). Equally relevant, this means that it would require 10^6 axons firing simultaneously and continuously at 100 Hz to double the normal amount of K^+ exiting each second. (This hypothetical firing rate is much higher than for normal CNS axons.) Only during seizure activity would that level of K^+ efflux become readily detectable above normal CSF turnover. Perhaps equally important, the $[K^+]_O$ of CSF is normally about half that of serum (Kandel et al., 2000), providing substantial instantaneous additional K^+ buffering capacity. We also call attention to other sources of K^+ efflux, including unmyeli-nated axons, but here especially, K^+ is thought to directly reenter axons, primarily via Na^+/K^+ATPase, as well as to enter nearby astrocyte processes by a variety of K^+ "pumps" and channels (Keep et al., 1999). Nevertheless, at low stimulation frequen-cies, we presume that Na^+/K^+ATPase in the axon plasma membrane, in innermost myelin, and in astrocytes would likely provide the primary mechanism for restoring normal $[K^+]_I$ in both myelinated and unmyelinated axons, whereas at high stimula-tion frequencies, alternate pathways are involved, primarily uptake by the innermost layer of myelin.

From another perspective, a 100 μm diameter alpha motor neuron in spinal cord or a similar diameter pyramidal cell in hippocampus contains ca. 10^{15} K^+ and 5 × 10^{13} Na^+. Thus, the number of K^+ exiting the axon during an action potential is five or six orders of magnitude smaller than the number remaining behind. This huge difference also means that repeated electrical activity of axons produces no significant change in intracellular ion concentration in the axon, but instead primar-ily changes the membrane electrical potential. However, this activity precipitously increases $[K^+]_O$ in the peri-internodal compartment/inner mesaxon.

If K^+ enters the inner mesaxon (which based on its volume, normally contains ca. 10^9 K^+ ions per internodal segment, and based on 10^4 Na^+ channels per node and entry of 10^4 Na^+ ions entering each channel), this would mean that stimulation for 1 s at 10 Hz would transiently double the $[K^+]_M$ and significantly raise the membrane potential in the inner mesaxon, possibly to the value of +75 mV, as measured by David and coworkers in the peri-internodal compartment (David et al., 1992, 1993). Moreover, such strong depolarization of the inner mesaxon would likely open Cx32-containing gap junctions linking successive paranodal loops of myelin, as detailed below.

8.11 PROPOSED ULTRASTRUCTURAL PATHWAYS FOR K+ FROM AXON TO MYELIN CYTOPLASM

Despite more than half a century of effort, the precise pathways for K^+ after efflux from myelinated axons are still unknown. However, K^+ cannot travel circumferen-tially in the myelin interlamellar extracellular space because of the abundant tight junction strands that link and isolate each successive turn of myelin (Figure 8.1B,D).

Nevertheless, K⁺ and associated osmotic water (at least 350 water molecules per potassium ion) ultimately exit from outermost myelin, directly into astrocytes, presumably via their O:A gap junctions (Kamasawa et al., 2005; Rash, 2010). At the opposite end of this pathway, excess K⁺ arising from high-frequency axonal activity must enter cytoplasmic myelin, either (1) indirectly from accumulating pools of K⁺ in the peri-internodal extracellular space, possibly via Kir4.1 K⁺ leak channels or via Na⁺/K⁺ATPase located in the innermost plasma membrane of myelin (Poopalasundaram et al., 2000), or (2) via axonal K_V1 channels linked directly to Cx29 hemichannels (as proposed in Rash, 2010). But if millions of potassium ions and associated osmotic water molecules enter the innermost layer of cytoplasmic myelin during high-frequency saltatory conduction, how is this excess K⁺ (and water) transported to the outermost myelin for removal via the astrocyte syncytium? After each action potential, what happens to the membrane potential of each layer of myelin, as well as to $[K^+]_I$ in the innermost cytoplasmic layer of oligodendrocyte myelin? Moreover, what happens to these two properties in each successive layer of myelin? And what would happen if this K⁺ and associated osmotic water were prevented from exiting from this innermost myelin compartment (as for example, by blocking Cx32-containing gap junctions between paranodal loops)? To begin to answer some of these types of questions, David and coworkers (David et al., 1992, 1993) used ultrasharp electrodes to show that during an action potential, the peri-internodal voltage rose by up to 75 mV (from 0 mV recorded with respect to bath ground). Likewise, by injection of K⁺ into the peri-internodal compartment, David and coworkers estimated that $[K+]_P$ rose 20–100 mM during repetitive firing. This apparent decreasing gradient of K⁺ from innermost to outermost myelin suggests that K⁺ rapidly builds up in the innermost compartment and that K⁺ is rapidly transported to the outermost cytoplasmic layer of myelin by an as yet unidentified mechanism. Although the precise pathway and mechanisms for radial transport of K⁺ within internodal myelin are not firmly established, it is now clear that a large number of neurological diseases result from disruptions of any one of these K⁺ and water transport pathways, with almost all having the same composite results: initial localized myelin swelling, blockade of saltatory conduction within those same fibers, and ultimately, sclerosis of white matter tracts over large regions of the CNS (proposed mechanisms described below).

8.12 OLIGODENDROCYTES MYELINATE MULTIPLE AXONS IN THE CNS, THUS SERVING AS A DISTRIBUTION "HUB"

Unlike myelin in the PNS, in which each segment of myelin is formed by a single Schwann cell (Mugnaini et al., 1977; Schnapp and Mugnaini, 1978), each oligodendrocyte typically myelinates from 20 to 60 axons (Matthews and Duncan, 1971; Peters et al., 1991), but occasionally as few as only one (Bunge, 1968). Moreover, only rarely does an oligodendrocyte myelinate two successive segments of the same axon (Bunge, 1968). As a consequence of the cytoplasmic processes linking the oligodendrocyte cell bodies to each internodal segment of myelin, the ionic and osmotic load on each oligodendrocyte soma is dispersed temporally and spatially, gathering K$^+$ and water from many different, and therefore necessarily asynchronously firing, axons. According to the K$^+$ gateway hypothesis, this means that under normal conditions, each individual oligodendrocyte would act as a central hub for K$^+$ sequestration, temporally and spatially averaging the ionic and osmotic load from multiple axons, and providing a central locus for gap junctional coupling to multiple astrocytes (Figure 8.1A).

8.13 GAP JUNCTIONS AND CONNEXINS OF THE PANGLIAL SYNCYTIUM

The glial syncytium (Mugnaini, 1986) or panglial syncytium (Rash et al., 1997; Kamasawa et al., 2005; Rash, 2010) is comprised of all three types of macroglial cells—i.e., oligodendrocytes, astrocytes, and ependymocytes—all of which are heavily interlinked by gap junctions, forming an ionic, osmotic, and electrical continuum. The concept of the panglial syncytium emphasizes that during repetitive, high-frequency activation, excess K$^+$ and osmotic water are actively transported via strong electrical potential driving forces, from the axon cytoplasm (which during an action potential, depolarizes to +55 mV), into the ill-defined peri-internodal compartment/innermost cytoplasmic layer of oligodendrocyte myelin (which reaches a localized electrical potential (driving force) of up to +75 mV (David et al., 1992, 1993)), through successive paranodal layers having an outwardly directed voltage gradient ranging from +75 mV (innermost myelin) to an unknown voltage (in outermost myelin) (David et al., 1992, 1993), and thence through gap junctions into surrounding astrocytes (nominally –85 mV), where K$^+$ and osmotic water are transported to and released at astrocyte endfeet (Kamasawa et al., 2005; Rash, 2010).

Astrocyte endfeet are greatly enriched in Kir4.1 potassium leak channels and AQP4 water channels (Nielsen et al., 1997; Nagelhus et al., 1998, 1999, 2004), the latter forming dense accumulations of AQP4 "square arrays," as identified by FRIL (Rash et al., 1998, 2004; Furman et al., 2003). Thus, the potassium gateway model identifies oligodendrocyte myelin as the "headwaters" of the panglial syncytium, with the innermost cytoplasmic expansion on both sides of the inner mesaxon serving to accept and store K$^+$ received from the axon, thereby creating the large electrical gradient (driving force) underlying long-distance "dynamic potassium siphoning." We have hypothesized (Rash, 2010) that the +40 mV to +75 mV potential created in the "peri-internodal space" during axon depolarization (David et al., 1992, 1993) is coupled via interlamellar paranodal gap junctions to the outermost layer of cytoplasmic myelin, and thence via oligodendrocyte-to-astrocyte gap junctions, to the strong negative voltage within astrocyte endfeet (–85 mV). This net 120–160 mV electrical driving force, along with a strong osmotic gradient, was hypothesized to provide for "voltage-augmented" or "dynamic" potassium siphoning (Kamasawa et al., 2005; Rash, 2010; also see similar proposals in Menichella et al., 2006 and Sargiannidou et al., 2010).

Gap junctions composed solely of Cx32 are present between successive paranodal loops in CNS myelin (Kamasawa et al., 2005), as in PNS myelin (Meier et al., 2004).

These interlamellar gap junctions provide a nearly direct radial route for ions and water, from innermost to outermost cytoplasmic myelin. This route was described as 1,000-fold shorter than the circumferential/spiral pathway around each successive cytoplasmic myelin layer (Balice-Gordon et al., 1998). (Actually, based on spacing of successive paranodal loops of 0.1–0.2 μm vs. 1–5 μm diameter of axons (multiplied by pi), the gap junction pathway from innermost to outermost myelin is 30- to 100-fold shorter, which is nevertheless, still significantly shorter than the circumferential pathway.)

At astrocyte-to-oligodendrocyte (O:A) gap junctions, oligodendrocyte Cx47 couples only with astrocyte Cx43, and oligodendrocyte Cx32 couples with astrocyte Cx30, and in white matter areas, also with astrocyte Cx26 (Altevogt and Paul, 2004), thus forming multiply heterotypic, heterologous gap junctions. Within the astrocyte syncytium, astrocytes strongly express Cx43 and Cx30 at their gap junctions (reviewed in Nagy et al. 2004), and in some white matter areas, they also express Cx26, including at O:A gap junctions (Nagy et al., 2001, 2003; Rash et al., 2001a). In A:A gap junctions, Cx43 couples only to Cx43, Cx30 to Cx30, and apparently, Cx26 to Cx26, with Cx30 to Cx26 also possible (Altevogt and Paul, 2004). Until recently, oligodendrocytes were thought not to form gap junctions directly with other oligodendrocytes (Mugnaini, 1986), but instead to communicate with other oligodendrocytes via gap junctions shared with an intervening astrocyte "intermediary" (Mugnaini, 1986; also Rash et al., 1997; Rash, 2010), a pathway that appears to be both electrically rectified (Kettenmann and Ransom, 1988; Pastor et al., 1998) and voltage gated (Orthmann-Murphy et al., 2007), with voltage-dependent gating potentially of direct relevance to potassium siphoning, as described above. However, in the corpus callosum, chains of oligodendrocyte somata are linked by gap junctions (Wasseff and Scherer, 2010) and show evidence for tracer coupling, even in the absence of astrocyte expression of Cx43 and Cx30 (Maglione, et al., 2010). These gap junnctions may provide an additional pathway for K$^+$ siphoning in dense white matter.

8.14   DISEASES OF PANGLIAL K$^+$ AND WATER TRANSPORT CAUSE DISTINCTIVE FORMS OF MS AND OTHER NEUROLOGICAL DISEASES WITH MS-LIKE SYMPTOMS

Multiple sclerosis represents a constellation of diseases affecting white matter integrity. MS is characterized by dysmyelination and the formation of multiple areas of myelin disruption or scars known as sclerotic plaques, or simply plaques. In MS, sclerotic plaques are usually detected by magnetic resonance imaging (MRI) or upon autopsy. Some forms of MS are undoubtedly autoimmune diseases directed against the most abundant myelin proteins (proteolipid protein (PLP) and myelin basic protein (MBP)) (Ota et al., 1990), but the causes of many forms of MS are mostly unknown. An emerging concept is that several forms of MS may arise from altered interactions of astrocytes with oligodendrocytes, and more specifically, from defects in K$^+$ siphoning and water transport pathways that result in local or widespread failure of saltatory conduction, disruption of CNS myelin, dysmyelination, and white matter sclerosis/leukodystrophy (Bergoffen et al., 1993; Senderek et al., 1999; Brenner et al., 2001; Menichella et al., 2003, 2006; Paznekas et al., 2003; Kleopa et al., 2004; Hsiao et al., 2005; Orthmann-Murphy et al., 2009; Rash, 2010).

Heterologous O:A gap junctions and autologous or "homocellular" O:O gap junctions (i.e., between successive myelin layers at paranodal loops and Schmidt-Lanterman incisures (Meier et al., 2004; Kamasawa et al., 2005; Rash et al., 2007)) are essential elements for voltage-augmented K^+ siphoning in myelinated axons of the CNS, and their disruptions provide new insights in MS as a constellation of diseases of the panglial syncytium. Interestingly, the original images of Waxman and Black (1984) emphasized that the astrocyte processes found near the nodes of Ranvier also formed multiple gap junctions with the outermost paranodal loops of myelin. Those images suggested a substantial role for O:A gap junctions in normal myelin physiology, but until recently, that role was unknown. Nor was the role of O:A and A:A gap junctions in maintenance of myelin widely recognized. The following are examples of MS and MS-like diseases resulting from disruption of molecular components of the K^+ siphoning pathway.

8.14.1 Diseases of Oligodendrocyte Connexins Block K^+ Siphoning

8.14.1.1 Mutations of Cx32 Damage Myelin and Cause X-Linked Charcot-Marie-Tooth Disease (CMTX)

CMTX, the most commonly inherited neurological disorder, is a demyelinating leukodystrophy that has been mapped to chromosome Xq13.1 (defining the gene for Cx32), with many different mutations identified (Bergoffen et al., 1993). In addition to forming gap junctions in the liver and other organs, Cx32 protein forms gap junctions between paranodal loops of myelin in the PNS. In CNS myelin, Cx32 is present, along with Cx47, in heterologous gap junctions linking oligodendrocytes to astrocytes (Kamasawa et al., 2005). In addition to decreased or absent saltatory conduction in the PNS, symptoms of CMTX apparently resulting from defects in Cx32 include myelin swelling and segmental demyelination (Bergoffen et al., 1993; Scherer et al., 1998), suggesting that in addition to an essential role in myelin compaction (Meier et al., 2004), Cx32 is required for additional aspects of myelin homeostasis (e.g., K^+ siphoning). However, because Balice-Gordon et al. (1998) found no difference in injected dye movement between outer and inner layers of myelin in normal vs. Cx32 knockout (KO) mice, they concluded that *radial diffusion* was unchanged, and therefore that there must be an additional unidentified connexin at paranodes and Schmidt-Lanterman incisures. This view can no longer be supported. With the subsequent completion of the human genome (Venter et al., 2001; Lander and Internat. Human Genome Seq. Cons., 2001) and identification of all 21 or 22 connexins expressed in mice and humans (Eiberger et al., 2001; Willecke et al., 2002); with evidence that oligodendrocytes express only three of those connexins (Cx47 > Cx32 > Cx29, with Cx29 *not* forming gap junctions) (Rash et al., 2001b; Altevogt et al., 2002; Nagy et al., 2003; Altevogt and Paul, 2004); and with neither Cx47 nor Cx29 present at gap junctions between cytoplasmic layers at paranodes or Schmidt-Lanterman incisures (Nagy et al., 2003; Li et al., 2004; Kamasawa et al., 2005), Cx32 appears to be the sole connexin in gap junctions between myelin layers in the PNS (Scherer et al., 1995; Bruzzone et al., 1996; Meier et al., 2004) and CNS (Kamasawa et al., 2005). Accordingly, radial interlamellar transport of K^+ and water would be completely blocked after functional deletion of Cx32, although circumferential transport would likely be *increased* due to additional pathways created within the incompletely compacted myelin (Scherer et al., 1998; Orthmann-Murphy et al., 2009), potentially accounting for the observations of Balice-Gordon et al. (1998) for maintained high dye diffusion within myelin of Cx32 KO mice.* Thus, instead of providing a major diffusion pathway, we suggest that the primary function of the few and tiny gap junctions at CNS paranodes and PNS Schmidt-Lanterman incisures is to provide an efficient pathway for *voltage-driven* interlammelar movement of the positively charged K^+ ions through connexon ion channels.

Regardless, in CMTX patients, the innermost and outermost layers of PNS myelin were swollen and separated (Senderek et al., 1999; Kleopa et al., 2004; Sargiannidou et al., 2010). According to the K^+ gateway hypothesis, in the absence of Cx32, K^+ and water could no longer be driven by voltage differences to flow from inner to outer myelin layers via those gap junctions, which would result in K^+ accumulation and osmotic swelling of cytoplasmic myelin, and ultimately, segmental demyelination. In addition, myelin may not compact during postnatal development in Cx32 KO mice due to absence of the Cx32-containing gap junctions that are normally found between the outermost two layers of myelin at the edge of the zone of myelin compaction (Meier et al., 2004). Finally, some CMTX patients also show strong CNS symptoms, including white matter alterations detected by MRI (summarized by Sargiannidou, 2010). Apparently, Cx47, normally present along with Cx32 at O:A gap junctions, is able to partially compensate for the lack of Cx32 at those heterologous heterotypic intercellular gap junctions. In contrast, in PNS myelin, there is no potential to compensate for missing or altered Cx32 because Schwann cell myelin segments are not coupled by gap junctions to any other cells. Thus, in CMTX, the

---

* The data collection methods and interpretations of Balice-Gordon et al. (1998) were predicated on their assertion that radial pathways in normal mouse myelin have a 1,000,000-fold diffusion advantage over circumferential pathways. Thus, they interpreted their images as showing radial diffusion occurring in 90–147 s, which if circumferential diffusion were 1,000,000-fold slower than radial diffusion, would have required 90,000,000–147,000,000 s, or 3–5 years—clearly an error of many orders of magnitude. Instead, the images of Balice-Gordon and coworkers are consistent with largely circumferential diffusion of dyes within the cytoplasm of the Schmidt-Lanterman incisures. That pathway also would account for the reported similar rate of dye movement in normal vs. Cx32 KO mice. Moreover, substantial radial diffusion could not have occurred through extensive interlamellar gap junctions, as had been proposed, because only one or two tiny gap junctions occur at each cytoplasmic expansion at paranodes and Schmidt-Lanterman incisures, and those gap junctions consist of only 5–25 connexons each (Meier et al., 2004; Kamasawa et al., 2005; Rash et al., 2007). For calculating maximum radial diffusion rates, the total connexon pore cross-sectional area is 7–40 nm^2 per interlamellar gap junction (the proposed radial diffusion pathway), whereas the cross-sectional area of each paranodal loop and Schmidt-Lanterman incisure cytoplasm (the circumferential pathways) is ca. 20,000 nm^2—a 500- to 3,000-fold greater cross-sectional area for circumferential diffusion over radial diffusion. After accounting for the 100-fold greater circumferential diffusion distance in mouse myelin over interlamellar radial diffusion (800–1,000 μm of unrolled myelin vs. 5–10 μm of stacked cytoplasmic expansions at Schmidt-Lanterman incisures (Mugnaini et al., 1977)), we calculate that there is perhaps a 10-fold to 100-fold advantage of radial diffusion over circumferential diffusion, rather than a 1,000,000-fold advantage. Accordingly, voltage-driven ion movement through tiny gap junctions linking successive layers of paranodal or Schmidt-Lanterman incisure cytoplasm apparently occurs in the tens to hundreds of milliseconds time range (see David et al., 1992, 1993 for possible temporal correlations) vs. the tens of seconds that we calculate to be required for purely circumferential diffusion.

failure to transport K⁺ and water between paranodal loops is proposed to cause osmotic disruption of successive layers of PNS myelin, high incidence of conduction block, and dysmyelination, whereas in the CNS, neurologic symptoms attributable to myelin disruption are more difficult to detect.

8.14.1.2   Mutations of Cx46.6 (Cx47) Cause Pelizaeus-Merzbacher-Like Diseases (PMLDs) and Hereditary Spastic Paraplegia (HSP)

Pelizaeus-Merzbacher disease (PMD) results from mutation of proteolipid protein-1, the primary protein of myelin, resulting in primary failure of myelination, recognized as a leukodystrophy (http://www.ninds.nih.gov/disorders/pelizaeus_merzbacher/pelizaeus_merzbacher.htm). In contrast, PMLDs are a group of debilitating autosomal recessive human diseases that affect formation, structure, and function of CNS myelin (hypomyelination). Primary symptoms of PMLD include reduced conduction velocity and complete block of saltatory conduction in some myelinated fibers, along with dysmyelination of long white matter tracts, nystagmus, progressive spasticity, and ataxia. PMLDs result from mutations of the gene for Cx46.6 (Cx47 in mice) on chromosome 1q41-q42 (Uhlenberg et al., 2004). Thus, the absence of Cx47:Cx43 at O:A gap junctions would be expected to result in significant disruption of normal K⁺ and water transport mechanisms from myelin to astrocytes, leading to progressive myelin disruption (as summarized in Rash, 2010).

Hereditary spastic paraplegia (HSP) is a second group of human neurological diseases characterized by recessive mutations of Cx46.6 (Orthmann-Murphy et al., 2009) that result in myelin disruption and disrupted axonal saltatory conduction. Because sequential CNS biopsies have not been obtained from PMLD and HSP patients, understanding these maladies required examination of tissues from animal models (i.e., mice) having single knockouts (KOs) for Cx32 and for Cx47, as well as double knockouts (dKOs) for Cx32/Cx47.

8.14.1.3   Cx32 and Cx47 Single-Knockout and Cx32/ Cx47 Double-Knockout Mice

In mice, single knockout for Cx32 is not fatal, and gross physical examination reveals no significant phenotype. In contrast, single KO for Cx47 is usually fatal within 100 days of birth, with progressive myelin destruction evident. Still more destructive is the dKO for Cx32/Cx47, with severe symptoms and deaths seen as early as postnatal day 5 (P5), and 100% fatalities observed by P31 (Menichella et al., 2006). By histological examination of Cx32/Cx47 dKO optic nerve axons, Menichella et al. (2006) showed that vacuolation of myelin became evident at about P13 (2 days after eye opening and concomitant with greatly increased activation of the optic nerve), with myelin disruption becoming increasingly severe by P15. As predicted by the K⁺ gateway hypothesis, optic nerve myelin became grossly expanded, particularly in innermost myelin, where axonal K⁺ initially enters myelin—at least during the repetitive, high-frequency activity that underlies normal vision. Moreover, blocking optic nerve axonal saltatory conduction by injecting tetrodotoxin (TTX) into the vitreous humor of postnatal mice reduced or eliminated vacuolization of myelin, consistent with the effects of reduced Na⁺ influx into the axon and the concomitant reduction of the efflux of K⁺ and water into the peri-internodal compartment. Thus, the onset of myelin destruction in the optic nerves of dKO mice was traced to increased axonal activity at P15, and this destruction of myelin was blocked by TTX, both consistent with ionic and osmotic damage proportional to K^+ efflux and to the subsequent failure to remove excess K^+ via Cx32- and Cx47-containing gap junctions.

8.14.2 DISEASES OF ASTROCYTE CONNEXINS BLOCK K^+ SIPHONING

8.14.2.1 Oculodentodigital Dysplasia (ODDD) Results from Mutations of the Gene for Cx43

Because Cx43 is widely expressed in all four tissue types (epithelia, muscle, connective, and neural tissue), more than 35 known mutations in Cx43 cause a pleiotropic (i.e., multifaceted) disease affecting, most noticeably, facial features (eyes and teeth) and connective tissue development (digital abnormalities). Although Cx43 is also the primary connexin in astrocyte gap junctions, it was not immediately obvious how defects in oligodendrocyte myelin formation and in the late developing dysmyelination might result from mutations in an astrocyte connexin (Paznekas et al., 2003). However, the potassium gateway hypothesis provides a rational basis for understanding the CNS alterations, which include hypomyelination and sclerotic changes in white matter tracts and cerebellar atrophy (Wolf et al., 2010), both presumably resulting from diminished K^+ siphoning through O/A and A/A gap junctions.

8.14.2.2 Double Knockout of Cx43 and Cx30 Causes Demyelination and White Matter Sclerosis

Conditional double knockout (cdKO) of Cx43 and Cx30 under control of the GFAP promoter in astrocytes (and expressed in few other cell types) results in failure to form gap junctions between astrocytes and between astrocytes and oligodendrocytes, but causes no alterations in other tissues. These cdKO animals had greatly reduced saltatory conduction (Wallraff et al., 2006) and large-scale disruptions of myelin (Lutz et al., 2009), virtually indistinguishable from those seen after disruption of oligodendrocyte connexin in the Cx32/Cx47 dKO mice. Thus, a common feature of genetic diseases that disrupt gap junctions of the pan glial syncytium, which are essential for K^+ siphoning and water transport, is the near-simultaneous blockade of saltatory conduction and the disruption of CNS myelin, consistent with the essentiality of integrated K^+ siphoning and water transport in the CNS.

8.14.3 DISEASES AFFECTING OTHER ASPECTS OF ASTROCYTE K^+ SIPHONING AND WATER TRANSPORT CAUSE MYELIN SCLEROSIS

8.14.3.1 Mutations and Knockout of Kir4.1 Cause Dysmyelination and Sclerosis

To date, the only human neurological disease thought to be caused by mutations in the Kir4.1 gene (KCNJ10) is one form of temporal lobe epilepsy (Heuser et al., 2010). Because Kir4.1 is widely expressed in all neurons and glial cells, where it helps to set the resting membrane potential, identifying the source of disease etiology has proven to be difficult. In addition to involvement in K^+ siphoning, Kir4.1 also governs the transmembrane gradients from the myelin sheath to astrocyte endfeet.

The first study of Kir4.1 KO mice (Neusch et al., 2001) reported downregulation of Kir4.1 in *oligodendrocyte somata*. Thus, they proposed an effect on oligodendrocyte K^+ siphoning that directly resulted in dysmyelination. That interpretation of oligodendrocyte abnormalities in Kir4.1 KO mice, as well as interpretations of proposed mechanisms for myelin disruption, was subsequently explored by Kalsi and coworkers and by Menichella and coworkers, who favored an interpretation of damaged myelin based on disruption of K^+ exit pathways in astrocyte endfeet (Kalsi et al., 2004), which is consistent with the K^+ gateway hypothesis. Moreover, it has been proposed that KO of Kir4.1 in mice disrupts K^+ exit pathways in astrocytes, causing "bystander" swelling and vacuolation in both innermost and outermost myelin layers in the optic nerve (Menichella et al., 2006), similar to that seen in the Cx32/Cx47 and Cx43/Cx30 dKO mouse. Those observations lend further support to the hypothesis that any substantial disruption in the K^+ or water transport pathways, from innermost myelin to the astrocyte endfeet, results in ultrastructural alterations of myelin, dysmyelination, and sclerosis, and as a direct consequence, also disrupts axonal saltatory conduction.

8.14.3.2 Essential Role of Astrocyte GFAP Filaments in K^+ Siphoning and Osmotic Water Transport

Astrocytic glial fibrillary acidic protein (GFAP) filaments form a strongly hydrophilic, negatively charged "wick" that appears to be required for efficient distribution of both water and K^+. At physiological pH, GFAP filaments are strongly negatively charged, thereby providing for strong interaction with positively charged K^+, potentially allowing for addition of K^+ at astrocyte contacts with myelin occurring simultaneously with exit of K^+ at far distant astrocyte endfeet at the glia limitans. The subcellular distribution of the GFAP distribution network has been beautifully characterized by confocal microscopy (see 2004 *Neuroscience* cover image by X. Wang (vol. 129, no. 4) and by Nedergaard and coworkers (Nedergaard et al., 2003)). The GFAP bundles (Figure 8.2C, black staining from DAB reaction product) extend from contacts with myelin (not visualized in this tri-stained image), beginning as small processes in contact with myelin, increasing in diameter and converging on the cell soma (blue DAPI-stained nuclei, left and bottom), then branching again toward the astrocyte endfeet (red fluorescence for AQP4; reproduced as pink) that surround and clearly delineate the otherwise unstained capillaries. Thus, GFAP filaments create a pervasive, filamentous hydrophilic bundle (or wick) network that closely resembles both ends of a complex water irrigation network. GFAP filament bundles not only branch widely to contact multiple myelin segments (thereby resembling the small coalescing streams that comprise the headwaters of a river), but after coalescing into the main tributary bundle that crosses the astrocyte cell body, the GFAP bundles then branch again, resembling a desert irrigation network that is able to distribute K^+ and water to the glia limitans and to endfeet surrounding every capillary within the CNS parenchyma. Overall, the biophysical properties and subcellular distributions of GFAP filaments are consistent with them having an essential role in rapidly and efficiently transferring water and K^+ over centimeter distances within the CNS. If this description is accurate, what would happen to K^+ siphoning if the GFAP network were to be disrupted?

8.14.3.2.1 Alexander Disease Results from Mutations of the Gene for GFAP

In humans, Alexander disease is a fatal demyelinating disease resulting from any one of at least 15 different point mutations of the gene for GFAP (Brenner et al., 2001; Hsiao et al., 2005), primarily replacing arginine with cysteine, which would be expected to result in increased cross-linking of GFAP filaments. Infants with Alexander disease develop a pronounced leukoencephalopathy with macrocephaly, both resulting from gross disruption of white matter. They have seizures and psychomotor retardation, and usually die before age 10. Autopsies reveal that all Alexander disease patients have Rosenthal fibers, cytoplasmic inclusions in astrocytes that consist of unusually large (up to 10×50 μm) bundles of GFAP filaments, with GFAP intermixed with several other strongly-bound filamentous proteins, including vimentin. Rosenthal fibers are deposited as particularly large masses in astrocyte endfeet that surround capillaries and form the glia limitans, often disrupting the endfoot processes, and presumably the delicate K^+ and water transport mechanisms, which rely on close apposition of astrocytes and capillaries, for example. These disruptions of GFAP result in pronounced formation of sclerotic myelin, apparently based on reduced water and K^+ transport through the astrocyte syncytium, with concomitant swelling of both astrocyte processes and inner and outer layers of oligodendrocyte myelin, as predicted by the K^+ gateway hypothesis.

8.14.3.2.2 GFAP Knockout Mice Have Sclerotic Myelin

As a possible model for understanding the role of GFAP in CNS function, a mouse GFAP knockout model lacks detectable GFAP filaments but expresses increased vimentin in astrocyte processes (Liedtke et al., 1996). However, myelin is disrupted and greatly reduced over wide areas, and older animals exhibit hydrocephalus, the latter providing evidence for reduced or absent functioning of AQP4 water transport at astrocyte endfeet (Papadopoulos and Verkman, 2010), also consistent with predictions of the K^+ gateway hypothesis.

8.14.3.3 Neuromyelitis Optica, an Autoimmune Disease Affecting Water Transport at Astrocyte Endfeet, Causes a Severe Form of Multiple Sclerosis

Complete knockout of AQP4 is not fatal in mice, and symptoms are minimal, presumably because of compensating mechanisms developed during early CNS development (Verbavatz et al., 1997; Zhou et al., 2008). However, in humans, autoimmune destruction of astrocyte AQP4 results in a severe form of multiple sclerosis called neuromyelitis optica (NMO). Because it preferentially affects Asians and Africans, NMO is also called Asian MS.

NMO was originally defined as an idiopathic, severe demyelinating disease that is restricted to CNS myelin but that does not affect PNS myelin (i.e., affects oligodendrocytes but not Schwann cells). NMO preferentially affects myelin of the most frequently activated axons, notably those in the optic nerves and spinal cord. After more than a century of false leads, NMO was recently demonstrated to be a primary autoimmune disease directed against AQP4 water channels (Wingerchuk et al., 2007), particularly those that are concentrated in astrocyte endfeet surrounding capillaries, as well as those at the glia limitans. Before recognition of the integrated roles of oligodendrocytes and astrocytes in the panglial syncytium, understanding the etiology of NMO was particularly problematic. This was because most areas of high AQP4 expression in the brain are within and normally protected by the blood–brain barrier. It now seems clear that autoimmune destruction of AQP4 water channels in astrocyte endfeet of NMO patients disrupts essential water exit pathways that would otherwise dispose of the large volume of intracellular osmotic water that accompanies K^+ during K^+ siphoning. According to the K^+ gateway hypothesis, if the osmotic load on myelin is not rapidly corrected by large-scale potassium siphoning, K^+ plus osmotic water build up in cytoplasmic myelin (i.e., paranodes, and innermost and outermost myelin), which then expands and ultimately ruptures, producing ultrastructural disruptions of myelin that are similar to those seen following double-knockout of Cx32/Cx47 (Menichella et al., 2006), in dKO of Cx43/Cx30 (Lutz et al., 2009), in KO of Kir4.1 (Menichella et al., 2006), and following disruption of astrocyte GFAP filaments, as in Alexander disease (Brenner et al., 2001; Hsiao et al., 2005). Again, these changes are directly predicted by the K^+ gateway hypothesis.

8.15 SUMMARY

In this chapter, we have introduced the potassium gateway hypothesis, which describes the pathway of excess K^+ (along with osmotic water) that is released during high-frequency axonal saltatory conduction, tracing its movement from juxtaparanode and internode, into the innermost layer of myelin, through the multiple layers of paranodal myelin, across heterologous O:A gap junctions, and into the astrocyte syncytium, for release at astrocyte endfeet surrounding capillaries and forming the glia limitans. Because of the essential role of the panglial syncytium in K^+ siphoning and water homeostasis, the K^+ gateway hypothesis provides a new paradigm for better understanding the complex roles of glia in CNS health and disease, and provides a mechanism for understanding of the etiology of many human demyelinating diseases. We propose that molecular disruptions of K^+ and water transport proteins result in disruption of saltatory conduction, dysmyelination, and ultimately the formation of sclerotic plaques, including those found in several forms of multiple sclerosis, Alexander disease, Pelizaeus-Merzbacher-like disease, neuromyelitis optica, and oculodentodigital dysplasia, among a growing list of diseases that disrupt CNS myelin.

REFERENCES

Ahn, M., Lee, J., Gustafsson, A., Enriquez, A., Lancaster, E., Sul, J.-Y., Haydon, P.G., Paul, D.L., Huang, Y., Abrams, C.K., and Scherer, S. (2008). Cx29 and Cx32, two connexins expressed by myelinating glia, do not interact and are functionally distinct. *J. Neurosci. Res.* 86, 992–1006.

Altevogt, B.M., Kleopas, K.A., Postma, F.R., Scherer, S.S., and Paul, D. (2002). Connexin29 is uniquely distributed within myelinating glial cells of the central and peripheral nervous systems. *J. Neurosci.* 22, 6458–6470.

Altevogt, B.M., and Paul, D.L. (2004). Four classes of intercellular channels between glial cells in the CNS. *J. Neurosci.* 24, 4313–4323.

Aschenbrenner, S., and Walz, B. (1998). Pleated septate junctions in leech photoreceptors: ultrastructure, arrangement of septa, gate and fence functions. *Cell Tiss. Res.* 293, 253–269.

Balice-Gordon, R.J., Bone, L.J., and Scherer, S.S. (1998). Functional gap junctions in the Schwann cell myelin sheath. *J. Cell Biol.* 142, 1095–1104.

Banerjee, S., Sousa, A., and Bhat, M. (2006). Organization and function of septate junctions. *Cell Biochem. Biophys.* 46, 65–77.

Bergoffen, J., Scherer, S.S., Wang, S., Oronzi-Scott, M., Paul, D., Chen, K., Lensch, M.W., Chance, P., and Fischbeck, K. (1993). Connexin mutations in X-linked Charcot-Marie-Tooth disease. *Science* 262, 2039–2042.

Bhat, M.A., Rios, J.C., Lu, Y., Garcia-Fresco, G.P., Ching, W., Martin, M.S., Li, J., Einheber, S., Chesler, M., Rosenbluth, J., Salzer, J.L., and Bellen, H.J. (2001). Axon-glia interactions and the domain organization of myelinated axons requires neurexin IV/caspr/paranodin. *Neuron* 30, 369–383.

Black, J.A., and Waxman, S.G. (1988). The perinodal astrocyte. *Glia* 1, 169–183.

Brenner, M., Johnson, A.B., Boespflug-Tanguy, O., Rodriguez, D., Goldman, J.E., and Messing, A. (2001). Mutations in GFAP, encoding glial fibrillary acidic protein, are associated with Alexander disease. *Nat. Genet.* 27, 117–120.

Brightman, M.W., and Reese, T.S. (1969). Junctions between intimately apposed cell membranes in the vertebrate brain. *J. Cell Biol.* 40, 648–677.

Brophy, P.J. (2001). Axoglial junctions: separate the channels or scramble the message. *Curr. Biol.* 11, R555–R557.

Brophy, P.J. (2003). Myelinated nerves: filling in the juxtaparanodal gap. *Curr. Biol.* 13, R956–R957.

Bruzzone, R., White, T.W., and Paul, D.L. (1996). Connections with connexins: the molecular basis of direct intercellular signaling. *Eur. J. Biochem.* 238, 1–27.

Bullock, T.H., Moore, J.K., and Fields, R.D. (1984). Evolution of myelin sheaths: both lamprey and hagfish lack myelin. *Neurosci. Lett.* 48, 145–148.

Bunge, R.P. (1968). Glial cells and the central myelin sheath. *Physiol. Rev.* 48, 197–251.

Catterall, W.A., Goldin, A.L., and Waxman, S.G. (2005). International Union of Pharmacology. XLVII. Nomenclature and structure-function relationships of voltage-gated sodium channels. *Pharmacol. Rev.* 57, 397–409.

Chiu, S.Y. (1991). Functions and distribution of voltage-gated sodium and potassium channels in mammalian Schwann cells. *Glia* 4, 541–558.

Chiu, S.Y., and Ritchie, J.M. (1980). Potassium channels in nodal and internodal axonal membrane of mammalian myelinated fibres. *Nature* 284, 170–171.

David, G., Barrett, J.N., and Barrett, E.F. (1992). Evidence that action potentials activate an internodal potassium conductance in lizard myelinated axons. *J. Physiol.* 445, 277–301.

David, G., Barrett, J.N., and Barrett, E.F. (1993). Activation of internodal potassium conductance in rat myelinated axons. *J. Physiol.* 472, 177–202.

Dudek, F.E., Shao, L.-R., and Rash, J.E. (2010). Possible roles of nonsynaptic mechanisms in synchronization of epileptic seizures: potential antiepileptic targets? In *Epilepsy: mechanisms, models, and translational perspectives*, ed. Jong M. Rho, Raman Sankar, and Carl E. Stafstrom, 209–228.

Eiberger, J., Degen, J., Romualdi, A., Deutsch, U., Willecke, K., and Söhl, G. (2001). Connexin genes in the mouse and human genome. *Cell Commun. Adhes.* 8, 163–165.

Einheber, S., Zanazzi, G., Ching, W., Scherer, S., Milner, T.A., Peles, E., and Salzer, J.L. (1997). The axonal membrane protein Caspr, a homologue of neurexin IV, is a component of the septate-like paranodal junctions that assemble during myelination. *J. Cell Biol.* 139, 1495–1506.

Furman, C.S., Gorelick-Feldman, D.A., Davidson, K.G.V., Yasumura, T., Neely, J.D., Agre, P., and Rash, J.E. (2003). Aquaporin-4 square array assembly: opposing actions of M1 and M23 isoforms. *Proc. Natl. Acad. Sci. USA* 100, 13609–13614.

Gould, R.M., Morrison, H.G., Gilland, E., and Campbell, R.K. (2005). Myelin tetraspan family proteins but no non-tetraspan family proteins are present in the ascidian (*Ciona intestinalis*) genome. *Biol. Bull.* 209, 49–66.

Gutman, G.A., Chandy, K.G., Grissmer, S., Lazdunski, M., McKinnon, D., Pardo, L.A., Robertson, G.A., Rudy, B., Sanguinetti, M.C., Stühmer, W., and Wang, X. (2005). International Union of Pharmacology. LIII. Nomenclature and molecular relationships of voltage-gated potassium channels. *Pharmacol. Rev.* 57, 473–508.

Hablitz, J.J., and Heinemann, U. (1989). Alterations in the microenvironment during spreading depression associated with epileptiform activity in the immature neocortex. *Dev. Brain Res.* 46, 243–252.

Hartline, D.K. (2008). What is myelin? *Neuron Glia Biol.* 4, 153–163.

Heuser, K., Nagelhus, E.A., Tauboll, E., Indahl, U., Berg, P.R., Lien, S., Nakken, S., Gjerstad, L., and Ottersen, O.P. (2010). Variants of the genes encoding AQP4 and Kir4.1 are associated with subgroups of patients with temporal lobe epilepsy. *Epilepsy Res.* 88, 55–64.

Hirano, A., and Dembitzer, H.M. (1982). Further studies on the transverse bands. *J. Neurocytol.* 11, 861–866.

Hodgkin, A.L. (1951). The ionic basis of electrical activity in nerve and muscle. *Biol. Rev.* 26, 339–409.

Hodgkin, A.L., and Huxley, A.F. (1952). A quantitative description of membrane current and its application to conduction and excitation in nerve. *J. Physiol.* 117, 500–544.

Hsiao, V.C., Tian, R., Long, H., Der Perng, M., Brenner, M., Quinlan, R.A., and Goldman, J.E. (2005). Alexander-disease mutation of GFAP causes filament disorganization and decreased solubility of GFAP. *J. Cell Sci.* 118, 2057–2065.

Huxley, A.F., and Stämpfli, R. (1949). Evidence for saltatory conduction in peripheral myelinated nerve fibres. *J. Physiol.* 108, 315–339.

Kalsi, A.S., Greenwood, K., Wilkin, G., and Butt, A.M. (2004). Kir4.1 expression by astrocytes and oligodendrocytes in CNS white matter: a developmental study in the rat optic nerve. *J. Anat.* 204, 475–485.

Kamasawa, N., Sik, A., Morita, M., Yasumura, T., Davidson, K.G.V., Nagy, J.I., and Rash, J.E. (2005). Connexin47 and connexin32 in gap junctions of oligodendrocyte somata, myelin sheaths, paranodal loops, and Schmidt-Lanterman incisures: implications for ionic homeostasis and potassium siphoning. *Neuroscience* 136, 65–86.

Kandel, E.R., Schwartz, J.H., and Jessell, T.M., eds. (2000). *Principles of neural science.* New York: Elsevier.

Keep, R.F., Ulanski II, L.J., Xiang, J., Ennis, S.R., and Lorris Betz, A. (1999). Blood-brain barrier mechanisms involved in brain calcium and potassium homeostasis. *Brain Res.* 815, 200–205.

Kettenmann, H., and Ransom, B.R. (1988). Electrical coupling between astrocytes and between oligodendrocytes studied in mammalian cell cultures. *Glia* 1, 64–73.

Kleopa, K.A., Orthmann, J.L., Enriquez, A., Paul, D.L., and Scherer, S.S. (2004). Unique distributions of the gap junction proteins connexin29, connexin32, and connexin47 in oligodendrocytes. *Glia* 47, 346–357.

Kofuji, P., and Newman, E.A. (2004). Potassium buffering in the central nervous system. *Neuroscience* 129, 1043–1054.

Lander, E.S., and Internat. Human Genome Seq. Cons. (2001). Initial sequencing and analysis of the human genome. *Nature* 409, 860–921.

Lennon, V.A., Wingerchuk, D.M., Kryzer, T.J., Pittock, S.J., Lucchinetti, C.F., Fujihara, K., Nakashima, I., and Weinshenker, B.G. (2004). A serum autoantibody marker of neuromyelitis optica: distinction from multiple sclerosis. *Lancet* 364, 2106–2112.

Li, H., and Richardson, W.D. (2008). The evolution of Olig genes and their roles in myelination. *Neuron Glia Biol.* 4, 129–135.

Li, X., Ionescu, A.-V., Lynn, B.D., Lu, S., Kamasawa, N., Morita, M., Davidson, K.G.V., Yasumura, T., Rash, J.E., and Nagy, J.I. (2004). Connexin47, connexin29 and connexin32 co-expression in oligodendrocytes and Cx47 association with zonula occludens-1 (ZO-1) in mouse brain. *Neuroscience* 126, 611–630.

Li, X., Lynn, B.D., Olson, C., Meier, C., Davidson, K.G.V., Yasumura, T., Rash, J.E., and Nagy, J.I. (2002). Connexin29 expression, immunocytochemistry and freeze-fracture replica immunogold labeling (FRIL) in sciatic nerve. *Eur. J. Neurosci.* 16, 795–806.

Liedtke, W., Edelmann, W., Bieri, P.L., Chiu, F.C., Cowan, N.J., Kucherlapati, R., and Raine, C.S. (1996). GFAP is necessary for the integrity of CNS white matter architecture and long-term maintenance of myelination. *Neuron* 17, 607–615.

Lutz, S.E., Zhao, Y., Gulinello, M., Lee, S.C., Raine, C.S., and Brosnan, C.F. (2009). Deletion of astrocyte connexins 43 and 30 leads to a dysmyelinating phenotype and hippocampal CA1 vacuolation. *J. Neurosci.* 29, 7743–7752.

Matthews, M.A., and Duncan, D. (1971). A quantitative study of morphological changes accompanying the initiation and progress of myelin production in the dorsal funiculus of the rat spinal cord. *J. Comp. Neurol.* 142, 1–22.

Meier, C., Dermietzel, R., Davidson, K.G.V., Yasumura, T., and Rash, J.E. (2004). Connexin32-containing gap junctions in Schwann cells at the internodal zone of partial myelin compaction and in Schmidt-Lanterman incisures. *J. Neurosci.* 24, 3186–3198.

Menichella, D.M., Goodenough, D.A., Sirkowski, E., Scherer, S.S., and Paul, D.L. (2003). Connexins are critical for normal myelination in the CNS. *J. Neurosci.* 23, 5963–5973.

Menichella, D.M., Majdan, M., Awatramani, R., Goodenough, D.A., Sirkowski, E., Scherer, S.S., and Paul, D.L. (2006). Genetic and physiological evidence that oligodendrocyte gap junctions contribute to spatial buffering of potassium released during neuronal activity. *J. Neurosci.* 26, 10984–10991.

Miller, R.G., and Pinto da Silva, P. (1977). Particle rosettes in the periaxonal Schwann cell membrane and particle clusters in the axolemma of rat sciatic nerve. *Brain Res.* 130, 135–141.

Morales, R., and Duncan, D. (1975). Specialized contacts of astrocytes with astrocytes and with other cell types in the spinal cord of the cat. *Anat. Rec.* 182, 255–266.

Mugnaini, E. (1986). Cell junctions of astrocytes, ependyma, and related cells in the mammalian central nervous system, with emphasis on the hypothesis of a generalized functional syncytium of supporting cells. In *Astrocytes*, ed. S. Fedoroff and A. Vernadakis, 329–371. Vol. I. New York: Academic Press.

Mugnaini, E., Osen, K.K., Schnapp, B., and Friedrich, V.L., Jr. (1977). Distribution of Schwann cell cytoplasm and plasmalemmal vesicles (caveolae) in peripheral myelin sheaths. An electron microscopic study with thin sections and freeze-fracturing. *J. Neurocytol.* 6, 647–668.

Nagelhus, E.A., Horio, Y., Inanobe, A., Fujita, A., Haug, F.-M., Nielsen, S., Kurachi, Y., and Ottersen, O.P. (1999). Immunogold evidence suggests that coupling of K+ siphoning and water transport in rat retinal Muller cells is mediated by coenrichment of Kir4.1 and AQP4 in specific membrane domains. *Glia* 26, 47–54.

Nagelhus, E.A., Mathiisen, T.M., and Ottersen, O.P. (2004). Aquaporin-4 in the central nervous system: cellular and subcellular distribution and coexpression with KIR4.1. *Neuroscience* 129, 905–913.

Nagelhus, E.A., Veruki, M.L., Torp, R., Haug, F.-M., Laake, J.H., Nielsen, S., Agre, P., and Ottersen, O.P. (1998). Aquaporin-4 water channel protein in the rat retina and optic nerve: polarized expression of Müller cells of fibrous astrocytes. *J. Neurosci.* 18, 2506–2519.

Nagy, J.I., Dudek, F.E., and Rash, J.E. (2004). Update on connexins and gap junctions in neurons and glia in the mammalian central nervous system. *Brain Res. Brain Res. Rev.* 47, 191–215.

Nagy, J.I., Ionescu, A.-V., Lynn, B.D., and Rash, J.E. (2003). Coupling of astrocyte connexins Cx26, Cx30, Cx43 to oligodendrocyte Cx29, Cx32, Cx47: implications from normal and connexin32 knockout mice. *Glia* 44, 205–218.

Nagy, J.I., Li, X., Rempel, J., Stelmack, G.L., Patel, D., Staines, W.A., Yasumura, T., and Rash, J.E. (2001). Connexin26 in adult rodent CNS: demonstration at astrocytic gap junctions and co-localization with connexin30 and connexin43. *J. Comp. Neurol.* 441, 302–323.

Nedergaard, M., Ransom, B., and Goldman, S.A. (2003). New roles for astrocytes: redefining the functional architecture of the brain. *Trends Neurosci. (TINS)* 26, 523–530.

Neusch, C., Rozengurt, N., Jacobs, R.E., Lester, H.A., and Kofuji, P. (2001). Kir4.1 potassium channel subunit is crucial for oligodendrocyte development and *in vivo* myelination. *J. Neurosci.* 21, 5429–5438.

Newman, E.A. (1986). High potassium conductance in astrocyte endfeet. *Science* 233, 453–454.

Nielsen, S., Nagelhus, E.A., Amiry-Moghaddam, M., Bourque, C., Agre, P., and Ottersen, O.P. (1997). Specialized membrane domains for water transport in glial cells: high-resolution immunogold cytochemistry of aquaporin-4 in rat brain. *J. Neurosci.* 17, 171–180.

Orkand, R.K., Nicholls, J.G., and Kuffler, S.W. (1966). Effect of nerve impulses on the membrane potential of glial cells in the central nervous system of amphibia. *J. Neurophysiol.* 29, 788–806.

Orthmann-Murphy, J.L., Freidin, M., Fischer, E., Scherer, S.S., and Abrams, C.K. (2007). Two distinct heterotypic channels mediate gap junction coupling between astrocyte and oligodendrocyte connexins. *J. Neurosci.* 27, 13949–13957.

Orthmann-Murphy, J.L., Salsano, E., Abrams, C.K., Bizzi, A., Uziel, G., Freidin, M.M., Lamantea, E., Zeviani, M., Scherer, S.S., and Pareyson, D. (2009). Hereditary spastic paraplegia is a novel phenotype for GJA12/GJC2 mutations. *Brain* awn328.

Ota, K., Matsui, M., Milford, E.L., Mackin, G.A., Weiner, H.L., and Hafler, D.A. (1990). T-cell recognition of an immuno-dominant myelin basic protein epitope in multiple sclerosis. *Nature* 346, 183–187.

Papadopoulos, M.C., and Verkman, A.S. (2010). Aquaporin-4 and brain edema. *Pediatr. Nephrol.* 22, 778–784.

Pastor, A., Kremer, M., Möller, T., Kettenmann, H., and Dermietzel, R. (1998). Dye coupling between spinal cord oligodendrocytes: differences in coupling efficiency between gray and white matter. *Glia* 24, 108–120.

Paznekas, W.A., Boyadjiev, S.A., Shapiro, R.E., Daniels, O., Wollnik, B., Keegan, C.E., Innis, J.W., Dinulos, M.B., Christian, C., Hannibal, M.C., and Jabs, E.W. (2003). Connexin 43 (GJA1) mutations cause the pleiotropic phenotype of oculodentodigital dysplasia. *Am. J. Hum. Genet.* 72, 408–418.

Peters, A., Palay, S.L., and Webster, H.D. (1991). *The fine structure of the nervous system: Neurons and their supporting cells*, 1–494. New York: Oxford University Press.

Poopalasundaram, S., Knott, C., Shamotienko, O.G., Foran, P.G., Dolly, J.O., Ghiani, C.A., Gallo, V., and Wilkin, G.P. (2000). Glial heterogeneity in expression of the inwardly rectifying K(+) channel, Kir4.1, in adult rat CNS. *Glia* 30, 362–372.

Rasband, M.N. (2004). It's "juxta" potassium channel! *J. Neurosci. Res.* 76, 749–757.

Rasband, M.N., Peles, E., Trimmer, J.S., Levinson, S.R., Lux, S.E., and Shrager, P. (1999). Dependence of nodal sodium channel clustering on paranodal axoglial contact in the developing CNS. *J. Neurosci.* 19, 7516–7528.

Rasband, M.N., and Shrager, P. (2000). Ion channel sequestration in central nervous system axons. *J. Physiol. (Lond.)* 525, 63–73.

Rash, J.E. (2010). Molecular disruptions of the panglial syncytium block potassium siphoning and axonal saltatory conduction: pertinence to neuromyelitis optica and other demyelinating diseases of the central nervous system. *Neuroscience* 168, 982–1008.

Rash, J.E., Davidson, K.G.V., Yasumura, T., and Furman, C.S. (2004). Freeze-fracture and immunogold analysis of aquaporin-4 (AQP4) square arrays, with models of AQP4 lattice assembly. *Neuroscience* 129, 915–934.

Rash, J.E., Duffy, H.S., Dudek, F.E., Bilhartz, B.L., Whalen, L.R., and Yasumura, T. (1997). Grid-mapped freeze-fracture analysis of gap junctions in gray and white matter of adult rat central nervous system, with evidence for a "panglial syncytium" that is not coupled to neurons. *J. Comp. Neurol.* 388, 265–292.

Rash, J.E., Olson, C.O., Pouliot, W.A., Davidson, K.G.V., Yasumura, T., Furman, C.S., Royer, S., Kamasawa, N., Nagy, J.I., and Dudek, F.E. (2007). Connexin36, miniature neuronal gap junctions, and limited electrotonic coupling in rodent suprachiasmatic nucleus. *Neuroscience* 149, 350–371.

Rash, J.E., Yasumura, T., Davidson, K., Furman, C.S., Dudek, F.E., and Nagy, J.I. (2001a). Identification of cells expressing Cx43, Cx30, Cx26, Cx32 and Cx36 in gap junctions of rat brain and spinal cord. *Cell Commun. Adhes.* 8, 315–320.

Rash, J.E., Yasumura, T., Dudek, F.E., and Nagy, J.I. (2001b). Cell-specific expression of connexins, and evidence for restricted gap junctional coupling between glial cells and between neurons. *J. Neurosci.* 21, 1983–2001.

Rash, J.E., Yasumura, T., Hudson, C.S., Agre, P., and Nielsen, S. (1998). Direct immunogold labeling of aquaporin-4 in "square arrays" of astrocyte and ependymocyte plasma membranes in rat brain and spinal cord. *Pro. Natl. Acad. Sci. USA* 95, 11981–11986.

Rios, J.C., Rubin, M., St. Martin, M., Downey, R.T., Einheber, S., Rosenbluth, J., Levinson, S.R., Bhat, M., and Salzer, J.L. (2003). Paranodal interactions regulate expression of sodium channel subtypes and provide a diffusion barrier for the node of Ranvier. *J. Neurosci.* 23, 7001–7011.

Roemer, S.F., Parisi, J.E., Lennon, V.A., Benarroch, E.E., Lassmann, H., Bruck, W., Mandler, R.N., Weinshenker, B.G., Pittock, S.J., Wingerchuk, D.M., and Lucchinetti, C.F. (2007). Pattern-specific loss of aquaporin-4 immunoreactivity distinguishes neuromyelitis optica from multiple sclerosis. *Brain* 130, 1194–1205.

Rosenbluth, J. (2009). Multiple functions of the paranodal junction of myelinated nerve fibers. *J. Neurosci. Res.* 87, 3250–3258.

Sargiannidou, I., Markoullis, K., and Kleopa, K.A. (2010). Molecular mechanisms of gap junction mutations in myelinating cells. *Histol. Histopath. Cell. Mol. Biol.* 25, 1191–1206.

Satir, B.H., and Satir, P. (1979). Partitioning of intramembrane particles during the freeze-fracture procedure. In *Freeze fracture: methods, artifacts, and interpretations*, ed. J.E. Rash and C.S. Hudson, 43–49. New York: Raven Press.

Scherer, S.S., Deschenes, S.M., Xu, Y.-T., Grinspan, J.P., Fischbeck, K.H., and Paul, D.L. (1995). Connexin32 is a myelin-related protein in the PNS and CNS. *J. Neurosci.* 15, 8281.

Scherer, S.S., Xu, Y.T., Nelles, E., Fischbeck, K., Willecke, K., and Bone, L.J. (1998). Connexin32-null mice develop demyelinating peripheral neuropathy. *Glia* 24, 8–20.

Schnapp, B., and Mugnaini, E. (1978). Membrane architecture of myelinated fibers as seen by freeze fracture. In *Physiology and pathobiology of axons*, ed. S. Waxman, 83–123. New York: Raven Press.

Segal, M.B. (1993). Extracellular and cerebrospinal fluids. *J. Inherit. Metab. Disord.* 16, 617–638.

Senderek, J., Hermanns, B., Bergmann, C., Boroojerdi, B., Bajbouj, M., Hungs, M., Ramaekers, V.T., Quasthoff, S., Karch, D., and Schroder, J.M. (1999). X-linked dominant Charcot-Marie-Tooth neuropathy: clinical, electrophysiological, and morphological phenotype in four families with different connexin32 mutations. *J. Neurol. Sci.* 167, 90–101.

Sherratt, R.M., Bostock, H., and Sears, T.A. (1980). Effects of 4-aminopyridine on normal and demyelinated mammalian nerve fibres. *Nature* 283, 570–572.

Stolinski, C., Breathnach, A.S., Martin, B., Thomas, P.K., King, R.H.M., and Gabriel, G. (1981). Associated particle aggregates in juxtaparanodal axolemma and adaxonal Schwann cell membrane of rat peripheral nerve. *J. Neurocytol.* 10, 679–691.

Stolinski, C., Breathnach, A.S., Thomas, P.K., Gabriel, G., and King, R.H.M. (1985). Distribution of particle aggregates in the internodal axolemma and adaxonal Schwann cell membrane of rodent peripheral nerve. *J. Neurol. Sci.* 67, 213–222.

Tasaki, I. (1939). The electro-saltatory transmission of the nerve impulse and the effect of narcosis upon the nerve fiber. *Am. J. Physiol.* 127, 211–227.

Uhlenberg, B., Schuelke, M., Ruschendorf, F., Rug, N., Kaindl, A.M., Henneke, M., Thiele, H., Stoltenburg-Didinger, G., Aksu, F., Topaloglu, H., Hubner, C., Weschke, B., and Gartner, J. (2004). Mutations in the gene encoding gap junction protein alpha 12 (connexin 46.6) cause Pelizaeus-Merzbacher-like disease. *Am. J. Hum. Genet.* 75, 251–260.

Venter, J.C., Adams, M.D., et al. (2001). The sequence of the human genome. *Science* 291, 1304–1351.

Verbavatz, J.M., Ma, T., Gobin, P., and Verkman, A.S. (1997). Absence of orthogonal arrays in kidney, brain, and muscle from transgenic knockout mice lacking aquaporin-4. *J. Cell Sci.* 110, 2855–2860.

Wallraff, A., Kohling, R., Heinemann, U., Theis, M., Willecke, K., and Steinhauser, C. (2006). The impact of astrocytic gap junctional coupling on potassium buffering in the hippocampus. *J. Neurosci.* 26, 5438–5447.

Wasseff, S.K., and Scherer, S.S. (2011). Cx32 and Cx47 mediate oligodendrocyte: astrocyte and oligodendrocyte: oligodendrocyte gap junction coupling. *Neurobiology of Disease* 42, 506–513.

Waxman, S.G., and Black, J.A. (1984). Freeze-fracture ultrastructure of the perinodal astrocyte and associated glial junctions. *Brain Res.* 308, 77–87.

Willecke, K., Eiberger, J., Degen, J., Eckardt, D., Romualdi, A., Gueldenagel, M., Deutsch, U., and Soehl, G. (2002). Structural and functional diversity of connexin genes in the mouse and human genome. *Biol. Chem.* 383, 725–737.

Wingerchuk, D.M., Lennon, V.A., Lucchinetti, C.F., Pittock, S.J., and Weinshenker, B.G. (2007). The spectrum of neuromyelitis optica. *Lancet Neurol.* 6, 805–815.

Wolf, N.I., Harting, I., Innes, A.M., Patzer, S., Zeitler, P., Schneider, A., Wolff, A., Baier, K., Zschocke, J., Ebinger, F., Boltshauser, E., and Rating, D. (2010). Ataxia, delayed dentition and hypomyelination: a novel leukoencephalopathy. *Neuropediatrics* 38, 64–70.

Zalc, B., Goujet, D., and Colman, D. (2008). The origin of the myelination program in vertebrates. *Curr. Biol.* 18, R511–R512.

Zhou, J., Kong, H., Hua, X., Xiao, M., Ding, J., and Hu, G. (2008). Altered blood-brain barrier integrity in adult aquaporin-4 knockout mice. *Neuroreport* 19, 1–5.

9 Microglia and Non-CNS Cells in Paracrine Signaling and CNS Immunity

Effects of Pathogens, Age, and Life History

Emma H. Wilson and Monica J. Carson
Division of Biomedical Sciences, Center for
Glial-Neuronal Interactions, University of
California Riverside, Riverside, California

CONTENTS

9.1   JUST ENOUGH IMMUNITY TO MAINTAIN A WORKING CNS

The central nervous system (CNS) is defined as an immune-privileged site (reviewed in Carson et al., 2006). This does not mean that the CNS is immune isolated or immune incompetent. Both CNS-intrinsic and CNS-extrinsic sources of immunity serve to defend the CNS from pathogens and to maintain optimal CNS function. CNS immune privilege refers to the tissue-specific regulation of both sources of immunity to provide sufficient immunity to limit the neurodestructive consequences of CNS pathogens without causing more CNS damage than the pathogen.

In this chapter, we will illustrate that neither CNS-intrinsic nor CNS-extrinsic immunity is by itself sufficient to defend *and* maintain optimal CNS function. We will also discuss the different roles played by each form of immunity and their regulation by paracrine (local) factors with particular focus on the regulatory roles of microglia and astrocytes. Finally, because CNS immunity is actively regulated, we will discuss how pathogens, age, and life history can alter or disrupt this regulation, sometimes with detrimental consequences for CNS function.

9.1.1   Too Much versus Too Little CNS Immunity

Too much immunity can be just as neurotoxic as a neurotrophic pathogen when it occurs within the CNS (reviewed in Carson et al., 2006). Too much immunity can even cause lethal brain damage after the pathogen has been effectively cleared! For example, acute postinfectious measles encephalomyelitis (APME) afflicts 0.1% of those infected with the measles virus (Poser, 1990; Huynh et al., 2008). APME is characterized by abrupt onset of fever, seizures, and multifocal neurological signs 5–14 days after appearance of the measles rash. These clinical features of APME correlate with robust influx of blood-derived macrophages and lymphocytes into the CNS. However, the measles virus is not highly neurotropic, nor can infectious virus be isolated from brain tissue or cerebral spinal fluid (CSF) of individuals with APME. Currently, the sum of the published data suggests that APME is an autoimmune (anti-CNS) disorder (Poser, 1990; Huynh et al., 2008). Specifically, it is presumed that the robust immune response that was effective in eliminating the measles virus outside of the CNS was also cross-reactive for molecules located in the CNS in 0.1% of individuals infected with measles. Thus, activation of a cross-reactive immune response during the clearance of the measles virus outside of the CNS can also lead to the influx of highly activated neurotoxic immune cells into the CNS. The extreme consequences of too much immunity are demonstrated by the 10–20% mortality rate in those with APME (Poser, 1990; Huynh et al., 2008).

Conversely, too little immunity can result in normally benign viruses, bacteria, and parasites lethally disrupting brain function. In this situation, brain damage or death can be due to competition for metabolic components, to disruption of CNS intracellular signaling by pathogen products, or to direct pathogen killing of CNS cells. For example, progressive infectious measles encephalitis can develop in children with acquired immunodeficiency due to chemotherapy or HIV-AIDS (Poser, 1990; Huynh et al., 2008). In these patients, measles virus replication is not sufficiently controlled and can be detected in brain tissue. Clinical presentation is associated with general progressive neurological deterioration, including seizures and altered mental status. Pathologic analysis of brain tissue reveals neuronal damage, astrogliosis, and inclusion bodies in both neurons and glia. In these cases, treatment is directed at halting viral replication with antiviral drug therapies (Poser, 1990; Huynh et al., 2008).

9.1.2 JUST ENOUGH CNS IMMUNITY

From this introductory discussion, the amount of immunity construed to be just enough might be articulated as robust enough to control pathogen spread to minimize damage to CNS neurons and glia, yet sufficiently limited in magnitude and time to minimize damage to CNS neurons and glia.

Chronic inflammation by definition is not limited in time, and low-grade chronic inflammation is often ascribed to be a causative or contributing factor to many neurodegenerative diseases. Consistent with the assumption that too much immunity also applies to the kinetics of CNS immunity is the example of subacute sclerosing measles panencephalitis (SSPE) (Poser, 1990; Huynh et al., 2008). SSPE is a very rare complication of measles infection and is associated with continual low-grade activation of the immune system. Clinical symptoms emerge 5–10 years after systemic clearance of the measles virus and include mental deterioration, appearance of myoclonus, optic atrophy, and akinetic mutism. Based on experimental models, it is speculated that the measles virus infects the brain at the time of the original infection in patients that develop SSPE. However, acute encephalitis does not occur because the viral clones infecting the CNS in SSPE patients have been shown to be defective, in that they are unable to generate budding viral particles from the surface of infected cells (Poser, 1990; Huynh et al., 2008). While chronic inflammation can still be documented in the CNS of SSPE patients, effective recruitment and reactivation of a measles-specific immune response within the CNS does not occur. Despite the persistent production of tumor necrosis factor (TNF) and interleukin (IL)-1β within the infected CNS and the accumulation of intrathecal oligoclonal IgG, the budding defective viral clones are able to spread and persist throughout the CNS (Poser, 1990; Huynh et al., 2008).

Although persistent chronic CNS inflammation is widely presumed to be the cause of many of the clinical symptoms of SSPE, this does not mean the CNS cannot tolerate some forms of well-regulated chronic inflammation. For example, approximately 30% of the world's population (and 10–20% of the U.S. population) is estimated to have chronic CNS infections of *Toxoplasma gondii* (Ferguson, 2009; Meerburg and Kijlstra, 2009). Data from experimental models and immune-deficient humans have clearly demonstrated that chronic CNS inflammation associated with production of interferon (IFN)-γ is absolutely required to prevent the obligate intracellular parasite *T. gondii* from causing necrotizing damage to the CNS (Noor et al., 2010; Harris et al., 2010; Wilson et al., 2006). While subtle changes in brain function may result from this type of chronic inflammation (see Section 9.4), the *T. gondii*–infected CNS of an immune-competent individual does not display overt clinical or histologic signs of neuropathology, such as those listed above for measles-associated diseases (Filisetti and Candolfi, 2004). In the following sections, we define the role of microglia and peripheral immune cells (non-CNS cells) in determining the types of acute and chronic CNS inflammation triggered by pathogens, injury, and tissue homeostasis. We also focus on defining the paracrine mechanisms within the CNS that regulate and trigger just enough of the right type of immunity to maintain CNS function in a nonsterile world.

9.2   THE NEUROIMMUNE CIRCUIT

CNS immunity is distinct from that occurring in most other tissues in the body (reviewed in Carson et al., 2006). Most notably, foreign grafts placed in the parenchyma of the CNS avoid acute immune rejection and exhibit prolonged survival compared to foreign grafts placed under the skin or in other non-CNS sites. This phenomenon is a feature only of the CNS parenchyma. Grafts placed in nonparenchymal sites such as in the ventricles or meninges are readily rejected by the immune system (reviewed in Carson et al., 2006).

For decades CNS immune privilege was viewed as a passive phenomenon resulting from

1. The absence of a draining lymphatic system
2. The absence of an effective tissue macrophage/immature dendritic cell able to traffic to draining lymph nodes and activate antigen-specific T cells
3. The presence of a blood–brain-barrier

However, the current scientific literature has demonstrated that the peripheral immune system can rapidly detect and accumulate at antigenic deposits in CNS parenchyma if the immune system is activated outside the CNS (reviewed in Carson et al., 2006). For example, if foreign grafts are placed at two sites, one inside and outside the CNS, both grafts will be rapidly attacked and rejected by the peripheral immune system. These and similar observations reveal that CNS immune privilege is not a passive process based on structural barriers. Instead, it is an actively regulated phenomenon with a CNS-intrinsic component that can be hijacked by pathogens, altered by environmental toxins, and by aged associated changes in paracrine regulatory mechanisms.

9.2.1   THE PLAYERS IN THE NEUROIMMUNE CIRCUIT

CNS immunity is a composite of interactions between CNS-resident cells (CNS-intrinsic immunity) and peripheral immune cells (CNS-extrinsic immunity). However, determination of which CNS-resident cells are immunologically active and whether a cell is truly CNS resident and part of the CNS-intrinsic immunity is a subject of ongoing debate.

9.2.1.1 CNS-Resident Cells

Neurons and macroglia (astrocytes and oligodendrocytes) are unquestionably CNS-resident cells. However, until recently neurons and macroglia were not recognized to play biologically significant roles in CNS immunity (Biber et al., 2007; Carson and Lo, 2007). Conversely, microglia are the primary tissue macrophage of the CNS (Graeber and Streit, 2010; Carson et al., 2007; Pocock and Kettenmann, 2007; Raivich, 2005). As such they are presumed to be immunologically active and to have the potential to develop effector functions common to other non-CNS macrophage populations (Graeber and Streit, 2010; Carson et al., 2007).

Microglia are found in all regions of the CNS by at least as early as embryonic day 13 in the rodent CNS (Graeber and Streit, 2010; Carson et al., 2007). Morphologically, microglia display a stellate morphology in the healthy noninflamed adult CNS, but can develop an ameboid morphology in response to pathogenic stimuli. In CNS tissue sections, it is impossible to conclusively identify a cell as a CNS-resident microglia vs. an acutely infiltrating blood-derived macrophage because microglia express the same macrophage cell-type-specific markers (such as Iba1, CD11b, and F4/80) as other macrophages. Adoptive transfer of labeled monocytes and macrophages has demonstrated that non-CNS macrophages can enter the CNS and even acquire the stellate morphology characteristic of microglia (Graeber and Streit, 2010; Carson et al., 2007). Conversely, activated microglia can acquire the amoeboid morphology often ascribed to infiltrating macrophages. Therefore, it is not uncommon for many studies to label all macrophage-like cells found in the CNS as microglia.

9.2.1.2 CNS-Infiltrating Peripheral Immune Cells: Macrophages

Two types of studies have revealed that fundamental differences exist between CNS-resident microglia and blood-derived CNS-infiltrating macrophages, and that they should be considered two distinct populations. First, flow cytometric analysis of CNS cell suspensions prepared from perfused CNS cleared of trapped blood has revealed that 95% of CNS cells expressing macrophage markers also express low levels of the molecule referred to as cluster of differentiation (CD)45. The low level of CD45 is abbreviated as CD45lo, also known as leukocyte common antigen (Carson et al., 2007). By contrast, all other macrophage populations in non-CNS sites, and indeed all other nucleated cells of hematopoietic origin, express an order of magnitude higher levels of the molecule referred to as cluster of differentiaion (CD)45. The low level of CD45 is abbreviated as CD45lo.

Second, irradiation chimeric studies have demonstrated that the CD45lo population in the CNS is not replaced by bone-marrow-derived cells with the same kinetics as other tissue macrophages (Byram et al., 2004; Garg et al., 2009; Hickey and Kimura, 1988). In these studies, rodents were lethally irradiated with doses sufficient to kill the bone marrow but low enough not to cause nonspecific radiation-induced systemic inflammation. Irradiated rodents were then supplemented with bone marrow from a genetically distinct source. For example, bone marrow from mice expressing the CD45.2 allele is transferred into irradiated mice expressing the CD45.1 allele. In these chimeric mice, non-bone-marrow-derived tissue is of the recipient genotype (in this example CD45.1), while bone-marrow-derived cells are of the donor genotype (in this example CD45.2). Within a few weeks all macrophage populations in nearly all tissues of the body are found to express the donor phenotype. The primary exception is the CNS. Here ~90–95% of the macrophages located in the CNS express the host genotype, indicating they were not replaced by bone-marrow-derived cells. The low percentage of macrophage cells found to be readily replaced by bone-marrow-derived cells was preferentially enriched in perivascular spaces and found to be preferentially of the CD45hi phenotype (Byram et al., 2004; Garg et al., 2009; Hickey and Kimura, 1988).

The failure of bone-marrow-derived cells to repopulate the CNS even after extended periods of time is not an artifact of irradiation or specific to rodent physiology. Parabiosis studies that link the vascular system of two congenic mice (for example, CD45.1 and CD45.2, GFP+ and GFP−) have confirmed the observations made in irradiation chimeric mice (Ajami et al., 2007). In parabiotically linked rodents, both macrophage genotypes were detected in all tissues with the exception of the CNS. In the CNS, tissue macrophages (microglia) retained their original genotype. When humans receive bone marrow transplant therapy, they must first receive lethal irradiation to kill their own host bone marrow (reviewed in Carson et al., 2006). Occasionally there is a gender mismatch such that female hosts receive male bone marrow. When organ tissue is analyzed at autopsy years after bone marrow therapy, male-derived macrophages can be detected in most tissues but not in the CNS parenchyma. These bone marrow transplant data replicate in humans the irradiation chimeric data generated in rodent models.

A major function of nearly all tissue macrophages/immature tissue dendritic cells as well as classically activated inflammatory macrophages is the capture of antigens within the tissue followed by emigration out of the tissue into the draining lymph node (Lo et al., 1999). Adoptive transfer studies of dye-labeled microglia and macrophages into the CNS of nonirradiated hosts have demonstrated that transferred CD45lo microglia can migrate great distances within the CNS. However, they cannot be detected to exit the CNS and travel to the draining cervical lymph nodes (Carson et al., 1999). Indeed, up to 2 weeks posttransfer, nearly all of the transferred CD45lo microglia are retained in the healthy murine CNS. By contrast, CD45hi macrophages transferred into healthy murine CNS can be detected within the draining cervical lymph nodes within 24 hours of transfer. Indeed, nearly all of the transferred CD45hi cells either migrate to draining lymph nodes or die (often within the CNS) within 72 hours postinjection into the healthy adult CNS (Carson et al., 1999).

Gene profiling studies confirmed that these two cell types display distinct cell-type-specific gene expression patterns even when isolated from the same inflamed CNS environment (Schmid et al., 2002, 2009; Carson et al., 2004). Molecules completely unique to either cell type have not yet been identified. Furthermore, microglial activation does increase microglial expression of many macrophage markers (MHC class II, CD40, FcR, B7.2), further confounding the ability to distinguish microglia from macrophages based on activation markers. However, activated microglia acutely isolated from lipopolysaccharide (LPS)-injected CNS do express higher levels of the anti-inflammatory receptor, triggering receptor expressed on myeloid cells-2 (TREM2), while CNS-infiltrating macrophages express higher levels of the pro-inflammatory receptor, TREM1 (Schmid et al., 2002, 2009; Carson et al., 2004). In addition, CNS-infiltrating macrophages express nearly 10-fold higher levels of golli-myelin basic protein (golli-MBP) (Papenfuss et al., 2007). As yet, the functional consequences of preferential golli-MBP expression in macrophages are unresolved; nevertheless, it may provide a useful cell-type-specific marker.

Taken together, these studies illustrate that two functionally distinct macrophage populations serve the CNS: a long-lived or self-renewing population of CD45lo CNS-resident microglia (the majority of macrophage-like cells in the CNS) and a short-lived population of macrophages that is readily replaced by bone marrow (Carson et al., 2006, Ajami et al., 2007). The CD45lo microglial population is larger, comprising at least 90% of the macrophage-like cells in the CNS, while the CD45hi macrophage population comprises only 10% of the total macrophage-like cells in the healthy CNS. It should be stressed that the enrichment of these CD45hi cells at perivascular spaces coupled with their ability to capture antigen and migrate to draining lymph nodes to activated antigen-specific T cells, allows the immune system to routinely survey CNS antigens during health, CNS injury, and pathogenic attack (Hickey and Kimura, 1988).

9.2.1.3   Additional CNS-Infiltrating Peripheral Immune Cells

In the healthy CNS, few lymphocytes are detected in the CNS. However, low numbers of T cells displaying a memory phenotype can be detected in the CSF of healthy individuals (Wilson et al., 2010). At present there is much speculation on the biological significance of these observations, but this phenomenon may represent ongoing immune surveillance of parenchymal CNS antigens that have drained into the CSF (Wilson et al., 2010).

In the injured or infected CNS, nearly all types of immune cells have been detected in the CNS (Müller et al., 2007; Odoardi et al., 2007). Lymphocytes (CD4 T cells, CD8 T cells, B cells), macrophages, neutrophils, mast cells, and NK T and NK cells are all present in specific forms of neuroinflammation (Wilson et al., 2010). Their entry into the CNS is dependent on activation-induced expression of adhesion molecules that allow for adhesion and extravasation. Lastly, it should be noted that eosinophils and basophils are only rarely observed in CNS inflammatory responses in human diseases and in animal experimental models.

Immune cell entry into the CNS is highly regulated (Wilson et al., 2010). Most naïve leukocytes do not express sufficiently high levels of adhesion molecules to bind and extravasate from the cerebral vasculature into the CNS parenchyma. In the case of T cells, there is a growing body of evidence that suggests that CD8 T cells primarily enter the CNS parenchyma using antigen-dependent mechanisms (Bajénoff et al., 2006). In particular, recent studies using multiphoton imaging have demonstrated both antigen-dependent scanning behavior by CD8 T cells when at the CNS pial membrane and exclusion of non-antigen-specific CD8+ T cells from the CNS parenchyma (Kim et al., 2010; Bartholomäus et al., 2009; Wilson et al., 2009, 2010).

At present, the bulk of current data indicates that antigen-dependent CNS infiltration may be restricted to CD8 T cells. Studies using CD4+ T cells with defined antigenic specificity (using either T cell receptor (TCR) transgenic mice or tetramer analysis of TCR specificity) have demonstrated that CD4 T cells readily gain entry into the CNS during ongoing inflammatory events, even if their target antigen is not in the CNS (McGavern and Truong, 2004). In part, this is a consequence of the multiple mechanisms able to induce elevated expression on CD4 T cells of adhesion molecules necessary for extravasation. Most notably C-C Motif Ligand (CCL) 21 drives homeostatic (antigen-independent) proliferation of CD4, but not CD8 T cells (Ploix et al., 2001). With each round of CCL21-driven homeostatic proliferation, CD4 T cells further increase their expression of Lenkocyte Function Associated Molecule 1 LFA-1 and Very late antigen-4 (VLA4), the two adhesion molecules required for CD4 extravasation and infiltration into the CNS. Strikingly, CCL21 expression is robustly induced during the early stages of nearly all neuroinflammatory disorders associated with CD4 T cell infiltration (reviewed in Ploix et al., 2010).

However, CCL21 expression by itself is unlikely to be sufficient for T cell entry into the CNS. Transgenic expression of CCL21 in non-CNS tissues is sufficient to recruit T cells into the CNS and initiate their organization into lymph node-like structures (Ploix et al., 2001). By contrast, T cells are not recruited to the CNS of transgenic mice that express CCL21 in oligodendrocytes (MBP-CCL21) or astrocytes (GFAP-CCL21) (Chen et al., 2002; Ploix et al., 2010). This failure to enter the CNS is unlikely due to the presence of the blood–brain barrier because neutrophils accumulate in the MBP-CCL21 CNS and damage the blood–brain barrier. These data are consistent with the unique failure of immune cells to organize in lymph-node-like structures in any human neuroinflammatory disorder, including the autoimmune disorder multiple sclerosis (reviewed in Carson et al., 2006).

9.2.2   THE PLASTICITY OF IMMUNE CHOICES

Once recruited to the CNS, immune cells can develop a broad array of pro-inflammatory or anti-inflammatory phenotypes (Goverman, 2009; Siffrin et al., 2007; Korn et al., 2007; Lo et al., 1999). A complete review of immune cell effector functions and lymphocytic differentiation lineages is beyond the scope of this chapter. Here we will focus on the primary phenotypes developed by CNS-resident microglia and CNS-infiltrating macrophages for the simple reason that these cells do more than provide innate first-response defenses. Microglia and macrophages also act as antigen-presenting cells able to direct the function of CNS-infiltrating T cells. In addition, one of the primary consequences of T cell activation is to direct the subsequent activation and function of microglia and macrophages.

Antigen presentation is therefore a critical paracrine regulatory step in CNS immunity (reviewed in Lo et al., 1999), but what is it? The T cell receptor on T cells cannot bind free antigen. Instead, an antigen-presenting cell such as a microglia, macrophage, or immature dendritic cell captures the free, debris-, cell-, or pathogen-associated antigen and then presents it to the T cell receptor in the pocket of MHC. CD4 T cells recognize the antigen in the pocket of MHC class II, while CD8 T cells recognize the antigen in the pocket of MHC class I. Antigens presented by MHC class II are captured during phagocytosis, while antigens presented by MHC class I are synthesized by the protein synthetic machinery within the antigen-presenting cell.

Most cells do not constitutively express high levels of MHC class I, unless they are infected by intracellular pathogens (for example, viruses and *T. gondii*), severely stressed or preapoptotic (reviewed in Lo et al., 1999). If infected by an intracellular pathogen, the protein synthetic machinery will be producing both self and pathogen peptides, both of which will be loaded into the pocket of the MHC class I molecule as it travels to the plasma membrane. Once activated by antigen presentation, CD8 T cells kill the presumed infected or severely damaged MHC class I–expressing cell.

While all cells, including neurons, can be induced to produce MHC class I, MHC class II expression is normally limited to cells with primary immune functions (reviewed in Lo et al., 1999). MHC class II–expressing cells include microglia, macrophages, dendritic cells, and B cells. Following antigen presentation, CD4 T cells can develop a broad array of pro-inflammatory, neuroprotective, or even anti-inflammatory phenotypes. The type of phenotype generated is largely dependent on the activation state of the antigen-presenting cell: classical, alternative, or regulatory (Gordon and Martinez, 2010; Colton and Wilcock, 2009).

9.2.2.1 Classical Pro-Inflammatory Pathogen Defense Activation States

Both CNS-resident microglia and CNS-infiltrating macrophages can become classically activated (Gordon and Martinez, 2010; Colton and Wilcock, 2009). The characteristics of classical activation are elevated expression of molecules required to present antigens to T cells and drive pro-inflammatory T helper (Th)-1 differentiation, namely, expression of high levels of MHC class II and pro-inflammatory costimulatory molecules, CD40 and B7.2. Pro-inflammatory cytokines produced by activated Th1 cells such as IFNγ further prime and enhance polarized classical microglia and macrophage activation characterized by production of high levels of cytotoxic products (free radicals, NO), pro-inflammatory cytokines (TNF, IL-1b, IL-6), and extracellular degrading enzymes serving to facilitate immune cell infiltration throughout the inflamed CNS. In addition, classically activated macrophages express high levels of TREM1 (Ford and McVicar, 2009; Lanier, 2009). TREM1 is a receptor that acts to amplify the intracellular signaling of Toll-like receptor (TLR)-4, the pro-inflammatory receptor for LPS (a component of gram-negative bacteria), and thus amplifies the magnitude of classical pro-inflammatory immune responses (Ford and McVicar, 2009; Lanier, 2009). Strikingly, even when displaying robust classical activation states, microglia isolated from adult murine CNS express low to undetectable levels of TREM1 on their cell surface (Schmid et al., 2009).

9.2.2.2 Alternative Wound Healing Activation States

As with classical activation, both microglia and macrophages can become alternatively activated (Gordon and Martinez, 2010; Colton and Wilcock, 2009). Alternative activation states are defined by high expression of macrophage mannose receptor (CD206), Fizz1 (found in inflammatory zone 1), YM1, as well as anti-inflammatory cytokines such as tumor growth factor (TGF)-β and IL-10. Alternatively activated microglia also express high levels of arginase I and TREM2. The high levels of arginase I expression decrease cytotoxic nitric oxide production while simultaneously increasing the production of polyamines, which are well described to promote cell proliferation and tissue growth. Alternatively, activated microglia and macrophages can still act as highly effective antigen-presenting cells. However, unlike classically activated antigen-presenting cells, alternatively activated cells serve to decrease

Th1 T cell activation and promote neuroprotective, growth-factor-producing T cell activation states (Gordon and Martinez, 2010; Colton and Wilcock, 2009).

Even when developing polarized activation states, CNS-infiltrating macrophages express 10- to 30-fold lower levels of TREM2 than similarly activated CNS-resident microglia (Schmid et al., 2009, Melchior et al., 2010). The significance of this differential TREM2 expression is illustrated by three observations. First, increasing TREM2 expression in microglia increases their phagocytic and antigen-presenting cell activity without increasing CD40 or T cell IFNγ production (Melchior et al., 2010). Second, introduction of TREM2 into the peripheral immune system did not alter the onset of experimentally induced autoimmune encephalomyelitis (EAE). However, overexpression of TREM2 in peripheral immune cells did dramatically increase the rate CNS inflammation resolved, as well as the rate and extent of CNS tissue repair. Third, using irradiation chimeric models and serial adoptive transfer of activated macrophages, peripheral antigen-presenting cells were essential to initiate anti-CNS responses following facial axotomy (Byram et al., 2004). However, TREM2+ microglia were absolutely essential to evoke and to sustain the neuroprotective CD 4 T cell responses that slowed CNS motoneuron death following facial axotomy (Byram et al., 2004).

In the CNS, only microglia express TREM2 (Schmid et al., 2002; Thrash et al., 2009). Humans lacking functional TREM2 pathways develop early-onset cognitive dementia called Nasu-Hakola disease characterized by emotional and suicidal impulses (Klünemann et al., 2005). Thus, a primary genetic defect in microglia leads to a form of classic neurologic and not neuroinflammatory neurodegenerative disease. It is tempting to speculate that the clinical neurologic symptoms result from the failure to generate neuroprotective T cell responses required to support optimal neuronal function during the "wear and tear" of normal aging. Consistent with this speculation, T-cell-deficient rodents have defects in hippocampal memory tasks and display greater stress-induced deficits in learning than immunocompetent rodents (Derecki et al., 2010; Kipnis et al., 2008; Brynskikh et al., 2008). These deficits can be reversed by transfer of CD4 T cell populations. Recent experiments demonstrate the neuroprotective role of CD4 T cells in learning is dependent on antigen presentation by alternatively activated microglia/macrophages.

Neuroprotection and promotion of tissue repair are not always the most appropriate responses to maintain optimal CNS function. For example, with respect to astrogliomas and glioblastoma, a high degree of microglial or macrophage influx is associated with poor prognosis for tumor eradication (Yang et al., 2010). When examined ex vivo, microglial/macrophage phenotypes of tumor-associated cells are of a mixed alternative, regulatory phenotype.

9.2.2.3 Regulatory Anti-Inflammatory/Tolerogenic Activation States

Antigen-presenting cells (APCs) can also develop regulatory phenotypes that direct T cell activation toward T cell death, T cell anergy (inactivation), or generation of anti-inflammatory T-regulatory (Treg) CD4 T cell functions (Lo et al., 1999; Carson et al., 2006). Regulatory APC function is characterized by expression of anti-inflammatory coinhibitory molecules such as programmed death ligand (PD-L)-1 and PD-L2, or by the absence of pro-inflammatory costimulatory molecules.

Recently, a screen designed to identify molecules enriched on macrophages in CNS tissue revealed that following CNS infiltration, blood-derived macrophages are induced to express the tolerance-related transcript, Tmem176b (Schmid et al., 2009). CNS microglia also increase Tmem176b expression during the self-resolving neuroinflammatory insults. Tmem176b is a tetraspan molecule located on a variety of intracellular membranes (Condamine et al., 2010; Louvet et al., 2005). Overexpression of Tmem176b prevents immature dendritic cells from acquiring mature pro-inflammatory antigen-presenting cell function and instead promotes their acquisition of regulatory, tolerogenic antigen-presenting cell function (Condamine et al., 2010; Louvet et al., 2005). Overexpression of Tmem176b in microglial and macrophage cell lines also promoted cell death of activated cells (Melchior et al., 2010). Thus, Tmem176b plays dual roles in driving anti-inflammatory activation states: promotion of regulatory macrophage function and the elimination of activated microglia and macrophages.

9.3   ASTROCYTES AS ACTIVE REGULATORS OF CNS IMMUNITY ± NEURONS

From the discussion so far, it is apparent that the CNS can modulate both CNS-intrinsic and -extrinsic immunity. While in this chapter we focus on the regulatory roles played by astrocytes and microglia, it is also important to note the dominant immunoregulatory played by CNS neurons. Neurons express a large number of molecules that inhibit microglia and macrophage activation (reviewed in Biber et al., 2007). For example, fractalkine, CD200, and CD22 are all expressed on the surface of CNS neurons, while their receptors are expressed on unactivated and activated microglia and macrophages. Deletion of these molecules or their receptors on microglia and macrophages primes both microglia and macrophages to develop classical activation states more rapidly than wild-type cells and in response to lower doses of pro-inflammatory stimuli.

The CD200 receptor is expressed at similar levels on both microglia and macrophages (Koning et al., 2009; Banerjee and Dick, 2004). Thus, neuronal regulation of both microglia and macrophages is similar and coordinate via this pathway. By contrast, CX3CR1, the fractalkine receptor, is expressed at higher levels on microglia (Cardona et al., 2008; Sunnemark et al., 2005), while CD45, the receptor for CD22, is expressed at higher levels on CNS-infiltrating macrophages (Mott et al., 2004). Thus, neurons also can simultaneously but separately regulate microglia and macrophage functions using cell-type-specific mechanisms. Lastly, microglial function is also regulated by neurotransmitter exposure. For example, glutamate amplifies classical activation responses while norepinephrine reduces these responses. These types of data demonstrate the potential for the types and degree of neuronal activity to regulate microglial immune responses (Pocock and Kettenmann, 2007).

9.3.1   CHEMOKINES AND LEUKOCYTE ENTRY INTO THE CNS

CNS-resident microglia enter the CNS early in development (reviewed in Carson et al., 2007). Irradiation chimeric and parabiosis models demonstrate that in the adult

CNS, peripheral monocytes and macrophages do not contribute to the replenishment or maintenance of CNS-resident microglia. It is unclear when in postnatal development these populations become distinct. However, in the adult, influx and retention of large numbers of peripheral monocytes and macrophages within the CNS are dependent on chemokine production. Simple disruption of the blood–brain barrier is by itself insufficient to cause robust recruitment and retention of blood-derived macrophages. The primary chemokine implicated in recruitment and retention of macrophages and monocytes is CCL2 (Ge et al., 2009). While microglia and CNS endothelial cells can produce CCL2 during neuroinflammation, astrocytes are the primary source of CCL2. Indeed, astrocytic production of CCL2 is essential for macrophage influx following acute bouts of systemic inflammation or following direct injury or infection of the CNS (Ge et al., 2009).

Recent studies investigating the migration of specific subsets of T cells have highlighted the important and specific role of chemokines and their receptors. Thus, Reboldi and colleagues demonstrated that the secretion of CCL20 by choroid plexus epithelium and the expression of its receptor, CCR6, were required for the entry of the pro-inflammatory CD4+ TH17 subset to initiate an autoimmune inflammatory response (Reboldi et al., 2009). This initial entry process into the CSF seems to be independent of astrocyte activation status. However, further migration across the astroglia limitans that form part of the blood–brain barrier is also required for entry into the actual CNS parenchyma.

A further level of control lies in the secretion of chemokines into the perivascular space and the presence of receptors on the surface of migrating cells. CXCR3 is the receptor for CXCL10. Several studies suggest that the secretion of CXCL10 by astrocytes at the blood–brain barrier acts to limit pathology during autoimmune inflammation by restricting CXCR3+ T cells to the perivascular space (Müller et al., 2007). Both T regulatory cells and effector T cells express CXCR3. Therefore, the effect of astrocytic expression of CXCL10 at the blood–brain barrier is synergistic. It allows increased interaction of the anti-inflammatory CD4 Treg population with the pro-inflammatory effector T cell population that they inhibit.

9.3.2 Leukocyte Migration into and within the CNS

Leukocyte infiltration of the CNS is a two-step procedure. As just discussed in the previous section, immune cells must first extravasate from the blood into the perivascular or extraparenchymal (submeningeal and ventricular) spaces of the CNS. Extravasated immune cells may then either remain and accumulate in these extraparenchymal sites or infiltrate the parenchyma. The regulation of this second step, migration from extraparenchymal spaces into and within the CNS parenchyma, is an area of active research and debate, but is likely multifactoral (reviewed in Wilson et al., 2010).

The interactive roles of astrocytes, microglia, and non-CNS in regulating the two-step nature of leukocyte infiltration and dissemination throughout the CNS have been demonstrated in two different models of brain infection. In response to injury, infection, and autoimmune triggers, astrocytes and microglia begin to express high levels of the monocyte/macrophage chemoattractant protein, CCL2.

In response to the same insults, astrocytes and neurons increase expression of the T cell chemoattractant, CCL21. In the first study, lymphocyte influx from perivascular spaces across the astrocytic glial limitans into the CNS parenchyma was studied in mice infected with a neurotropic coronavirus (Sauvin et al., 2010). Here the authors demonstrated that in the absence of CCL2 expression within the CNS, monocytes were not recruited to the infected brain. In the absence of CNS-infiltrating monocytes, lymphocytes could not cross the glial limitans and migrate into the infected CNS. Instead, lymphocytes accumulated in the perivascular regions of infected mice.

In the second study, lymphocyte influx from perivascular spaces across the astrocytic glial limitans into the CNS parenchyma was studied in mice infected with a *Toxoplasma gondii* (Ploix et al., 2010). Using CCL21 deletion and CNS-overexpressing mouse lines, this study demonstrated that CNS-expressed CCL21 was not required for extravasation of lymphocytes into perivascular spaces. Although monocytes, macrophages, and CD8+ T cells did infiltrate the CNS parenchyma of infected CCL21-deficient mice, CD4+ T cells could not. CD4+ T cells accumulated in perivascular, ventricular, and submeningeal sites of *Toxoplasma*-infected CCL21-deficient brains. Another notable feature of this study was the observation that despite a continual influx and accumulation of lymphocytes into the infected brain, constitutive overexpression of CCL21 failed to lead to the organization of inflammatory infiltrates into neolymphoid structures observed in other inflamed CCL21-expressing tissues. Considered together, these two studies reveal the existence of multiple levels of CNS-intrinsic regulation controlling the influx of distinct populations of immune cells into the CNS.

Multiphoton microscopy of inflamed brain tissue has provided suggestive data of one additional level of regulation. In response to pathogen infection or autoimmunity, once lymphocytes have crossed the glial limitans and entered the CNS parenchyma, they do not migrate randomly throughout the CNS. Lymphocyte migration within the CNS parenchyma is closely associated with a reticular network of fibers visualized by two-photon microscopy. The molecular identity of this network is unknown, but it is not present in the noninflamed brain. Similar reticular lymphocyte migratory fibers have been detected in lymph nodes by two-photon microscopy. Collagen has been identified as the primary component of this fiber network in the lymph node. The T cell chemokines, CCL19 and CCL21, are bound to this network and promote lymphocyte migration along the collagen fiber network.

Within the CNS, CCL21 immunoreactivity is associated with the reticular fiber network visualized by two-photon microscopy. However, collagen is not a component of this reticular lymphocyte migratory pathway in the CNS. Although the molecular identify of the CCL21-coated fiber network is unknown, the secreted protein acidic rich in cysteine (SPARC) has been identified as a candidate molecule. SPARC is well defined to bind extracellular molecules and is currently being tested for its ability to bind and present CCL21 to lymphocytes (Noor and Wilson, studies in progress). In normal development SPARC is required for CNS cell migration in developing pre- and postnatal brain (Weaver et al., 2010; Sullivan et al., 2008). SPARC is also associated with increased metastasis of cancerous cells and angiogenesis (Clark and Sage, 2008; Zhang et al., 2009). Importantly, glia upregulate SPARC expression in

Toxoplasma-infected brains in the regions associated with lymphocyte migration (Wilson and Carson, unpublished observations).

9.3.3 MICROGLIAL AND MACROPHAGE ACTIVATION: TREM2 AND TMEM176B

Astrocytes also contribute to the regulation of CNS-infiltrating T cells by regulating the phenotype of microglia and macrophages. As noted in Section 9.2.2.2, macrophages infiltrating the CNS are induced to express the tolerance-related transcript, Tmem176b, at levels similar to that of microglia. Coculture experiments with macrophages and glia revealed that activated astrocytes are the cell types inducing Tmem176b expression in both microglia and CNS-infiltrating macrophages (Schmid et al., 2009).

TREM2 expression is highly induced in amyloid-associated microglia. However, even with extended exposure, neither soluble nor aggregated amyloid was sufficient to increase microglial expression of TREM2 (Melchior et al., 2010). Similar to the studies on Tmem176b, coculture studies revealed that astrocytes were essential to mediate amyloid-associated increases in TREM2 expression by microglia (Melchior et al., 2010). Once TREM2 expression is induced, activation of microglia via the TREM2 pathway also appears to be largely astrocyte dependent. Although TREM2 is an orphan receptor, putative TREM2 ligands are detected using a fusion protein of the human IgG Fc region with the extracellular ligand-binding domain of TREM2 (TREM2-Fc). TREM2-Fc only detects TREM2 binding activity on activated astrocytes in CNS histologic sections (Melchior et al., 2010).

9.4 HIJACKED IMMUNITY

The prior discussions in this chapter demonstrate the ability of the CNS to titrate and direct the type of immunity required to limit pathogen damage to the CNS and promote CNS repair and function. In limiting CNS immunity to just enough, the result is that chronic infections are not uncommon in the CNS. For example, ~70% of individuals in the United States are estimated to be infected with the John Cunningham (JC) polyoma virus. In immunocompetent individuals, JC virus causes no problems. Similarly, recent estimates indicate that *Toxoplasma gondii* is present in at least 10% of those in North America. Up to 80% of the population are seropositive in areas of Europe and South America (Filisetti and Candolfi, 2004; Ferguson, 2009). *Toxoplasma* leads to lifelong infection because it cannot be eliminated from the brain. Chronic infection appears well controlled and well tolerated in immune-competent individuals. Reactivation of latent infection is a major risk for individuals with acquired immune deficiency. Indeed, even just decreasing the ability of T cells to enter the CNS due to biotherapies such as Tysabri, which are directed at blocking VLA4 function, can trigger lethal JC-mediated destruction of CNS neurons and glia (reviewed in Carson et al., 2006). These types of observations stress the importance of continuous immune surveillance of the CNS.

Despite the presence of chronic inflammation and chronic IFNγ in the *T. gondii*–infected CNS, no neuropathology is detected in the immune-competent host. However, the consequences of continuous accumulation of peripheral immune cells in the brain parenchyma do have the potential to induce as yet unknown disease.

There has been a growing body of correlative data that suggests that infection with *T. gondii* and other inflammatory events may be an underlying trigger for mental health disorders such as depression and schizophrenia and even chronic pain (Torrey and Yolken, 2003; Yolken et al., 2009; Prandota, 2010).

In support of these correlative data, clear changes in behavior of *T. gondii–*infected mice are well documented. The parasite's definitive host is the cat, and it undergoes sexual replication only in the epithelial cells of the cat's gastrointestinal tract. As an evolved behavior, mice avoid the scent of cat urine; however, a mouse with *Toxoplasma* cysts in its brain will run toward cat urine, thereby likely facilitating parasite transmission to its definitive host (Vyas et al., 2007). *T. gondii* cysts are found intracellularly in neurons in the brain and predominantly in the frontal cortex and amygdala, regions associated with control of the mouse behaviors just discussed (Ferguson, 2009). However, the mechanisms mediating *T. gondii–*induced behavioral changes are unknown. Whether this is a direct change in the function of neurons due to their infection or as a secondary consequence of glial activation and *T. gondii–*directed inflammation is still to be addressed.

9.5 THE CONSEQUENCES OF ENVIRONMENT, EXPERIENCE, AND AGE

This chapter has reviewed multiple paracrine interactions that direct and titrate CNS immunity, focusing on effector functions of microglia and non-CNS cells, such as CNS-infiltrating macrophages and lymphocytes. While the focus of this chapter was on pathogens and injury, environmental factors can also alter the normal balanced regulation of CNS immunity. For example, inhalation exposure to ultra-fine particulate matter isolated from actual Southern Californian highway pollution was sufficient to promote long-term classical activation in adult rodent CNS (Gerlofs-Nijland et al., 2010; Herbert, 2010; Campbell et al., 2009; Block and Calderón-Garcidueñas, 2009). Aberrant neuronal activity can also lead to long-term changes in CNS immunity. For example, triggering either acute or chronic epilepsy by kainic acid injections surprisingly promotes high TREM2 expression and alternative activation of microglia of the adult murine CNS (Melchior et al., 2006).

CNS immunity is also not constant during normal aging. In young postnatal rodent CNS at ages equivalent to human infancy, microglia display highly activated morphologies and express at least 30-fold higher levels of TREM2 than microglia in young adult CNS (Thrash et al., 2009). However, only very low levels of TREM2 binding activity can be detected, suggesting that cells are primed to respond rapidly to TREM2 signals in the developing young CNS, perhaps to facilitate repair during this critical developmental window. Conversely, in aged (>24 months old) mice, TREM2 binding activity can be seen throughout the CNS, suggesting chronic TREM2-mediated activation of microglia (Melchior et al., 2010). Perhaps this type of chronic microglial activation is necessary to maintain optimal function of aging neurons. Alternatively or in addition, chronic TREM2 activation is required to offset the increasing potential for microglia to develop classical activation states and to become nonfunctional (senescent) in the noninflamed, aged CNS (Sierra et al., 2007; Graeber and Streit, 2010).

9.6 CONCLUSIONS AND SPECULATIONS

The data discussed throughout this chapter have demonstrated that the CNS is not immune isolated, immune inactive, or immune incompetent. Instead, the data discussed illustrate active regulation and tailoring of CNS immunity by neurons and glia to support optimal CNS function. Thus, rather than view the CNS as a *non*-immune-privileged site because it is now recognized to be immune active, we conclude the CNS *is* immune privileged because of the unique manner in which it supports and directs CNS immunity during pathogen defense and tissue maintenance/regeneration.

9.6.1 CONCLUSIONS: THE ROAD TO JUST ENOUGH CNS IMMUNITY

Much of what has been presented in this chapter is the evaluation of what is involved in immune privilege. It is not isolation or segregation from the immune system. Like all tissues in the body, the CNS is in ongoing contact with immune cells. Notably, it is populated by a resident population of tissue macrophages (microglia) and is defended by blood-derived CNS-infiltrating immune cells during pathogen defense. However, there are many tissue-specific features for the resident macrophage population, as well as many tissue-specific mechanisms regulating leukocyte influx and activation in the CNS (Carson et al., 2006; Wilson et al., 2010).

Among the former CNS-specific features, the most notable are that microglia are a terminally differentiated macrophage that can still proliferate but are also long lived. Furthermore, unlike all other tissue macrophages, they are not detected to emigrate from the CNS to the draining lymph nodes during health or disease. Among the latter CNS-specific regulatory mechanisms, it is notable that influx of blood-derived monocytes/macrophages, CD8+ T cells, and CD4+ T cells is separately controlled. Furthermore, their migration and activation within the CNS is also regulated in a CNS-specific manner, with the CNS being refractory to organization of inflammatory infiltrates into lymphoid-like structures (Aloisi and Pujol-Borrell 2006; Ploix et al., 2010). One consequence of the inability to support organization of inflammatory infiltrates into lymphoid-like structures is that the regulatory influences of neurons and glia on inflammatory cells are not as diluted as in other tissues.

9.6.2 SPECULATION: COULD ASTROCYTES, NEURONS, AND MICROGLIA ACT TOGETHER TO PROVIDE A CNS FORM OF INNATE IMMUNE MEMORY?

Traditionally, the peripheral immune system is viewed as comprised of two complementary parts:

1. An innate immune system comprised of myeloid cells and other nonlymphocytes that serve as the first responders and has no immunologic memory of pathogens previously encountered. Innate immune cells have no memory because they are short lived and do not generate progeny that are prepolarized to respond to a previously encountered pathogen. In humans, most myeloid cells live days to a couple of weeks (Lo et al., 1999).

2. An adaptive immune system (comprised of lymphocytes) that serves as a slow response that has immunologic memory of specific pathogens encountered by an individual. Adaptive immune cells generate memory because only the cells that recognize and respond to a pathogen generate multiple long-lived progeny prepolarized to respond immediately upon pathogen reencounter. In humans, memory lymphocytes have life spans extending into decades (Lo et al., 1999).

We end this chapter with a highly speculative and provocative hypothesis aimed at provoking discussion and future experimental approaches to CNS immunity (Figure 9.1). Specifically, we propose that the CNS can generate a type of innate immune memory. Microglia are long-lived cells able to acquire multiple classical, alternative, and regulatory immune functions. Furthermore, because microglial phenotype and function are actively determined by paracrine interactions with neurons and macroglia, long-term changes in these cells will lead to long-term changes in microglial function. Indeed, microglial phenotype and function change simply as a function of normal development and aging. For example, nearly all murine microglia express the anti-inflammatory receptor TREM2 at birth, but

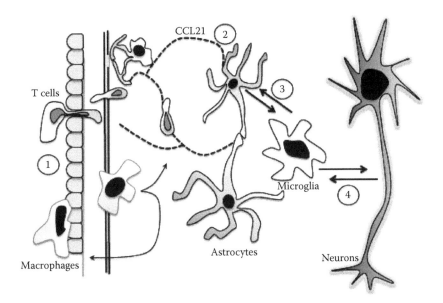

FIGURE 9.1 Interactions of CNS-resident cells and infiltrating peripheral leukocytes. Steps during CNS immune interactions: (1) T cells and peripheral macrophages extravasate across the blood–brain barrier into perivascular spaces. Extravasated immune cells, but not microglia, can leave the CNS and return to peripheral lymphoid organs. (2) CCL21 production by astrocytes and neurons provides a guidance signal for T cell migration from extraparenchmal sites into and within the brain parenchyma. (3) Ongoing long-term interactions between microglia and astrocytes and (4) between microglia and neurons define the immunoregulatory properties and immune functions of the CNS microenvironment.

microglia become increasingly heterogeneous in their level of TREM2 expression after mice begin to open their eyes in the second week of life (Carson et al., 2006; Thrash et al., 2009). Unlike, peripheral innate immune cells that enter the CNS and subsequently leave and die, microglia with polarized phenotypes remain in the CNS. We speculate that with age, pathogen encounter, and perhaps even experience-associated changes in neuronal activity, microglial phenotypes (and thus function) become an increasingly heterogeneous and specialized record of the life history of the CNS.

REFERENCES

Ajami B, Bennett JL, Krieger C, Tetzlaff W, Rossi FMV. 2007. Local self-renewal can sustain CNS microglia maintenance and function throughout adult life. *Nature Neuroscience* 10: 1538–1543.

Aloisi F, Pujol-Borrell R. 2006. Lymphoid neogenesis in chronic inflammatory diseases. *Nature Reviews: Immunology* 6: 205–217.

Bajénoff M, Egen JG, Koo LY, Laugier JP, Brau F, Glaichenhaus N, Germain RN. 2006. Stromal cell networks regulate lymphocyte entry, migration, and territoriality in lymph nodes. *Immunity* 25: 989–1001.

Banerjee D, Dick AD. 2004. Blocking CD200-CD200 receptor axis augments NOS-2 expression and aggravates experimental autoimmune uveoretinitis in Lewis rats. *Ocular Immunology and Inflammation* 12: 115–125.

Bartholomäus I, Kawakami N, Odoardi F, Schläger C, Miljkovic D, Ellwart JW, Klinkert WEF, et al. 2009. Effector T cell interactions with meningeal vascular structures in nascent autoimmune CNS lesions. *Nature* 462: 94–98.

Biber K, Neumann H, Inoue K, Boddeke HWGM. 2007. Neuronal 'on' and 'off' signals control microglia. *Trends in Neurosciences* 30: 596–602.

Block ML, Calderón-Garcidueñas L. 2009. Air pollution: mechanisms of neuroinflammation and CNS disease. *Trends in Neurosciences* 32: 506–516.

Brynskikh A, Warren T, Zhu J, Kipnis J. 2008. Adaptive immunity affects learning behavior in mice. *Brain, Behavior, and Immunity* 22: 861–869.

Byram SC, Carson MJ, DeBoy CA, Serpe CJ, Sanders VM, Jones KJ. 2004. CD4-positive T cell-mediated neuroprotection requires dual compartment antigen presentation. *Journal of Neuroscience* 24: 4333–4339.

Campbell A, Araujo JAA, Li H, Sioutas C, Kleinman M. 2009. Particulate matter induced enhancement of inflammatory markers in the brains of apolipoprotein E knockout mice. *Journal of Nanoscience and Nanotechnology* 9: 5099–5104.

Cardona AE, Sasse ME, Liu L, Cardona SM, Mizutani M, Savarin C, Hu T, Ransohoff RM. 2008. Scavenging roles of chemokine receptors: chemokine receptor deficiency is associated with increased levels of ligand in circulation and tissues. *Blood* 112: 256–263.

Carson MJ, Bilousova TV, Puntambekar SP, Melchior B, Doose JM, Ethell IM. 2007. A rose by any other name? The potential consequences of microglial heterogeneity during CNS health and disease. *Neurotherapeutics* 4: 571–579.

Carson MJ, Doose JM, Melchior B, Schmid CD, Ploix CC. 2006. CNS immune privilege: hiding in plain sight. *Immunological Reviews* 213: 48–65.

Carson MJ, Lo DD. 2007. Perspective is everything: an irreverent discussion of CNS-immune system interactions as viewed from different scientific traditions. *Brain, Behavior, and Immunity* 21: 367–373.

Carson MJ, Reilly CR, Sutcliffe JG, Lo D. 1999. Disproportionate recruitment of CD8+ T cells into the central nervous system by professional antigen-presenting cells. *American Journal of Pathology* 154: 481–494.

Carson MJ, Thrash JC, Lo D. 2004. Analysis of microglial gene expression: identifying targets for CNS neurodegenerative and autoimmune disease. *American Journal of Pharmacogenomics* 4: 321–330.

Chen S-C, Leach MW, Chen Y, Cai X-Y, Sullivan L, Wiekowski M, Dovey-Hartman BJ, Zlotnik A, Lira SA. 2002. Central nervous system inflammation and neurological disease in transgenic mice expressing the CC chemokine CCL21 in oligodendrocytes. *Journal of Immunology* 168: 1009–1017.

Clark CJ, Sage EH. 2008. A prototypic matricellular protein in the tumor microenvironment—where there's SPARC, there's fire. *Journal of Cellular Biochemistry* 104: 721–732.

Colton CA, Wilcock DM. 2009. Assessing activation states in microglia. *CNS & Neurological Disorders Drug Targets* (http://www.ncbi.nlm.nih.gov/pubmed/20015027).

Condamine T, Le Texier L, Howie D, Lavault A, Hill M, Halary F, Cobbold S, Waldmann H, Cuturi M, Chiffoleau E. 2010. Tmem176B and Tmem176A are associated with the immature state of dendritic cells. *Journal of Leukocyte Biology*, doi:10.1189/jlb.1109738, in press.

Ferguson DJP. 2009. *Toxoplasma gondii*: 1908–2008, homage to Nicolle, Manceaux and Splendore. *Memórias Do Instituto Oswaldo Cruz* 104: 133–148.

Filisetti D, Candolfi E. 2004. Immune response to *Toxoplasma gondii*. *Annali dell'Istituto Superiore Di Sanità* 40: 71–80.

Ford JW, McVicar DW. 2009. TREM and TREM-like receptors in inflammation and disease. *Current Opinion in Immunology* 21: 38–46.

Garg S, Nichols JR, Esen N, Liu S, Phulwani NK, Syed MM, Wood WH, et al. 2009. MyD88 expression by CNS-resident cells is pivotal for eliciting protective immunity in brain abscesses. *ASN Neuro* 1, no. 2. doi:10.1042/AN20090004.

Ge S, Murugesan N, Pachter JS. 2009. Astrocyte- and endothelial-targeted CCL2 conditional knockout mice: critical tools for studying the pathogenesis of neuroinflammation. *Journal of Molecular Neuroscience: MN* 39: 269–283.

Gerlofs-Nijland ME, van Berlo D, Cassee FR, Schins RPF, Wang K, Campbell A. 2010. Effect of prolonged exposure to diesel engine exhaust on proinflammatory markers in different regions of the rat brain. *Particle and Fibre Toxicology* 7: 12.

Gordon S, Martinez FO. 2010. Alternative activation of macrophages: mechanism and functions. *Immunity* 32: 593–604. doi:10.1016/j.immuni.2010.05.007.

Goverman J. 2009. Autoimmune T cell responses in the central nervous system. *Nature Reviews: Immunology* 9: 393–407.

Graeber MB, Streit WJ. 2010. Microglia: biology and pathology. *Acta Neuropathologica* 119: 89–105.

Harris TH, Wilson EH, Tait ED, Buckley M, Shapira S, Caamano J, Arti D, Hunter CA. 2010. NF-kappaB1 contributes to T cell-mediated control of *Toxoplasma gondii* in the CNS. *Journal of Neuroimmunology* 222: 19–28.

Herbert HR. 2010. Contributions of the environment and environmentally vulnerable physiology to autism spectrum disorders. *Current Opinion in Neurology* 23: 103–110.

Hickey WF, Kimura H. 1988. Perivascular microglial cells of the CNS are bone marrow-derived and present antigen *in vivo*. *Science* 239: 290–292.

Huynh W, Cordato DJ, Kehdi E, Masters LT, Dedousis C. 2008. Post-vaccination encephalomyelitis: literature review and illustrative case. *Journal of Clinical Neuroscience* 15: 1315–1322.

Kim JV, Jiang NE, Tadokoro CE, Liu L, Ransohoff RM, Lafaille JJ, Dustin ML. 2010. Two-photon laser scanning microscopy imaging of intact spinal cord and cerebral cortex reveals requirement for CXCR6 and neuroinflammation in immune cell infiltration of cortical injury sites. *Journal of Immunological Methods* 352: 89–100.

Kipnis J, Derecki NC, Yang C, Scrable H. 2008. Immunity and cognition: what do age-related dementia, HIV-dementia and "chemo-brain" have in common? *Trends in Immunology* 29: 455–463.

Klünemann HH, Ridha BH, Magy L, Wherrett JR, Hemelsoet DM, Keen RW, De Bleecker JL, et al. 2005. The genetic causes of basal ganglia calcification, dementia, and bone cysts: DAP12 and TREM2. *Neurology* 64: 1502–1507.

Koning N, Swaab DF, Hoek RM, Huitinga I. 2009. Distribution of the immune inhibitory molecules CD200 and CD200R in the normal central nervous system and multiple sclerosis lesions suggests neuron-glia and glia-glia interactions. *Journal of Neuropathology and Experimental Neurology* 68: 159–167.

Korn T, Anderson AC, Bettelli E, Oukka M. 2007. The dynamics of effector T cells and Foxp3+ regulatory T cells in the promotion and regulation of autoimmune encephalomyelitis. *Journal of Neuroimmunology* 191: 51–60.

Lanier LL. 2009. DAP10- and DAP12-associated receptors in innate immunity. *Immunological Reviews* 227: 150–160.

Lo D, Feng L, Li L, Carson MJ, Crowley M, Pauza M, Nguyen A, Reilly CR. 1999. Integrating innate and adaptive immunity in the whole animal. *Immunological Reviews* 169: 225–239.

Louvet C, Chiffoleau E, Heslan M, Tesson L, Heslan JM, Brion R, Bériou G, et al. 2005. Identification of a new member of the CD20/FcepsilonRIbeta family overexpressed in tolerated allografts. *American Journal of Transplantation* 5: 2143–2153.

McGavern DB, Truong P. 2004. Rebuilding an immune-mediated central nervous system disease: weighing the pathogenicity of antigen-specific versus bystander T cells. *Journal of Immunology* 173: 4779–4790.

Meerburg BG, Kijlstra A. 2009. Changing climate-changing pathogens: *Toxoplasma gondii* in North-Western Europe. *Parasitology Research* 105: 17–24.

Melchior B, Garcia AE, Hsiung BK, Lo KM, Doose JM, Thrash JC, Stalder AK, Staufenbiel M, Neumann H, Carson MJ. 2010. Dual induction of TREM2 and tolerance related transcript, Tmem176b in amyloid transgenic mice: implications for vaccine-based therapies for Alzheimer's disease. *ASN Neuro* 2: e00037.

Melchior B, Puntambekar SP, Carson MJ. 2006. Microglia and the control of autoreactive T cell responses. *Neurochemistry International* 49: 145–153.

Mott RT, Ait-Ghezala G, Town T, Mori T, Vendrame M, Zeng J, Ehrhart J, Mullan M, Tan J. 2004. Neuronal expression of CD22: novel mechanism for inhibiting microglial proinflammatory cytokine production. *Glia* 46: 369–379.

Müller M, Carter SL, Hofer MJ, Manders P, Getts DR, Getts MT, Dreykluft A, et al. 2007. CXCR3 signaling reduces the severity of experimental autoimmune encephalomyelitis by controlling the parenchymal distribution of effector and regulatory T cells in the central nervous system. *Journal of Immunology* 179: 2774–2786.

Noor S Habashy AS, Nance JP, Clark RT, Nemati K, Carson MJ, Wilson EH. 2010. CCR7-dependent immunity during acute *Toxoplasma gondii* infection. *Infection and Immunity* 78: 2257–2263.

Odoardi F, Kawakami N, Li Z, Cordiglieri C, Streyl K, Nosov M, Klinkert WEF, et al. 2007. Instant effect of soluble antigen on effector T cells in peripheral immune organs during immunotherapy of autoimmune encephalomyelitis. *PNAS* 104: 920–925.

Papenfuss TL, Thrash JC, Danielson PE, Foye PE, Hllbrush BS, Sutcliffe JG, Whitacre CC, Carson MJ. 2007. Induction of Golli-MBP expression in CNS macrophages during acute LPS-induced CNS inflammation and experimental autoimmune encephalomyelitis (EAE). *Scientific World Journal* 7: 112–120.

Ploix CC, Lo D, Carson MJ. 2001. A ligand for the chemokine receptor CCR7 can influence the homeostatic proliferation of CD4 T cells and progression of autoimmunity. *Journal of Immunology* 167: 6724–6730.

Ploix CC, Noor S, Crane J, Masek K, Carter W, Lo DD, Wilson EH, Carson MJ. 2010. CNS-derived CCL21 is both sufficient to drive homeostatic CD4+ T cell proliferation and necessary for efficient CD4+ T cell migration into the CNS parenchyma following *Toxoplasma gondii* infection. *Brain Behavior and Immunity*, in press.

Pocock JM, Kettenmann H. 2007. Neurotransmitter receptors on microglia. *Trends in Neurosciences* 30: 527–535.

Poser CM. 1990. Notes on the pathogenesis of subacute sclerosing panencephalitis. *Journal of the Neurological Sciences* 95: 219–224.

Prandota J. 2010. Migraine associated with patent foramen ovale may be caused by reactivation of cerebral toxoplasmosis triggered by arterial blood oxygen desaturation. *International Journal of Neuroscience* 120: 81–87.

Raivich G. 2005. Like cops on the beat: the active role of resting microglia. *Trends in Neurosciences* 28: 571–573.

Reboldi A, Coisne C, Baumjohann D, Benvenuto F, Bottinelli D, Lira S, Uccelli A, Lanzavecchia A, Engelhardt B, Sallusto F. 2009. C-C chemokine receptor 6-regulated entry of TH-17 cells into the CNS through the choroid plexus is required for the initiation of EAE. *Nature Immunology* 10: 514–523.

Sauvin Stohlman SA, Atkinson R, Ransohoff RM, Bergmann CC. 2010. Monocytes regulate T cell migration through the glia limitans during acute viral encepalitis. *Journal of Virology* 84: 4878–4888.

Schmid CD, Melchior B, Masek K, Puntambekar SP, Danielson PE, Lo DD, Sutcliffe JG, Carson MJ. 2009. Differential gene expression in LPS/IFNγamma activated microglia and macrophages: *in vitro* versus *in vivo*. *Journal of Neurochemistry* 109 (Suppl 1): 117–125.

Schmid CD, Sautkulis LN, Danielson PE, Cooper J, Hasel KW, Hilbush BS, Sutcliffe JG, Carson MJ. 2002. Heterogeneous expression of the triggering receptor expressed on myeloid cells-2 on adult murine microglia. *Journal of Neurochemistry* 83: 1309–1320.

Sierra A, Gottfried-Blackmore AC, McEwen BS, Bulloch K. 2007. Microglia derived from aging mice exhibit an altered inflammatory profile. *Glia* 55: 412–424.

Siffrin V, Brandt AU, Herz J, Zipp F. 2007. New insights into adaptive immunity in chronic neuroinflammation. *Advances in Immunology* 96: 1–40.

Sullivan MM, Puolakkainen PA, Barker TH, Funk SE, Sage EH. 2008. Altered tissue repair in hevin-null mice: inhibition of fibroblast migration by a matricellular SPARC homolog. *Wound Repair and Regeneration* 16: 310–319.

Sunnemark D, Eltayeb S, Nilsson M, Wallström E, Lassmann H, Olsson T, Berg A, Ericsson-Dahlstrand A. 2005. CX3CL1 (fractalkine) and CX3CR1 expression in myelin oligodendrocyte glycoprotein-induced experimental autoimmune encephalomyelitis: kinetics and cellular origin. *Journal of Neuroinflammation* 2: 17.

Thrash JC, Torbett BE, Carson MJ. 2009. Developmental regulation of TREM2 and DAP12 expression in the murine CNS: implications for Nasu-Hakola disease. *Neurochem Res.* 34:38–45.

Torrey EF, Yolken RH. 2003. *Toxoplasma gondii* and schizophrenia. *Emerging Infectious Diseases* 9: 1375–1380.

Vyas A, Kim S, Giacomini N, Boothroyd JC, Sapolsky RM. 2007. Behavioral changes induced by *Toxoplasma* infection of rodents are highly specific to aversion of cat odors. *PNAS* 104: 6442–6447.

Weaver MS, Workman G, Cardo-Vila M, Arap W, Pasqualini R, Sage EH. 2010. Processing of the matricellular protein hevin in mouse brain is dependent on ADAMTS4. *Journal of Biological Chemistry* 285: 5868–5877.

Wilson EH, Harris TH, Mrass P, John B, Tait ED, Wu GF, Pepper M, et al. 2009. Behavior of parasite-specific effector CD8+ T cells in the brain and visualization of a kinesis-associated system of reticular fibers. *Immunity* 30: 300–311.

Wilson EH, Weninger W, Hunter CA. 2010. Trafficking of immune cells in the central nervous system. *Journal of Clinical Investigation* 120: 1368–1379.

Wilson EH, Zaph C, Mohrs M, Welcher A, Siu J, Artis D, Hunter CA. 2006. B7RP-1-ICOS interactions are required for optimal infection-induced expansion of CD4+ Th1 and Th2 responses. *Journal of Immunology* 177: 2365–2372.

Yang I, Han SJ, Kaur G, Crane C, Parsa. 2010. AT. The role of microglia in central nervous system immunity and glioma immunology. *Journal of Clinical Neuroscience* 17: 6–10.

Yolken RH, Dickerson FB, Torrey EF. 2009. *Toxoplasma* and schizophrenia. *Parasite Immunology* 31: 706–715.

Zhang J, Gratchev A, Riabov V, Mamidi S, Schmuttermaier C, Krusell L, Kremmer E, et al. 2009. A novel GGA-binding site is required for intracellular sorting mediated by stabilin-1. *Molecular and Cellular Biology* 29: 6097–6105.

Section III

CNS Pathology: Disruption
of Astrocyte Connectivity:
Introduction

The chapters in this section deal with pathological changes in astrocyte physiology and in their connections, and consider whether the disrupted astrocyte networks could be the cause of the pathology. It is a general hypothesis of this section that multiple mechanisms may trigger disruption of astrocytes, and that these differences in causation may allow therapeutic targeting of the individual pathologies. There is evidence for morphological, functional, and biochemical changes in astrocytes in all neuropsychiatric diseases. While it is currently difficult to assign these changes as either cause or effect of the illness, understanding mechanisms leading to changes in astrocytes could yield either diagnostic or novel therapeutic strategies. The fundamental issue for the chapters of this section is: What are the triggering molecules that induce astrocytes to undergo changes in diseases?

In the first part (Chapter 10) in this section, Christian Steinhäuser and Gerald Seifert discuss the dysfunctions of several astrocyte biomarkers in epilepsy. For instance, the dysregulation of K^+ channels, mislocalization and loss of AQP4 from astrocyte endfeet, decreased expression of glutamate transporters, and loss of glutamine synthase have been associated with epilepsy. It is considered likely that these biomarkers could be novel targets for preventing or diminishing the substantial changes in neuronal excitability in these diseases.

In Chapter 11, Sarah Lutz, Cedric Raine, and Celia Brosnan describe astrocyte involvement in neuroinflammatory demyelinating disease. In both animal models (such as EAE) and human disease (MS), astrocytes contribute to the brain lesions by astrocyte-astrocyte, astrocyte-neuron, astrocyte–blood–brain barrier, and astrocyte-oligodendrocyte interactions. A major concept of this chapter is that astrocytes release cytokines, which can be both bad (inducing cell death) and good (repairing the lesion by providing a scaffold that promotes remyelination and as potential progenitor cells).

Chapter 12, by José Rodríguez Arellano, Carlos Matute, and Alexei Verkhratsky, considers the effects of Alzheimer's disease and representative animal models on glia. All three types of glia are affected in this disease, leading to weakening synaptic connectivity, loss of myelin due to excitotoxicity, and microglial activation leading to sustained inflammation. Remodeling of glial networks may cause cognitive impairment. Astrocytes in dementia and Alzheimer's disease show differences in cell density, molecular expression, and differences in the neurovascular unit due to changes in astrocyte endfoot coverage; oligodendrocytes show myelination changes/loss due to imbalance of Ca^{2+} and excitotoxicity; and microglia in Alzheimer's disease exhibit changes in inflammation due to amyloid plaques. The thrust of this chapter is to understand changes in the glial cells in Alzheimer's disease and to identify the cause of these changes.

Chapter 13, by Rachel Kneeland, Stephanie Liesch, Timithy Folsom, and Hossein Fatemi, details brain pathology in autism, which is characterized by abnormalities of the immune system, altered gliogenesis, and changes in the astrocyte markers GFAP, AQP4, and Cx43. This chapter also discusses the association between genetic mutations that are found in mental retardation and in animal models of autism that are relevant to gliogenesis. It also highlights the fact that many drugs used to treat autism are targets of astroglial proteins.

Chapter 14, by Markus Schwarz and Aye-Mu Myint, discusses evidence that in major depression there is loss of astrocytes as a consequence of kynurenine imbalance. Such loss of astrocytes leads to disturbances in GABAergic and glutamatergic transmission. The authors point out that the therapeutic aspects of antidepressant therapies involve restoring the health of astrocytes. In schizophrenia, a role of astrocytes may be in disrupting the glutamatergic control of the dopaminergic neurotransmission due to enhanced expression and activity of the glutamate transporter and of the glutamate-degrading enzyme glutamine synthetase.

10 Astrocytes in Epilepsy

Christian Steinhäuser and Gerald Seifert
Institute of Cellular Neurosciences, Medical
Faculty, University of Bonn, Bonn, Germany

CONTENTS

10.1 INTRODUCTION

Epilepsy is a condition of the brain characterized by the periodic and unpredictable occurrence of seizures. Even with optimal current antiepileptic drug (AED) therapy, about one-third of patients have poor seizure control and become medically refractory. Many AEDs act as central nervous system (CNS) depressants and must be taken chronically for seizure suppression, which often leads to marked side effects on cognition. Thus, there is a need for the development of more specific AEDs that may target functional abnormalities responsible for epilepsy but avoid unwanted side effects. In this regard, recent developments in the understanding of glial (especially astrocytic) changes in epilepsy can potentially provide novel therapeutic targets.

Recent work has identified glial cells and astrocytes in particular as active partners in neural information processing. Application of advanced electrophysiological and

Ca^{2+} imaging techniques unravelled that astrocytes in acute brain slices or after fresh isolation from the tissue express a spectrum of functional ion channels and transmitter receptors that is similarly broad, as in neurons (Verkhratsky and Steinhäuser 2000). The presence of ionotropic and metabotropic neurotransmitter receptors led to the conclusion that astrocytes are endowed with the machinery to sense and respond to neuronal activity. In 1994 two groups discovered that elevation of the intracellular Ca^{2+} concentration ($[Ca^{2+}]_i$) in cultured astrocytes upon membrane receptor activation can induce glial release of glutamate (Nedergaard 1994; Parpura et al. 1994). This astonishing finding for the first time demonstrated that astrocytes are sensing neuronal activity and feed back to neurons to modulate CNS signaling (Pasti et al. 1997). Later studies corroborated the view that astrocytes are direct communication partners of neurons and dynamically interact with synapses through uptake of neurotransmitters, receptor-mediated Ca^{2+} signaling, and subsequent gliotransmitter release. The intimate morphological and physiological interconnection between both of these types gave rise to the term *tripartite synapse*, which comprises not only pre- and postsynaptic elements but also the astrocytic processes (Araque et al. 1999; Halassa et al. 2007).

According to a long-standing concept, astrocytes supply neurons with nutrition metabolites and oxygen. Fundamental new insight into this aspect of astrocyte function was gained through the discovery that astrocytes control in an activity-dependent manner cerebral blood flow, by releasing vasoactive substances such as polyunsaturated fatty acids, adenosine, and prostaglandins (Koehler et al. 2009; Mulligan and MacVicar 2004; Zonta et al. 2003). In addition to the only recently discovered modulatory actions on brain signaling and circulation, astrocytes have been known for decades to serve homeostatic functions, including the clearance of neuronally released K^+ and glutamate from the extracellular space.

Despite the fact that the pathways enabling activation of these cells under physiological conditions are still ill-determined, evidence is emerging suggesting a critical role of astrocyte dysfunction in the pathogenesis of neurological disorders (Seifert et al. 2006). In this chapter, we focus on alterations of astrocytes that are currently thought to be associated with human or experimental epilepsy. Most of these data were obtained from focal epilepsies. While different types of cells with astroglial properties have been described to coexist in the brain (Matthias et al. 2003), in this chapter only alterations of bona fide astrocytes are discussed.

10.2   IMPAIRED K^+ BUFFERING IN TEMPORAL LOBE EPILEPSY

10.2.1   REGULATION OF INWARDLY RECTIFYING K^+ CHANNELS

Neuronal activity, propagation of action potentials, and synaptic activity after local depolarization lead to fast fluctuations of the extracellular K^+ concentration ($[K^+]_o$) because of the restricted extracellular space volume (Nicholson and Syková 1998). If increases of $[K^+]_o$ would remain uncorrected, the resting potential would become more positive and affect activation of transmembrane ion channels, receptors, and transporters. During neuronal hyperactivity in vivo, $[K^+]_o$ may increase from 3 mM to a ceiling level of 10–12 mM (Heinemann and Lux 1977). Such high $[K^+]_o$ levels can generate epileptiform activity in acute brain slices. Two different mechanisms are thought to balance $[K^+]_o$ during neuronal activity: K^+ uptake and spatial K^+ buffering (for review see Kofuji and Newman 2004). K^+ uptake, mediated by Na,K-Adenosine triphosphatase (AT) or Na-K-Cl cotransporters, is accompanied by cell swelling and local depolarization of astrocytes. Spatial K^+ buffering instead is driven by the glial syncytium membrane potential and the local K^+ equilibrium potential. This allows transfer of K^+ from regions of elevated $[K^+]_o$, through the syncytium, to regions of lower $[K^+]_o$. Spatial buffering depends on proper distribution and function of astrocytic K^+ channels, water channels, and gap junctions. In astrocytes, the inward rectifying K^+ channel Kir4.1, which is activated by intracellular ATP, is thought to allow for K^+ influx at negative membrane potentials (for review see Hibino et al. 2010; Reimann and Ashcroft 1999).

Because of its presumed role in K^+ homeostasis, properties of astroglial Kir channels have been investigated in experimental and human epilepsy. Measurements of $[K^+]_o$ with ion-sensitive microelectrodes and patch-clamp studies suggested that impaired K^+ buffering in sclerotic human hippocampus resulted from altered Kir channel expression. Differences were observed in the effect of Ba^{2+} on stimulus-induced changes in $[K^+]_o$ in the CA1 region of hippocampal brain slices obtained from temporal lobe epilepsy (TLE) patients with hippocampal sclerosis (HS) or without sclerosis (non-HS). In non-HS tissue, Ba^{2+} application significantly enhanced $[K^+]_o$, while this effect was not observed in HS specimens. Since Ba^{2+} is a blocker of Kir channels in astrocytes of the hippocampus (Seifert et al. 2009), this finding suggested impaired function of these channels in the sclerotic tissue (Kivi et al. 2000; Jauch et al. 2002). The hypothesis was confirmed with patch-clamp analyses demonstrating downregulation of Kir currents in the sclerotic human CA1 region of TLE patients (Bordey and Sontheimer 1998; Hinterkeuser et al. 2000). Accordingly, in HS, impaired K^+ buffering and enhanced seizure susceptibility result from reduced expression of Kir channels.

Fine mapping of a locus on mouse chromosome 1 identified KCNJ10 as a candidate gene exhibiting a potentially important polymorphism with regard to fundamental aspects of seizure susceptibility (Ferraro et al. 2004). Similarly, variations in KCNJ10 in the human genome associate with multiple seizure phenotypes. Missense mutations of KCNJ10 influence the risk of acquiring forms of human epilepsy (Buono et al. 2004). Mutations in the KCNJ10 gene encoding Kir4.1 channels cause a multiorgan disorder in patients with clinical features of epilepsy, sensorineural deafness, ataxia, and electrolyte imbalance (Bockenhauer et al. 2009; Scholl et al. 2009). These patients suffer from generalized tonic-clonic seizures and focal seizures since childhood. Single-nucleotide mutations in the KCNJ10 gene cause missense or nonsense mutations on the protein level in the pore region, transmembrane helices, or the C-terminus, the latter resulting in deletion of a Postsynaptic density protein 95/Disc large tumor suppressor/Zonula occludens-1 protein (PDZ) binding domain and proper membrane localization of Kir4.1. Experiments with heterologous expression systems demonstrated that the mutations indeed affect channel function and lead to reduced transmembrane currents.

Genetic downregulation of Kir4.1, the main Kir channel subunit in astrocytes (Kofuji et al. 2000; Neusch et al. 2006; Olsen et al. 2006; Seifert et al. 2009), profoundly reduced the ability of astrocytes to remove glutamate and K^+ from the extracellular space, both in cell culture (Kucheryavykh et al. 2007) and in vivo

(Djukic et al. 2007). General knockout of Kir4.1 leads to early postnatal lethality and mice with astrocytic deletion of the channel developed a pronounced behavioral phenotype, including seizures (Kofuji et al. 2000; Djukic et al. 2007).

10.2.2 IMPAIRED K⁺ BUFFERING AFTER DISRUPTION OF THE BLOOD–BRAIN BARRIER

The blood–brain barrier (BBB) is composed of a complex of endothelial cells, basement membrane, pericytes, and astrocytic endfeet. It represents a diffusion barrier maintaining the extracellular brain environment constant and impeding diffusion of large hydrophilic molecules and proteins from the peripheral bloodstream into the brain. Cells of the BBB are endowed with specific transporters allowing the transfer of nutrients such as glucose and amino acids across the barrier. Pathophysiological conditions such as trauma, stroke, and epilepsy result in the release of mediators, e.g., glutamate, Adenosine triphosphate (ATP), Tumor necrosis factor (TNF)α endothelin-1, and IL-1 interleukin-β, which increase BBB permeability. This leads to influx of serum albumin into the brain and induces further inflammatory responses that may exacerbate the disease (reviewed by Ballabh et al. 2004).

Dysfunction of the BBB seems to be involved in seizure generation. Transient opening of the BBB is sufficient for induction of focal epileptogenesis in rat (Seiffert et al. 2004). BBB lesion is also a primary event in human TLE leading to the extravasation of serum albumin, which is taken up by neurons, astrocytes, and microglia (Seiffert et al. 2004; Van Vliet et al. 2007). In astrocytes, albumin is taken up through a TGFβ receptor-mediated mechanism, resulting in downregulation of Kir4.1 and Kir2.3 channels, impaired buffering of extracellular K⁺, and N-Methyl-D-aspartate (NMDA) receptor-mediated hyperexcitability (Perillan et al. 2002; Tomkins et al. 2007; Ivens et al. 2007; review by Friedman et al. 2009). Transcript analysis after albumin and bile-salt-treated brains confirmed the downregulation of the inwardly rectifying K⁺ channels Kir4.1 and Kir3.1 and connexins Cx26, Cx30, and Cx43 (David et al. 2009). Most of these genes are predominantly expressed by astrocytes, suggesting a key role of these cells in epileptogenesis.

10.2.3 GAP JUNCTIONS AND K⁺ BUFFERING

As already mentioned in Section 10.2.1, a prerequisite for the operation of spatial K⁺ buffering is the presence of Kir channels and connexins forming gap junctions (Kofuji and Newman 2004; Walz 2000). According to this concept, K⁺ entry into the astrocytic network occurs at sites of maximal extracellular K⁺ accumulation, driven by the difference between glial syncytium membrane potential and the local K⁺ equilibrium potential. Since K⁺ diffuses freely through the glial network, at sites distant to elevated $[K^+]_o$, a driving force for K⁺ efflux results because here local depolarization exceeds the K⁺ equilibrium potential. Surprisingly, clearance and redistribution of K⁺ were still preserved in the hippocampal stratum radiatum (but not in the lacunosum moleculare) of mice with coupling-deficient astrocytes, indicating that gap-junction-independent mechanisms add to K⁺ homeostasis in the brain (e.g., *indirect coupling*; Figure 10.1; Wallraff et al. 2006). Nevertheless, genetic deletion of astrocyte gap

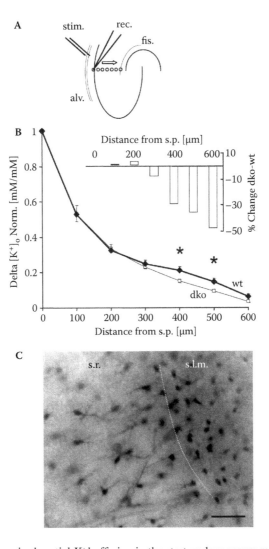

FIGURE 10.1 Impaired spatial K[+] buffering in the stratum lacunosum moleculare of mice with deletion of Cx43 and Cx30 in astrocytes (dko). (A) Experimental setup for the analysis of laminar profiles of changes in $[K^+]_o$. Stimulation was performed in the alveus (20 Hz), while the recording electrode was stepped from the stratum pyramidale to the hippocampal fissure (100 µm step size). (B) Mean rises in $[K^+]_o$ (normalized to rise at the stratum pyramidale) plotted against the distance from stratum pyramidale (thin line, wild type (wt); thick line, dko). Normalized $[K^+]_o$ in dko mice reached lower levels at 400, 500, and 600 µm from the stratum pyramidale as compared to wt. See inset for relative changes. (C) Illustration of the difference in astrocyte morphology and orientation in the stratum radiatum vs. stratum lacunosum moleculare, as obtained after biocytin injection into a stratum radiatum astrocyte proximal to the stratum lacunosum moleculare. Note the small size and random orientation of cells in the stratum lacunosum moleculare. Scale bar: 50 µm. alv., alveus; fis., fissure; stim., stimulation electrode; rec., recording electrode; s.r., stratum radiatum; s.l.m., stratum lacunosum moleculare. (From Wallraff, A., et al., *J. Neurosci.*, 26, 5438–5447, 2006. With permission.)

junctions leads to impaired K⁺ buffering, spontaneous epileptiform activity, and a decreased threshold for eliciting seizure activity (Wallraff et al. 2006).

10.2.4 INTERPLAY BETWEEN KIR CHANNELS, AQP4, AND K⁺ BUFFERING

Ultrastructural analyses in rat demonstrated spatial overlap of Kir4.1 and the water channel aquaporin 4 (AQP4) in astroglial endfeet contacting the capillaries (Higashi et al. 2001; Nielsen et al. 1997). This finding gave rise to the hypothesis that K^+ clearance through Kir channels might critically depend on concomitant transmembrane flux of water in a given cell, to dissipate osmotic imbalances due to K^+ redistribution. Subsequent functional work corroborated this idea by showing that in mice the clearance of extracellular K^+ is compromised if the number of perivascular AQP4 channels is decreased (Amiry-Moghaddam et al. 2003b). Similarly, impaired K^+ buffering and prolonged seizure duration was observed in AQP4 knockout mice (Binder et al. 2006).

Epileptic rats show mislocalization of AQP4 in astrocytic endfeet contacting blood vessels in the hippocampus. Eight weeks after status epilepticus, immunohistochemistry revealed loss of AQP4 in vacuolized astrocytes of the hippocampus. Instead, AQP1 was found in astrocytes, a protein not expressed by these cells under physiological conditions. Nonvacuolized astrocytes still contained AQP4, even at higher levels, and in addition expressed AQP9 (Kim et al. 2009). In TLE patients presenting with HS, immunostaining indicated loss of AQP4 in vasculature-associated astrocyte endfeet, compared with specimens from non-HS patients (Eid et al. 2005). The decrease of perivascular AQP4 channels might be secondary, following the disruption of the dystrophin complex that is essential for anchoring of AQP4 in the plasma membrane (Amiry-Moghaddam et al. 2003a). Together, these findings suggest that in HS, dislocation of water channels in concert with decreased expression of Kir channels in astrocytes might underlie impaired K^+ buffering and increased seizure propensity. The functional consequence of the upregulation of AQP subunits other than AQP4 in astrocytes of epileptic tissue remains unclear yet.

In addition to spatial buffering, transient K^+ accumulations can be counterbalanced by net K^+ uptake through Na-K-ATPase and the Na-K-Cl cotransporter NKCC1, at the cost of cell swelling due to concomitant water influx (reviewed by Kofuji and Newman 2004). In rodent hippocampus, the Na-K-ATPase was reported to have a potential role in maintaining low $[K^+]_o$ levels and to clear elevations in $[K^+]_o$ after epileptiform activity (D'Ambrosio et al. 2002; Xiong and Stringer 2000). However, whether alterations in net K^+ uptake contribute to the enhanced $[K^+]_o$ levels seen in epileptic tissue has still to be elucidated.

10.3 AMBIGUOUS ROLE OF GAP JUNCTIONS IN EPILEPTOGENESIS

The abundant expression of gap junctions in astrocytes and their formation as a functional syncytium enables long-range intercellular exchange of ions, nutritional metabolites, amino acids, and nucleotides. The permeability of gap junctions is regulated by endogenous membrane receptors, second messengers, and pH (see review by Rouach et al. 2002). Thus, trafficking of nutritional metabolites such as glucose-6-phosphate, lactate, and glutamine through astrocytes is controlled by endogeneous compounds that are released by endothelial cells, astrocytes, and neurons in an activity-dependent manner (Giaume et al. 1997). In addition to the aforementioned homeostic functions, astrocytic gap junctions may affect neuronal migration and proliferation (Elias and Kriegstein 2008; Giaume and Theis 2010; Kunze et al. 2009). Recent work revealed that inhibition of gap junction coupling not only enhances glucose uptake, synthesis of nucleic acids, and proliferation of astrocytes (Tabernero et al. 2006), but also allows intercellular trafficking of glucose through the astrocytic network and the delivery energetic metabolites from blood vessels to neurons to maintain synaptic transmission in the murine brain (Rouach et al. 2008). Glucose uptake and trafficking was shown to be dependent on synaptic transmission: trafficking was increased during epileptiform activity, and in turn glucose delivery through the astrocytic network is needed to sustain epileptiform activity. Neuronal activity was sustained by the transport of nutrition metabolites through the astrocytic network even under conditions of transient limited substrate availability (see also Giaume et al. 2010).

Cx43 and Cx30 comprise the main connexins forming gap junctions in astrocytes of the CNS (Nagy et al. 2004), and as discussed above, its cell-type-specific deletion in mice led to the generation of spontaneous epileptiform activity and a decreased threshold for evoking seizure activity (Wallraff et al. 2006). Disruption of the BBB and albumin-dependent generation of epilepsy in rat is accompanied by a transient decrease in transcripts of both connexins (Cacheaux et al. 2009; David et al. 2009). These findings are in line with the long-standing concept that astrocyte gap junctions are essential for proper K^+ regulation (Orkand et al. 1966) and help counteract the generation of epileptiform activity. However, an opposite effect was observed in organotypic hippocampal slice cultures where presumed long-term block of gap junctions through Cx43 mimetic peptides attenuated spontaneous seizure-like events (but not evoked epileptiform responses; Samoilova et al. 2008). The authors also observed that serum deprivation strongly reduced spontaneous recurrent network activity and assigned this effect to a neuroprotective role of gap junctional communication. Hence, a decrease of gap junction permeability seems to exert opposite effects on excitability: a fast-onset, proconvulsive effect due to impaired K^+ redistribution, but a delayed antiepileptic effect because of disruption of neuronal energy supply.

Despite these intriguing new insights into astrocyte function, the role of gap junctions in human epilepsy is still unresolved. Published data are not always consistent, reflecting that (1) human epilepsy cannot be considered a uniform condition, (2) most of the currently available gap junction blockers do not distinguish between neuronal and glial gap junctions, and (3) these blockers usually have dramatic side effects (Steinhäuser and Seifert 2002). Moreover, analysis of tissue samples is restricted to the chronic phase of the disorder and is likely to be affected by patients' long-lasting treatment with different AEDs, a problem inherent to experiments with neurosurgically resected specimens. Increased expression of Cx43 protein was observed in low-grade tumors and reactive astrocytes of human epileptic cortical tissue surrounding tumors, although high-grade gliomas exhibited great variations in Cx43 (Aronica et al. 2001). Specimens from pharmacoresistant TLE patients with HS showed strongly enhanced Cx43 immunoreactivity and transcript levels (Collignon et al. 2006; Fonseca et al. 2002; Naus et al. 1991). The authors speculated that upregulation of connexins might represent a compensatory response of astrocytes to cope with the enhanced K$^+$ release during seizure activity. However, in the light of the aforementioned findings, enhanced coupling could also serve to fuel hyperactivity and thereby exacerbate generalized seizures. Importantly, it has to be emphasized that any functional evidence of enhanced gap junction coupling in human epilepsy is yet missing, which considerably limits conclusions that can be drawn from the above studies.

10.4   GLUTAMATE UPTAKE IN EPILEPSY

10.4.1   Glutamate Transporter in Astrocytes

The uptake of glutamate that helps to terminate the action of this neurotransmitter at CNS synapses is mainly mediated by transporters localized at the astrocytic membrane. The high efficiency of these glial transporters ensures the maintenance of low concentrations of extracellular glutamate to prevent excitotoxic cell death (Choi 1992; Danbolt 2001). They are densely packed and keep the extracellular glutamate concentration in the nM range, preventing significant receptor activation (Herman and Jahr 2007; Le Meur et al. 2007; Cavelier and Attwell 2005). Fine-tuning of extrasynaptic glutamate through glial glutamate transporters is important for proper synaptic function and plasticity (Arnth-Jensen et al. 2002; Filosa et al. 2009; Huang and Bordey 2004; review by Tzingounis and Wadiche 2007). Downregulation of glial glutamate transporters or metabolic inhibition of the glutamate-to-glutamine converting enzyme, glutamine synthetase (GS), immediately causes extracellular accumulation of the transmitter that, when reaching µM concentrations, causes depolarization, compromises synaptic transmission, and induces neuronal cell death (Herman and Jahr 2007; Jabaudon et al. 1999; Sah et al. 1989; Zorumski et al. 1996). It is therefore not surprising that dysfunction of the astrocytic glutamate transporters, EAAT1 and EAAT2, is observed under various pathological conditions, including epilepsy (Seifert et al. 2006).

Excessive extracellular glutamate is found in human epileptogenic tissue and can induce recurrent seizures and neuronal cell death (Glass and Dragunow 1995). Different reports exist about the regulation of glial glutamate transporters in patients presenting with pharmacoresistant TLE. Employing in situ hybridization and Western blot analyses in specimens from patients with HS, Tessler et al. (1999) did not find changes of EAAT1 or EAAT2. In contrast, other groups reported downregulation of EAAT2 immunoreactivity in the CA1 region displaying profound neuronal loss in human HS (Mathern et al. 1999; Proper et al. 2002). EAAT1 was found increased in the sclerotic CA2/3 region (Mathern et al. 1999). Later work showed downregulation of EAAT1 and EAAT2 in the CA1 region in HS and emphasized that currently it is still unclear whether this reduction is causative of the condition or rather represents a compensatory effect (Sarac et al. 2009). The latter hypothesis would be in line with recent work reporting that expression of EAAT2 is critically dependent on synaptic activity. In this study, EAAT2-mediated uptake was found to be decreased after nerve fiber transection or neurodegeneration in a mouse model of amyotrophic lateral sclerosis (ALS) (Yang et al. 2009). β-Lactam antibiotics increased glutamate uptake in primary human astrocytes through Nuclear factor "kappa-light-chain-enhancer" of activated κ-cells (NF-κκ)-mediated EAAT2 promoter activation (Lee et al. 2008). Hence, the antibiotics might represent a therapeutical tool to counteract glutamate transporter dysfunction in neurological disorders such as ALS, epilepsy, stroke, or hepatic encephalopathy (Rothstein et al. 2005).

Depolarization of astrocytes by inadequate K^+ buffering leads to compromised functioning of the glial transporters (Djukic et al. 2007; Kucheryavykh et al. 2007). In a rat model of cortical dysplasia, pharmacological inhibition of glial glutamate transporters in the lesion area led to opening of neuronal NMDA receptors, prolonged synaptic currents, and decreased the threshold for the induction of epileptiform activity (Campbell and Hablitz 2008). This enhanced activity of NMDA receptors also triggered dephosphorylation of Kv2.1 K^+ channels produced a negative shift of its voltage-dependent activation, and hence modulated excitability and neuronal plasticity in mice (Mulholland et al. 2008).

10.4.2 CONVERSION OF GLUTAMATE TO GLUTAMINE BY ASTROCYTES

For effective removal of excess extracellular glutamate, the transmitter must be converted by GS into the receptor-inactive substrate glutamine, under consumption of ATP and ammonia. Increasing evidence indicates a loss of this astrocyte-specific enzyme in epilepsy. In patients presenting with HS, loss of GS in the hippocampus was accompanied by elevated extracellular glutamate levels (Figure 10.2, Eid et al.

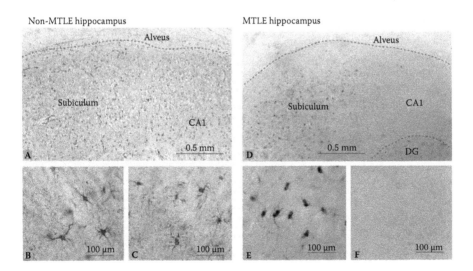

FIGURE 10.2 Glutamine synthetase immunoreactivity of hippocampal specimens obtained from nonmesial temporal lobe epilepsy (non-MTLE) (A–C) and MTLE patients (D–F). (Modified from Eid, T., et al., *Lancet*, 363, 28–37, 2004. With permission.)

2004; van der Hel et al. 2005; reviewed by Eid et al. 2008b). In experimental epilepsy, upregulation of Glutamine synthetase (GS) and Glial fibrillary acidic protein (GFAP) was observed in the latent phase, prior to recurrent seizure onset, while in the chronic phase GS was downregulated with elevated GFAP immunoreactivity persisting (Hammer et al. 2008). By contrast, glutamate dehydrogenase, another glutamate-degrading enzyme, remained unaltered in this rat model. Compatible with a potential causative role of GS loss in initiating epilepsy was the finding that pharmacological inhibition of GS produced recurrent seizure activity and rat brain pathology resembling human HS (Eid et al. 2008a).

Inhibition of GS in astrocytes or glutamine transporters in neurons reduced amplitudes of evoked inhibitory postsynaptic currents (IPSCs) and Gamma-aminobutyric acid (GABA) release from interneurons in the hippocampus. Hence, in the rat the glial glutamate-glutamine cycle is a major contributor to synaptic GABA release and regulates inhibitory synaptic strength (Liang et al. 2006), while inhibition of GS did not significantly affect glutamatergic transmission in the same species (Kam and Nicoll 2007). However, for periods of intense neuronal activity or epileptiform activity, glial supply of glutamine and the transport into neurons are required (Tani et al. 2010). The latter findings do not support the current view that protecting GS function might represent a promising therapeutic strategy to prevent seizures.

10.5   ASTROCYTE CA^{2+} SIGNALING IN EPILEPSY: RELEASE OF GLIOTRANSMITTERS AND SYNCHRONIZATION OF NEURONAL ACTIVITY

Initially, glutamate and bradykinin were shown to induce Ca^{2+} rises in astrocytes that propagated as Ca^{2+} waves through the syncytium and evoked Ca^{2+} elevations in adjacent neurons (Nedergaard 1994; Parpura et al. 1994). This astrocyte-to-neuron signaling is now thought to be mediated by release of the gliotransmitters glutamate, ATP, and D-serine from astrocytes (for review see Hamilton and Attwell 2010). Transmitters released from astrocytes may synchronize synaptic activity (Angulo et al. 2004; Fellin et al. 2004), regulate nearby synapses in an activity-dependent manner (Pascual et al. 2005; Serrano et al. 2006; Zhang et al. 2003; see review by Haydon and Carmignoto 2006), and control NMDA-receptor-dependent synaptic plasticity (Henneberger et al. 2010). Thus, disturbed release of transmitters from glial sources may potentially result in hyperactivity and excitotoxicity.

10.5.1   RELEASE OF GLUTAMATE

In rodents gliotransmitter release is stimulated by neuronal activity-dependent elevation of intracellular Ca^{2+} (Halassa et al. 2009; Petzold et al. 2008; Schummers et al. 2008; Wang et al. 2006) and may occur through Ca^{2+}-dependent exocytosis (Bezzi et al. 2004; Chen et al. 2005). In the normal brain, Ca^{2+} elevations in astrocytes are mediated mainly by metabotropic receptors. Indeed, a large body of evidence has shown that the activation of astrocytes by neuronal activity-derived glutamate is due to activation of metabotropic glutamate receptors (mGluRs) mGluR3 and mGluR5. Activation of these receptors affects 5'-3'-cyclic Adenosine monophosphate (cAMP) accumulation and leads to increases in intracellular Ca^{2+}, respectively. The Ca^{2+} rise may oscillate and initiate Ca^{2+} wave propagation within the astrocyte network, activate Ca^{2+}-dependent ion channels, and induce glutamate release from astrocytes (cf. above). In epilepsy models, elevated mGluR3, mGluR5, and mGluR8 protein levels have been found (Steinhäuser and Seifert 2002). High-resolution analysis of hippocampal specimens from TLE patients detected mGluR2/3, mGluR4, and mGluR8 in reactive astrocytes, suggesting an involvement of these mGluRs in gliosis (Tang and Lee 2001). Enhanced levels of astroglial mGluR2/3 and mGluR5 were also observed in epileptic specimens from patients with focal cortical dysplasia (Aronica et al. 2003b). Since their activation affects expression of EAAT1 and EAAT2 (Aronica et al. 2003a) and elevates $[Ca^{2+}]_i$, astrocytic mGluRs might contribute to the generation of seizure foci.

The aforementioned observation that astrocytes exhibit Ca^{2+}-induced release of glutamate providing direct excitation to neighboring neurons is of particular interest in the context of epileptogenesis. Accordingly, alterations in this glia-derived excitatory pathway, in concert with impaired glutamate uptake, might increase excitability of the neuron-astrocyte network, favor neuronal synchronization, and ultimately predispose neurons to seizures. Recent work suggested that in chemically induced acute epilepsy models astrocytic Ca^{2+} oscillations and glutamate release generate paroxysmal depolarizing shifts, i.e., interictal spikes (Kang et al. 2005; Tian et al. 2005), although this conclusion is disputed by others (Fellin and Haydon 2005). Nevertheless, glutamate released from rodent astrocytes during status epilepticus (SE) contributes to neuronal death. This process could be averted by inhibiting glianeuron signaling and by applying mGluR5 and NR2B NMDA receptor antagonists, which suppress astrocytic Ca^{2+} increase and block extrasynaptic NMDA receptors activated by glia-derived glutamate (Ding et al. 2007).

To better understand the role of gliotransmission in epilepsy, a number of important issues remain to be clarified. Glutamate from astrocytes can modulate synaptic transmission by affecting presynaptic receptors and regulate exocytosis, or by acting postsynaptically on extrasynaptic NMDA receptors, which can counteract or promote seizure activity (reviewed by Halassa et al. 2007; Wetherington et al. 2008). Moreover, recent evidence suggests that the increase in intracellular Ca^{2+} per se might not be sufficient for the release of astroglial glutamate. Rather, other factors, such as spatial relations between Ca^{2+} increase and release sites or different modes of exocytosis, might be of critical importance (Agulhon et al. 2008; Shigetomi et al. 2008).

10.5.2 Release of ATP

ATP can be released from rodent astrocytes in response to the same stimuli that trigger release of glutamate from astrocytes (Montana et al. 2006; Zhang et al. 2007). The potential impact of astroglial ATP release on seizure activity is now gradually emerging. Under physiological conditions, released ATP is rapidly degraded to adenosine through ectonucleotidases. By acting on presynaptic A1 receptors adenosine leads to inhibition of transmitter release and heterosynaptic depression (Pascual et al. 2005; Serrano et al. 2006; Zhang et al. 2003). The ambient adenosine level is controlled by the activity of the astrocytic enzyme adenosine kinase, which phosphorylates adenosine to 5'-AMP (Boison 2008; Martin et al. 2007). In experimental epilepsy, seizure induction results in upregulation of adenosine kinase and decreases ambient adenosine concentration, while genetic reduction of adenosine kinase prevents seizures (Li et al. 2008). These findings gave rise to the adenosine kinase hypothesis of epileptogenesis, which considers this enzyme both a diagnostic marker and a potential therapeutic target to prevent epileptogenesis (Boison 2008).

Hence, by releasing glutamate and ATP, astrocytes may potentially exert pro- or anticonvulsive actions, respectively. Full clarification of the functional significance of these different astrocyte-to-neuron signaling pathways represents an intriguing challenge. Notably, use of molecular genetics is providing important tools that allow us to selectively manipulate the different signaling pathways. Gliotransmission may soon be recognized as a target for developing new antiepileptic therapies.

10.5.3 Release of GABA

In addition to the above-mentioned gliotransmitters, astrocytes may release GABA through inverse operation of transporters. Recent data suggested that activation of glutamate transporters, leading to local increase in $[Na^+]_i$, can fuel transporter-mediated release of GABA from astrocytes. GABA release was enhanced at low $[Mg^{2+}]_o$, i.e., conditions generating epileptiform activity in brain slices (Heja et al. 2009). GABA release was mediated by the transporters rGAT2/3, which in the brain are mainly located on astrocytes. It was suggested that astrocytic GABA release might activate extrasynaptic GABA receptors on neurons, producing tonic GABA receptor currents. Hence, astrocytes might represent important regulators of tonic inhibition. In human hippocampus, hGAT-3 transporters are weakly expressed by astrocytes. An upregulation was observed in sclerotic specimens, but not in the CA3 and CA1 regions showing neuronal loss (Lee et al. 2006). In contrast, another study found decreased hGAT-3 immunoreactivity in human hippocampal sclerosis (Mathern et al. 1999), suggesting that glutamate-induced GABA release is impaired in TLE because of decreased glial GABA transporter expression (During et al. 1995).

GABA released by astrocytes activates extrasynaptic GABA receptors on neurons and glial cells. Activation of these receptors in neurons may well result in seizure generation in epileptic tissue, e.g., due to a switch of GABA receptor action from inhibition to excitation, and altered subunit combination, pharmacology, or subcellular localization (reviewed by Cossart et al. 2005). Thus, modified GABA release from astrocytes as observed in epileptic tissue may significantly affect the excitability of the neuronal network.

10.6   ASTROCYTE IMMUNE RESPONSES AND EPILEPSY

Astrocytes contribute to the inflammatory environment of the CNS by producing and responding to immunologically relevant molecules, including a variety of chemokines and cytokines (Dong and Benveniste 2001). Fever provokes seizures in

1 out of 20–50 children, termed febrile seizures, the most common form of pathological brain activity during early development. Prolonged or focal febrile seizures predispose to the development of intractable TLE (Dube et al. 2007). Both in rodents and humans, such febrile seizures have been shown to be associated with increased IL-1β levels in CNS and blood plasma (Dube et al. 2005; Virta et al. 2002), which enhances neuronal excitability, in part by augmenting glutamate receptor function and inhibiting GABAergic transmission (Vezzani et al. 2008). However, IL-1β and other cytokines can also act on astrocytic receptors. Thus, in culture or acute rat brain slices, IL-1β and TNFα were reported to inhibit glutamate reuptake and increase glial glutamate release (Bezzi et al. 2001; Hu et al. 2000; Ye and Sontheimer 1996), which can be expected to produce hyperactivity. TNFα released from rodent astrocytes controls synaptic efficiency by increasing surface expression of α-Amino-3-hydroxy-5-methyl-4-isoxazolepropanoic acid (AMPA) receptors (Beattie et al. 2002) and causing endocytosis of $GABA_A$ receptors, which decreases inhibitory synaptic strength (Stellwagen et al. 2005). Hence, excess TNFα release as observed during inflammation may entail overexcitation and neurodegeneration. In addition, Ca^{2+} increase and glutamate release from astrocytes might be mediated through other stimuli of the NFκB pathway in HS (Steinhäuser et al. 2007), which would result in uncoupling and promote hyperexcitability.

IL-1β and TNFα also interfere with astrocyte gap junction communication. In cell culture, release of these proinflammatory cytokines from activated microglia leads to a downregulation of Cx43 and closure of gap junctions in astrocytes (John et al. 1999; Meme et al. 2006). Since in mice status epilepticus (SE) induces massive microglial activation in vivo (Avignone et al. 2008), inhibition of gap junction communication in astrocytes appears to represent a very early alteration in the process of epileptogenesis. On the other hand, proinflammatory cytokines have been reported to open Cx43 hemichannels in cultured astrocytes, which might represent an alternative pathway of glucose entry under pathological condition when intercellular astrocytic junctional communication is impaired (Retamal et al. 2007). It has still to be clarified whether the opposite effects of cytokines on gap junction coupling and disruption of neuronal energy supply (see Section 10.3) or hemichannel activity have a pro- or anticonvulsive outcome.

Inflammation causes loss of polarized localization of AQP4, as recently shown in a mouse model of experimental autoimmune encephalomyelitis (Wolburg-Buchholz et al. 2009), which might contribute to increased seizure propensity, as discussed in Section 10.2.4. It is tempting to speculate that inflammatory mediators released after seizures (such as IL-1β and TNFα) may account for changes in the distribution and function of several other astrocyte membrane channels or receptors.

Infections of the brain, e.g., through meningitis, human herpes virus 6, or herpes simplex virus, have also been associated with TLE, and again astrocytes might be important in this context. Human astrocytes express Toll-like receptors (TLRs) (Bsibsi et al. 2002), which belong to the innate immune system and sense microbial infections (Medzhitov 2001). Several TLRs are expressed by astrocytes, including those that recognize and respond to viruses and bacteria. Activation of TLRs, similar to IL-1 receptors, induces signal transduction pathways leading to stimulation of the transcription factor NF-κB, which accordingly might add to enhanced release of astroglial glutamate in the sclerotic hippocampus (Steinhäuser et al. 2007). Inflammatory responses occur after neurotrauma, stroke, and seizures and harbor the risk of developing epilepsy. Molecules released by injured tissue, the so-called damage-associated molecular patterns (DAMPs), induce and activate TLRs in astrocytes, which has proconvulsant effects (Maroso et al. 2010). The latter study describes the release of high-mobility group box 1 (HMGB1), a chromatin component belonging to DAMPs, from neurons and glial cells after acute and chronic seizures. HMGB1 activates TLR4 and in turn contributes to generation and perpetuation of seizures. Hence, HMBG1, liberated by tissue damage and degeneration, is a promising target for pharmaceutical intervention to relieve recurrent seizures.

10.7   CONCLUDING REMARKS

The novel view that considers astrocytes as communication partners of neurons rather than "brain glue" has rekindled the question regarding the role of these cells in neurological disorders such as epilepsy. Indeed, an increasing body of evidence has documented astroglial dysfunction, and even dysregulation of astroglia-specific functions in human and experimental epilepsy. However, a number of key questions need to be addressed before a unifying picture can be proposed. For example, it is still unclear whether the reported glial alterations are causative of the condition or rather represent a consequence. In addition, difficulties arise from the fact that the term *astrocyte* covers a heterogeneous group of cells, which complicates comparison of individual studies. It is worthwhile, however, underlining that the molecular, functional, and structural characterization of astroglial heterogeneity is a rapidly evolving field that may soon lead to a better definition of astroglial subtypes. In a comprehensive approach that uses modern molecular genetics and in vivo models we may now have the opportunity to clarify the specific roles of astroglia in epilepsy and to develop novel therapeutic approaches to cure this disorder.

ACKNOWLEDGMENTS

Work of the authors is supported by Deutsche Forschungsgemeinschaft (grants SPP 1172 SE 774/3, SFB/TR3 C1, C9) and European Commission (FP7–202167 NeuroGLIA). We thank Dr. I. Nauroth for comments on the manuscript, and apologize to all those whose work could not be discussed due to space constraints.

REFERENCES

Agulhon, C., Petravicz, J., McMullen, A. B., et al. 2008. What is the role of astrocyte calcium in neurophysiology? *Neuron* 59: 932–946.
Amiry-Moghaddam, M., Otsuka, T., Hurn, P. D., et al. 2003a. An alpha-syntrophin-dependent pool of AQP4 in astroglial end-feet confers bidirectional water flow between blood and brain. *Proc. Natl. Acad. Sci. USA* 100: 2106–2111.
Amiry-Moghaddam, M., Williamson, A., Palomba, M., et al. 2003b. Delayed K^+ clearance associated with aquaporin-4 mislocalization: phenotypic defects in brains of alpha-syntrophin-null mice. *Proc. Natl. Acad. Sci. USA* 100: 13615–13620.

Angulo, M. C., Kozlov, A. S., Charpak, S., et al. 2004. Glutamate released from glial cells synchronizes neuronal activity in the hippocampus. *J. Neurosci.* 24: 6920–6927.

Araque, A., Parpura, V., Sanzgiri, R. P., et al. 1999. Tripartite synapses: glia, the unacknowledged partner. *Trends Neurosci.* 22: 208–215.

Arnth-Jensen, N., Jabaudon, D., and Scanziani, M. 2002. Cooperation between independent hippocampal synapses is controlled by glutamate uptake. *Nat. Neurosci.* 5: 325–331.

Aronica, E., Gorter, J. A., Ijlst-Keizers, H., et al. 2003a. Expression and functional role of mGluR3 and mGluR5 in human astrocytes and glioma cells: opposite regulation of glutamate transporter proteins. *Eur. J. Neurosci.* 17: 2106–2118.

Aronica, E., Gorter, J. A., Jansen, G. H., et al. 2001. Expression of connexin 43 and connexin 32 gap-junction proteins in epilepsy-associated brain tumors and in the perilesional epileptic cortex. *Acta Neuropathol. (Berl.)* 101: 449–459.

Aronica, E., Gorter, J. A., Jansen, G. H., et al. 2003b. Expression and cell distribution of group I and group II metabotropic glutamate receptor subtypes in Taylor-type focal cortical dysplasia. *Epilepsia* 44: 785–795.

Avignone, E., Ulmann, L., Levavasseur, F., et al. 2008. Status epilepticus induces a particular microglial activation state characterized by enhanced purinergic signaling. *J. Neurosci.* 28: 9133–9144.

Ballabh, P., Braun, A., and Nedergaard, M. 2004. The blood-brain barrier: an overview: structure, regulation, and clinical implications. *Neurobiol. Dis.* 16: 1–13.

Beattie, E. C., Stellwagen, D., Morishita, W., et al. 2002. Control of synaptic strength by glial TNFalpha. *Science* 295: 2282–2285.

Bezzi, P., Domercq, M., Brambilla, L., et al. 2001. CXCR4-activated astrocyte glutamate release via TNFalpha: amplification by microglia triggers neurotoxicity. *Nat. Neurosci.* 4: 702–710.

Bezzi, P., Gundersen, V., Galbete, J. L., et al. 2004. Astrocytes contain a vesicular compartment that is competent for regulated exocytosis of glutamate. *Nat. Neurosci.* 7: 613–620.

Binder, D. K., Yao, X., Zador, Z., et al. 2006. Increased seizure duration and slowed potassium kinetics in mice lacking aquaporin-4 water channels. *Glia* 53: 631–636.

Bockenhauer, D., Feather, S., Stanescu, H. C., et al. 2009. Epilepsy, ataxia, sensorineural deafness, tubulopathy, and KCNJ10 mutations. *N. Engl. J. Med.* 360: 1960–1970.

Boison, D. 2008. The adenosine kinase hypothesis of epileptogenesis. *Prog. Neurobiol.* 84: 249–262.

Bordey, A., and Sontheimer, H. 1998. Properties of human glial cells associated with epileptic seizure foci. *Epilepsy Res.* 32: 286–303.

Bsibsi, M., Ravid, R., Gveric, D., et al. 2002. Broad expression of Toll-like receptors in the human central nervous system. *J. Neuropathol. Exp. Neurol.* 61: 1013–1021.

Buono, R. J., Lohoff, F. W., Sander, T., et al. 2004. Association between variation in the human KCNJ10 potassium ion channel gene and seizure susceptibility. *Epilepsy Res.* 58: 175–183.

Cacheaux, L. P., Ivens, S., David, Y., et al. 2009. Transcriptome profiling reveals TGF-ß signaling involvement in epileptogenesis. *J. Neurosci.* 29: 8927–8935.

Campbell, S. L., and Hablitz, J. J. 2008. Decreased glutamate transport enhances excitability in a rat model of cortical dysplasia. *Neurobiol. Dis.* 32: 254–261.

Cavelier, P., and Attwell, D. 2005. Tonic release of glutamate by a DIDS-sensitive mechanism in rat hippocampal slices. *J. Physiol.* 564: 397–410.

Chen, X., Wang, L., Zhou, Y., et al. 2005. "Kiss-and-run" glutamate secretion in cultured and freshly isolated rat hippocampal astrocytes. *J. Neurosci.* 25: 9236–9243.

Choi, D. W. 1992. Excitotoxic cell death. *J. Neurobiol.* 23: 1261–1276.

Collignon, F., Wetjen, N. M., Cohen-Gadol, A. A., et al. 2006. Altered expression of connexin subtypes in mesial temporal lobe epilepsy in humans. *J. Neurosurg.* 105: 77–87.

Cossart, R., Bernard, C., and Ben-Ari, Y. 2005. Multiple facets of GABAergic neurons and synapses: multiple fates of GABA signalling in epilepsies. *Trends Neurosci.* 28: 108–115.

D'Ambrosio, R., Gordon, D. S., and Winn, H. R. 2002. Differential role of KIR channel and Na^+/K^+-pump in the regulation of extracellular K^+ in rat hippocampus. *J. Neurophysiol.* 87: 87–102.

Danbolt, N. C. 2001. Glutamate uptake. *Prog. Neurobiol.* 65: 1–105.

David, Y., Cacheaux, L. P., Ivens, S., et al. 2009. Astrocytic dysfunction in epileptogenesis: consequence of altered potassium and glutamate homeostasis? *J. Neurosci.* 29: 10588–10599.

Ding, S., Fellin, T., Zhu, Y., et al. 2007. Enhanced astrocytic Ca^{2+} signals contribute to neuronal excitotoxicity after status epilepticus. *J. Neurosci.* 27: 10674–10684.

Djukic, B., Casper, K. B., Philpot, B. D., et al. 2007. Conditional knock-out of Kir4.1 leads to glial membrane depolarization, inhibition of potassium and glutamate uptake, and enhanced short-term synaptic potentiation. *J. Neurosci.* 27: 11354–11365.

Dong, Y., and Benveniste, E. N. 2001. Immune function of astrocytes. *Glia* 36: 180–190.

Dube, C., Vezzani, A., Behrens, M., et al. 2005. Interleukin-1beta contributes to the generation of experimental febrile seizures. *Ann. Neurol.* 57: 152–155.

Dube, C. M., Brewster, A. L., Richichi, C., et al. 2007. Fever, febrile seizures and epilepsy. *Trends Neurosci.* 30: 490–496.

During, M. J., Ryder, K. M., and Spencer, D. D. 1995. Hippocampal GABA transporter function in temporal-lobe epilepsy. *Nature* 376: 174–177.

Eid, T., Ghosh, A., Wang, Y., et al. 2008a. Recurrent seizures and brain pathology after inhibition of glutamine synthetase in the hippocampus in rats. *Brain* 131: 2061–2070.

Eid, T., Lee, T. S., Thomas, M. J., et al. 2005. Loss of perivascular aquaporin 4 may underlie deficient water and K^+ homeostasis in the human epileptogenic hippocampus. *Proc. Natl. Acad. Sci. USA* 102: 1193–1198.

Eid, T., Thomas, M. J., Spencer, D. D., et al. 2004. Loss of glutamine synthetase in the human epileptogenic hippocampus: possible mechanism for raised extracellular glutamate in mesial temporal lobe epilepsy. *Lancet* 363: 28–37.

Eid, T., Williamson, A., Lee, T. S et al. 2008b. Glutamate and astrocytes—key players in human mesial temporal lobe epilepsy? *Epilepsia* 49 (Suppl. 2): 42–52.

Elias, L. A., and Kriegstein, A. R. 2008. Gap junctions: multifaceted regulators of embryonic cortical development. *Trends Neurosci.* 31: 243–250.

Fellin, T., and Haydon, P. G. 2005. Do astrocytes contribute to excitation underlying seizures? *Trends Mol. Med.* 11: 530–533.

Fellin, T., Pascual, O., Gobbo, S., et al. 2004. Neuronal synchrony mediated by astrocytic glutamate through activation of extrasynaptic NMDA receptors. *Neuron* 43: 729–743.

Ferraro, T. N., Golden, G. T., Smith, G. G., et al. 2004. Fine mapping of a seizure susceptibility locus on mouse chromosome 1: nomination of Kcnj10 as a causative gene. *Mamm. Genome* 15: 239–251.

Filosa, A., Paixao, S., Honsek, S. D., et al. 2009. Neuron-glia communication via EphA4/ephrin-A3 modulates LTP through glial glutamate transport. *Nat. Neurosci.* 12: 1285–1292.

Fonseca, C. G., Green, C. R., and Nicholson, L. F. 2002. Upregulation in astrocytic connexin 43 gap junction levels may exacerbate generalized seizures in mesial temporal lobe epilepsy. *Brain Res.* 929: 105–116.

Friedman, A., Kaufer, D., and Heinemann, U. 2009. Blood–brain barrier breakdown-inducing astrocytic transformation: novel targets for the prevention of epilepsy. *Epilepsy Res.* 85: 142–149.

Giaume, C., Koulakoff, A., Roux, L., et al. 2010. Astroglial networks: a step further in neuroglial and gliovascular interactions. *Nat. Rev. Neurosci.* 11: 87–99.

Giaume, C., Tabernero, A., and Medina, J. M. 1997. Metabolic trafficking through astrocytic gap junctions. *Glia* 21: 114–123.

Giaume, C., and Theis, M. 2010. Pharmacological and genetic approaches to study connexin-mediated channels in glial cells of the central nervous system. *Brain Res. Rev.* 63: 160–176.

Glass, M., and Dragunow, M. 1995. Neurochemical and morphological changes associated with human epilepsy. *Brain Res. Rev.* 21: 29–41.

Halassa, M. M., Fellin, T., and Haydon, P. G. 2007. The tripartite synapse: roles for gliotransmission in health and disease. *Trends Mol. Med.* 13: 54–63.

Halassa, M. M., Florian, C., Fellin, T., et al. 2009. Astrocytic modulation of sleep homeostasis and cognitive consequences of sleep loss. *Neuron* 61: 213–219.

Hamilton, N. B., and Attwell, D. 2010. Do astrocytes really exocytose neurotransmitters? *Nat. Rev. Neurosci.* 11: 227–238.

Hammer, J., Alvestad, S., Osen, K. K., et al. 2008. Expression of glutamine synthetase and glutamate dehydrogenase in the latent phase and chronic phase in the kainate model of temporal lobe epilepsy. *Glia* 56: 856–868.

Haydon, P. G., and Carmignoto, G. 2006. Astrocyte control of synaptic transmission and neurovascular coupling. *Physiol Rev.* 86: 1009–1031.

Heinemann, U., and Lux, H. D. 1977. Ceiling of stimulus induced rises in extracellular potassium concentration in the cerebral cortex of cat. *Brain Res.* 120: 231–249.

Heja, L., Barabas, P., Nyitrai, G., et al. 2009. Glutamate uptake triggers transporter-mediated GABA release from astrocytes. *PLoS. ONE* 4: e7153.

Henneberger, C., Papouin, T., Oliet, S. H., et al. 2010. Long-term potentiation depends on release of D-serine from astrocytes. *Nature* 463: 232–236.

Herman, M. A., and Jahr, C. E. 2007. Extracellular glutamate concentration in hippocampal slice. *J. Neurosci.* 27: 9736–9741.

Hibino, H., Inanobe, A., Furutani, K., et al. 2010. Inwardly rectifying potassium channels: their structure, function, and physiological roles. *Physiol Rev.* 90: 291–366.

Higashi, K., Fujita, A., Inanobe, A., et al. 2001. An inwardly rectifying K(+) channel, Kir4.1, expressed in astrocytes surrounds synapses and blood vessels in brain. *Am. J. Physiol. Cell. Physiol.* 281: C922–C931.

Hinterkeuser, S., Schröder, W., Hager, G., et al. 2000. Astrocytes in the hippocampus of patients with temporal lobe epilepsy display changes in potassium conductances. *Eur. J. Neurosci.* 12: 2087–2096.

Hu, S., Sheng, W. S., Ehrlich, L. C., et al. 2000. Cytokine effects on glutamate uptake by human astrocytes. *Neuroimmunomodulation* 7: 153–159.

Huang, H., and Bordey, A. 2004. Glial glutamate transporters limit spillover activation of presynaptic NMDA receptors and influence synaptic inhibition of Purkinje neurons. *J. Neurosci.* 24: 5659–5669.

Ivens, S., Kaufer, D., Flores, L. P., et al. 2007. TGF-beta receptor-mediated albumin uptake into astrocytes is involved in neocortical epileptogenesis. *Brain* 130: 535–547.

Jabaudon, D., Shimamoto, K., Yasuda-Kamatani, Y., et al. 1999. Inhibition of uptake unmasks rapid extracellular turnover of glutamate of nonvesicular origin. *Proc. Natl. Acad. Sci. USA* 96: 8733–8738.

Jauch, R., Windmüller, O., Lehmann, T. N., et al. 2002. Effects of barium, furosemide, ouabaine and 4,4'-diisothiocyanatostilbene-2,2'-disulfonic acid (DIDS) on ionophoretically-induced changes in extracellular potassium concentration in hippocampal slices from rats and from patients with epilepsy. *Brain Res.* 925: 18–27.

John, G. R., Scemes, E., Suadicani, S. O., et al. 1999. IL-1beta differentially regulates calcium wave propagation between primary human fetal astrocytes via pathways involving P2 receptors and gap junction channels. *Proc. Natl. Acad. Sci. USA* 96: 11613–11618.

Kam, K., and Nicoll, R. 2007. Excitatory synaptic transmission persists independently of the glutamate-glutamine cycle. *J. Neurosci.* 27: 9192–9200.

Kang, N., Xu, J., Xu, Q., et al. 2005. Astrocytic glutamate release-induced transient depolarization and epileptiform discharges in hippocampal CA1 pyramidal neurons. *J. Neurophysiol.* 94: 4121–4130.

Kim, J. E., Ryu, H. J., Yeo, S. I., et al. 2009. Differential expressions of aquaporin subtypes in astroglia in the hippocampus of chronic epileptic rats. *Neuroscience* 163: 781–789.

Kivi, A., Lehmann, T. N., Kovacs, R., et al. 2000. Effects of barium on stimulus-induced rises of [K+]o in human epileptic non-sclerotic and sclerotic hippocampal area CA1. *Eur. J. Neurosci.* 12: 2039–2048.

Koehler, R. C., Roman, R. J., and Harder, D. R. 2009. Astrocytes and the regulation of cerebral blood flow. *Trends Neurosci.* 32: 160–169.

Kofuji, P., Ceelen, P., Zahs, K. R., et al. 2000. Genetic inactivation of an inwardly rectifying potassium channel (Kir4.1 subunit) in mice: phenotypic impact in retina. *J. Neurosci.* 20: 5733–5740.

Kofuji, P., and Newman, E. A. 2004. Potassium buffering in the central nervous system. *Neuroscience* 129: 1045–1056.

Kucheryavykh, Y. V., Kucheryavykh, L. Y., Nichols, C. G., et al. 2007. Downregulation of Kir4.1 inward rectifying potassium channel subunits by RNAi impairs potassium transfer and glutamate uptake by cultured cortical astrocytes. *Glia* 55: 274–281.

Kunze, A., Congreso, M. R., Hartmann, C., et al. 2009. Connexin expression by radial glia-like cells is required for neurogenesis in the adult dentate gyrus. *Proc. Natl. Acad. Sci. USA* 106: 11336–11341.

Le Meur K., Galante, M., Angulo, M. C., et al. 2007. Tonic activation of NMDA receptors by ambient glutamate of non-synaptic origin in the rat hippocampus. *J. Physiol.* 580: 373–383.

Lee, S. G., Su, Z. Z., Emdad, L., et al. 2008. Mechanism of ceftriaxone induction of excitatory amino acid transporter-2 expression and glutamate uptake in primary human astrocytes. *J. Biol. Chem.* 283: 13116–13123.

Lee, T. S., Bjornsen, L. P., Paz, C., et al. 2006. GAT1 and GAT3 expression are differently localized in the human epileptogenic hippocampus. *Acta Neuropathol.* 111: 351–363.

Li, T., Ren, G., Lusardi, T., et al. 2008. Adenosine kinase is a target for the prediction and prevention of epileptogenesis in mice. *J. Clin. Invest.* 118: 571–582.

Liang, S. L., Carlson, G. C., and Coulter, D. A. 2006. Dynamic regulation of synaptic GABA release by the glutamate-glutamine cycle in hippocampal area CA1. *J. Neurosci.* 26: 8537–8548.

Maroso, M., Balosso, S., Ravizza, T., et al. 2010. Toll-like receptor 4 and high-mobility group box 1 are involved in ictogenesis and can be targeted to reduce seizures. *Nat. Med.* 16: 413–419.

Martin, E. D., Fernandez, M., Perea, G., et al. 2007. Adenosine released by astrocytes contributes to hypoxia-induced modulation of synaptic transmission. *Glia* 55: 36–45.

Mathern, G. W., Mendoza, D., Lozada, A., et al. 1999. Hippocampal GABA and glutamate transporter immunoreactivity in patients with temporal lobe epilepsy. *Neurology* 52: 453–472.

Matthias, K., Kirchhoff, F., Seifert, G., et al. 2003. Segregated expression of AMPA-type glutamate receptors and glutamate transporters defines distinct astrocyte populations in the mouse hippocampus. *J. Neurosci.* 23: 1750–1758.

Medzhitov, R. 2001. Toll-like receptors and innate immunity. *Nat. Rev. Immunol.* 1: 135–145.

Meme, W., Calvo, C. F., Froger, N., et al. 2006. Proinflammatory cytokines released from microglia inhibit gap junctions in astrocytes: potentiation by beta-amyloid. *FASEB J.* 20: 494–496.

Montana, V., Malarkey, E. B., Verderio, C., et al. 2006. Vesicular transmitter release from astrocytes. *Glia* 54: 700–715.

Mulholland, P. J., Carpenter-Hyland, E. P., Hearing, M. C., et al. 2008. Glutamate transporters regulate extrasynaptic NMDA receptor modulation of Kv2.1 potassium channels. *J. Neurosci.* 28: 8801–8809.

Mulligan, S. J., and MacVicar, B. A. 2004. Calcium transients in astrocyte endfeet cause cerebrovascular constrictions. *Nature* 431: 195–199.

Nagy, J. I., Dudek, F. E., and Rash, J. E. 2004. Update on connexins and gap junctions in neurons and glia in the mammalian nervous system. *Brain Res. Rev.* 47: 191–215.

Naus, C. C. G., Bechberger, J. F., and Paul, D. L. 1991. Gap junction gene expression in human seizure disorder. *Exp. Neurol.* 111: 198–203.

Nedergaard, M. 1994. Direct signaling from astrocytes to neurons in cultures of mammalian brain cells. *Science* 263: 1768–1771.

Neusch, C., Papadopoulos, N., Müller, M., et al. 2006. Lack of the Kir4.1 channel subunit abolishes K+ buffering properties of astrocytes in the ventral respiratory group: impact on extracellular K+ regulation. *J. Neurophysiol.* 95: 1843–1852.

Nicholson, C., and Syková, E. 1998. Extracellular space structure revealed by diffusion analysis. *Trends Neurosci.* 21: 207–215.

Nielsen, S., Nagelhus, E. A., Amiry-Moghaddam, M., et al. 1997. Specialized membrane domains for water transport in glial cells: high-resolution immunogold cytochemistry of aquaporin-4 in rat brain. *J. Neurosci.* 17: 171–180.

Olsen, M. L., Higashimori, H., Campbell, S. L., et al. 2006. Functional expression of Kir4.1 channels in spinal cord astrocytes. *Glia* 53: 516–528.

Orkand, R. K., Nicholls, J. G., and Kuffler, S. W. 1966. Effect of nerve impulses on the membrane potential of glial cells in the central nervous system of amphibia. *J Neurophysiol* 29: 788–806.

Parpura, V., Basarsky, T. A., Liu, F., et al. 1994. Glutamate-mediated astrocyte-neuron signalling. *Nature* 369: 744–747.

Pascual, O., Casper, K. B., Kubera, C., et al. 2005. Astrocytic purinergic signaling coordinates synaptic networks. *Science* 310: 113–116.

Pasti, L., Volterra, A., Pozzan, T., et al. 1997. Intracellular calcium oscillations in astrocytes: a highly plastic, bidirectional form of communication between neurons and astrocytes *in situ*. *J. Neurosci.* 17: 7817–7830.

Perillan, P. R., Chen, M., Potts, E. A., et al. 2002. Transforming growth factor-beta 1 regulates Kir2.3 inward rectifier K+ channels via phospholipase C and protein kinase C-delta in reactive astrocytes from adult rat brain. *J. Biol. Chem.* 277: 1974–1980.

Petzold, G. C., Albeanu, D. F., Sato, T. F., et al. 2008. Coupling of neural activity to blood flow in olfactory glomeruli is mediated by astrocytic pathways. *Neuron* 58: 897–910.

Proper, E. A., Hoogland, G., Kappen, S. M., et al. 2002. Distribution of glutamate transporters in the hippocampus of patients with pharmaco-resistant temporal lobe epilepsy. *Brain* 125: 32–43.

Reimann, F., and Ashcroft, F. M. 1999. Inwardly rectifying potassium channels. *Curr. Opin. Cell Biol.* 11: 503–508.

Retamal, M. A., Froger, N., Palacios-Prado, N., et al. 2007. Cx43 hemichannels and gap junction channels in astrocytes are regulated oppositely by proinflammatory cytokines released from activated microglia. *J. Neurosci.* 27: 13781–13792.

Rothstein, J. D., Patel, S., Regan, M. R., et al. 2005. Beta-lactam antibiotics offer neuroprotection by increasing glutamate transporter expression. *Nature* 433: 73–77.

Rouach, N., Avignone, E., Meme, W., et al. 2002. Gap junctions and connexin expression in the normal and pathological central nervous system. *Biol. Cell* 94: 457–475.

Rouach, N., Koulakoff, A., Abudara, V., et al. 2008. Astroglial metabolic networks sustain hippocampal synaptic transmission. *Science* 322: 1551–1555.

Sah, P., Hestrin, S., and Nicoll, R. A. 1989. Tonic activation of NMDA receptors by ambient glutamate enhances excitability of neurons. *Science* 246: 815–818.

Samoilova, M., Wentlandt, K., Adamchik, Y., et al. 2008. Connexin 43 mimetic peptides inhibit spontaneous epileptiform activity in organotypic hippocampal slice cultures. *Exp. Neurol.* 210: 762–775.

Sarac, S., Afzal, S., Broholm, H., et al. 2009. Excitatory amino acid transporters EAAT-1 and EAAT-2 in temporal lobe and hippocampus in intractable temporal lobe epilepsy. *APMIS* 117: 291–301.

Scholl, U. I., Choi, M., Liu, T., et al. 2009. Seizures, sensorineural deafness, ataxia, mental retardation, and electrolyte imbalance (SeSAME syndrome) caused by mutations in KCNJ10. *Proc. Natl. Acad. Sci. USA* 106: 5842–5847.

Schummers, J., Yu, H., and Sur, M. 2008. Tuned responses of astrocytes and their influence on hemodynamic signals in the visual cortex. *Science* 320: 1638–1643.

Seifert, G., Hüttmann, K., Binder, D. K., et al. 2009. Analysis of astroglial K^+ channel expression in the developing hippocampus reveals a predominant role of the Kir4.1 subunit. *J. Neurosci.* 29: 7474–7488.

Seifert, G., Schilling, K., and Steinhäuser, C. 2006. Astrocyte dysfunction in neurological disorders: a molecular perspective. *Nat. Rev. Neurosci.* 7: 194–206.

Seiffert, E., Dreier, J. P., Ivens, S., et al. 2004. Lasting blood-brain barrier disruption induces epileptic focus in the rat somatosensory cortex. *J. Neurosci.* 24: 7829–7836.

Serrano, A., Haddjeri, N., Lacaille, J. C., et al. 2006. GABAergic network activation of glial cells underlies hippocampal heterosynaptic depression. *J. Neurosci.* 26: 5370–5382.

Shigetomi, E., Bowser, D. N., Sofroniew, M. V., et al. 2008. Two forms of astrocyte calcium excitability have distinct effects on NMDA receptor-mediated slow inward currents in pyramidal neurons. *J. Neurosci.* 28: 6659–6663.

Steinhäuser, C., de Lanerolle N. C., and Haydon, P. G. 2007. Astroglial mechanisms in epilepsy. In *Epilepsy: A comprehensive textbook*, ed. Engel, J., and Pedley, T. A., 277–288. 2nd ed. Philadelphia: Lippincott Williams & Wilkins.

Steinhäuser, C., and Seifert, G. 2002. Glial membrane channels and receptors in epilepsy: impact for generation and spread of seizure activity. *Eur. J. Pharmacol.* 447: 227–237.

Stellwagen, D., Beattie, E. C., Seo, J. Y., et al. 2005. Differential regulation of AMPA receptor and GABA receptor trafficking by tumor necrosis factor-alpha. *J. Neurosci.* 25: 3219–3228.

Tabernero, A., Medina, J. M., and Giaume, C. 2006. Glucose metabolism and proliferation in glia: role of astrocytic gap junctions. *J. Neurochem.* 99: 1049–1061.

Tang, F. R., and Lee, W. L. 2001. Expression of the group II and III metabotropic glutamate receptors in the hippocampus of patients with mesial temporal lobe epilepsy. *J. Neurocytol.* 30: 137–143.

Tani, H., Dulla, C. G., Huguenard, J. R., et al. 2010. Glutamine is required for persistent epileptiform activity in the disinhibited neocortical brain slice. *J. Neurosci.* 30: 1288–1300.

Tessler, S., Danbolt, N. C., Faull, R. L. M., et al. 1999. Expression of the glutamate transporters in human temporal lobe epilepsy. *Neuroscience* 88: 1083–1091.

Tian, G. F., Azmi, H., Takano, T., et al. 2005. An astrocytic basis of epilepsy. *Nat. Med.* 11: 973–981.

Tomkins, O., Friedman, O., Ivens, S., et al. 2007. Blood-brain barrier disruption results in delayed functional and structural alterations in the rat neocortex. *Neurobiol. Dis.* 25: 367–377.

Tzingounis, A. V., and Wadiche, J. I. 2007. Glutamate transporters: confining runaway excitation by shaping synaptic transmission. *Nat. Rev. Neurosci.* 8: 935–947.

van der Hel, W. S., Notenboom, R. G., Bos, I. W., et al. 2005. Reduced glutamine synthetase in hippocampal areas with neuron loss in temporal lobe epilepsy. *Neurology* 64: 326–333.

Van Vliet, E. A., da Costa, A. S., Redeker, S., et al. 2007. Blood-brain barrier leakage may lead to progression of temporal lobe epilepsy. *Brain* 130: 521–534.

Verkhratsky, A., and Steinhäuser, C. 2000. Ion channels in glial cells. *Brain Res. Rev.* 32: 380–412.

Vezzani, A., Balosso, S., and Ravizza, T. 2008. The role of cytokines in the pathophysiology of epilepsy. *Brain Behav. Immun.* 22: 797–803.

Virta, M., Hurme, M., and Helminen, M. 2002. Increased plasma levels of pro- and anti-inflammatory cytokines in patients with febrile seizures. *Epilepsia* 43: 920–923.

Wallraff, A., Köhling, R., Heinemann, U., et al. 2006. The impact of astrocytic gap junctional coupling on potassium buffering in the hippocampus. *J. Neurosci.* 26: 5438–5447.

Walz, W. 2000. Role of astrocytes in the clearance of excess extracellular potassium. *Neurochem. Int.* 36: 291–300.

Wang, X., Lou, N., Xu, Q., et al. 2006. Astrocytic Ca(2+) signaling evoked by sensory stimulation *in vivo. Nat. Neurosci.* 9: 816–823.

Wetherington, J., Serrano, G., and Dingledine, R. 2008. Astrocytes in the epileptic brain. *Neuron* 58: 168–178.

Wolburg-Buchholz, K., Mack, A. F., Steiner, E., et al. 2009. Loss of astrocyte polarity marks blood-brain barrier impairment during experimental autoimmune encephalomyelitis. *Acta Neuropathol.* 118: 219–233.

Xiong, Z. Q., and Stringer, F. L. 2000. Sodium pump activity, not glial spatial buffering, clears potassium after epileptiform activity induced in the dentate gyrus. *J. Neurophysiol.* 83: 1443–1451.

Yang, Y., Gozen, O., Watkins, A., et al. 2009. Presynaptic regulation of astroglial excitatory neurotransmitter transporter GLT1. *Neuron* 61: 880–894.

Ye, Z. C., and Sontheimer, H. 1996. Cytokine modulation of glial glutamate uptake: a possible involvement of nitric oxide. *NeuroReport* 7: 2181–2185.

Zhang, J., Wang, H., Ye, C., et al. 2003. ATP released by astrocytes mediates glutamatergic activity-dependent heterosynaptic suppression. *Neuron* 40: 971–982.

Zhang, Z., Chen, G., Zhou, W., et al. 2007. Regulated ATP release from astrocytes through lysosome exocytosis. *Nat. Cell Biol.* 9: 945–953.

Zonta, M., Angulo, M. C., Gobbo, S., et al. 2003. Neuron-to-astrocyte signaling is central to the dynamic control of brain microcirculation. *Nat. Neurosci.* 6: 43–50.

Zorumski, C. F., Mennerick, S., and Que, J. 1996. Modulation of excitatory synaptic transmission by low concentrations of glutamate in cultured rat hippocampal neurons. *J. Physiol.* 494: 465–477.

11 Astrocyte Involvement in the Acquired Demyelinating Diseases

Sarah E. Lutz
Department of Neuroscience, Albert Einstein
College of Medicine, Bronx, New York

Cedric S. Raine
Departments of Neuroscience, Pathology, and Neurology,
Albert Einstein College of Medicine, Bronx, New York

Celia F. Brosnan
Departments of Neuroscience, and Pathology, Albert
Einstein College of Medicine, Bronx, New York

CONTENTS

11.1 INTRODUCTION

11.1.1 THE ACQUIRED DEMYELINATING DISEASES OF MAN

The acquired demyelinating diseases of the human central nervous system (CNS) are a heterogeneous group of disorders characterized by selective damage to myelin (reviewed in Raine, 1984; Lassmann, 1999; Wingerchuk and Weinshenker, 2008). The underlying cause of myelin loss may be infectious (e.g., progressive multifocal leukoencephalopathy, post-infectious encephalitis), or inflammatory/ immune mediated (multiple sclerosis, neuromyelitis optica), or may result from exposure to toxic factors, chronic ischemia of the deep white matter associated with aging, or nutritional deficiencies (vitamin B12 deficiency). It has also been postulated that iatrogenic autoimmune encephalomyelitis may have occurred following immunization with CNS homogenates, as, for example, when several patients treated with Pasteur's vaccine for rabies developed encephalitis, polyneuritis, and perivascular demyelination (Baxter, 2007). In contrast to the acquired demyelinating diseases listed above, myelin diseases with a genetic basis usually develop early in life, display symmetric and extensive involvement of the hemispheric white matter, and are associated with widespread loss of AXONS and high morbidity.

The most prevalent of the acquired demyelinating disorders is multiple sclerosis (MS), which affects more than 2 million individuals worldwide (www.nmss.org). The majority of patients in the early stages of disease exhibit a relapsing-remitting disease pattern that gradually converts to a chronic secondary progressive disease state in which periods of remission are not as pronounced. In a minor population of patients, the disease is continually progressive from onset, and is referred to as primary progressive. Onset of disease usually occurs between the 20th and 25th year, and women are affected more than men. Optic neuritis and internuclear ophthalmoplegia are frequent early signs. Although MS is rarely fatal in the early stages, the burden of disease becomes progressively worse with time, likely due to the accumulating loss of axons (Filippi et al., 1996; Losseff et al., 1996; Lassmann, 1999; Wujek et al., 2002). Imaging studies have shown that lesions are particularly common in the periventricular white matter (see Figure 11.1A), although any structures that contain myelin may harbor MS plaques (Ludwin and Raine, 2008). Geographically, MS shows a distinct increase in prevalence in more temperate zones (Compston, 1990), which suggests the involvement of environmental factors in disease pathogenesis, such as infectious agents. However, most current hypotheses support the conclusion

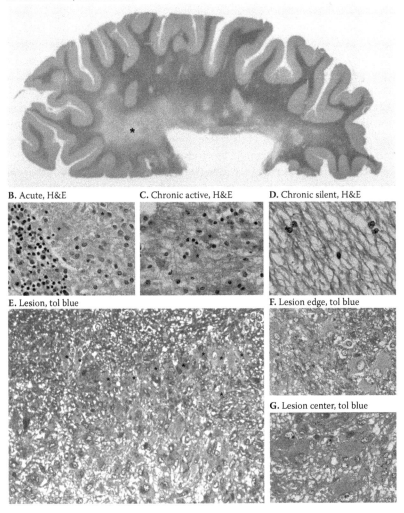

FIGURE 11.1 **(See color insert.)** Astrocytic reactivity in MS lesions. (A) A section of a hemisphere from the brain of a patient with chronic MS stained with Luxol-fast-blue for myelin. Note the many large periventricular lesions lacking myelin, some of which present with diffuse edges (*) suggestive of ongoing activity, whereas others possess sharper borders, indicative of chronic myelin loss. (B) An H&E preparation from an acute MS lesion shows hypertrophic astrocytes (*) in the CNS parenchyma adjacent to a perivascular accumulation of plasma cells and lymphocytes (at left). (C) An H&E-stained section from the center of a chronic active MS lesion reveals many hypertrophic astrocytes in the demyelinated parenchyma. (D) The center of a chronic silent lesion shows a network of fibrous astrocytes. (E) A one-micron epoxy section from the margin of an acute MS lesion shows a broad zone of hypertrophic astrocytes (asterisks) between myelinated white matter (upper left) and lesion center (below). Note how the same cells extend into the lesion proper. (F, G) Detail of lesion edge and center, respectively, to show hypertrophic astrocytes. Note how the cells are more globoid at the edge and more process bearing within the lesion.

that an autoimmune attack directed against components of CNS myelin represents the final common pathway of disease induction.

Pathological staging of MS lesions remains controversial, but they are usually categorized as active/acute, chronic active, and chronic inactive/silent (Figure 11.1). In acute and chronic active lesions, perivascular cuffs of lymphocytes and macrophages are associated with areas of primary demyelination, with chronic active lesions often showing a chronic core and an active edge. Chronic silent lesions display loss of oligodendrocytes, myelin, and axons, with little evidence of ongoing inflammation (Ludwin and Raine, 2008). Remarkably, all three types of lesions may be found within adjacent areas of the same brain, suggesting that it is the microenvironment of the lesion itself that influences lesion outcome.

Oligodendrocytes and myelin are thought to be the primary targets of the immune response, but pathological changes are also noted in neuronal elements, astrocytes, microglia, and the vasculature. Axonal and neuronal loss within the lesion correlates with chronic dysfunction in MS patients (Trapp et al., 1998; Kornek et al., 2000). A reactive astrogliosis is also prevalent in many chronic lesions and is responsible for the hallmark *sclérose en plaques* for which the disease is named.

11.1.2 ANIMAL MODELS OF THE DEMYELINATING DISEASES

Animal models that have proved of value for studying the pathogenesis of MS can be divided into three broad categories: those that are autoimmune in origin, those mediated by an infectious agent, and those mediated by gliotoxins. With respect to autoimmune models, it is now well recognized that specific peptides of most myelin proteins can cause a paralytic disease that is associated with inflammation and demyelination of the CNS when administered to the right host in the presence of adjuvant. This process involves association of peptide with major histocompatibility complex (MHC) class II and expression of costimulatory molecules by antigen-presenting cells, leading to the activation of T lymphocytes expressing either interferon-gamma (Th1 cells) or interleukin (IL)-17 (Th17 cells) that drive the inflammatory process. Passive (adoptive) transfer of disease to naïve animals can also be mediated by activated myelin-peptide-specific T cells expressing these cytokines (Paterson, 1960; Stromnes and Goverman, 2006; Miller et al., 2010). Transgenic animals expressing T cell receptors (TCRs) that recognize specific myelin peptides have also been developed and have helped to define the role of regulatory cells in disease expression, as well as to define distinct clinical features of experimental autoimmune encephalomyelitis (EAE) mediated by specific myelin peptides (Goverman et al., 1993; Lafaille et al., 1994; Mendel et al., 1995; Waldner et al., 2000; Bettelli et al., 2003). Taken together, these studies have shown that depending upon the antigen, species, and strain of animal chosen, different patterns of clinical expression of disease can be obtained that mimic in their complexity the variability in disease expression found in MS patients. The common patterns of disease pathogenesis that have been observed from mice to primates have provided a fertile testing ground for the development of new modes of therapy (Raine and Genain, 2008).

Several infectious agents have also been tested for their potential role in initiating inflammatory demyelinating lesions in animals, but the two that have been most actively pursued are Theiler's murine encephalomyelitis virus (TMEV) and Semliki Forest virus (SFV) (Olson et al., 2005). Interestingly, in both of these models, stimulation of autoreactive T cells is thought to contribute to disease pathogenesis either through molecular mimicry between viral and host peptides, or via activation of autoreactive T cells at sites of viral-induced tissue damage and inflammation in the CNS, a phenomenon known as bystander activation. Canine distemper encephalomyelitis is a naturally occurring, paramyxovirus demyelinating condition with many similarities to EAE (Wisniewski et al., 1972). Although astrocytes are infected by several viruses and may function as reservoirs for others (e.g., HHV6), the involvement of astrocytes in disease expression in viral models of the inflammatory demyelinating diseases has received less attention than in EAE, and thus the major focus of studies addressed below will be on MS and EAE.

In addition to a direct role for viruses in inducing demyelination, other infectious agents have also been tested for their potential role as adjuvants. The important contribution that ambient infectious agents can play in disease expression has been amply demonstrated in transgenic mice expressing TCRs specific for myelin peptides, where it was found that mice developed spontaneous disease when kept in a conventional "dirty" housing facility but not when maintained in a specific pathogen-free environment (Goverman et al., 1993; Farez et al., 2009). Activation of T cells, which then readily gain access to the CNS, is thought to be a contributing mechanism. Taken together, these data support a role for infections, particularly upper respiratory tract infections, in precipitating attacks in MS through an adjuvant effect.

Gliotoxins that have provided useful demyelinating models in vivo include injection into white matter tracts of lysolecithin or ethidium bromide, or the addition of the copper chelating agent cuprizone to food pellets (reviewed in Blakemore and Franklin, 2008). These procedures result in a reversible demyelinating event that proceeds with well-defined kinetics following a single course of treatment, or a chronic irreversible injury following multiple exposures (Carlton, 1966; Blakemore, 1972). They have proven to be great value in defining factors that protect against toxicity, such as prolactin (Raine, 1984), as well as factors that promote or inhibit repair (Woodruff and Franklin, 1999). They have also proven valuable in testing the merits of the involvement of or treatment with various glial precursor cell populations in mediating repair (Gallo and Armstrong, 2008).

11.2   THE PATHOGENIC RESPONSE OF ASTROCYTES IN MS AND EAE

11.2.1   ASTROCYTE PATHOLOGY IN MS

Classically, neuropathologists have used the appearance of the astrocyte as a major indicator in the assessment of activity and age in the multiple sclerosis lesion (Figure 11.1). It is well known that the astrocyte is the first element of the CNS to display reactivity during lesion formation in MS, and that it provides the initial structural barrier in the CNS parenchyma to the entry of immune cells and their products (D'Amelio et al., 1990; Ludwin and Raine, 2008; Owens et al., 2008).

Around the early (acute) inflammatory lesion, astrocyte reactivity is widespread (Figure 11.1) and there is a gradient of reactivity from modestly swollen cells with processes in normal adjacent white matter to grossly enlarged globoid (hypertrophic) astrocytes in the lesion center (Figure 11.1E–G). Astrocyte hypertrophy begins at the lesion edge, where the cells show somewhat retracted processes and are sometimes almost arranged as a wall (Figure 11.1E, F). Deeper areas display much enlarged globoid astrocytes suspended in an edematous parenchyma containing swollen astrocyte processes, macrophages, and infiltrating cells (Figure 11.1B, C, G and Figure 11.2). These astrocytes display diffuse immunoreactivity for glial fibrillary acidic protein (GFAP) (Wu and Raine, 1992), corresponding to the abundance of fragmented (depolymerized) glial filaments seen ultrastructurally, and contribute to the fleshy edematous state of the tissue (Figure 11.2). Sometimes, these enlarged cells are multinucleated (a.k.a. Creutzfeldt astrocytes; see Figure 11.2D), exceed 100 μm in diameter, and share some similarities to cells encountered in gliomas (reviewed in Sofroniew and Vinters, 2010). Such reactive astrocytes within the acute lesion are unusual inasmuch as they are capable of internalizing oligodendrocytes (Figure 11.2E) and lymphocytes, a phenomenon known as emperipolesis (Ghatak, 1992; Wu and Raine, 1992; Nguyen and Pender, 1998). Hypertrophic astrocytes are also capable of phagocytosing myelin fragments (Lee et al., 1990), taking up IgG from the surrounding tissue, expressing molecules of the complement cascade, and showing staining for proinflammatory and regulatory cytokines and chemokines (Brosnan et al., 1995), in addition to growth factors, inhibitory molecules, nitric oxide (NO) (Bo et al., 1994; Brosnan et al., 1994; Liu et al., 2001), glutamate, and other neurotransmitters (reviewed in Seifert et al., 2006; Sofroniew and Vinters, 2010).

In some acute MS lesions, astrocytic damage is extensive. In such cases, the perivascular layer of astrocyte processes forming the glia limitans is disrupted, leaving an intact basal lamina with scattered astroglial endfeet still attached (Figure 11.2A). Individual astrocyte processes are swollen and contain a mixture of polymerized and depolymerized filaments (Figure 11.2B). Some astrocyte processes display gap junctions (Figure 11.2B). The occurrence of disruptions in the perivascular glia limitans in such acute MS lesions raises the possibility that the astrocyte may be a target of the inflammatory response, thus contributing pathogenic significance to this phenomenon.

As the lesion ages, hypertrophic astrocytes persist but begin to accumulate GFAP, and develop processes packed with dense bundles of glial filaments (Figures 11.1C,D and 11.2F). GFAP immunoreactivity is intense, tissue edema decreases, and collectively, these cells provide a scaffold for surviving axons and other cellular elements within the chronic demyelinated lesion. By this time, the lesion is usually devoid of oligodendrocytes and inflammatory cells are much reduced in number, although macrophage activity continues well into the chronic phase.

Ultimately, macrophages (foamy macrophages) are engorged with lipid, the end product of myelin destruction, and congregate around blood vessels, from which site they are believed to exit the tissue, leaving the fibrous astrocyte, now much reduced in size, as the major cellular component of the gliotic, chronic demyelinated lesion (Figure 11.3). Older, inactive lesions are intensely glassy and

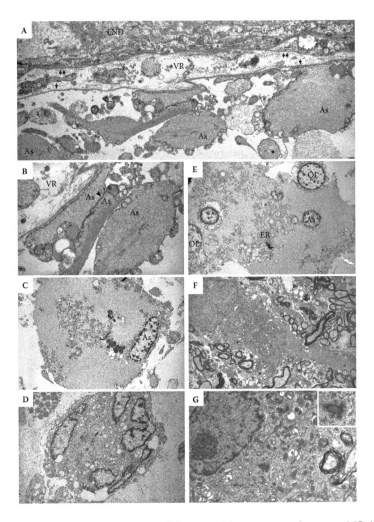

FIGURE 11.2 Electron microscopy of hypertrophic astrocytes in acute MS lesions. (A) A perivascular lesion shows loss of continuity of the astrocytic layer (the glia limitans). The endothelial basal lamina (double arrows) and the glia limitans basal lamina (arrows) are intact. Astrocytic processes, some of which remain attached to their basal lamina, lie within an edematous matrix. End, endothelial cell; VR, Virchow-Robin space; As, astrocyte. (B) Detail of (A) to show retention of membrane specialization between astrocyte and its basal lamina and the presence of junctional complexes between astrocytes (arrowheads). Note also the existence of both polymerized and depolymerized filaments within the processes. (C) A large hypertrophic astrocyte lies within an edematous parenchyma. Note depolymerized glial filaments within the cytosol. (D) A multinucleated hypertrophic astrocyte lies within an edematous matrix. (E) Another hypertrophic astrocyte filled with depolymerized filaments and endoplasmic reticulum (ER) displays emperipolesis of one oligodendrocyte (*) and close association with two others on the cell surface (OL). The astrocytic nucleus is shown at AS. (F) In the white matter adjacent to another acute lesion, a large reactive astrocyte is shown. (G) A reactive astrocyte at the margin of an acute lesion displays a centriole (inset), evidence of recent mitotic activity.

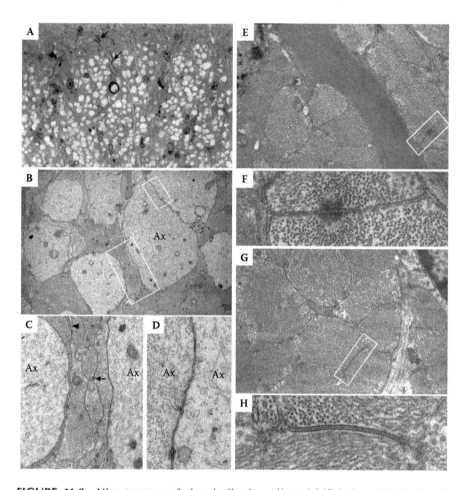

FIGURE 11.3 Ultrastructure of chronically demyelinated MS lesions. (A) A chronic silent MS lesion of the anterior column displays almost total loss of myelin and presence of surviving axons in cross section within an intensely gliotic parenchyma. The interstitial astrogliosis contains a population of astrocytes with dark thread-like processes (arrows). One-micron epoxy section stained with toluidine blue. (B) An electron micrograph from a chronic silent lesion shows naked axons (A) in cross section surrounded by a dense network of fibrous astrogliotic processes. (C) Detail of the interstitial gliosis to show tightly packed fibrous astrocyte processes between two axons. Note the presence of a desmosome (arrow) and a gap junction (arrowhead). (D) Detail of B showing lack of astrocytic processes between two adjacent axons. (E) Higher magnification from the lesion edge to show a cell process on the dense astrocyte traversing a field of fibrous astrocytes. (F) Detail of the desmosomes indicated in (E). (G) High magnification of another area of the margin of a chronic lesion to show astroglial scar tissue in cross section. The gap junction (boxed) is shown in (H).

shrunken in gross specimens, and consist histologically almost exclusively of fibrous astrocytes in a dense array of glial filaments, junctional complexes, and extracellular matrix elements containing scattered microglia and demyelinated axons (Figures 11.1D and 11.3A–H). Fibrous astrogliosis invariably extends for some distance into the normal adjacent white matter. Typically, the chronic gliotic lesion is devoid of oligodendrocytes and shows little prospect for myelin repair. Relapsing disease activity is accompanied by recurrent inflammatory activity and a fresh wave of astroglial reactivity and hypertrophy at the lesion edge (chronic active lesions).

Remyelination occurs in MS and is usually associated with the margins of chronic active lesions displaying ongoing activity, where the tissue contains oligodendrocytes and oligodendrocyte progenitor cells (OPCs), which are capable of remyelinating axons in a narrow rim around the edge of the lesion (Raine et al., 1981). Frequently in acute MS lesions, there is a prolific oligodendrocyte response, believed to derive from the differentiation of oligodendrocyte progenitors (rather than mitosis of mature cells), which is accompanied by extensive remyelination, even in the face of ongoing inflammation, demyelination, and astrocytic gliosis (Raine et al., 1981; Prineas et al., 1989, 2002b; Patrikios et al., 2006; Patani et al., 2007; Goldschmidt et al., 2009). However, with the relentless progression of the disease, this remyelination is usually transient and abortive and eventually falls victim to the ongoing expansion of the lesion, which once more becomes demyelinated; the oligodendrocytes also degenerate in the process.

An interesting question that remains to be more fully explored is whether an astrogliotic response ever fully resolves. Data suggest that, at least in MS, this does not occur. Notable, for example, are images from shadow plaques in MS where remyelinated axons appear embedded in highly filamentous astrocytic processes (Prineas et al., 2002a). Perhaps in the future, a comparison at the molecular level of astrogliotic responses in lesions of differing pathology may provide insight into the factors present in the microenvironment that might account for differences in the outcome of the inflammatory event.

11.2.2 Astrocyte Pathology in Experimental Autoimmune Encephalomyelitis (EAE)

The response of astrocytes in animals sensitized to develop EAE has been examined predominantly in the context of effects on the expression of glial fibrillary acidic protein (GFAP), an 8–9 nm intermediate filament found in mature astrocytes and some multipotent neural precursor cells. GFAP was first isolated ~40 years ago from the center of a chronic-silent MS plaque (reviewed in Eng et al., 2000). Almost universally, using western blotting, immunohistochemistry, and in situ hybridization, EAE studies have shown that GFAP upregulation is an early event in lesion formation and fluctuates in concert with disease activity (see, for example, Aquino et al., 1990; Kothavale et al., 1995; Tani et al., 1996). However, these early studies were limited in that they did not permit analysis over time in a single animal. The myelin oligodendrocyte glycoprotein (MOG) model of EAE in C57Bl6 mice has provided an opportunity to study the response of astrocytes to immune cell invasion of the CNS using transgenic animals. What these recent studies have shown is that activation of GFAP in EAE occurs earlier than has previously been appreciated and occurs prior to the onset of the acute clinical episode. So, for example, one study used a transgenic mouse in which the gene for luciferase was placed under the transcriptional control of the mouse GFAP promoter (GFAP-luciferase mice; Luo et al., 2008). Following sensitization for EAE, mice were injected intraperitoneally with D-luciferin at varying times in the disease process, and later subjected to in vivo imaging for bioluminescence. In these mice, clinical signs of disease first occurred approximately 10 days postsensitization with MOG, and peaked around day 14. Injection of D-luciferin demonstrated a small increase in luciferase activity in the CNS as early as day 3, with significant increases detected by day 7. Luciferase activity then fluctuated in intensity in conjunction with clinical signs, and partially resolved during the chronic phase of the disease. Of particular interest was the observation that the intensity of the signal was a good predictor of the maximum clinical score noted for the animals. The authors suggested that monitoring GFAP activity using bioluminescence might be a useful noninvasive tool for determining the severity of EAE. Another study examined the early astrocytic response during the preclinical to the onset phase (days 6–12) in MOG-induced EAE, particularly as it related to evidence of axonal damage in the optic nerve (Wang et al., 2005). Evidence of axonal injury was detected during the preonset phase that was associated with astrocyte hypertrophy, and which occurred over a wide area in regions where evidence of immune cell inflammation was sparse. These studies support a role for astrocytes in neurodegenerative effects associated with inflammation. Taken together, these works showed that astrocyte activation precedes clinical disease by several days, and that the extent of activation correlated with subsequent severity of disease, thus supporting an important role for astrocytes in the initial events in EAE.

Persuasive as these data might be that astrocytes are intimately involved in the development of the EAE lesion and that studying their role in EAE has relevance to the pathogenesis of MS, it is important to point out that several aspects of the pathological changes noted in MS lesions, such as multinucleated astrocytes and astrocytic uptake of oligodendrocytes and lymphocytes, have not yet been described in EAE. In the human brain, the ratio of astrocytes to neurons is greater than that found in any other species, and human astrocytes are also larger, structurally more complex, and more diverse than those of rodents (Oberheim et al., 2009). It is also important to point out that species-specific responses of astrocytes, particularly between mouse and man, to inflammatory and immune activators have been amply documented (John et al., 2003; Farina et al., 2005), indicating that EAE in mice is a good but limited model to study astrocyte responses to autoimmune events directed against oligodendrocytes and myelin.

11.2.3 Astrogliosis and Proliferation

Astrocytes react to many different stimuli with a series of morphologic and biochemical changes that are broadly referred to as activation or astrogliosis. These include upregulation of GFAP protein and mRNA, and condensation of GFAP fila-ments resulting in the formation of a more compact "stellate" shape. For many years it was thought that activated astrocytes proliferated at sites of injury, based in large part upon the marked increase in GFAP immunoreactive cells near focal wounds. It is now appreciated that in most models of acute neuroinflammation, the apparent increase in GFAP-positive cells results not from the generation of new astrocytes, but rather from an increase in GFAP synthesis and a condensation of glial filaments in preexisting cells, resulting in more readily detectable GFAP by immunostaining (reviewed in Sofroniew, 2009; Sofroniew and Vinters, 2010). Incorporation of labeled nucleotide analogs in GFAP-positive cells has also been used as indicative for astrocyte proliferation in severe reactive astrogliosis in the lesioned CNS (Latov et al., 1979). However, it is difficult to resolve if newly gen-erated GFAP-positive cells in the lesioned CNS derive from GFAP-positive pro-genitor cells, from GFAP-negative cells that divide and subsequently upregulate GFAP, or if these cells originate from fully differentiated, mature astrocytes that undergo mitosis. Most (but not all) lineage-tracing experiments have suggested that newly generated astrocytes derive from populations of cells with progenitor characteristics, but that the identity of the cells giving rise to new astrocytes may differ depending on the nature of the insult and the location within the CNS. In the cerebrum, progenitor activity is found in ependymal cells in the lateral ventricle wall (Carlen et al., 2009), NG2 proteoglycan or olig2-positive cells (Magnus et al., 2008), and endothelin-1-expressing astrocytes (Gadea et al., 2008). Severing the spinal cord of mice induces proliferation of ciliated ependymal cells of the cen-tral canal, a small percentage of which exhibit radial morphology and upregulate GFAP (although in the resting state these cells are GFAP-negative) (Meletis et al., 2008). In acute EAE, mitotic activity is present in up to 25% of adult radial-glia cells immunoreactive for brain lipid-binding protein and GFAP in focal inflamma-tory lesions (Bannerman et al., 2007).

In MS, it is not uncommon in chronic lesions to observe astrocytes contain-ing centrioles, indicative of recent mitotic activity (Figure 11.2G), and a survey of human biopsy material demonstrated immunoreactivity for a mitotic marker in as many as 1% of GFAP-positive cells in five of six cases of demyelinating lesions (Colodner et al., 2005). Besides GFAP, activated astrocytes also upregulate other structural and adhesion molecules, extracellular matrix components, and inflam-matory chemokines and cytokines (see below). This can result in a positive feed-forward loop, because the secretion of inflammatory cytokines further activates local cells.

11.3 ASTROCYTES AS CONTRIBUTORS TO LESION DEVELOPMENT AND REPAIR

11.3.1 LESION DEVELOPMENT

Astrocytes contribute to lesion development by promoting the downregulation of tight junction proteins in endothelial cells, by expressing cytokines and chemo-kines that attract leukocytes to sites of injury in the CNS, by upregulating adhesion molecules and matrix metalloproteases required for leukocyte migration into the parenchyma, by producing factors toxic for oligodendrocytes and neurons, by inhibiting oligodendrocyte precursor cell maturation to a mature myelinating phenotype, and in late stages, by forming a dense gliotic scar that blocks the regeneration of axons and the migration of oligodendrocytes into the tissue. The end result is a lesion consisting of a mesh of fibrotic astrocytic processes containing a few scattered axons (Figures 11.1D, 3). Many of the factors involved in each of these events have been identified, at least in part.

11.3.1.1   Astrocytes and the Blood–Brain Barrier (BBB)

In the brain, astrocytes critically regulate permeability to solutes (Owens et al., 2008), but the extent of glial coverage of cerebral microvessels has been debated. Resolution of this issue has important implications for the movement of water and ions, and also for the trafficking of inflammatory lymphocytes into the CNS parenchyma in the context of autoimmune demyelination. Estimates extrapolated from electron or light microscopy have indicated that astrocytic endfeet cover anywhere from 80%–100% of the endothelial basal lamina, and that microglial processes cover up to 13% (Lassmann et al., 1991). Ottersen and colleagues have helped resolve this issue with a combination of quantitative morphometric analysis of serial two-dimensional electron micrographs (EMs) and three-dimensional reconstructions of capillaries in the hippocampal CA1 stratum moleculaire of 7-week-old rats (Mathiisen et al., 2010). Their morphometric analysis of serial EMs shows that the basal lamina underlying capillary endothelial cells is completely (100%) covered by astrocytic endfoot processes, with considerable overlap between adjacent endfeet, such that no gaps occur between endfeet. Interestingly, astrocytic coverage of pericytes embedded in the common endothelial/pericyte basal lamina was slightly less complete (99%). Note that these authors did not investigate postcapillary venules, which are the primary site of extravasation of inflammatory cells into the CNS parenchyma during EAE (D'Amelio et al., 1990).

The single basal lamina of brain capillaries is composed of laminins 8 and 10. However, at the level of the postcapillary venule there is a second basal lamina that is produced by astrocytes and leptomeningeal cells, called the parenchymal basement membrane (BM) composed of laminins 1 and 2. Astrocytic endfeet are attached to this parenchymal BM via dystroglycan interactions with the endothelial isoform of the heparin sulfate proteoglycan, agrin (Noell et al., 2007). Using EAE as a model, Sorokin and colleagues (Agrawal et al., 2006) have shown that encephalitogenic T cells breach these barriers in a step-wise fashion that involves the expression of specific adhesion molecules and matrix metalloproteases. Somewhat surprisingly, their data support the conclusion that whereas activated T cells can effectively cross the endothelial cell and its associated basal lamina, metalloproteases (MMPs) secreted by macrophages are required for T cell migration across the parenchymal (astrocytic) barrier. The authors refer to this as a double-barrier migration process, each step of which involves distinct molecular mechanisms, and they further suggest that it is the second step that is disease relevant (see also D'Amelio et al., 1990; Owens et al., 2008). Additional studies showed that this process involved gelatinase activity of MMP2 and 9, did not lead to a general digestion of parenchymal BM components but was relatively selective for beta-dystroglycan, and was reversible (Graesser et al., 2000; Agrawal et al., 2006). Expression of MMP7 and 9 is upregulated in MS lesions (Lindberg et al., 2001), and broad inhibition of MMP activity has been found to reduce immune cell trafficking into the CNS and attenuate EAE (Gijbels et al., 1994; Liedtke et al., 1998b; Dubois et al., 1999). Tissue inhibitors of metalloproteinases (TIMPs) are endogenous inhibitors of MMPs and are found in astrocytes in active demyelinating lesions in EAE (Pagenstecher et al., 1998). Inactivation of TIMP-1 in mice led to enhanced inflammation in the CNS parenchyma, accompanied by increased microglia/macrophage activity and prolonged myelin injury following sensitization for EAE (Crocker et al., 2006).

Disruption of the dystroglycan-agrin complex in astrocytic endfeet associated with inflamed vessels causes loss of polarized expression of aquaporin4 (AQP4) (Wolburg-Buchholz et al., 2009), the channel primarily responsible for movement of water across astrocytic membranes. AQP4 is normally anchored to the laminin-dystroglycan-dystrophin-agrin complex via interactions with alpha-syntrophin (Neely et al., 2001; Amiry-Moghaddam et al., 2003). The disruption of AQP4 expression at perivascular astrocytic endfeet directly contributes to development of intracerebral edema (Manley et al., 2000; Amiry-Moghaddam et al., 2003). Astrocytic expression of dystroglycan is also likely to have an important role in organizing the pial invaginations of the brain, as evidenced by the finding that hGFAP-Cre:dystroglycan$^{F/F}$ mice fail to develop an appropriate glia limitans/meningeal barrier, resulting in fusion of the cerebral hemispheres and cerebellar folia with consequent neuronal migration deficits (Moore et al., 2002).

Loss of BBB function and edema formation are early and clinically significant events in both MS and EAE. Activated T cells, macrophages, mast cells, and neutrophils have all been implicated in loss of barrier function (Cannella and Raine, 1995; Christy and Brown, 2007), but the contribution of astrocytes to barrier breakdown in inflammation has been less extensively explored. Recent studies have implicated IL-1, a macrophage-derived cytokine expressed at high levels in MS and EAE lesions, in the activation of the transcription factor hypoxia-inducible factor 1 (HIF-1), and its target vascular endothelial growth factor A (VEGF-A), in astrocytes in loss of barrier function (Argaw et al., 2006). Production of VEGF-A by astrocytes acts on endothelial cells via its receptor VEGFR2/flk-1, resulting in downregulation or loss of endothelial tight junction proteins claudin-5 (CLN5) and occludin (OCLN), culminating in focal loss of BBB function in lesioned tissue (Argaw et al., 2009). Reconstitution with recombinant CLN5, but not OCLN, restored endothelial barrier properties and rescued the BBB permeability phenotype.

11.3.1.2 Astrocyte Involvement at Other Sites of Inflammation in the CNS

The glia limitans refers to the continuous layer of astrocytic endfeet that separate the brain parenchyma from the perivascular, ventricular, and meningeal spaces. It has been well appreciated for many years that although the anatomic substrate of BBB impermeability to molecules and cells exists at the level of the endothelial cell tight junctions, astrocytic endfeet provide an additional barrier. In contrast, the glia limitans forms the primary barrier between the circumventricular organs and the parenchyma (Wolburg and Paulus, 2010).

The choroid plexus consists of a single layer of cuboidal, epithelial-like cells that are continuous with the ependymal cells lining the lateral, third, and fourth ventricles (Del Bigio, 1995). The choroid plexus produces the majority of the cerebral spinal fluid (CSF), and choroid plexus epithelial cells are specialized for this secretory function. These cells are interconnected by tight junctions at their apical side and exhibit extensive microvilli and cilia at their apical aspect (facing the CSF and the interior of the ventricles). A network of fenestrated capillaries and venules permeable to polar molecules and cells is sequestered behind the choroid plexus epithelial barrier.

Recent data suggest that the choroid plexus blood-CSF barrier is an early and essential site of entry of T cells into the CNS in EAE. Interaction of T cells bearing CCR6 with choroid plexus epithelial cells expressing the CCR6 ligand CCL20 is necessary for EAE initiation (Reboldi et al., 2009). CCR6 null mice sensitized for EAE exhibit greater accumulation of CD45-positive cells in the choroid plexus and dramatically decreased dissemination of CD45-positive cells in the CNS parenchyma. Notably, Reboldi et al. (2009) suggest that many of the CD45-positive cells in these mice appeared to be trapped between the basement membrane of the choroid plexus epithelial cells and the laminin-positive endothelial basal lamina of the capillaries and venules inside the choroid plexus. In addition to CCL20, other studies have shown that transmigration at this site is regulated by CD73 and the formation of adenosine, signaling through the A2A adenosine receptor (Mills et al., 2008). Following the normal flow of CSF, CCR6-positive T cells then disseminate into the subarachnoid space, where they may be reactivated by antigen-presenting dendritic cells. Astrocytes do not contribute to this initial step in EAE, but at a later stage, CCL20 production by activated astrocytes may contribute to the recruitment of CCR6 T cells into the brain parenchyma. In tissues from patients with MS, high CCL20 expression is detected in inflamed areas in astrocytes positive for GFAP as well as within the choroid plexus (Reboldi et al., 2009).

Other organs of the circumventricular system, including the area postrema, subfornical organ, organum vasculosum of the lamina terminalis, and median eminence, are also potential gateways of CD45-positive cell infiltration during EAE (Schulz and Engelhardt, 2005). Despite the fact that EAE is traditionally considered to be a disease of the spinal cord (owing to the prevalence of inflammatory demyelinating lesions at that site), these data indicate that the choroid plexus is a critical site of lymphocyte transmigration for disease initiation.

11.3.1.3 Astrocytes, Cytokines, and Chemokines

Numerous studies have shown that astrocytes can be activated in vitro to produce inflammatory cytokines and chemokines that can recruit lymphocytes as well as activate both glial and endothelial cells, resulting in the upregulation of integrins that promote lymphocyte adhesion to vascular walls (reviewed in John et al., 2003). Further studies targeting production of these cytokines to astrocytes in vivo using a murine GFAP genomic vector have demonstrated that astrocytes can be induced to express biologically active cytokines in a dose- and time-dependent fashion, that these cytokines activate appropriate signaling pathways, and that the pathologic effects of transgene expression in astrocytes may be distinct from that which occurs following transgene expression in other cell types, for example, neurons (Campbell et al., 2009).

The extent to which astrocytes contribute to cytokine and chemokine production at sites of inflammation within the CNS has been more difficult to determine. To address this question directly, Bethea and colleagues induced EAE in transgenic mice with a dominant-negative NF-kB super-repressor under control of the GFAP promoter (Brambilla et al., 2009). Remarkably, these mice showed significantly less severe clinical signs of disease and reduced evidence of demyelination and neuronal loss compared to wild type (WT), with the group difference particularly pronounced during the chronic disease phase. Transgenic animals displayed significantly decreased levels of TNFalpha and the chemokines CCL5, CXCL9, and CXCL10 during preclinical, acute, and chronic time points. Immunohistochemical analysis indicated that TNFalpha immunoreactivity was noted in microglia and macrophages, not astrocytes, indicating cross talk between astrocytes and microglia/monocytes in the regulation of cytokine expression in the lesion. IFNgamma, a cytokine well known for its role in defining the Th1 proinflammatory phenotype, was also decreased in transgenic animals compared to wild-type in the acute disease phase, likely reflecting differences in T cell infiltration into the CNS. Both transgenic and wild-type EAE animals had extensive infiltration of T cells and B cells into the CNS parenchyma. Perplexingly, however, mice expressing the astrocyte-directed NF-kB super-repressor actually exhibited increased leukocyte infiltration, despite having less severe disease and decreased production of the chemokine CXCL10 (IP10). The authors speculated that the population of infiltrating leukocytes might differ between mice expressing the NF-kB super-repressor and wild-type mice. Indeed, they went on to show that transgenic animals had greater numbers of protective, CD8+/CD122+ regulatory T cells and better relative preservation of neuronal structures. The numbers of CD4+ regulatory T cells were similar in the transgenic and wild-type animals. Given that the NF-kB super-repressor astrocytes had decreased expression of most chemokines studied, it remains to be determined the molecular mechanism that results in increased recruitment of protective CD8+/CD122+ T cells. Nevertheless, the data generated by Brambilla et al. (2009) indicate that astrocytic activation of the NF-kB pathway regulates TNFalpha, IFNgamma, and chemokine expression in the CNS, thus regulating directly and indirectly the severity and the inflammatory cell profile of the lesion.

Astrocytes associated with the BBB have also been reported to be the first cells in the CNS to make proinflammatory cytokines and chemokines essential for the induction of EAE following activation by MOG-reactive Th17 cells (Kang et al., 2010). Somewhat unexpectedly, studies in vitro showed that the pattern of cytokines induced in astrocytes by IL-17 plus TNFalpha (i.e., CXCL1, CXCL2, CCL20) differed from those activated in astrocytes by the Th1 cytokine IFNgamma plus TNF-alpha (i.e., CXCL9, CXCL10, CXCL11). This again demonstrates that chemokines secreted by astrocytes may regulate the nature of T cells and other leukocytes that infiltrate the CNS.

11.3.1.4   Astrocytes as a Source of Cytotoxic Factors

Activated astrocytes also secrete compounds with potentially direct toxic effects on neurons/axons and oligodendrocytes/myelin, such as reactive oxygen and nitrogen species, glutamate, and adenosine triphosphate (ATP). In rodents, treatment with IFNgamma or lipopolysaccharide causes a subset of astrocytes to produce inducible nitric oxide synthase (iNOS; also known as NOS-2). Combination treatment with transforming growth factor TGFbeta plus IFNgamma increased the percentage of astrocytes secreting NO (Hamby et al., 2006). Human astrocytes do not respond to lipopolysaccharide (LPS), but upregulate iNOS following activation with IL-1beta plus IFNgamma or the Toll-like receptor 3 (TLR3) ligand polyI:C. In situ hybridization and immunohistochemistry showed extensive iNOS reactivity in hypertrophic astrocytes in acute but not chronic MS lesions (Liu et al., 2001). Nitrotyrosine, a footprint for peroxynitrite, was also detected in parenchymal and perivascular regions in these acute lesions. Decreased uptake of glutamate by astrocytic or oligodendrocytic glutamate transporters could also contribute to pathologically elevated levels of extracellular glutamate, which is also directly toxic to oligodendrocytes and neurons (Matute et al., 1997). Glutamate excitotoxicity is thought to contribute to oligodendrocyte and axonal loss in MS (Werner et al., 2001), and glutamate receptor antagonists ameliorate EAE (Pitt et al., 2000). Oligodendrocytes in human, rodent, and rabbit optic nerve also express the ATP-sensitive purinergic receptor $P2X_7$, and excessive $P2X_7$ receptor activation has been implicated in oligodendroglial cell death and axonal injury in EAE and MS (Matute et al., 2007).

11.3.1.5   Astrocytes as a Source of Growth Inhibitory Compounds

Historically, the glial scar has been considered to have a negative impact on recovery from injury due to the classic role of the glial scar in inhibiting axonal regrowth. Prominent biochemical components of the gliotic scar include secreted macromolecules such as chondroitin sulfate proteoglycans (CSPGs). CSPGs, including aggrecan, brevican, neurocan, and phosphacan, inhibit neurite outgrowth and axonal regeneration in vitro and in vivo (reviewed in Silver and Miller, 2004; Fawcett, 2006; Rolls et al., 2009). In rodent models of spinal cord and cortical injury, GFAP-positive astrocytes are a significant source of neurocan (McKeon et al., 1999; Jones et al., 2003). At the edges of active MS plaques, versican, aggrecan, and neurocan are elevated in association with astrogliosis (Sobel and Ahmed, 2001) and may contribute to the failure of plaque repair. Enzymes that degrade CSPGs have shown promise in animal models of spinal cord injury (Bradbury et al., 2002). Interestingly, the production and degradation of CSPGs is a regulated process: astrocytes in acute lesions upregulate CSPGs, whereas astrocytes in chronic MS lesions express high levels of ADAMTS (a disintegrin and metalloproteinase with thrombospondin motifs)-4, a protease that metabolizes CSPGs (Haddock et al., 2006).

11.3.2   Astrocytes as Positive Regulators of Repair

11.3.2.1   Revascularizaton

In Section 11.3.1.1, we discussed the HIF-VEGF axis through which astrocytes, activated by inflammation, downregulate endothelial tight junction proteins CLN5 and OCLN, resulting in increased BBB permeability. Downregulation of endothelial tight junctions is also required, however, for endothelial plasticity and revascularization of injured tissue. The scaffold created by the glial scar also provides structural support for revascularization. In a model of guinea pig spinal cord contusion, the mean number of capillary profiles was twice that in unaffected spinal cord, and this increased vascularization extended to white matter regions as far as 400 μm from the lesion (Blight, 1991). The hypervascularization of injured tissue may facilitate repair by providing metabolic, trophic, and structural support for regenerating axons and myelin.

11.3.2.2 Structural Support Provided by the Glial Scar

Despite the fact that astrocytic responses can contribute to lesion development, it is also clear that astrocytes play an important role in limiting lesion development and promoting repair (Figure 11.4). As previously discussed, a prominent feature of CNS inflammation is the accumulation of a dense network of glial filaments and extracellular matrix (ECM) molecules termed the glial scar that forms in association with the inflammatory process. The pro- and antilesion effects of astrocytes may be due to the dual roles of specific astroglial functions. For example, despite the clear inhibitory role for CSPGs in attenuating axonal regrowth and remyelination, the dense network of CSPGs in the glial scar comprises a diffusion barrier that might protect the surrounding tissue from secondary degeneration caused by elevated levels of extracellular potassium, glutamate, purines, etc. (Roitbak and Sykova, 1999).

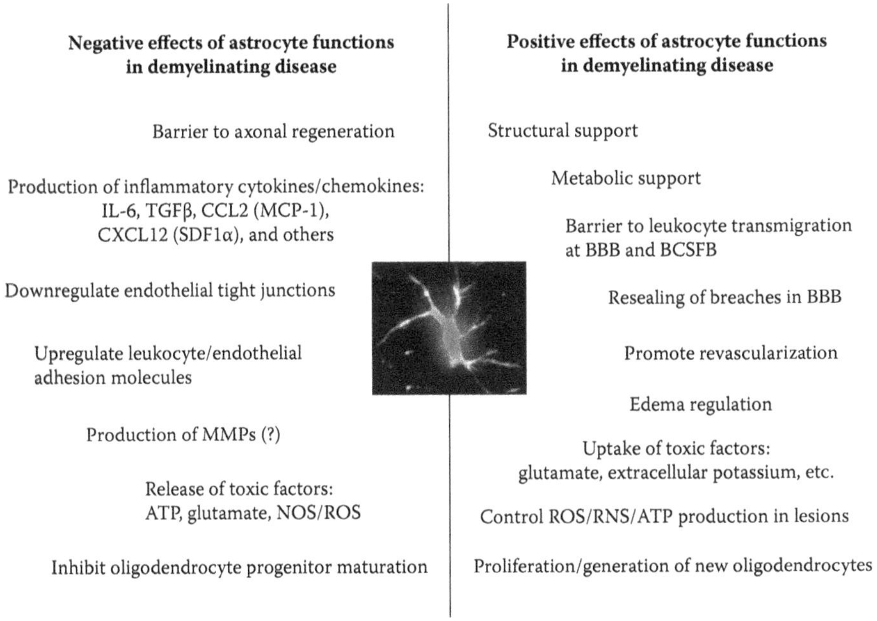

Negative effects of astrocyte functions in demyelinating disease	Positive effects of astrocyte functions in demyelinating disease
Barrier to axonal regeneration	Structural support
Production of inflammatory cytokines/chemokines: IL-6, TGFβ, CCL2 (MCP-1), CXCL12 (SDF1α), and others	Metabolic support
	Barrier to leukocyte transmigration at BBB and BCSFB
Downregulate endothelial tight junctions	Resealing of breaches in BBB
Upregulate leukocyte/endothelial adhesion molecules	Promote revascularization
	Edema regulation
Production of MMPs (?)	Uptake of toxic factors: glutamate, extracellular potassium, etc.
Release of toxic factors: ATP, glutamate, NOS/ROS	Control ROS/RNS/ATP production in lesions
Inhibit oligodendrocyte progenitor maturation	Proliferation/generation of new oligodendrocytes

FIGURE 11.4   Schematic depiction of some of the functions of the astrocyte relevant to demyelinating disease.

Extracellular matrix molecules also bind to and concentrate cytokines, chemokines, and growth factors, as has been recently reviewed (Rolls et al., 2009; Sofroniew and Vinters, 2010).

Studies in GFAP null mice have shown that following spinal cord transection or brain stab wounds, both GFAP and vimentin are required for proper glial scar formation (Pekny et al., 1999), and that CNS injury in animals lacking these intermediate filaments developed less dense scars that were frequently accompanied by signs of hemorrhage. GFAP expression was also found to be essential for normal white matter development and for the integrity of the BBB (Liedtke et al., 1996). Clinical signs of EAE were also more severe in these animals and were suggested to be related to the reduced ability of astrocytes to "wall-off" the lesion area (Liedtke et al., 1998a). However, in mice lacking GFAP, axonal sprouting and regeneration were not altered following spinal cord injury (De Keyser et al., 2010), consistent with some degree of functional redundancy between GFAP and vimentin.

Our appreciation of the contribution that astrocytes make to the repair process within the CNS has been greatly enhanced by the development of an animal model in which reactive astrocytes can be selectively ablated at sites of injury (Bush et al., 1999). This has been achieved by placing the gene for the herpes simplex thymidine kinase under the control of the full-length mouse GFAP promoter. This renders activated astrocytes vulnerable to the antiviral drug ganciclovir, which can be administered at specific time points following injury to the CNS. Thus, this model system permits both temporal and cell-type-specific control of astrocyte ablation. The effects of this procedure have now been tested in mice subjected to a stab wound (Bush et al., 1999), spinal cord injury (Faulkner et al., 2004; Myer et al., 2006), and EAE (Voskuhl et al., 2009). In these models, administration of ganciclovir led to the ablation of a population of astrocytes that surrounded the lesioned areas of the CNS, resulting in a disruption of scar formation, increased spread and persistence of inflammatory cells, more persistent loss of BBB function, increased tissue damage associated with increased neuronal loss, and increased demyelination. These data clearly show the benefits of astrocytic scar formation to neural protection and repair. Reactive astrocytes have also been shown to protect against the spread of infections (Drogemuller et al., 2008). Studies into possible mechanisms involved in astrocyte-mediated protection against tissue damage using the spinal injury model have clearly implicated a role for the uptake of potentially excitotoxic neurotransmitters such as glutamate, and protection from oxidative stress via the production of glutathione (Sofroniew, 2009).

The contribution of activated astrocytes to lesion development and repair has also been tested in mice in which deletion of STAT3 has been targeted to astrocytes using a mouse GFAP-Cre-loxP system creating a conditional knockout (cKO). STAT3 is activated by a number of cytokines implicated in the injury response, several of which, including IL-6, ciliary neurotrophic factor (CNTF), leukemia inhibitory factor (LIF), epidermal growth factor, and TGFalpha, have been implicated as triggers of a reactive astrogliosis (Takeda et al., 1998; Sofroniew, 2009). Thus, these studies, in contrast to those cited above with the NF-kB super-repressor that looked at the production of cytokines by astrocytes, examined the role of astrocytes as targets of the inflammatory cascade. Following spinal cord injury in these mice, the data showed that astrocyte hypertrophy and upregulation of GFAP were attenuated in STAT3 astrocyte cKO mice (Herrmann et al., 2008). Lesions in these animals showed a pronounced disruption of glial scar formation and displayed increased extravasation of CD45-positive cells into the peri-lesioned area, increased myelin loss, increased volume of necrotic tissue, and reduced functional recovery. Studies in vitro supported an important role for regulation of reactive oxygen species and ATP production in this process (Sarafian et al., 2010). Thus, these studies identified a beneficial role for activated astrocytes and development of the astrocytic scar in lesion resolution (Herrmann et al., 2008), and complement the diversity of responses noted following astrocyte-targeted expression of cytokines such as IL-6, TGFbeta, TNFalpha, and IL-12 (Campbell et al., 2009).

11.3.3 ASTROCYTES AS PROGENITOR CELLS

11.3.3.1 Astrocytes and Regulation of Regeneration

Astrocytes may also contribute to lesion repair through their diverse roles in the regeneration process that has now been recognized to occur within the adult CNS. So, for example, in rodents ribbons of astrocytes form a network along which neural progenitor cells migrate to the olfactory bulb forming the rostral migratory stream (Prineas et al., 2002a). Astrocytes may also function as support cells within the specialized niches where neural progenitor cells reside, such as in the subventricular zone (SVZ) (Raine and Genain, 2008). Astrocytes express growth factor and signaling molecules that regulate stem cell proliferation and differentiation, such as members of the Jagged/Notch and WNT signaling pathways. Finally, a subpopulation of astrocytes may themselves function as pluri- or unipotential progenitor cells, although the extent to which this occurs in the adult human CNS remains unclear at the present time (see Section 11.2.3). The composition of the extracellular matrix, much of which is astrocyte derived, also plays a critical role at sites thought to control stem cell fate, maturation, and survival. So, for example, chondroitin sulfate and dermatan sulfate proteoglycans are abundantly expressed in neurogenic niches such as the SVZ and contribute to niche maintenance and cell fate (Oohira et al., 2000; Rolls et al., 2009). Proteoglycans synthesized by astrocytes, neurons, and microglia control proliferation, migration, and survival of stem cell niche populations (Sirko et al., 2007, 2010; Ishii and Maeda, 2008; Jen et al., 2009). How injury, particularly repeated injury, to the CNS modulates these pathways is an area of intensive research at the present time.

It is now clear that cytokines, such as TGFbeta, influence many aspects of the astrocytic scar. In MS lesions, TGFbeta was found predominantly associated with extracellular matrix material around blood vessels (Cannella and Raine, 1995). Recent studies have implicated this cytokine in inducing the Notch ligand Jagged1 in astrocytes in MS lesions. During development, Notch receptor signaling regulates oligodendrocyte progenitor differentiation and maturation. Reactivation of this developmental pathway can be detected in active MS lesions with Jagged1 expression noted in astrocytes and its receptor Notch1 in OPCs (John et al., 2002). In demyelinating diseases such as MS, OPCs are present in considerable numbers, particularly at the edge of active as well as chronic-active MS lesion, but remyelination frequently fails at these sites (Wolswijk, 1998). This may reflect a failure of OPCs to migrate from the edge of the lesion into the demyelinated center or to differentiate into mature myelinating oligodendrocytes within the lesion center. Conversely, later in the disease course, OPC number may act as a limiting factor. Differential manipulation of notch signaling may provide a potential avenue for manipulating oligodendrocyte proliferation vs. differentiation in the adult CNS, with signaling through the Jagged/Notch pathway promoting OPC proliferation, and inhibition of this pathway promoting OPC maturation. In this regard, it is of interest to note that in contrast to signaling via canonical ligands such as Jagged1, that maintain the undifferentiated state, activation of Notch via noncanonical ligands, such as contactin (which is expressed on demyelinated axons in MS lesions), may promote OPC differentiation, which then fails because of failure of the Notch intracellular domain to locate to the nucleus (Nakahara et al., 2009). Thus, TGFbeta1-triggered Jagged1-Notch1 interactions may represent just one of many pathways that regulate the size and differentiation of the OPC pool. Included in these potential negative regulators of OPC differentiation (and axonal regeneration) in MS are components of the WNT signaling pathway (Fancy et al., 2010) and pathways activated by myelin debris (McKerracher et al., 1994; Mukhopadhyay et al., 1994; GrandPre et al., 2000; Kottis et al., 2002; Wang et al., 2002; reviewed in Domeniconi and Filbin, 2005; Giger et al., 2008). However, astrocytes have not as yet been implicated in contributing to either of these other possible negative regulators of myelination.

11.4   CONCLUDING REMARKS

Although the astrocyte was briefly considered by some to be the major cell type targeted in MS in early descriptions of the lesion (Müller, 1904), this view was rapidly eclipsed by the identification of the oligodendrocyte as the myelinating cell of the CNS. This observation, together with the development of the animal model of EAE, led to the focus of attention shifting to oligodendrocytes and myelin as targets of an autoimmune response. This was followed by renewed recognition of the contribution of neuronal and axonal damage to the pathogenesis of MS (Lassmann, 1999; Wingerchuk and Weinshenker, 2008). Now we have almost come full circle with the identification of the astrocyte-specific protein AQP4 as the primary target of the immune response in patients with neuromyelitis optica (NMO) (Liedtke et al., 1998b; Lennon et al., 2005; Vincent et al., 2008; Parratt and Prineas, 2010), a condition that had previously been considered by most to be a subtype of MS. In NMO, the primary event is loss or damage to the perivascular astrocyte, with the development of demyelination occurring as a secondary effect. This new knowledge of selective astrocyte pathology, together with the increasing availability of molecular and immunological tools that permit specific targeting of astrocytes in vivo, renders this a particularly exciting time for those interested in the biology of astrocytes, the most abundant cell type in the human CNS. Nevertheless, the task of understanding what these cells actually do in the normal and injured CNS is daunting. As can readily be appreciated from the present and other recent reviews (Williams et al., 2007; Nair et al., 2008; Sofroniew and Vinters, 2010), new techniques have illustrated just how complex this response is likely to be, with both injurious and reparative responses probably occurring within the same cell, and with subtle and transient danger signals, such as the release of ATP and other neurotransmitters like norepinephrine and glutamate (De Keyser et al., 2010), or changes in expression of heat shock proteins, tipping the balance one way or the other. However, the contribution that this knowledge could make to improvements in human health and hopefully one day, recovery from devastating injuries and chronic neurodegenerative conditions, is a worthy goal and we look forward to substantial progress in this area.

ACKNOWLEDGMENTS

Supported by the National Multiple Sclerosis Society USA grants RG 3827A5/1 to CFB and RG1001-K-11 to CSR, NINDS-NIH NS52245 to SEL. and USPHS grant NS11920. We thank Dr. G.R. Wayne Moore (Vancouver, British Columbia, Canada) for Figure 11.1A. CSR is the Wollowick Family Foundation Professor for Multiple Sclerosis Research.

REFERENCES

Agrawal S, Anderson P, Durbeej M, van Rooijen N, Ivars F, Opdenakker G, Sorokin LM. (2006). Dystroglycan is selectively cleaved at the parenchymal basement membrane at sites of leukocyte extravasation in experimental autoimmune encephalomyelitis. *J Exp Med* 203:1007–1019.

Amiry-Moghaddam M, Otsuka T, Hurn PD, Traystman RJ, Haug FM, Froehner SC, Adams ME, Neely JD, Agre P, Ottersen OP, Bhardwaj A. (2003). An alpha-syntrophin-dependent pool of AQP4 in astroglial end-feet confers bidirectional water flow between blood and brain. *Proc Natl Acad Sci USA* 100:2106–2111.

Aquino DA, Shafit-Zagardo B, Brosnan CF, Norton WT. (1990). Expression of glial fibrillary acidic protein and neurofilament mRNA in gliosis induced by experimental autoimmune encephalomyelitis. *J Neurochem* 54:1398–1404.

Argaw AT, Gurfein BT, Zhang Y, Zameer A, John GR. (2009). VEGF-mediated disruption of endothelial CLN-5 promotes blood-brain barrier breakdown. *Proc Natl Acad Sci USA* 106:1977–1982.

Argaw AT, Zhang Y, Snyder BJ, Zhao ML, Kopp N, Lee SC, Raine CS, Brosnan CF, John GR. (2006). IL-1beta regulates blood-brain barrier permeability via reactivation of the hypoxia-angiogenesis program. *J Immunol* 177:5574–5584.

Bannerman P, Hahn A, Soulika A, Gallo V, Pleasure D. (2007). Astrogliosis in EAE spinal cord: derivation from radial glia, and relationships to oligodendroglia. *Glia* 55:57–64.

Baxter AG. (2007). The origin and application of experimental autoimmune encephalomyelitis. *Nat Rev Immunol* 7:904–912.

Bettelli E, Pagany M, Weiner HL, Linington C, Sobel RA, Kuchroo VK. (2003). Myelin oligodendrocyte glycoprotein-specific T cell receptor transgenic mice develop spontaneous autoimmune optic neuritis. *J Exp Med* 197:1073–1081.

Blakemore WF. (1972). Observations on oligodendrocyte degeneration, the resolution of status spongiosus and remyelination in cuprizone intoxication in mice. *J Neurocytol* 1:413–426.

Blakemore WF, Franklin RJ. (2008). Remyelination in experimental models of toxin-induced demyelination. *Curr Top Microbiol Immunol* 318:193–212.

Blight AR. (1991). Morphometric analysis of blood vessels in chronic experimental spinal cord injury: hypervascularity and recovery of function. *J Neurol Sci* 106:158–174.

Bo L, Dawson TM, Wesselingh S, Mork S, Choi S, Kong PA, Hanley D, Trapp BD. (1994). Induction of nitric oxide synthase in demyelinating regions of multiple sclerosis brains. *Ann Neurol* 36:778–786.

Bradbury EJ, Moon LD, Popat RJ, King VR, Bennett GS, Patel PN, Fawcett JW, McMahon SB. (2002). Chondroitinase ABC promotes functional recovery after spinal cord injury. *Nature* 416:636–640.

Brambilla R, Persaud T, Hu X, Karmally S, Shestopalov VI, Dvoriantchikova G, Ivanov D, Nathanson L, Barnum SR, Bethea JR. (2009). Transgenic inhibition of astroglial NF-kappa B improves functional outcome in experimental autoimmune encephalomyelitis by suppressing chronic central nervous system inflammation. *J Immunol* 182:2628–2640.

Brosnan CF, Battistini L, Raine CS, Dickson DW, Casadevall A, Lee SC. (1994). Reactive nitrogen intermediates in human neuropathology: an overview. *Dev Neurosci* 16:152–161.

Brosnan CF, Cannella B, Battistini L, Raine CS. (1995). Cytokine localization in multiple sclerosis lesions: correlation with adhesion molecule expression and reactive nitrogen species. *Neurology* 45:S16–S21.

Bush TG, Puvanachandra N, Horner CH, Polito A, Ostenfeld T, Svendsen CN, Mucke L, Johnson MH, Sofroniew MV. (1999). Leukocyte infiltration, neuronal degeneration, and neurite outgrowth after ablation of scar-forming, reactive astrocytes in adult transgenic mice. *Neuron* 23:297–308.

Campbell IL, Hofer MJ, Pagenstecher A. (2010). Transgenic models for cytokine-induced neurological disease. *Biochim Biophys Acta* 1802: 903–17.

Cannella B, Raine CS. (1995). The adhesion molecule and cytokine profile of multiple sclerosis lesions. *Ann Neurol* 37:424–435.

Carlen M, Meletis K, Goritz C, Darsalia V, Evergren E, Tanigaki K, Amendola M, Barnabe-Heider F, Yeung MS, Naldini L, Honjo T, Kokaia Z, Shupliakov O, Cassidy RM, Lindvall O, Frisen J. (2009). Forebrain ependymal cells are Notch-dependent and generate neuroblasts and astrocytes after stroke. *Nat Neurosci* 12:259–267.

Carlton WW. (1966). Response of mice to the chelating agents sodium diethyldithiocarbamate, alpha-benzoinoxime, and biscyclohexanone oxaldihydrazone. *Toxicol Appl Pharmacol* 8:512–521.

Christy AL, Brown MA. (2007). The multitasking mast cell: positive and negative roles in the progression of autoimmunity. *J Immunol* 179:2673–2679.

Colodner KJ, Montana RA, Anthony DC, Folkerth RD, De Girolami U, Feany MB. (2005). Proliferative potential of human astrocytes. *J Neuropathol Exp Neurol* 64:163–169.

Compston A. (1990). Risk factors for multiple sclerosis: race or place? *J Neurol Neurosurg Psychiatry* 53:821–823.

Crocker SJ, Whitmire JK, Frausto RF, Chertboonmuang P, Soloway PD, Whitton JL, Campbell IL. (2006). Persistent macrophage/microglial activation and myelin disruption after experimental autoimmune encephalomyelitis in tissue inhibitor of metalloproteinase-1-deficient mice. *Am J Pathol* 169:2104–2116.

D'Amelio FE, Smith ME, Eng LF. (1990). Sequence of tissue responses in the early stages of experimental allergic encephalomyelitis (EAE): immunohistochemical, light microscopic, and ultrastructural observations in the spinal cord. *Glia* 3:229–240.

De Keyser J, Laureys G, Demol F, Wilczak N, Mostert J, Clinckers R. (2010). Astrocytes as potential targets to suppress inflammatory demyelinating lesions in multiple sclerosis. *Neurochem Int* 57:446–450.

Del Bigio MR. (1995). The ependyma: a protective barrier between brain and cerebrospinal fluid. *Glia* 14:1–13.

Domeniconi M, Filbin MT. (2005). Overcoming inhibitors in myelin to promote axonal regeneration. *J Neurol Sci* 233:43–47.

Drogemuller K, Helmuth U, Brunn A, Sakowicz-Burkiewicz M, Gutmann DH, Mueller W, Deckert M, Schluter D. (2008). Astrocyte gp130 expression is critical for the control of *Toxoplasma* encephalitis. *J Immunol* 181:2683–2693.

Dubois B, Masure S, Hurtenbach U, Paemen L, Heremans H, van den Oord J, Sciot R, Meinhardt T, Hammerling G, Opdenakker G, Arnold B. (1999). Resistance of young gelatinase B-deficient mice to experimental autoimmune encephalomyelitis and necrotizing tail lesions. *J Clin Invest* 104:1507–1515.

Eng LF, Ghirnikar RS, Lee YL. (2000). Glial fibrillary acidic protein: GFAP-thirty-one years (1969–2000). *Neurochem Res* 25:1439–1451.

Fancy SP, Kotter MR, Harrington EP, Huang JK, Zhao C, Rowitch DH, Franklin RJ. (2010). Overcoming remyelination failure in multiple sclerosis and other myelin disorders. *Exp Neurol* 225: 18–23.

Farez MF, Quintana FJ, Gandhi R, Izquierdo G, Lucas M, Weiner HL. (2009). Toll-like receptor 2 and poly(ADP-ribose) polymerase 1 promote central nervous system neuroinflammation in progressive EAE. *Nat Immunol* 10:958–964.

Farina C, Krumbholz M, Giese T, Hartmann G, Aloisi F, Meinl E. (2005). Preferential expression and function of Toll-like receptor 3 in human astrocytes. *J Neuroimmunol* 159:12–19.

Faulkner JR, Herrmann JE, Woo MJ, Tansey KE, Doan NB, Sofroniew MV. (2004). Reactive astrocytes protect tissue and preserve function after spinal cord injury. *J Neurosci* 24:2143–2155.

Fawcett JW. (2006). Overcoming inhibition in the damaged spinal cord. *J Neurotrauma* 23:371–383.

Filippi M, Campi A, Colombo B, Pereira C, Martinelli V, Baratti C, Comi G. (1996). A spinal cord MRI study of benign and secondary progressive multiple sclerosis. *J Neurol* 243:502–505.

Gadea A, Schinelli S, Gallo V. (2008). Endothelin-1 regulates astrocyte proliferation and reactive gliosis via a JNK/c-Jun signaling pathway. *J Neurosci* 28:2394–2408.

Gallo V, Armstrong RC. (2008). Myelin repair strategies: a cellular view. *Curr Opin Neurol* 21:278–283.

Ghatak NR. (1992). Occurrence of oligodendrocytes within astrocytes in demyelinating lesions. *J Neuropathol Exp Neurol* 51:40–46.

Giger RJ, Venkatesh K, Chivatakarn O, Raiker SJ, Robak L, Hofer T, Lee H, Rader C. (2008). Mechanisms of CNS myelin inhibition: evidence for distinct and neuronal cell type specific receptor systems. *Restor Neurol Neurosci* 26:97–115.

Gijbels K, Galardy RE, Steinman L. (1994). Reversal of experimental autoimmune encephalomyelitis with a hydroxamate inhibitor of matrix metalloproteases. *J Clin Invest* 94:2177–2182.

Goldschmidt T, Antel J, Konig FB, Bruck W, Kuhlmann T. (2009). Remyelination capacity of the MS brain decreases with disease chronicity. *Neurology* 72:1914–1921.

Goverman J, Woods A, Larson L, Weiner LP, Hood L, Zaller DM. (1993). Transgenic mice that express a myelin basic protein-specific T cell receptor develop spontaneous autoimmunity. *Cell* 72:551–560.

Graesser D, Mahooti S, Madri JA. (2000). Distinct roles for matrix metalloproteinase-2 and alpha4 integrin in autoimmune T cell extravasation and residency in brain parenchyma during experimental autoimmune encephalomyelitis. *J Neuroimmunol* 109:121–131.

GrandPre T, Nakamura F, Vartanian T, Strittmatter SM. (2000). Identification of the Nogo inhibitor of axon regeneration as a reticulon protein. *Nature* 403:439–444.

Haddock G, Cross AK, Plumb J, Surr J, Buttle DJ, Bunning RA, Woodroofe MN. (2006). Expression of ADAMTS-1, -4, -5 and TIMP-3 in normal and multiple sclerosis CNS white matter. *Mult Scler* 12:386–396.

Hamby ME, Hewett JA, Hewett SJ. (2006). TGF-beta1 potentiates astrocytic nitric oxide production by expanding the population of astrocytes that express NOS-2. *Glia* 54:566–577.

Herrmann JE, Imura T, Song B, Qi J, Ao Y, Nguyen TK, Korsak RA, Takeda K, Akira S, Sofroniew MV. (2008). STAT3 is a critical regulator of astrogliosis and scar formation after spinal cord injury. *J Neurosci* 28:7231–7243.

Ishii M, Maeda N. (2008). Oversulfated chondroitin sulfate plays critical roles in the neuronal migration in the cerebral cortex. *J Biol Chem* 283:32610–32620.

Jen YH, Musacchio M, Lander AD. (2009). Glypican-1 controls brain size through regulation of fibroblast growth factor signaling in early neurogenesis. *Neural Dev* 4:33.

John GR, Lee SC, Brosnan CF. (2003). Cytokines: powerful regulators of glial cell activation. *Neuroscientist* 9:10–22.

John GR, Shankar SL, Shafit-Zagardo B, Massimi A, Lee SC, Raine CS, Brosnan CF. (2002). Multiple sclerosis: re-expression of a developmental pathway that restricts oligodendrocyte maturation. *Nat Med* 8:1115–1121.

Jones LL, Margolis RU, Tuszynski MH. (2003). The chondroitin sulfate proteoglycans neurocan, brevican, phosphacan, and versican are differentially regulated following spinal cord injury. *Exp Neurol* 182:399–411.

Kang Z, Altuntas CZ, Gulen MF, Liu C, Giltiay N, Qin H, Liu L, Qian W, Ransohoff RM, Bergmann C, Stohlman S, Tuohy VK, Li X. (2010). Astrocyte-restricted ablation of interleukin-17-induced Act1-mediated signaling ameliorates autoimmune encephalomyelitis. *Immunity* 32:414–425.

Kornek B, Storch MK, Weissert R, Wallstroem E, Stefferl A, Olsson T, Linington C, Schmidbauer M, Lassmann H. (2000). Multiple sclerosis and chronic autoimmune encephalomyelitis: a comparative quantitative study of axonal injury in active, inactive, and remyelinated lesions. *Am J Pathol* 157:267–276.

Kothavale A, Di Gregorio D, Somera FP, Smith ME. (1995). GFAP mRNA fluctuates in synchrony with chronic relapsing EAE symptoms in SJL/J mice. *Glia* 14:216–224.

Kottis V, Thibault P, Mikol D, Xiao ZC, Zhang R, Dergham P, Braun PE. (2002). Oligodendrocyte-myelin glycoprotein (OMgp) is an inhibitor of neurite outgrowth. *J Neurochem* 82:1566–1569.

Lafaille JJ, Nagashima K, Katsuki M, Tonegawa S. (1994). High incidence of spontaneous autoimmune encephalomyelitis in immunodeficient anti-myelin basic protein T cell receptor transgenic mice. *Cell* 78:399–408.

Lassmann H. (1999). Pathology of multiple sclerosis. In *McAlpine's multiple sclerosis*, ed. Compston A, Ebers G, Lassmann H, McDonald I, Matthews B, Wekerle H, 323–385. 3rd ed. London: Churchill Livingstone.

Lassmann H, Zimprich F, Vass K, Hickey WF. (1991). Microglial cells are a component of the perivascular glia limitans. *J Neurosci Res* 28:236–243.

Latov N, Nilaver G, Zimmerman EA, Johnson WG, Silverman AJ, Defendini R, Cote L. (1979). Fibrillary astrocytes proliferate in response to brain injury: a study combining immunoperoxidase technique for glial fibrillary acidic protein and radioautography of tritiated thymidine. *Dev Biol* 72:381–384.

Lee SC, Moore GR, Golenwsky G, Raine CS. (1990). Multiple sclerosis: a role for astroglia in active demyelination suggested by class II MHC expression and ultrastructural study. *J Neuropathol Exp Neurol* 49:122–136.

Lennon VA, Kryzer TJ, Pittock SJ, Verkman AS, Hinson SR. (2005). IgG marker of optic-spinal multiple sclerosis binds to the aquaporin-4 water channel. *J Exp Med* 202:473–477.

Liedtke W, Cannella B, Mazzaccaro RJ, Clements JM, Miller KM, Wucherpfennig KW, Gearing AJ, Raine CS. (1998b). Effective treatment of models of multiple sclerosis by matrix metalloproteinase inhibitors. *Ann Neurol* 44:35–46.

Liedtke W, Edelmann W, Bieri PL, Chiu FC, Cowan NJ, Kucherlapati R, Raine CS. (1996). GFAP is necessary for the integrity of CNS white matter architecture and long-term maintenance of myelination. *Neuron* 17:607–615.

Liedtke W, Edelmann W, Chiu FC, Kucherlapati R, Raine CS. (1998a). Experimental autoimmune encephalomyelitis in mice lacking glial fibrillary acidic protein is characterized by a more severe clinical course and an infiltrative central nervous system lesion. *Am J Pathol* 152:251–259.

Lindberg RL, De Groot CJ, Montagne L, Freitag P, van der Valk P, Kappos L, Leppert D. (2001). The expression profile of matrix metalloproteinases (MMPs) and their inhibitors (TIMPs) in lesions and normal appearing white matter of multiple sclerosis. *Brain* 124:1743–1753.

Liu JS, Zhao ML, Brosnan CF, Lee SC. (2001). Expression of inducible nitric oxide synthase and nitrotyrosine in multiple sclerosis lesions. *Am J Pathol* 158:2057–2066.

Losseff NA, Webb SL, O'Riordan JI, Page R, Wang L, Barker GJ, Tofts PS, McDonald WI, Miller DH, Thompson AJ. (1996). Spinal cord atrophy and disability in multiple sclerosis. A new reproducible and sensitive MRI method with potential to monitor disease progression. *Brain* 119 (Pt 3):701–708.

Ludwin SK, Raine CS. (2008). The neuropathology of MS. In *Multiple sclerosis: a comprehensive text*, ed. Raine CS, McFarland HF, Hohlfeld R, 151–177. New York: Saunders/Elsevier.

Luo J, Ho P, Steinman L, Wyss-Coray T. (2008). Bioluminescence *in vivo* imaging of autoimmune encephalomyelitis predicts disease. *J Neuroinflammation* 5:6.

Magnus T, Carmen J, Deleon J, Xue H, Pardo AC, Lepore AC, Mattson MP, Rao MS, Maragakis NJ. (2008). Adult glial precursor proliferation in mutant SOD1G93A mice. *Glia* 56:200–208.

Manley GT, Fujimura M, Ma T, Noshita N, Filiz F, Bollen AW, Chan P, Verkman AS. (2000). Aquaporin-4 deletion in mice reduces brain edema after acute water intoxication and ischemic stroke. *Nat Med* 6:159–163.

Mathiisen TM, Lehre KP, Danbolt NC, Ottersen OP. (2010). The perivascular astroglial sheath provides a complete covering of the brain microvessels: an electron microscopic 3D reconstruction. *Glia* 58:1094–1103.

Matute C, Sanchez-Gomez MV, Martinez-Millan L, Miledi R. (1997). Glutamate receptor-mediated toxicity in optic nerve oligodendrocytes. *Proc Natl Acad Sci USA* 94:8830–8835.

Matute C, Torre I, Perez-Cerda F, Perez-Samartin A, Alberdi E, Etxebarria E, Arranz AM, Ravid R, Rodriguez-Antiguedad A, Sanchez-Gomez M, Domercq M. (2007). P2X(7) receptor blockade prevents ATP excitotoxicity in oligodendrocytes and ameliorates experimental autoimmune encephalomyelitis. *J Neurosci* 27:9525–9533.

McKeon RJ, Jurynec MJ, Buck CR. (1999). The chondroitin sulfate proteoglycans neurocan and phosphacan are expressed by reactive astrocytes in the chronic CNS glial scar. *J Neurosci* 19:10778–10788.

McKerracher L, David S, Jackson DL, Kottis V, Dunn RJ, Braun PE. (1994). Identification of myelin-associated glycoprotein as a major myelin-derived inhibitor of neurite growth. *Neuron* 13:805–811.

Meletis K, Barnabe-Heider F, Carlen M, Evergren E, Tomilin N, Shupliakov O, Frisen J. (2008). Spinal cord injury reveals multilineage differentiation of ependymal cells. *PLoS Biol* 6:e182.

Mendel I, Kerlero de Rosbo N, Ben-Nun A. (1995). A myelin oligodendrocyte glycoprotein peptide induces typical chronic experimental autoimmune encephalomyelitis in H-2b mice: fine specificity and T cell receptor V beta expression of encephalitogenic T cells. *Eur J Immunol* 25:1951–1959.

Miller SD, Karpus WJ, Davidson TS. (2010). Experimental autoimmune encephalomyelitis in the mouse. *Curr Protoc Immunol*, chap. 15, unit 15, 11.

Mills JH, Thompson LF, Mueller C, Waickman AT, Jalkanen S, Niemela J, Airas L, Bynoe MS. (2008). CD73 is required for efficient entry of lymphocytes into the central nervous system during experimental autoimmune encephalomyelitis. *Proc Natl Acad Sci USA* 105:9325–9330.

Moore SA, Saito F, Chen J, Michele DE, Henry MD, Messing A, Cohn RD, Ross-Barta SE, Westra S, Williamson RA, Hoshi T, Campbell KP. (2002). Deletion of brain dystroglycan recapitulates aspects of congenital muscular dystrophy. *Nature* 418:422–425.

Mukhopadhyay G, Doherty P, Walsh FS, Crocker PR, Filbin MT. (1994). A novel role for myelin-associated glycoprotein as an inhibitor of axonal regeneration. *Neuron* 13:757–767.

Müller E. (1904). Pathologische Anatomie und Pathogenese. In *Die multiple Sklerose des Gehirns und Rückenmarks*, 300–344. Jena, Germany: Gustav Fischer.

Myer DJ, Gurkoff GG, Lee SM, Hovda DA, Sofroniew MV. (2006). Essential protective roles of reactive astrocytes in traumatic brain injury. *Brain* 129:2761–2772.

Nair A, Frederick TJ, Miller SD. (2008). Astrocytes in multiple sclerosis: a product of their environment. *Cell Mol Life Sci* 65:2702–2720.

Nakahara J, Kanekura K, Nawa M, Aiso S, Suzuki N. (2009). Abnormal expression of TIP30 and arrested nucleocytoplasmic transport within oligodendrocyte precursor cells in multiple sclerosis. *J Clin Invest* 119:169–181.

Neely JD, Amiry-Moghaddam M, Ottersen OP, Froehner SC, Agre P, Adams ME. (2001). Syntrophin-dependent expression and localization of aquaporin-4 water channel protein. *Proc Natl Acad Sci USA* 98:14108–14113.

Nguyen KB, Pender MP. (1998). Phagocytosis of apoptotic lymphocytes by oligodendrocytes in experimental autoimmune encephalomyelitis. *Acta Neuropathol* 95:40–46.

Noell S, Fallier-Becker P, Beyer C, Kroger S, Mack AF, Wolburg H. (2007). Effects of agrin on the expression and distribution of the water channel protein aquaporin-4 and volume regulation in cultured astrocytes. *Eur J Neurosci* 26:2109–2118.

Oberheim NA, Takano T, Han X, He W, Lin JH, Wang F, Xu Q, Wyatt JD, Pilcher W, Ojemann JG, Ransom BR, Goldman SA, Nedergaard M. (2009). Uniquely hominid features of adult human astrocytes. *J Neurosci* 29:3276–3287.

Olson JK, Ercolini AM, Miller SD. (2005). A virus-induced molecular mimicry model of multiple sclerosis. *Curr Top Microbiol Immunol* 296:39–53.

Oohira A, Matsui F, Tokita Y, Yamauchi S, Aono S. (2000). Molecular interactions of neural chondroitin sulfate proteoglycans in the brain development. *Arch Biochem Biophys* 374:24–34.

Owens T, Bechmann I, Engelhardt B. (2008). Perivascular spaces and the two steps to neuroinflammation. *J Neuropathol Exp Neurol* 67:1113–1121.

Pagenstecher A, Stalder AK, Kincaid CL, Shapiro SD, Campbell IL. (1998). Differential expression of matrix metalloproteinase and tissue inhibitor of matrix metalloproteinase genes in the mouse central nervous system in normal and inflammatory states. *Am J Pathol* 152:729–741.

Parratt JD, Prineas JW. (2010). Neuromyelitis optica: a demyelinating disease characterized by acute destruction and regeneration of perivascular astrocytes. *Mult Scler* 16: 1156–72.

Patani R, Balaratnam M, Vora A, Reynolds R. (2007). Remyelination can be extensive in multiple sclerosis despite a long disease course. *Neuropathol Appl Neurobiol* 33:277–287.

Paterson PY. (1960). Transfer of allergic encephalomyelitis in rats by means of lymph node cells. *J Exp Med* 111:119–136.

Patrikios P, Stadelmann C, Kutzelnigg A, Rauschka H, Schmidbauer M, Laursen H, Sorensen PS, Bruck W, Lucchinetti C, Lassmann H. (2006). Remyelination is extensive in a subset of multiple sclerosis patients. *Brain* 129:3165–3172.

Pekny M, Johansson CB, Eliasson C, Stakeberg J, Wallen A, Perlmann T, Lendahl U, Betsholtz C, Berthold CH, Frisen J. (1999). Abnormal reaction to central nervous system injury in mice lacking glial fibrillary acidic protein and vimentin. *J Cell Biol* 145:503–514.

Pitt D, Werner P, Raine CS. (2000). Glutamate excitotoxicity in a model of multiple sclerosis. *Nat Med* 6:67–70.

Prineas JW, McDonald WI, Franklin RJ. (2002a). Demyelinating diseases. In *Greenfield's neuropathology*, ed. Graham DI, Lantos PL, 487. 7th ed. London: Arnold.

Prineas JW, McDonald WI, Franklin RJ. (2002b). Demyelinating diseases. In *Greenfield's neuropathology*, ed. Graham DI, Lantos PL, 471–535. 7th ed. London: Arnold.

Prineas JW, Kwon EE, Goldenberg PZ, Ilyas AA, Quarles RH, Benjamins JA, Sprinkle TJ. (1989). Multiple sclerosis. Oligodendrocyte proliferation and differentiation in fresh lesions. *Lab Invest* 61:489–503.

Raine CS. (1984). *The neuropathology of myelin diseases*. In Myelin, ed. Morell P, 259–310. 2nd ed. New York: Plenum Press.

Raine CS, Genain CP. (2008). Models of chronic relapsing experimental autoimmune encephalomyelitis. In *Multiple sclerosis: a comprehensive text*, ed. Raine CS, McFarland HF, Hohlfeld R, 237–260. Edinburgh: Saunders Elsevier.

Raine CS, Scheinberg L, Waltz JM. (1981). Multiple sclerosis. Oligodendrocyte survival and proliferation in an active established lesion. *Lab Invest* 45:534–546.

Reboldi A, Coisne C, Baumjohann D, Benvenuto F, Bottinelli D, Lira S, Uccelli A, Lanzavecchia A, Engelhardt B, Sallusto F. (2009). C-C chemokine receptor 6-regulated entry of TH-17 cells into the CNS through the choroid plexus is required for the initiation of EAE. *Nat Immunol* 10:514–523.

Roitbak T, Sykova E. (1999). Diffusion barriers evoked in the rat cortex by reactive astrogliosis. *Glia* 28:40–48.

Rolls A, Shechter R, Schwartz M. (2009). The bright side of the glial scar in CNS repair. *Nat Rev Neurosci* 10:235–241.

Sarafian TA, Montes C, Imura T, Qi J, Coppola G, Geschwind DH, Sofroniew MV. (2010). Disruption of astrocyte STAT3 signaling decreases mitochondrial function and increases oxidative stress *in vitro*. *PLoS One* 5:e9532.

Schulz M, Engelhardt B. (2005). The circumventricular organs participate in the immunopathogenesis of experimental autoimmune encephalomyelitis. *Cerebrospinal Fluid Res* 2:8.

Seifert G, Schilling K, Steinhauser C. (2006). Astrocyte dysfunction in neurological disorders: a molecular perspective. *Nat Rev Neurosci* 7:194–206.

Silver J, Miller JH. (2004). Regeneration beyond the glial scar. *Nat Rev Neurosci* 5:146–156.

Sirko S, von Holst A, Weber A, Wizenmann A, Theocharidis U, Gotz M, Faissner A. (2010). Chondroitin sulfates are required for fibroblast growth factor-2-dependent proliferation and maintenance in neural stem cells and for epidermal growth factor-dependent migration of their progeny. *Stem Cells* 28:775–787.

Sirko S, von Holst A, Wizenmann A, Gotz M, Faissner A. (2007). Chondroitin sulfate glycosaminoglycans control proliferation, radial glia cell differentiation and neurogenesis in neural stem/progenitor cells. *Development* 134:2727–2738.

Sobel RA, Ahmed AS. (2001). White matter extracellular matrix chondroitin sulfate/dermatan sulfate proteoglycans in multiple sclerosis. *J Neuropathol Exp Neurol* 60:1198–1207.

Sofroniew MV. (2009). Molecular dissection of reactive astrogliosis and glial scar formation. *Trends Neurosci* 32:638–647.

Sofroniew MV, Vinters HV. (2010). Astrocytes: biology and pathology. *Acta Neuropathol* 119:7–35.

Stromnes IM, Goverman JM. (2006). Passive induction of experimental allergic encephalomyelitis. *Nat Protoc* 1:1952–1960.

Takeda K, Kaisho T, Yoshida N, Takeda J, Kishimoto T, Akira S. (1998). Stat3 activation is responsible for IL-6-dependent T cell proliferation through preventing apoptosis: generation and characterization of T cell-specific Stat3-deficient mice. *J Immunol* 161:4652–4660.

Tani M, Glabinski AR, Tuohy VK, Stoler MH, Estes ML, Ransohoff RM. (1996). *In situ* hybridization analysis of glial fibrillary acidic protein mRNA reveals evidence of biphasic astrocyte activation during acute experimental autoimmune encephalomyelitis. *Am J Pathol* 148:889–896.

Trapp BD, Peterson J, Ransohoff RM, Rudick R, Mork S, Bo L. (1998). Axonal transection in the lesions of multiple sclerosis. *N Engl J Med* 338:278–285.

Vincent T, Saikali P, Cayrol R, Roth AD, Bar-Or A, Prat A, Antel JP. (2008). Functional consequences of neuromyelitis optica-IgG astrocyte interactions on blood-brain barrier permeability and granulocyte recruitment. *J Immunol* 181:5730–5737.

Voskuhl RR, Peterson RS, Song B, Ao Y, Morales LB, Tiwari-Woodruff S, Sofroniew MV. (2009). Reactive astrocytes form scar-like perivascular barriers to leukocytes during adaptive immune inflammation of the CNS. *J Neurosci* 29:11511–11522.

Waldner H, Whitters MJ, Sobel RA, Collins M, Kuchroo VK. (2000). Fulminant spontaneous autoimmunity of the central nervous system in mice transgenic for the myelin proteolipid protein-specific T cell receptor. *Proc Natl Acad Sci USA* 97:3412–3417.

Wang D, Ayers MM, Catmull DV, Hazelwood LJ, Bernard CC, Orian JM. (2005). Astrocyte-associated axonal damage in pre-onset stages of experimental autoimmune encephalomyelitis. *Glia* 51:235–240.

Wang KC, Koprivica V, Kim JA, Sivasankaran R, Guo Y, Neve RL, He Z. (2002). Oligodendrocyte-myelin glycoprotein is a Nogo receptor ligand that inhibits neurite outgrowth. *Nature* 417:941–944.

Werner P, Pitt D, Raine CS. (2001). Multiple sclerosis: altered glutamate homeostasis in lesions correlates with oligodendrocyte and axonal damage. *Ann Neurol* 50:169–180.

Williams A, Piaton G, Lubetzki C. (2007). Astrocytes—friends or foes in multiple sclerosis? *Glia* 55:1300–1312.

Wingerchuk DM, Weinshenker BG. (2008). Unusual presentations and variants of idiopathic central nervous system demyelinating diseases. In *Multiple sclerosis: a comprehensive text*, ed. Raine CS, McFarland HF, Hohlfeld R, 24–42. Edinburgh: Saunders Elsevier.

Wisniewski H, Raine CS, Kay WJ. (1972). Observations on viral demyelinating encephalomyelitis. Canine distemper. *Lab Invest* 26:589–599.

Wolburg H, Paulus W. (2010). Choroid plexus: biology and pathology. *Acta Neuropathol* 119:75–88.

Wolburg-Buchholz K, Mack AF, Steiner E, Pfeiffer F, Engelhardt B, Wolburg H. (2009). Loss of astrocyte polarity marks blood-brain barrier impairment during experimental autoimmune encephalomyelitis. *Acta Neuropathol* 118:219–233.

Wolswijk G. (1998). Chronic stage multiple sclerosis lesions contain a relatively quiescent population of oligodendrocyte precursor cells. *J Neurosci* 18:601–609.

Woodruff RH, Franklin RJ. (1999). Demyelination and remyelination of the caudal cerebellar peduncle of adult rats following stereotaxic injections of lysolecithin, ethidium bromide, and complement/anti-galactocerebroside: a comparative study. *Glia* 25:216–228.

Wu E, Raine CS. (1992). Multiple sclerosis. Interactions between oligodendrocytes and hypertrophic astrocytes and their occurrence in other, nondemyelinating conditions. *Lab Invest* 67:88–99.

Wujek JR, Bjartmar C, Richer E, Ransohoff RM, Yu M, Tuohy VK, Trapp BD. (2002). Axon loss in the spinal cord determines permanent neurological disability in an animal model of multiple sclerosis. *J Neuropathol Exp Neurol* 61:23–32.

12 Neuroglia in Alzheimer's Disease

José Julio Rodríguez Arellano
IKERBASQUE, Basque Foundation for Science, Bilbao, Spain
Department of Neurosciences, University
of the Basque Country, Leioa, Spain
Instituto de Investigacion Sanitano Biodonostia,
Hospital Donostia, Sans Sebastian, Spain
Institute of Experimental Medicine,
ASCR, Prague, Czech Republic

Carlos Matute
Department of Neurosciences, University
of the Basque Country, Leioa, Spain
Neurotek-UPV/EHU, Zamudio, Spain
Centro de Investigación Biomédica en Red en Enfermedades
Neurodegenerativas (CiBERNED), Leioa, Spain

Alexei Verkhratsky
Faculty of Life Sciences, University of
Manchester, Manchester, United Kingdom
IKERBASQUE, Basque Foundation for Science, Bilbao, Spain
Department of Neurosciences, University
of the Basque Country, Leioa, Spain
Institute of Experimental Medicine,
ASCR, Prague, Czech Republic

CONTENTS

12.1   GLIA IN NEUROLOGICAL DISEASES

Neuroglia comprise the main cellular homeostatic element of the central nervous system. Glial cells are classically divided into macroglia (astrocytes, oligodendrocytes, and NG2 cells, all of ectodermal origin) and microglia (the resident Central Nervous System (CNS) macrophages of myeloid origin). The functions of these cells are many (for review see Iadecola and Nedergaard 2007; Magistretti 2006; Nedergaard et al. 2003; Newman et al. 1984; Simard and Nedergaard 2004; Zonta et al. 2003; Kettenmann and Verkhratsky 2008; Verkhratsky 2009, 2010; Verkhratsky and Kirchhoff 2007); however, all of them are specifically involved in control of brain homeostasis at various levels, from control over ion and neurotransmitter concentrations in brain interstitium to regulation of morphological plasticity of neural networks, including synapto- and neurogenesis.

Neuroglia form the principal defense system of the brain and the spinal cord. This system is of a fundamental importance, because the CNS, being separated from the rest of the body by the blood–brain barrier, requires special arrangements to oppose various types of insults and infectious invasions. Lesions of the CNS, regardless of their nature, trigger evolutionarity-conserved cascades of neuroglial "damage" responses, generally known as reactive gliosis. This "gliotic" response is characterized by morphological and functional remodeling of glial cells with a defined pathological context. The astrocytes subjected to the damage signals undergo reactive astrogliosis (Li et al. 2008; Pekny and Nilsson 2005), which is critical for both sealing the damaged area through scar formation (anisomorphic astrogliosis) and postinsult remodeling of neural circuits (isomorphic astrogliosis). The pathologically challenged oligodendrocytes embark into a complex process of Wallerian degeneration (Koeppen 2004; Vargas and Barres 2007), and finally microglial cells react to CNS damage by multistage and complex processes of microglial activation (Hanisch and Kettenmann 2007) that result in immune and phagocytic responses of the nervous tissue.

The complement of homeostatic and defensive cascades determines the pathological potential of neuroglia, which controls the progression and outcome of virtually all neurological diseases (Giaume et al. 2007; Halassa et al. 2007; Rossi and Volterra 2009; Heneka et al. 2010). Conceptually, neuroglia play fundamental roles in both acute and chronic neuropathologies. Astrocytes are specifically important for the progression of acute brain insults, such as trauma and stroke (Nedergaard and Dirnagl 2005; Nedergaard et al. 2010), where astroglial performance is of critical importance for limiting the site of damage. Astrocytes are also intimately involved in chronic neurological diseases. For example, astroglial functional and morphological remodeling are prominent in various forms of epilepsy and epileptiform disorders (Jabs et al. 2008; Tian et al. 2005). Similarly, dystrophic changes and loss of astrocytes accompany various forms of psychiatric disorders, including depression (Rajkowska et al. 1999), while altered release of the gliotransmitter D-serine can be involved in pathological glutamatergic transmission in schizophrenia (Halassa et al. 2007; Tsai and Coyle 2002).

The oligodendrocytes are central to diseases of white matter; these diseases are specifically important for human brain, where the white matter accounts for >50% of the overall volume (Fields 2008). The oligodendroglia are critical for both proper function and survival of axons, and therefore ischemic lesions that trigger oligoden-rocytic death (Matute 2008; Matute et al. 2007) result in severe neurological deficits and dementia, such as observed in periventricular leucomalacia (Blumenthal 2004) or in Binswanger's disease (Akiguchi et al. 1997), both being characterized by profound damage to white matter. The damage to oligodendroglia is also central for demyelinating disorders, including multiple sclerosis (Antel and Arnold 2005; Ercolini and Miller 2006; Jessen and Mirsky 2008; van der Valk and Amor 2009).

The microglial cells, being the innate immune and phagocytic cellular elements of the CNS, are invariably involved in every type of neuropathology. Through a complex and multistage process of activation, microglial cells grade the reaction to brain lesion, and shape the course of neuroinflammatory responses (Hanisch and Kettenmann 2007; Ransohoff and Perry 2009).

12.2 NEUROGLIA IN NEURODEGENERATION

Neurodegenerative diseases that destroy the intricate web of connections within neural networks, thus altering cognitive function and ultimately leading to the death of neural cells and atrophy of brain parenchyma, are specific for humans, being generally absent in other mammals. Neurodegenerative processes can be triggered by various etiological factors, from acute trauma, chemical poisoning, or infectious attacks to sporadic accumulation of genetic/biochemical errors of unknown nature.

Conceptually, the neurodegenerative process starts from altering synaptic and nonsynaptic contacts between neural cells, which cause initial cognitive deficits. This early stage of neurodegeneration is of specific importance, because the pathological process can be arrested or even reversed, thus offering the hope for arresting or reversing the cognitive decline. The neuroglia, which provide for the birth, maintenance, and demise of contacts in neural circuitry, are most likely the main cellular element shaping the early stages of neurodegenerative processes.

Astroglial reaction in neurodegenerative processes is complex and includes both reactive astrogliosis and astrodegeneration or atrophy. As a rule, astrodegeneration precedes astrogliosis, the latter being regarded as a general neuroinflammatory reaction at later stages of the disease. In amyotrophic lateral sclerosis (ALS), for example, astrodegeneration and astroglial atrophy occur before neuronal death and the appearance of manifest clinical symptoms in the Tg(SOD1*G93A)1Gur transgenic model of the disease (Rossi et al. 2008; Rossi and Volterra 2009). At the later stages of the ALS reactive astrogliosis becomes evident (McGeer and McGeer 2002), although the atrophic astrocytes are still present around lesion sites (Rossi et al. 2008; Rossi and Volterra 2009). Importantly, the silencing of ALS-related mutant SOD1 gene in astrocytes delayed the appearance of clinical symptoms in transgenic mouse model (Yamanaka et al. 2008).

The signs of astroglial atrophy were also observed in the thiamine-deficient rat model of Wernicke encephalopathy; these atrophic changes included decreased expression of GFAP, astrocytic glutamine synthetase, and astrocytic GAT-3 gamma aminobutiric acid (GABA) transporter. These changes were also accompanied with significant downregulation of astroglial glutamate transporters EAAT1 and EAAT2 that have a key pathogenic role; the compromised glutamate homeostasis induces neuronal toxicity and severe brain lesions characteristic for the disease (Hazell 2009; Hazell et al. 2009). The role for astroglia in another major neurodegenerative disease; Parkinson's disease has not been characterized in detail. Nonetheless, it is known that the substantia nigra has a relatively low density of astrocytes, and that astrocytes have a clear neuroprotective action on dopaminergic neurons in vitro in neuronal-glial cocultures (Mena et al. 1996, 2002; Mena and Garcia de Yebenes 2008). In addition, astrocytes are critically important for L-3,4-dihydroxypheny-lalanine (Levodopa or L-Dopa) substitute therapy, because these are astrocytes, which prevent L-DOPA direct neurotoxicity and convert L-DOPA into a trophic agent by utilizing yet unknown mechanisms (Mena et al. 1996; Mena and Garcia de Yebenes 2008). Incidentally, there are some reports that show additional beneficial roles of astrocytes in the context of Parkinson's disease (PD), because astroglial cells may support differentiation of stem cells into dopaminergic neurones and their functional integration into neuronal circuitry (Mena and Garcia de Yebenes 2008).

12.3   EXPERIMENTAL ANIMAL MODELS OF ALZHEIMER'S DISEASE

Alzheimer's disease (AD), as every other dementia, is a sole suffering of humans; no animal acquires (Toledano and Alvarez 2004). Substantial efforts therefore were invested in producing relevant animal models of AD to reproduce single or multiple subsets of neuropathological, behavioral, and biochemical alterations resembling those seen in typical AD (Cassel et al. 2008; Gotz et al. 2004).

12.3.1   AGED ANIMALS

Initially, aged animals were utilized as AD models (Bertoni-Freddari et al. 1986; Biegon et al. 1986; Toledano and Alvarez 2004). The cell loss and atrophy of basal forebrain neurones expressing choline acetyltransferase or nerve growth factor were observed in several old animals from rodents to primates (Sani et al. 2003; Toledano and Alvarez 2004; Decker 1987; Fischer et al. 1992). In monkeys the cholinergic degeneration was also associated with β-amyloid depositions (Toledano and Alvarez 2004). Furthermore, as frequently happens in aging humans with or without AD, normal old animals showed not only a cholinergic dysfunction but also a concomitant alteration of other neurotransmitter systems, such as the aminergic, peptidergic, and serotonergic (Toledano and Alvarez 2004; Arranz et al. 1996; McEntee and Crook 1991; Noristani et al. 2010).

12.3.2 LESIONED ANIMALS

Loss of cholinergic neurones is a prominent feature of AD (Bartus et al. 1982). This has led to the design of several cholinergic models of the disease. Among these, the most relevant were the rodent models with lesions in the nucleus basalis magnocellularis (nbm), which is the equivalent of the nucleus of Meynert in humans (Pepeu and Marconcini Pepeu 1994; Wellman and Pelleymounter 1999) and in the diagonal band of Broca and the septum (Toledano and Alvarez 2004) (Table 12.1).

12.3.2.1 Global Lesions

The global lesion models (which are nonselective for the cholinergic system) initially used electrolytic lesions that affected not only the intended lesion area (e.g., as in the case of AD the basal forebrain), but other distal and not cognitive associated areas due to the parallel effect produced on passing fibers (Lescaudron and Stein 1999; Toledano and Alvarez 2004). In the majority of global lesion models, however, the nonselective excitotoxic toxins such as N-Methyl-D-aspartic acid (NMDA), ibotenic acid, quisqualic acid, quinolic acid, colchicine, and other alkaloid substances, were used (Toledano and Alvarez 2004). Administration of these toxins triggers various forms of neuronal dysfunction with subsequent cell death (Olney 1994; Toledano and Alvarez 2004).

Alternative approaches included injections of alcohol, which is toxic to cholinergic neurons (Arendt 1994; Toledano and Alvarez 2004), or direct administration of β-amyloid peptides, which produces multiple alterations of basocortical neurones affecting acetylcholine release and cholinergic receptors (Giovannini et al. 2002; Pavia et al. 2000; Toledano and Alvarez 2004).

TABLE 12.1
Summary of Lesion Models of Alzheimer's Disease

Lesion	Cholinergic	Noncholinergic	Neuropathology	Reference
Electrolytic	Yes	Yes	Neuronal death	Lescaudron and Stein 1999; Vale-Martinez et al. 2002
Excitotoxins (NMDA, ibotenic acid, quisqualic acid)	Yes	Yes	Neuronal death	Dunnett et al. 1991; Winkler and Thal 1995
Quinolic acid	Yes	Yes	Neuronal death	Boegman et al. 1985
Colchicine	Yes	Yes	Neuronal death	Shaughnessy et al. 1994
Alkaloids	Yes	Yes	Neuronal death	Di Patre et al. 1989
AF64A	Yes	No	Neuronal death	Chrobak et al. 1988; Hanin 1996; Waite et al. 1995
192Ig-G saporin	Yes	No	Neuronal death	Wiley 1992; Wiley et al. 1991
Alcohol	Yes	Yes	Neuronal death	Arendt 1994
β-Amyloid	No	No	Cholinergic dysfunction	Giovannini et al. 2002; Pavia et al. 2000

12.3.2.2   Specific Cholinergic Lesions

More refined lesion models employed specific toxins that affected only cholinergic neurones in the relevant areas, such as septum, nbm, and the diagonal band of Broca, but preserved noncholinergic neurones (Toledano and Alvarez 2004; Wiley 1992). These models were produced by using initially AF64A cholinotoxin, which binds to the high-affinity choline uptake system. Subsequently the immunotoxin 192 IgG-saporin that binds selectively and irreversibly to low-affinity nerve growth factor receptor interrupting cholinergic neuronal protein synthesis was employed. Both techniques lead to selective impairment and death of cholinergic neurones (Toledano and Alvarez 2004).

12.3.3   Transgenic Animals

None of the models mentioned above, however, produced the histopathology (plaques and tangles) and mimicked the progression of AD. The recent approach concentrated on developing transgenic animals that replicate various neuropathological features of AD (Gotz et al. 2004; Rodriguez et al. 2009) (Table 12.2). Initially, simple transgenic models were created expressing single mutated β-amyloid-related proteins (APPs or presenilines (PSs)) or mutated tau protein. The first APP transgenic animal demonstrating an AD-like pathology was developed in 1995 (Games et al. 1995). This model was named Tg(APPV717F)109Ili (PDAPP) and showed many pathological hallmarks of AD, including extracellular β-amyloid deposits, neuritic dystrophy, astrogliosis, and memory impairments. The latter, however, did not correlate with the degree of β-amyloid load (Games et al. 1995; Gotz et al. 2004). Subsequently, transgenic mice carrying APP_{swe} mutation Tg(APPSWE)2576Kha (Tg2576 mice) were developed (Hsiao et al. 1996). The Tg2576 mice had many Aβ plaques associated with memory and learning impairment, which started from 9 months of age (Hsiao et al. 1996). At the next step, the double APP mutation, the Tg(Thy1-APP)3Somm (APP23) mouse, was created. This model showed a 14% loss of hippocampal CA1 pyramidal neurones (Eriksen and Janus 2007; Sturchler-Pierrat et al. 1997).

Mutations in PS proteins are observed in the majority of Familial Alzheimer's disease (FAD) cases. Therefore double transgenic mice harboring PS and APP mutations have been created. Coexpression of $PS1_{dE9}$ with APP resulted in a viable model characterized with accelerated Aβ deposition and memory deficits without tangle formation (Savonenko et al. 2005).

The pathological tau animals were developed in parallel, the first model being created in 1995 (Gotz et al. 1995); hyperphosphorlated tau was concentrated in somata and dendrites, but the animals never developed neurofibrillary tangle (NFT) pathology. After identification of the pathogenic mutations of tau in FTDP-17, different transgenic models with neuronal NFT were produced. The Tau_{P301L} mutation is the most common mutation linked with FTDP-17 (Lewis et al. 2000). Transgenic mice overexpressing Tau_{P301L} exhibit neurofibrillary tangles without Aβ pathology or neuronal loss, except in the spinal cord (Lewis et al. 2000).

Recently, triple transgenic mice Tg(APPSwe, tauP301L)1Lfa (3xTg-AD), which harbor the mutant genes for amyloid precursor protein (APP_{Swe}), for presenilin

TABLE 12.2
Transgenic Mouse and Rat Models of Alzheimer's Disease

Symbol	Synonyms	Neuropathology	Reference
Tg(Thy1-APP^{PSI})28L-pr	Thy-1 APP, Thy1-APP751SL	Plaques	Blanchard et al. 2003
Tg(Thy1-APPLon)2Vln	APP/Ld/2, APP/V717I, Tg(Thy1-APP*V642I)2Vln	Plaques	Moechars et al. 1996
Tg(Thy1-APP)3Somm	APP 23, APP23, Tg(Thy1-APPK670N;M671L), Tg(Thy1APP)23Sdz, TgAPP23	Plaques	Sturchler-Pierrat et al. 1997
Tg(PDGFB-PSEN1M146L)2Jhd	PDGF-hPS1, Ps1 line 6.2, PS1(M146L), PS1M146L, PS1mut, PSIM146L	Diffused plaques	Blanchard et al. 2003
APP$_{751SL}$/PS1$_{M146L}$		Plaques	Blanchard et al. 2003
Tg(APPswe,PSEN1dE9)85Dbo	APdE9, APP/PS1, APPswe/PS1dE9, Mo/Hu APPswe PS1dE9	Plaques	Savonenko et al. 2005
APPSwedish and PS1^{M146L}		Plaques	Janus et al. 2000
Tg(APPSWE)2576Kha	APP695swe, APPK670N, M671L, APPSw, APPswe Tg2576, hAPP, K670N/M671L, Tg(APPSWE)2576HKahs, Tg(APPSWE)2576Kahs, Tg(HuAPP695. K670-M671L)2576, Tg2576, TgN(APPSWE)2576	Plaques	Hsiao et al. 1996
Tg(APPV717F)109Ili	APP^{V717F}, APPInd, hAPP+, hbeta-APP, PD-APP transgenic, PDAPP, PDAPP+, PDAPP-109, PDGF-hAPP(V717F)	Plaques	Dodart et al. 2000
Tg(APPSwLon)96Btla	APP^{K670NM671L} and APP^{V717Im}, J1.96, K670N/M671L + V717I	Plaques	Janus et al. 2000

(Continued)

TABLE 12.2 (CONTINUED)
Transgenic Mouse and Rat Models of Alzheimer's Disease

Symbol	Synonyms	Neuropathology	Reference
Tg(PDGFB-APPSwInd)20Lms	APP Tg, APP/J20, J20, PDGF-APPSwInd, PDGF-hAPP695,751,770V171F, KM670/671NL	Plaques	Eriksen and Janus 2007
Tg(PDGFB-APPSwInd)J9Lms	APPSw,Ind, Tg(PDGFB-APP*)J 9Lms	Plaques	Chishti et al. 2001
Tg(Thy1-MAPT*V337M)1Godt	TauV337M, Tg(Thy1-MAPT*V337M)1Migo	Tangles	Tanemura et al. 2002
Tg(CamK2a-MAPT*R406W)748Atak	—	Tangles	Tatebayashi et al. 2002
Tg(Prnp-MAPT*P301L)JNPL3Hlmc	JNPL3, P301L, P301L tau 4R, tau (P301L), tau^{P301L}, Tg(Prnp-MAPT*P301L)3Hlmc	Tangles	Lewis et al. 2000
Tg(Thy1-MAPT)183Gotz	P301L tau	Tangles	Gotz et al. 2001
Tg(Thy1-MAPT)22Schd	THY-Tau22	Tangles	Schindowski et al. 2006
Tg(Thy1-MAPT*P301S)2541Godt	TauP301S	Tangles	Allen et al. 2002
Tg(Thy1-MAPT*)1Avil	tau^{VLW}	Tangles	Eriksen and Janus 2007
Endogenous tau KO		Tangles	Andorfer et al. 2003
P301L TET-off		Tangles	Ramsden et al. 2005
7TauTg		Tangles	Ishihara et al. 2001
Tg2576xJNPL3 (APP$_{SWE}$)		Plaques and tangles	Lewis et al. 2001
Tg2576 and VLW		Plaques and tangles	Ribe et al. 2005
Tg(APPSwe, tauP301L)1LFa	Tg2576/PSEN1	Plaques and tangles	Oddo et al. 2003b
Tg478		None	Flood et al. 2009
Tg1116		None	Flood et al. 2009
Tg478/Tg1116		Plaques	Flood et al. 2009
Tg478/1116/11587		Plaques	Flood et al. 2009

1PS1$_{M146V}$ and tau$_{P301L}$ were developed (Oddo et al. 2003a, 2003b). These mice show temporal- and region-specific Aβ and tau pathology that closely resembles that seen in the human AD brain. As well as progressively developing plaques and tangles, the 3xTg-AD animals also show clear functional and cognitive impairments, including reduced Long-term potentiation (LTP), as well as deficient spatial and long-term memory (Oddo et al. 2003a, 2003b). These pathological changes progress in an age-related manner; most importantly, functional deficits precede the appearance of histological hallmarks (Oddo et al. 2003a, 2003b). Cognitive deficits in the 3xTg-AD model correlate with the accumulation of intraneuronal Aβ (Carroll et al. 2007; McKee et al. 2008).

12.4 ASTROCYTES IN DEMENTIA AND ALZHEIMER'S DISEASE

12.4.1 Non-AD Dementia

Similarly to other types of neurodegeneration outlined in the previous chapter, astroglial cells display pathological changes of both atrophic and reactive character in various forms of non-AD dementia.

In frontotemporal dementia (which includes several forms of distinct pathological processes such as Pick's disease and frontotemporal lobar degeneration) astrocytes undergo early apoptotic death and dystrophy (Broe et al. 2004); incidentally, the degree of glial atrophy correlates with the severity of dementia. These findings remain controversial because alternative study found, in the postmortem tissues, prominent astrogliosis and significant (four or five times) increase in astroglial cell density (Kersaitis et al. 2004). Severe astrogliosis was also found in thalamic dementia; the pathological remodeling was specifically observed in perivascular and perineuronal astroglial processes (Potts and Leech 2005). Both astrogliosis and astroglial apoptosis and astrodegeneration were observed in immunodeficiency virus-1 (HIV-1)–associated dementia (HAD) (Kaul et al. 2001; Vanzani et al. 2006). Interestingly, most profound astrodegeneration and astroglial cell loss was observed in subjects with rapidly progressing cognitive deficits (Thompson et al. 2001).

12.4.2 Alzheimer's Disease

AD (Alzheimer 1907) appears in both genetic (familia AD, FAD) and sporadic forms. The AD is characterized by progressive neurodegeneration and an occurrence of specific histopathological markers represented by focal extracellular deposits of fibrillar β-amyloid (also called neuritic or senile plaques) in brain parenchyma and in the wall of blood vessels and intraneuronal accumulation of neurofibrillary tangles composed from abnormal hyperphosphorylated tau filaments. The initial neurodegenerative events in AD appear in the transentorhinal cortex, which subsequently spread to the entorhinal cortex and hippocampus. At the later stages of the disease the neurodegenerative process disseminates through the temporal, frontal, and parietal lobes (Thompson et al. 2003, 2007). At these late stages the grey matter undergoes severe damage manifested by a profound loss of neurones and synaptic contacts.

The role of astrocytes in the pathogenesis and progression of AD remains obscure, primarily because of the lack of longitudinal studies assessing the status of astroglia at different stages of the disease. It is generally agreed that at the late stages of the disease, the analysis of postmortem tissues demonstrates prominent reactive astrogliosis and inclusion of astrocytes into senile plaques (Heneka et al. 2010; Nagele et al. 2004; Rodriguez et al. 2009). This pathological role of neuroglia was initially suggested by Alois Alzheimer (1910), who described glial cells as an integral part of the neuritic plaque (Figure 12.1).

The nature of the molecules that trigger astrogliosis in the AD is yet to be fully characterized, although β-amyloid peptide (Aβ) by itself presents an activating signal. Exposure of cultured glial cells to aggregated β-amyloid or to amyloid plaques isolated from human AD brains was shown to trigger reactive astrogliosis (DeWitt et al. 1998). Simultaneously, the Aβ may induce calcium (Ca^{2+}) signals (represented by both Ca^{2+} transients and Ca^{2+} oscillations) in astrocytes in vitro, in astroglial-neuronal cocultures (Abramov et al. 2003). These Ca^{2+} oscillations were somehow involved in Aβ-induced neurotoxicity, probably through the release of reactive oxygen species from Aβ-stimulated astrocytes (Abramov et al. 2004). The abnormal Ca^{2+} signaling was also observed in vivo, in transgenic AD mouse

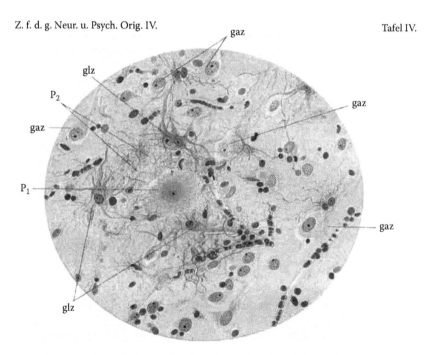

Z. f. d. g. Neur. u. Psych. Orig. IV.   Tafel IV.

FIGURE 12.1 Alois Alzheimer's drawing illustrating the glial reaction (astrogliosis and hypertrophy) in a pathological brain containing senile plaques. gaz, neurone; glz, glial cell, P central part of the plaque; P_2, peripheral part of the plaque. (From Alzheimer, A., in *Histologische und histopathologische Arbeiten über die Grosshirnrinde mit besonderer Berücksichtigung der pathologischen Anatomie der Geisteskrankheiten*, ed. F. Nissl and A. Alzheimer, Gustav Fischer, Jena, Germany, 1910.)

models. In this preparation spontaneous Ca^{2+} waves with much extended propagation distance were recorded from astrocytes located in the vicinity of neuritic plaques (Kuchibhotla et al. 2009). The Aβ was also reported to decrease expression and capacity of GLAST- and GLT-1-mediated glutamate uptake in cultured astrocytes (Matos et al. 2008).

The question of whether astrocytes participate in production/sequestration of Aβ in AD remains debatable. Astrocytes from mice expressing double-mutated Tg(APPSWE)2576Kha (K670N-M671L APP) were reported to express β-secretase, thus being able to produce Aβ (Hartlage-Rubsamen et al. 2003; Heneka et al. 2005). At the same time there are some indications that astrocytes may accumulate (by phagocytosis?) and degrade Aβ reducing, therefore, Aβ load (Nagele et al. 2004). Accumulation of Aβ was observed in astrocytes from the entorhinal cortex of AD patients (Nagele et al. 2003). Experiments in vitro have sown that astrocytes isolated from healthy brains and plated onto slices isolated from mutant APP transgenic mice migrated toward the plaques and accumulated Aβ. The astrocytes isolated from the APP mice, however, were unable to accumulate and remove Aβ (Wyss-Coray et al. 2003).

Similarly in astrocytes from the 3xTg-AD mice very little (if any) Aβ accumulation by reactive astrocytes surrounding the plaques was observed (Olabarria et al. 2010; Rodriguez et al. 2009). Therefore, astroglial cells undergo complex remodeling during AD progression, which may affect their defensive functions and hence modify the development of the disease.

In the 3x-TG-AD mice astrocytes undergo complex morphological changes (Figures 12.2 and 12.3). At the early (i.e., preplaque) stages of AD, astrocytes in the hippocampus and entorhinal cortex (from 6 months and 1 month of age, respectively) demonstrate signs of atrophy and astrodegeneration (Olabarria et al. 2010; Rodriguez et al. 2009; Yeh et al. 2011) (Figure 12.2). These changes comprised reduced expression of GFAP-rich cytoskeleton (surface and volume coverage) and decreased somatas volume, as well as number and branching of cell processes (Figure 12.2E–H); accounting for at least a 40%–50% decrease and being subfield and layer specific; DG and CA1 in the hippocampus and layers II and VI in the entorhinal cortex (Olabarria et al. 2010; Rodriguez et al. 2009; Yeh et al. 2011) (Figure 12.2A–D, I–J). However, there is no decrease in the overall number of GFAP-containing astrocytes either in the hippocampus or in the entorhinal cortex.

At the later stages the plaque formation and accumulation (12 and 18 months; when also the cells start to show evident neurofibrillary tangles) trigger a clear and specific astrogliotic reaction. Numerous hypertrophic astrocytes gather exclusively around neuritic plaques and β-amyloid-infested blood vessels (Olabarria et al. 2010; Rodriguez et al. 2009) (Figure 12.3A, B). This hypertrophy is characterized by an increased volume and surface of both astrocyte somata and processes, which can increase their size up to 70% (Figure 12.3).

The early atrophy of astroglial cells may have important functional consequences. Decrease in astroglial complexity may affect synaptic coverage and functional performance of neuronal-glial-vascular units. This in turn can affect connectivity in the neural network, reduce synaptic strength, and disturb synaptic plasticity with obvious cognitive consequences, which could result from an astroglial-mediated early imbalance in mnesic areas.

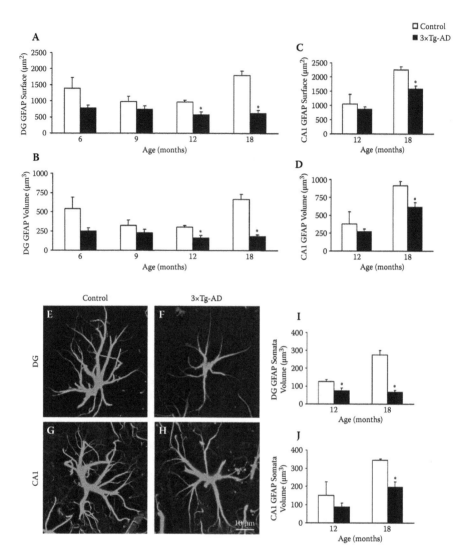

FIGURE 12.2 Bar graphs showing the significant decrease in GFAP surface, volume, and soma volume within both the DG (A, B, I) and the CA1 (C, D, J) of the hippocampus of the 3xTg-AD mice when compared with control animals. Bars represent mean ± SEM ($p < 0.05$) (G–J). Confocal micrographs illustrating the astrocytic atrophy in 3xTg-AD mice in the DG (F) and CA1 (H) compared to control animals (E and G). (Reproduced with permission from Olabarria, M., et al., *Glia*, 58, 831–38, 2010.)

12.5   OLIGODENDROGLIA AND MYELIN IN DEMENTIA AND ALZHEIMER'S DISEASE

Oligodendrocytes are the major cell type of CNS white matter, which in humans comprises about 50% of brain volume (Fields 2008). In addition, oligodendroglia are also present in grey matter, in particular throughout the cerebral cortex. Primary or

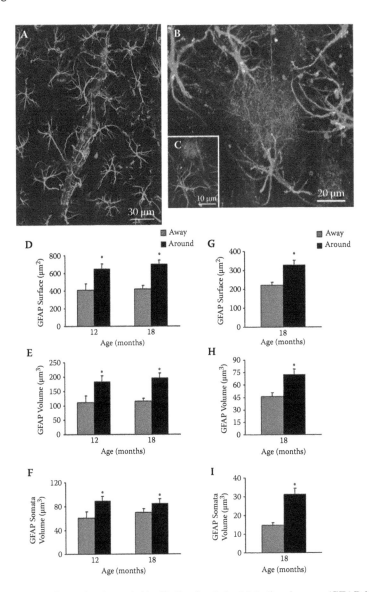

FIGURE 12.3 **(See color insert)** (A–C) Confocal dual labeling images (GFAP in green and Aβ in red) showing the accumulation of astrocytes around the Aβ plaques and vascular Aβ deposits in the hippocampus of 3xTg-AD mice. Astrocytes surrounding Aβ plaques (B, C) and Aβ-loaded blood vessel (A) undergo astrogliosis. (D–F) Bar graphs showing GFAP-positive astrocytic surface (D), volume (E), and soma volume (F) differences between astrocytes located around the amyloid plaques (Aβ) and those distant to the plaques in the CA1 of 3xTg-AD animals. (G–I) Similar astrocytic surface (G), volume (H), and soma volume (I) differences are observed in the DG at 18 months of age. Bars represent mean ± SEM ($p <$ 0.05). (Modified from Olabarria, M., et al., *Glia*, 58, 831–38, 2010.)

secondary oligodendrocyte death and myelin damage occur in most, if not all, CNS diseases, including stroke, perinatal ischemia, multiple sclerosis, psychiatric disorders, traumatic injury, and Alzheimer's disease (Matute 2010).

The main risk factor of dementia is aging. Intriguingly, myelin loss occurs along with aging in the cerebral cortex, especially in areas related to cognition and memory, including the frontal lobes (reviewed in Bartzokis 2009). Thus, intracortical and subcortical myelin levels diminish drastically as observed in both magnetic resonance imaging (MRI) of normal individuals and postmortem studies. The myelin profile reveals an inverted U profile of myelination, which increases during postnatal development, peaks at around 45 years, and subsequently decreases to infancy levels as age approaches 100 years (Bartzokis 2009). The molecular mechanisms leading to myelin loss with aging have not been elucidated yet and may include oxidative stress, hypoperfusion, increased cortisol levels, and excitotoxicity together with neuroinflammation. It is also unclear whether myelin decline is a consequence of neuronal demise or if it contributes primarily to dysfunction in the aging brain.

A key event to oligodendrocyte and myelin pathophysiology is Ca^{2+} dyshomeostasis, which is caused by aberrant signaling by extracellular molecules such as neurotransmitters, or alterations in the control of cytosolic Ca^{2+} levels by intracellular stores. Oligodendrocytes and myelin express ligand-gated channels that are permeable to Ca^{2+}, including glutamate and ATP receptors. Prolonged activation of ionotropic glutamate and ATP receptors triggers oligodendrocyte death and myelin destruction as a consequence of cytosolic Ca^{2+} overload, accumulation of this cation within mitochondria, increases in the production of radical oxygen species, and release of proapoptotic factors that activate caspases (Matute 2010; Salter and Fern 2005). Glutamate and ATP excitotoxicity constitute a relevant disease mechanism since glutamate and ATP are among the most abundant molecules in brain tissue and acute transient alterations in their homeostasis, or during chronic pathological states can overactivate their receptors and damage oligodendroglia and myelin (Matute 2010; Matute et al. 2007). However, the extent to which excitotoxicty contributes to demyelination in aging and dementia is not known.

12.5.1 Aging and Non-AD Dementia

The effects of age on oligodendrocytes and myelin have been recently examined experimentally in the primary visual cortices of rhesus monkeys that had been behaviorally tested (Peters et al. 2008). It was found that alterations in myelin sheaths increase with age, and that there is an age-related increase in the frequency of profiles of nerve fibers sectioned through paranodes, indicating that shorter lengths of myelin are being produced by remyelination. In addition, alterations in myelin correlate with the number of oligodendrocytes, suggesting that with age newly generated oligodendrocytes are required to remyelinate nerve fibers whose sheaths are altered. Importantly, the most cognitively impaired monkeys had the greatest numbers of oligodendrocytes, reflecting the fact that altered myelin slows down the rate of conduction along nerve fibers (Peters et al. 2008).

These data suggest that structural integrity of myelin sheaths deteriorates during aging and leads to impaired efficiency in information flow along neural networks, and thus to cognitive decline. Indeed, MRI studies of the human corpus callosum indicate that breakdown of white matter structural integrity occurs throughout its entire extent (Bartzokis et al. 2004). Notably, myelin decay is more prominent in the genu than in the splenium of the corpus callosum, which constitute the more rostral and caudal, respectively, parts of this structure, and indicates that former myelinating regions are more vulnerable to aging. In addition, these observations imply that interhemispheric connectivity of the frontal lobes is more severely affected, and consequently, cognition-related areas may malfunction (Bartzokis et al. 2004).

Early MRI studies of patients with non-Alzheimer's dementia have also shown diffuse, patchy white matter lesions in all cases examined (Brant-Zawadzki et al. 1985). The grade of severity of the changes and the prevalence observed were higher than in cognitively normal subjects. Interestingly, calcium binding proteins calmodulin and calbindin have altered expression and function in the white matter of non-Alzheimer's dementia patients (McLachlan et al. 1987). Thus, calmodulin has reduced efficacy in activating 3',5' cyclic nucleotide phosphodiesterase, which is exclusively and constitutively expressed in oligodendrocytes, whereas calbindin has reduced levels in white matter (McLachlan et al. 1987).

Recent studies have provided more details about the nature of the damage to white matter and the type of non-Alzheimer's dementia. Thus, patients with frontotemporal dementia show ischemic-like primary incomplete infarction of frontal white matter which shows gliosis and demyelination (Larsson et al. 2000), and those with amyotrophic lateral sclerosis-related dementia display diffuse fibrous gliosis in the frontotemporal white matter (Yoshida 2004). In turn, depressed patients with mild non-Alzheimer's dementia have changes in white matter MRI hyperintensity volumes in white matter (Steffens et al. 2007), a feature that also occurs in vascular dementia (Staekenborg et al. 2009). Importantly, these changes in hyperintensity have predictive value for the onset and progression of dementia. (Staekenborg et al. 2009; Steffens et al. 2007).

12.5.2 Alzheimer's Disease

It is well known that white matter is altered in Alzheimer's disease (AD). Thus, damage in specific locations appears to impair those cognitive functions that rely on networks involving those regions, and the extent of this damage is associated with dementia severity in AD (Bronge 2002). A high percentage of AD patients exhibit evidence of white matter degeneration with severe loss of oligodendrocytes by apoptosis (Brown et al. 2000). Although the mechanisms underlying this damage are not well understood, there is evidence that they involve amyloidoisis and Ca^{2+} dyshomeostasis (Mattson and Chan 2003; Matute 2010).

β-Amyloid peptides can damage oligodendrocytes in vitro (Xu et al. 2001) and increase their vulnerability to glutamate toxicity (Pak et al. 2003). This effect is not an in vitro artifact because injection of amyloid β_{1-42} into white matter causes axon disruption and myelin damage, as well as oligodendrocyte loss and profound gliosis (Jantaratnotai et al. 2003). On the other hand, oligodendrocytes from presenilin-1 mutant knock-in mice are more susceptible to glutamate excitotoxicity and exhibit an abnormality in Ca^{2+} regulation, which is responsible for their demise (Pak et al.

2003). Moreover, when exposed to the demyelinating agent cuprizone, presenilin-1 mutant mice exhibit enhanced white matter damage and learning and memory deficits that are not seen in wild-type mice exposed to cuprizone (Pak et al. 2003). These findings reveal that a specific presenilin-1 mutation in oligodendrocytes can have detrimental effects leading to disease, and indicate that white matter damage may well contribute to cognitive dysfunction in AD.

In the 3xTg-AD mice significant region-specific alterations in myelination patterns and in oligodendrocyte marker expression profiles were observed at time points preceding the appearance of amyloid and tau pathology (Desai et al. 2009). These findings indicate that myelin and oligodendrocyte defects in AD occur before the onset of symptoms and may be key players in the development of this disease. Indeed, lesions are prevalent in early-stage AD at periventricular and deep white matter (Burns et al. 2005). Notably, the burden of both types of lesions is associated with reduced global cognition as assessed by evaluating visual memory, processing speed, and executive function.

12.5.3 Relevance of Oligoprotection to Dementia

White and grey matter oligodendrocytes are key cells for information flow in myelinated axons. Cognitive deficits associated to AD and non-AD dementia are paralleled by alterations in myelin patterns. Consequently, novel, effective oligo- and myelin-protective strategies ought to be developed to treat more efficiently these disorders. A feasible approach is to explore the therapeutic potential of specific drugs aiming at restoring Ca^{2+} homeostasis.

12.6 MICROGLIA IN ALZHEIMER'S DISEASE

Neuroinflammation is now considered a leading pathogenic component of AD. Conceptually, activation of microglia occurs simultaneously with the formation of neuritic plaques, this being specific for late stages of AD (Heneka et al. 2010; McGeer and McGeer 1999). The activated microglial cells together with astrocytes are closely associated with senile plaques, and they fuel the neuroinflammatory process by secreting multiple pro-inflammatory factors (Heneka and O'Banion 2007). Furthermore, activated microglial cells are considered to phagocytose Aβ, thus clearing the Aβ load (Bolmont et al. 2008; Frautschy et al. 1998).

The factors triggering microglial activation in AD brains are most likely of multiple origin and nature and may include amyloid peptides and a variety of molecules released by damaged neurones and activated astrocytes. Indeed, amyloid peptides and APP were reported to activate microglia (Barger and Harmon 1997; Schubert et al. 2000). It remains however, unclear whether it is soluble Aβ or consolidated senile plaques that act as an ultimate activating signal (Heneka et al. 2010). Recent studies using in vivo multiphoton microscopy on 5- to 6-month-old B6C3-YFP transgenic mice (harboring APP_{swe} and PS1d9x-YFP genes) demonstrated that microglia are activated and recruited to Aβ plaques only after the latter have been formed (Meyer-Luehmann et al. 2008). In contrast, the focal activation of microglia was observed in 3-month-old APP V717I transgenic mice

(Heneka et al. 2005), in which senile plaques appear substantially later (at ~10–12 months of age).

The mechanisms of microglial activation by Aβ and senile plaques may involve multiple receptors and signaling cascades. For example, AD-associated activation of microglia requires P2X$_7$ purinoceptors and Ca^{2+} signaling (Sanz et al. 2009). The role of P2X$_7$ receptors was specifically highlighted by recent experiments in which intrahippocampal injection of Aβ$_{1-42}$ failed to induce microglial activation in animals deficient in P2X$_7$ receptors (Sanz et al. 2009). Microglial activation in AD also involves Toll-like receptors of TLR4 and TLR2 type (Okun et al. 2009; Lotz et al. 2005). These receptors are upregulated in AD animal models and in postmortem AD brains (Okun et al. 2009). Incidentally, a spontaneous loss-of-function mutation in the TLR4 gene markedly decreased microglial activation induced by Aβ (Walter et al. 2007).

The longitudinal AD-induced reactions of microglia are complex. In the 3xTg-AD mice we have found a very substantial increase in the density of resting (tomato-lectin-positive) microglia at both the early (i.e., preplaque) and late stages of the disease (Rodríguez et al. 2010) (Figures 12.4A and 12.5A, B). At 9 months of age (when the senile plaques are virtually absent) the density of resting microglia in the hippocampus of 3xTg-AD was ~105% larger than in control mice. This increased density of resting microglia remains at older ages (at 12 months it was 54% higher and at 18 months 131% higher compared to the controls) (Figure 12.5A). The appearance of plaques triggered microglial activation (Figure 12.4B). We observed a significant increase in the density of activated microglia in the CA1 hippocampal area of 3xTg-AD mice at 12 and 18 months, which correlate with the age of the appearance and development of Aβ plaques (Rodríguez et al. 2010) (Figures 12.5C–E). The early increase in the density of resting microglia may represent the generalized response of the brain defense system to the developing AD pathology.

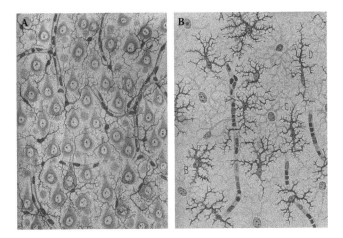

FIGURE 12.4 Drawings of Pío del Río-Hortega showing the morphological differences between resting and activated microglial cells in the rodent brain.

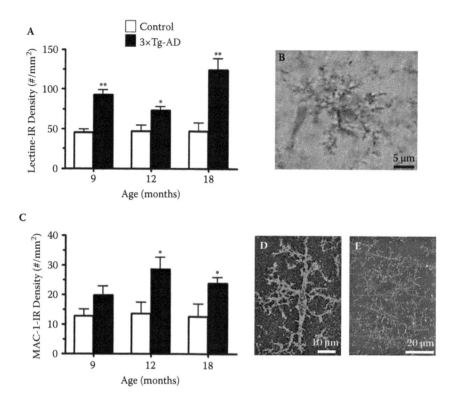

FIGURE 12.5 **(See color insert.)** (A, B) Visualization and quantification of resting and activated microglia in the hippocampi of 3xTg-AD animals. (A) Bar graph showing the area density of resting microglia (Sv; number/mm^2) in the hippocampal CA1 of 3xTg-AD mice and non-Tg control mice at 9, 12, and 18 months. (B) Brightfield micrographs of typical resting TL-IR microglia with small cell body showing thin to medium ramified processes extending to the surrounding neuropil in the CA1 subfield, which is not modified either by age or by Aβ-amyloid plaques. (CE) Microglial activation in plaque-infested hippocampal tissue. (C) Bar graph showing the effect of age-related changes on MAC-1-IR reactive microglia within the CA1 subfield of the hippocampus at 9, 12, and 18 months between non-Tg control and 3xTg-AD mice. (D) Fluorescence micrograph illustrating the characteristic morphology of reactive microglia within the CA1 subfield of the hippocampus of an 18-month-old 3xTg-AD mouse. Reactive microglia appear with most enlarged cell bodies from which a greater number of numerous processes emanated, but with an enlarged and thicker appearance. (E). Confocal image showing recruitment of MAC-1- IR microglia (green) in the vicinity of Aβ-amyloid aggregates (red) in the CA1 subfield of the hippocampus of an 18-month-old 3xTg-AD mouse. (Modified from Rodríguez, J. J., et al., *Cell Death Dis*, 1, e1, 2010.)

12.7   CONCLUSIONS

Neuroglial cells are the central element of the brain homeostatic and defense systems. All three types of neuroglia are involved in progression of Alzheimer's disease. The initial atrophy of astrocytes, observed at the early stages of the AD, reduces the synaptic coverage and support, leading thus to weakening of synaptic connectivity responsible for early cognitive deficits. At the later stages appearance of β-amyloid plaques triggers reactive astrogliosis, which promotes further remodeling of brain circuitry and disruption of neuronal networks. The reactive astrocytes also contribute to neurotoxicity and neuroinflammation through the compromised brain homeostasis, release of pro-inflammatory factors, and neurotoxic agents. AD-specific alterations of oligodendrocytes affect white matter, which contributes to cognitive decline. Finally, activation of microglia initiates and sustains the neuroinflammatory process that results in neural cell death and brain atrophy.

ACKNOWLEDGMENTS

We thank Mrs. Markel Olabarria and Harun N. Noristani for their help and assistance in the figures and table preparation. Authors' research was supported by Alzheimer's Research Trust (UK) Programme Grant (ART/PG2004A/1) to AV and JJR, by National Institute of Health (NIH) grant to AV, and by the Grant Agency of the Czech Republic (GACR 309/09/1696) to JJR and (GACR 305/08/1381 and GACR 305/08/1384) to AV.

REFERENCES

Abramov, A. Y., L. Canevari, and M. R. Duchen. 2003. Changes in intracellular calcium and glutathione in astrocytes as the primary mechanism of amyloid neurotoxicity. *J Neurosci* 23 (12):5088–95.

Abramov, A. Y., L. Canevari, and M. R. Duchen. 2004. b-Amyloid peptides induce mitochondrial dysfunction and oxidative stress in astrocytes and death of neurons through activation of NADPH oxidase. *J Neurosci* 24 (2):565–75.

Akiguchi, I., H. Tomimoto, T. Suenaga, H. Wakita, and H. Budka. 1997. Alterations in glia and axons in the brains of Binswanger's disease patients. *Stroke* 28 (7):1423–29.

Allen, B., E. Ingram, M. Takao, M. J. Smith, R. Jakes, K. Virdee, H. Yoshida, M. Holzer, M. Craxton, P. C. Emson, C. Atzori, A. Migheli, R. A. Crowther, B. Ghetti, M. G. Spillantini, and M. Goedert. 2002. Abundant tau filaments and nonapoptotic neurodegeneration in transgenic mice expressing human P301S tau protein. *J Neurosci* 22 (21):9340–51.

Alzheimer, A. 1907. Über eine eigenartige Erkrankung der Hirnrinde. *Allg. Z. Psychiat. Psych.-Gericht. Med.* 64:146–148.

Alzheimer, A. 1910. Beiträge zur Kenntnis der pathologischen Neuroglia und ihrer Beziehungen zu den Abbauvorgängen im Nervengewebe. In *Histologische und histopathologische Arbeiten über die Grosshirnrinde mit besonderer Berücksichtigung der pathologischen Anatomie der Geisteskrankheiten*, ed. F. Nissl and A. Alzheimer. Jena, Germany: Gustav Fischer.

Andorfer, C., Y. Kress, M. Espinoza, R. de Silva, K. L. Tucker, Y. A. Barde, K. Duff, and P. Davies. 2003. Hyperphosphorylation and aggregation of tau in mice expressing normal human tau isoforms. *J Neurochem* 86 (3):582–90.

Antel, J., and D. Arnold. 2005. Multiple sclerosis. In *Neuroglia*, ed. H. Kettenmann and B. Ransom. Oxford: Oxford University Press.

Arendt, T. 1994. Impairment in memory function and neurodegenerative changes in the cholinergic basal forebrain system induced by chronic intake of ethanol. *J Neural Transm Suppl* 44:173–87.

Arranz, B., K. Blennow, R. Ekman, A. Eriksson, J. E. Mansson, and J. Marcusson. 1996. Brain monoaminergic and neuropeptidergic variations in human aging. *J Neural Transm* 103 (1–2):101–15.

Barger, S. W., and A. D. Harmon. 1997. Microglial activation by Alzheimer amyloid precursor protein and modulation by apolipoprotein E. *Nature* 388 (6645):878–881.

Bartus, R. T., R. L. Dean 3rd, B. Beer, and A. S. Lippa. 1982. The cholinergic hypothesis of geriatric memory dysfunction. *Science* 217 (4558):408–14.

Bartzokis, G. In press. Alzheimer's disease as homeostatic responses to age-related myelin breakdown. doi: 10.1016/j.neurobiologing. 2009.08.007.

Bartzokis, G., D. Sultzer, P. H. Lu, K. H. Nuechterlein, J. Mintz, and J. L. Cummings. 2004. Heterogeneous age-related breakdown of white matter structural integrity: implications for cortical "disconnection" in aging and Alzheimer's disease. *Neurobiol Aging* 25 (7):843–51.

Bertoni-Freddari, C., C. Giuli, C. Pieri, and D. Paci. 1986. Quantitative investigation of the morphological plasticity of synaptic junctions in rat dentate gyrus during aging. *Brain Res* 366 (1–2):187–92.

Biegon, A., V. Greenberger, and M. Segal. 1986. Quantitative histochemistry of brain acetyl-cholinesterase and learning rate in the aged rat. *Neurobiol Aging* 7 (3):215–17.

Blanchard, V., S. Moussaoui, C. Czech, N. Touchet, B. Bonici, M. Planche, T. Canton, I. Jedidi, M. Gohin, O. Wirths, T. A. Bayer, D. Langui, C. Duyckaerts, G. Tremp, and L. Pradier. 2003. Time sequence of maturation of dystrophic neurites associated with Abeta deposits in APP/PS1 transgenic mice. *Exp Neurol* 184 (1):247–63.

Blumenthal, I. 2004. Periventricular leucomalacia: a review. *Eur J Pediatr* 163 (8):435–42.

Boegman, R. J., S. R. el-Defrawy, K. Jhamandas, R. J. Beninger, and S. K. Ludwin. 1985. Quinolinic acid neurotoxicity in the nucleus basalis antagonized by kynurenic acid. *Neurobiol Aging* 6 (4):331–36.

Bolmont, T., F. Haiss, D. Eicke, R. Radde, C. A. Mathis, W. E. Klunk, S. Kohsaka, M. Jucker, and M. E. Calhoun. 2008. Dynamics of the microglial/amyloid interaction indicate a role in plaque maintenance. *J Neurosci* 28 (16):4283–92.

Brant-Zawadzki, M., G. Fein, C. Van Dyke, R. Kiernan, L. Davenport, and J. de Groot. 1985. MR imaging of the aging brain: patchy white-matter lesions and dementia. *AJNR Am J Neuroradiol* 6 (5):675–82.

Broe, M., J. Kril, and G. M. Halliday. 2004. Astrocytic degeneration relates to the severity of disease in frontotemporal dementia. *Brain* 127 (Pt 10):2214–20.

Bronge, L. 2002. Magnetic resonance imaging in dementia. A study of brain white matter changes. *Acta Radiol Suppl* 428:1–32.

Brown, W. R., D. M. Moody, C. R. Thore, and V. R. Challa. 2000. Cerebrovascular pathology in Alzheimer's disease and leukoaraiosis. *Ann NY Acad Sci* 903:39–45.

Burns, J. M., J. A. Church, D. K. Johnson, C. Xiong, D. Marcus, A. F. Fotenos, A. Z. Snyder, J. C. Morris, and R. L. Buckner. 2005. White matter lesions are prevalent but differentially related with cognition in aging and early Alzheimer disease. *Arch Neurol* 62 (12):1870–76.

Carroll, J. C., E. R. Rosario, L. Chang, F. Z. Stanczyk, S. Oddo, F. M. LaFerla, and C. J. Pike. 2007. Progesterone and estrogen regulate Alzheimer-like neuropathology in female 3xTg-AD mice. *J Neurosci* 27 (48):13357–65.

Cassel, J. C., C. Mathis, M. Majchrzak, P. H. Moreau, and J. C. Dalrymple-Alford. 2008. Coexisting cholinergic and parahippocampal degeneration: a key to memory loss in dementia and a challenge for transgenic models? *Neurodegener Dis* 5 (5):304–17.

Chishti, M. A., D. S. Yang, C. Janus, A. L. Phinney, P. Horne, J. Pearson, R. Strome, N. Zuker, J. Loukides, J. French, S. Turner, G. Lozza, M. Grilli, S. Kunicki, C. Morissette, J. Paquette, F. Gervais, C. Bergeron, P. E. Fraser, G. A. Carlson, P. S. George-Hyslop, and D. Westaway. 2001. Early-onset amyloid deposition and cognitive deficits in transgenic mice expressing a double mutant form of amyloid precursor protein 695. *J Biol Chem* 276 (24):21562–70.

Chrobak, J. J., I. Hanin, D. E. Schmechel, and T. J. Walsh. 1988. AF64A-induced working memory impairment: behavioral, neurochemical and histological correlates. *Brain Res* 463 (1):107–17.

Decker, M. W. 1987. The effects of aging on hippocampal and cortical projections of the forebrain cholinergic system. *Brain Res* 434 (4):423–38.

Desai M. K., K. L. Sudol, M. C. Janelsins, M. A. Mastrangelo, M. E. Frazer, and W. J. Bowers. 2009. Triple-transgenic Alzheimer's disease mice exhibit region-specific abnormalities in brain myelination patterns prior to appearance of amyloid and tau pathology. *Glia* 57 (1): 54–65

DeWitt, D. A., G. Perry, M. Cohen, C. Doller, and J. Silver. 1998. Astrocytes regulate microglial phagocytosis of senile plaque cores of Alzheimer's disease. *Exp Neurol* 149 (2):329–40.

Di Patre, P. L., A. Abbamondi, L. Bartolini, and G. Pepeu. 1989. GM1 ganglioside counteracts cholinergic and behavioral deficits induced in the rat by intracerebral injection of vincristine. *Eur J Pharmacol* 162 (1):43–50.

Dodart, J. C., C. Mathis, J. Saura, K. R. Bales, S. M. Paul, and A. Ungerer. 2000. Neuroanatomical abnormalities in behaviorally characterized APP(V717F) transgenic mice. *Neurobiol Dis* 7 (2):71–85.

Dunnett, S. B., B. J. Everitt, and T. W. Robbins. 1991. The basal forebrain-cortical cholinergic system: interpreting the functional consequences of excitotoxic lesions. *Trends Neurosci* 14 (11):494–501.

Ercolini, A. M., and S. D. Miller. 2006. Mechanisms of immunopathology in murine models of central nervous system demyelinating disease. *J Immunol* 176 (6):3293–98.

Eriksen, J. L., and C. G. Janus. 2007. Plaques, tangles, and memory loss in mouse models of neurodegeneration. *Behav Genet* 37 (1):79–100.

Fields, R. D. 2008. White matter in learning, cognition and psychiatric disorders. *Trends Neurosci* 31 (7):361–70.

Fischer, W., K. S. Chen, F. H. Gage, and A. Bjorklund. 1992. Progressive decline in spatial learning and integrity of forebrain cholinergic neurons in rats during aging. *Neurobiol Aging* 13 (1):9–23.

Flood, D. G., Y. G. Lin, D. M. Lang, S. P. Trusko, J. D. Hirsch, M. J. Savage, R. W. Scott, and D. S. Howland. 2009. A transgenic rat model of Alzheimer's disease with extracellular Abeta deposition. *Neurobiol Aging* 30 (7):1078–90.

Frautschy, S. A., F. Yang, M. Irrizarry, B. Hyman, T. C. Saido, K. Hsiao, and G. M. Cole. 1998. Microglial response to amyloid plaques in APPsw transgenic mice. *Am J Pathol* 152 (1):307–17.

Games, D., D. Adams, R. Alessandrini, R. Barbour, P. Berthelette, C. Blackwell, T. Carr, J. Clemens, T. Donaldson, F. Gillespie, and et al. 1995. Alzheimer-type neuropathology in transgenic mice overexpressing V717F beta-amyloid precursor protein. *Nature* 373 (6514):523–27.

Giaume, C., F. Kirchhoff, C. Matute, A. Reichenbach, and A. Verkhratsky. 2007. Glia: the fulcrum of brain diseases. *Cell Death Differ* 14 (7):1324–35.

Giovannini, M. G., C. Scali, C. Prosperi, A. Bellucci, M. G. Vannucchi, S. Rosi, G. Pepeu, and F. Casamenti. 2002. Beta-amyloid-induced inflammation and cholinergic hypofunction in the rat brain *in vivo*: involvement of the p38MAPK pathway. *Neurobiol Dis* 11 (2):257–74.

Götz J. et al. 2001. Tau and transgenic animal models. *Brain Res Rev* 35 (3): 266–86

Götz, J., A. Probst, M. G. Spillantini, T. Schafer, R. Jakes, K. Burki, and M. Goedert. 1995. Somatodendritic localization and hyperphosphorylation of tau protein in transgenic mice expressing the longest human brain tau isoform. *Embo J* 14 (7):1304–13.

Götz, J., J. R. Streffer, D. David, A. Schild, F. Hoerndli, L. Pennanen, P. Kurosinski, and F. Chen. 2004. Transgenic animal models of Alzheimer's disease and related disorders: histopathology, behavior and therapy. *Mol Psychiatry* 9 (7):664–83.

Halassa, M. M., T. Fellin, and P. G. Haydon. 2007. The tripartite synapse: roles for gliotransmission in health and disease. *Trends Mol Med* 13 (2):54–63.

Hanin, I. 1996. The AF64A model of cholinergic hypofunction: an update. *Life Sci* 58 (22):1955–64.

Hanisch, U. K., and H. Kettenmann. 2007. Microglia: active sensor and versatile effector cells in the normal and pathologic brain. *Nat Neurosci* 10 (11):1387–94.

Hartlage-Rubsamen, M., U. Zeitschel, J. Apelt, U. Gartner, H. Franke, T. Stahl, A. Gunther, R. Schliebs, M. Penkowa, V. Bigl, and S. Rossner. 2003. Astrocytic expression of the Alzheimer's disease b-secretase (BACE1) is stimulus-dependent. *Glia* 41 (2):169–79.

Hazell, A. S. 2009. Astrocytes are a major target in thiamine deficiency and Wernicke's encephalopathy. *Neurochem Int* 55 (1–3):129–35.

Hazell, A. S., D. Sheedy, R. Oanea, M. Aghourian, S. Sun, J. Y. Jung, D. Wang, and C. Wang. 2009. Loss of astrocytic glutamate transporters in Wernicke encephalopathy. *Glia* 58:148–156.

Heneka, M. T., and M. K. O'Banion. 2007. Inflammatory processes in Alzheimer's disease *J Neuroimmunol* 184 (1–2):69–91.

Heneka, M. T., J. J. Rodriguez, and A. Verkhratsky. 2010. Neuroglia in neurodegeneration. *Brain Res Rev*, 63(1–2): 189–211.

Heneka, M. T., M. Sastre, L. Dumitrescu-Ozimek, I. Dewachter, J. Walter, T. Klockgether, and F. Van Leuven. 2005. Focal glial activation coincides with increased BACE1 activation and precedes amyloid plaque deposition in APP[V717I] transgenic mice. *J Neuroinflam* 2 (1):22.

Hsiao, K., P. Chapman, S. Nilsen, C. Eckman, Y. Harigaya, S. Younkin, F. Yang, and G. Cole. 1996. Correlative memory deficits, Abeta elevation, and amyloid plaques in transgenic mice. *Science* 274 (5284):99–102.

Iadecola, C., and M. Nedergaard. 2007. Glial regulation of the cerebral microvasculature. *Nat Neurosci* 10 (11):1369–76.

Ishihara, T., M. Higuchi, B. Zhang, Y. Yoshiyama, M. Hong, J. Q. Trojanowski, and V. M. Lee. 2001. Attenuated neurodegenerative disease phenotype in tau transgenic mouse lacking neurofilaments. *J Neurosci* 21 (16):6026–35.

Jabs, R., G. Seifert, and C. Steinhauser. 2008. Astrocytic function and its alteration in the epileptic brain. *Epilepsia* 49 (Suppl 2):3–12.

Jantaratnotai, N., J. K. Ryu, S. U. Kim, and J. G. McLarnon. 2003. Amyloid beta peptide-induced corpus callosum damage and glial activation *in vivo*. *Neuroreport* 14 (11):1429–33.

Janus, C., J. Pearson, J. McLaurin, P. M. Mathews, Y. Jiang, S. D. Schmidt, M. A. Chishti, P. Horne, D. Heslin, J. French, H. T. Mount, R. A. Nixon, M. Mercken, C. Bergeron, P. E. Fraser, P. St. George-Hyslop, and D. Westaway. 2000. A beta peptide immunization reduces behavioural impairment and plaques in a model of Alzheimer's disease. *Nature* 408 (6815):979–82.

Jessen, K. R., and R. Mirsky. 2008. Negative regulation of myelination: relevance for development, injury, and demyelinating disease. *Glia* 56 (14):1552–65.

Kaul, M., G. A. Garden, and S. A. Lipton. 2001. Pathways to neuronal injury and apoptosis in HIV-associated dementia. *Nature* 410 (6831):988–94.

Kersaitis, C., G. M. Halliday, and J. J. Kril. 2004. Regional and cellular pathology in frontotemporal dementia: relationship to stage of disease in cases with and without Pick bodies. *Acta Neuropathol* 108 (6):515–23.

Kettenmann, H., and A. Verkhratsky. 2008. Neuroglia: the 150 years after. *Trends Neurosci* 31 (12):653–59.

Koeppen, A. H. 2004. Wallerian degeneration: history and clinical significance. *J Neurol Sci* 220 (1–2):115–17.

Kuchibhotla, K. V., C. R. Lattarulo, B. T. Hyman, and B. J. Bacskai. 2009. Synchronous hyperactivity and intercellular calcium waves in astrocytes in Alzheimer mice. *Science* 323 (5918):1211–15.

Larsson, E., U. Passant, P. C. Sundgren, E. Englund, A. Brun, A. Lindgren, and L. Gustafson. 2000. Magnetic resonance imaging and histopathology in dementia, clinically of fronto-temporal type. *Dement Geriatr Cogn Disord* 11 (3):123–34.

Lescaudron, L., and D. G. Stein. 1999. Differences in memory impairment and response to GM1 ganglioside treatment following electrolytic or ibotenic acid lesions of the nucleus basalis magnocellularis. *Restor Neurol Neurosci* 15 (1):25–37.

Lewis, J., D. W. Dickson, W. L. Lin, L. Chisholm, A. Corral, G. Jones, S. H. Yen, N. Sahara, L. Skipper, D. Yager, C. Eckman, J. Hardy, M. Hutton, and E. McGowan. 2001. Enhanced neurofibrillary degeneration in transgenic mice expressing mutant tau and APP. *Science* 293 (5534):1487–91.

Lewis, J., E. McGowan, J. Rockwood, H. Melrose, P. Nacharaju, M. Van Slegtenhorst, K. Gwinn-Hardy, M. Paul Murphy, M. Baker, X. Yu, K. Duff, J. Hardy, A. Corral, W. L. Lin, S. H. Yen, D. W. Dickson, P. Davies, and M. Hutton. 2000. Neurofibrillary tangles, amyotrophy and progressive motor disturbance in mice expressing mutant (P301L) tau protein. *Nat Genet* 25 (4):402–5.

Li, L., A. Lundkvist, D. Andersson, U. Wilhelmsson, N. Nagai, A. C. Pardo, C. Nodin, A. Stahlberg, K. Aprico, K. Larsson, T. Yabe, L. Moons, A. Fotheringham, I. Davies, P. Carmeliet, J. P. Schwartz, M. Pekna, M. Kubista, F. Blomstrand, N. Maragakis, M. Nilsson, and M. Pekny. 2008. Protective role of reactive astrocytes in brain ischemia. *J Cereb Blood Flow Metab* 28 (3):468–81.

Lotz, M., S. Ebert, H. Esselmann, A. I. Iliev, M. Prinz, N. Wiazewicz, J. Wiltfang, J. Gerber, and R. Nau. 2005. Amyloid b peptide 1–40 enhances the action of Toll-like receptor-2 and -4 agonists but antagonizes Toll-like receptor-9-induced inflammation in primary mouse microglial cell cultures. *J Neurochem* 94 (2):289–98.

Magistretti, P. J. 2006. Neuron-glia metabolic coupling and plasticity. *J Exp Biol* 209 (Pt 12):2304–11.

Matos, M., E. Augusto, C. R. Oliveira, and P. Agostinho. 2008. Amyloid-beta peptide decreases glutamate uptake in cultured astrocytes: involvement of oxidative stress and mitogen-activated protein kinase cascades. *Neuroscience* 156 (4):898–910.

Mattson, M. P., and S. L. Chan. 2003. Neuronal and glial calcium signaling in Alzheimer's disease. *Cell Calcium* 34 (4–5):385–97.

Matute, C. 2008. P2X$_7$ receptors in oligodendrocytes: a novel target for neuroprotection. *Mol Neurobiol* 38 (2):123–28.

Matute, C. 2010. Calcium dyshomeostasis in white matter pathology. *Cell Calcium* 47 (2):150–57.

Matute, C., E. Alberdi, M. Domercq, M. V. Sanchez-Gomez, A. Perez-Samartin, A. Rodriguez-Antiguedad, and F. Perez-Cerda. 2007. Excitotoxic damage to white matter. *J Anat* 210 (6):693–702.

McEntee, W. J., and T. H. Crook. 1991. Serotonin, memory, and the aging brain. *Psychopharmacology (Berl)* 103 (2):143–49.

McGeer, E. G., and P. L. McGeer. 1999. Brain inflammation in Alzheimer disease and the therapeutic implications. *Curr Pharm Des* 5 (10):821–36.

McGeer, P. L., and E. G. McGeer. 2002. Inflammatory processes in amyotrophic lateral sclerosis. *Muscle Nerve* 26 (4):459–70.

McKee, A. C., I. Carreras, L. Hossain, H. Ryu, W. L. Klein, S. Oddo, F. M. LaFerla, B. G. Jenkins, N. W. Kowall, and A. Dedeoglu. 2008. Ibuprofen reduces Abeta, hyperphosphorylated tau and memory deficits in Alzheimer mice. *Brain Res* 1207:225–36.

McLachlan, D. R., L. Wong, C. Bergeron, and K. G. Baimbridge. 1987. Calmodulin and calbindin D28K in Alzheimer disease. *Alzheimer Dis Assoc Disord* 1 (3):171–79.

Mena, M. A., M. J. Casarejos, A. Carazo, C. L. Paino, and J. Garcia de Yebenes. 1996. Glia conditioned medium protects fetal rat midbrain neurones in culture from L-DOPA toxicity. *Neuroreport* 7 (2):441–45.

Mena, M. A., S. de Bernardo, M. J. Casarejos, S. Canals, and E. Rodriguez-Martin. 2002. The role of astroglia on the survival of dopamine neurons. *Mol Neurobiol* 25 (3):245–63.

Mena, M. A., and J. Garcia de Yebenes. 2008. Glial cells as players in parkinsonism: the "good," the "bad," and the "mysterious" glia. *Neuroscientist* 14 (6):544–60.

Meyer-Luehmann, M., T. L. Spires-Jones, C. Prada, M. Garcia-Alloza, A. de Calignon, A. Rozkalne, J. Koenigsknecht-Talboo, D. M. Holtzman, B. J. Bacskai, and B. T. Hyman. 2008. Rapid appearance and local toxicity of amyloid-beta plaques in a mouse model of Alzheimer's disease. *Nature* 451 (7179):720–24.

Moechars, D., K. Lorent, B. De Strooper, I. Dewachter, and F. Van Leuven. 1996. Expression in brain of amyloid precursor protein mutated in the alpha-secretase site causes disturbed behavior, neuronal degeneration and premature death in transgenic mice. *Embo J* 15 (6):1265–74.

Nagele, R. G., M. R. D'Andrea, H. Lee, V. Venkataraman, and H. Y. Wang. 2003. Astrocytes accumulate A b 42 and give rise to astrocytic amyloid plaques in Alzheimer disease brains. *Brain Res* 971 (2):197–209.

Nagele, R. G., J. Wegiel, V. Venkataraman, H. Imaki, and K. C. Wang. 2004. Contribution of glial cells to the development of amyloid plaques in Alzheimer's disease. *Neurobiol Aging* 25 (5):663–74.

Nedergaard, M., and U. Dirnagl. 2005. Role of glial cells in cerebral ischemia. *Glia* 50 (4):281–86.

Nedergaard, M., B. Ransom, and S. A. Goldman. 2003. New roles for astrocytes: redefining the functional architecture of the brain. *Trends Neurosci* 26 (10):523–30.

Nedergaard, M., J. J. Rodriguez, and A. Verkhratsky. 2010. Glial calcium and diseases of the nervous system. *Cell Calcium* 47 (2):140–49.

Newman, E. A., D. A. Frambach, and L. L. Odette. 1984. Control of extracellular potassium levels by retinal glial cell K+ siphoning. *Science* 225 (4667):1174–75.

Noristani, H. N., M. Olabarria, A. Verkhratsky, and J. J. Rodríguez. 2010. Serotonin fibre sprouting and increase in serotonin transporter immunoreactivity in the CA1 area of hippocampus in a triple transgenic mouse model of Alzheimer's disease. *Cell Death and Disease*, in press. (doi:10.1038/cddis.2011.79).

Oddo, S., A. Caccamo, M. Kitazawa, B. P. Tseng, and F. M. LaFerla. 2003a. Amyloid deposition precedes tangle formation in a triple transgenic model of Alzheimer's disease. *Neurobiol Aging* 24 (8):1063–70.

Oddo, S., A. Caccamo, J. D. Shepherd, M. P. Murphy, T. E. Golde, R. Kayed, R. Metherate, M. P. Mattson, Y. Akbari, and F. M. LaFerla. 2003b. Triple-transgenic model of Alzheimer's disease with plaques and tangles: intracellular Abeta and synaptic dysfunction. *Neuron* 39 (3):409–21.

Okun, E., K. J. Griffioen, J. D. Lathia, S. C. Tang, M. P. Mattson, and T. V. Arumugam. 2009. Toll-like receptors in neurodegeneration. *Brain Res Rev* 59 (2):278–92.

Olabarria, M., H. N. Noristani, A. Verkhratsky, and J. J. Rodriguez. 2010. Concomitant astroglial atrophy and astrogliosis in a triple transgenic animal model of Alzheimer's disease. *Glia* 58:831–38.

Olney, J. W. 1994. New mechanisms of excitatory transmitter neurotoxicity. *J Neural Transm Suppl* 43:47–51.

Pak, K., S. L. Chan, and M. P. Mattson. 2003. Presenilin-1 mutation sensitizes oligodendrocytes to glutamate and amyloid toxicities, and exacerbates white matter damage and memory impairment in mice. *Neuromolecular Med* 3 (1):53–64.

Pavia, J., J. Alberch, I. Alvarez, A. Toledano, and M. L. de Ceballos. 2000. Repeated intracerebroventricular administration of beta-amyloid(25–35) to rats decreases muscarinic receptors in cerebral cortex. *Neurosci Lett* 278 (1–2):69–72.

Pekny, M., and M. Nilsson. 2005. Astrocyte activation and reactive gliosis. *Glia* 50 (4):427–34.

Pepeu, G., and I. Marconcini Pepeu. 1994. Dysfunction of the brain cholinergic system during aging and after lesions of the nucleus basalis of Meynert. *J Neural Transm Suppl* 44:189–94.

Peters, A., A. Verderosa, and C. Sethares. 2008. The neuroglial population in the primary visual cortex of the aging rhesus monkey. *Glia* 56 (11):1151–61.

Potts, R., and R. W. Leech. 2005. Thalamic dementia: an example of primary astroglial dystrophy of Seitelberger. *Clin Neuropathol* 24 (6):271–75.

Rajkowska, G., J. J. Miguel-Hidalgo, J. Wei, G. Dilley, S. D. Pittman, H. Y. Meltzer, J. C. Overholser, B. L. Roth, and C. A. Stockmeier. 1999. Morphometric evidence for neuronal and glial prefrontal cell pathology in major depression. *Biol Psychiatry* 45 (9):1085–98.

Ramsden, M., L. Kotilinek, C. Forster, J. Paulson, E. McGowan, K. SantaCruz, A. Guimaraes, M. Yue, J. Lewis, G. Carlson, M. Hutton, and K. H. Ashe. 2005. Age-dependent neurofibrillary tangle formation, neuron loss, and memory impairment in a mouse model of human tauopathy (P301L). *J Neurosci* 25 (46):10637–47.

Ransohoff, R. M., and V. H. Perry. 2009. Microglial physiology: unique stimuli, specialized responses. *Annu Rev Immunol* 27:119–45.

Ribe, E. M., M. Perez, B. Puig, I. Gich, F. Lim, M. Cuadrado, T. Sesma, S. Catena, B. Sanchez, M. Nieto, P. Gomez-Ramos, M. A. Moran, F. Cabodevilla, L. Samaranch, L. Ortiz, A. Perez, I. Ferrer, J. Avila, and T. Gomez-Isla. 2005. Accelerated amyloid deposition, neurofibrillary degeneration and neuronal loss in double mutant APP/tau transgenic mice. *Neurobiol Dis* 20 (3):814–22.

Rodriguez, J. J., M. Olabarria, A. Chvatal, and A. Verkhratsky. 2009. Astroglia in dementia and Alzheimer's disease. *Cell Death Differ* 16 (3):378–85.

Rodríguez, J. J., J. Witton, M. Olabarria, H. N. Noristani, and A. Verkhratsky. 2010. Increase in the density of resting microglia precedes neuritic plaques formation and microglial activation in a transgenic model of Alzheimer's disease. *Cell Death Dis* 1:e1.

Rossi, D., L. Brambilla, C. F. Valori, C. Roncoroni, A. Crugnola, T. Yokota, D. E. Bredesen, and A. Volterra. 2008. Focal degeneration of astrocytes in amyotrophic lateral sclerosis. *Cell Death Differ* 15 (11):1691–700.

Rossi, D., and A. Volterra. 2009. Astrocytic dysfunction: insights on the role in neurodegeneration. *Brain Res Bull* 80:224–32.

Salter, M. G., and R. Fern. 2005. NMDA receptors are expressed in developing oligodendrocyte processes and mediate injury. *Nature* 438 (7071):1167–71.

Sani, S., D. Traul, A. Klink, N. Niaraki, A. Gonzalo-Ruiz, C. K. Wu, and C. Geula. 2003. Distribution, progression and chemical composition of cortical amyloid-beta deposits in aged rhesus monkeys: similarities to the human. *Acta Neuropathol* 105 (2):145–56.

Sanz, J. M., P. Chiozzi, D. Ferrari, M. Colaianna, M. Idzko, S. Falzoni, R. Fellin, L. Trabace, and F. Di Virgilio. 2009. Activation of microglia by amyloid b requires $P2X_7$ receptor expression. *J Immunol* 182 (7):4378–85.

Savonenko, A., G. M. Xu, T. Melnikova, J. L. Morton, V. Gonzales, M. P. Wong, D. L. Price, F. Tang, A. L. Markowska, and D. R. Borchelt. 2005. Episodic-like memory deficits in the APPswe/PS1dE9 mouse model of Alzheimer's disease: relationships to beta-amyloid deposition and neurotransmitter abnormalities. *Neurobiol Dis* 18 (3):602–17.

Schindowski, K., A. Bretteville, K. Leroy, S. Begard, J. P. Brion, M. Hamdane, and L. Buee. 2006. Alzheimer's disease-like tau neuropathology leads to memory deficits and loss of functional synapses in a novel mutated tau transgenic mouse without any motor deficits. *Am J Pathol* 169 (2):599–616.

Schubert, P., T. Morino, H. Miyazaki, T. Ogata, Y. Nakamura, C. Marchini, and S. Ferroni. 2000. Cascading glia reactions: a common pathomechanism and its differentiated control by cyclic nucleotide signaling. *Ann NY Acad Sci* 903:24–33.

Shaughnessy, L. W., S. Barone, Jr., W. R. Mundy, D. W. Herr, and H. A. Tilson. 1994. Comparison of intracranial infusions of colchicine and ibotenic acid as models of neurodegeneration in the basal forebrain. *Brain Res* 637 (1–2):15–26.

Simard, M., and M. Nedergaard. 2004. The neurobiology of glia in the context of water and ion homeostasis. *Neuroscience* 129 (4):877–96.

Staekenborg, S. S., E. L. Koedam, W. J. Henneman, P. Stokman, F. Barkhof, P. Scheltens, and W. M. van der Flier. 2009. Progression of mild cognitive impairment to dementia: contribution of cerebrovascular disease compared with medial temporal lobe atrophy. *Stroke* 40 (4):1269–74.

Steffens, D. C., G. G. Potter, D. R. McQuoid, J. R. MacFall, M. E. Payne, J. R. Burke, B. L. Plassman, and K. A. Welsh-Bohmer. 2007. Longitudinal magnetic resonance imaging vascular changes, apolipoprotein E genotype, and development of dementia in the neurocognitive outcomes of depression in the elderly study. *Am J Geriatr Psychiatry* 15 (10):839–49.

Sturchler-Pierrat, C., D. Abramowski, M. Duke, K. H. Wiederhold, C. Mistl, S. Rothacher, B. Ledermann, K. Burki, P. Frey, P. A. Paganetti, C. Waridel, M. E. Calhoun, M. Jucker, A. Probst, M. Staufenbiel, and B. Sommer. 1997. Two amyloid precursor protein transgenic mouse models with Alzheimer disease-like pathology. *Proc Natl Acad Sci USA* 94 (24):13287–92.

Tanemura, K., M. Murayama, T. Akagi, T. Hashikawa, T. Tominaga, M. Ichikawa, H. Yamaguchi, and A. Takashima. 2002. Neurodegeneration with tau accumulation in a transgenic mouse expressing V337M human tau. *J Neurosci* 22 (1):133–41.

Tatebayashi, Y., T. Miyasaka, D. H. Chui, T. Akagi, K. Mishima, K. Iwasaki, M. Fujiwara, K. Tanemura, M. Murayama, K. Ishiguro, E. Planel, S. Sato, T. Hashikawa, and A. Takashima. 2002. Tau filament formation and associative memory deficit in aged mice expressing mutant (R406W) human tau. *Proc Natl Acad Sci USA* 99 (21):13896–901.

Thompson, K. A., J. C. McArthur, and S. L. Wesselingh. 2001. Correlation between neurological progression and astrocyte apoptosis in HIV-associated dementia. *Ann Neurol* 49 (6):745–52.

Thompson, P. M., K. M. Hayashi, G. de Zubicaray, A. L. Janke, S. E. Rose, J. Semple, D. Herman, M. S. Hong, S. S. Dittmer, D. M. Doddrell, and A. W. Toga. 2003. Dynamics of gray matter loss in Alzheimer's disease. *J Neurosci* 23 (3):994–1005.

Thompson, P. M., K. M. Hayashi, R. A. Dutton, M. C. Chiang, A. D. Leow, E. R. Sowell, G. De Zubicaray, J. T. Becker, O. L. Lopez, H. J. Aizenstein, and A. W. Toga. 2007. Tracking Alzheimer's disease. *Ann NY Acad Sci* 1097:183–214.

Tian, G. F., H. Azmi, T. Takano, Q. Xu, W. Peng, J. Lin, N. Oberheim, N. Lou, X. Wang, H. R. Zielke, J. Kang, and M. Nedergaard. 2005. An astrocytic basis of epilepsy. *Nat Med* 11 (9):973–81.

Toledano, A., and M. I. Alvarez. 2004. Lesions and dysfunctions of the nucleus basalis as Alzheimer's disease models: general and critical overview and analysis of the long-term changes in several excitotoxic models. *Current Alzheimer Res* 1(3):189–214.

Tsai, G., and J. T. Coyle. 2002. Glutamatergic mechanisms in schizophrenia. *Annu Rev Pharmacol Toxicol* 42:165–79.

Vale-Martinez, A., G. Guillazo-Blanch, M. Marti-Nicolovius, R. Nadal, R. Arevalo-Garcia, and I. Morgado-Bernal. 2002. Electrolytic and ibotenic acid lesions of the nucleus basalis magnocellularis interrupt long-term retention, but not acquisition of two-way active avoidance, in rats. *Exp Brain Res* 142 (1):52–66.

van der Valk, P., and S. Amor. 2009. Preactive lesions in multiple sclerosis. *Curr Opin Neurol* 22 (3):207–13.

Vanzani, M. C., R. F. Iacono, R. L. Caccuri, A. R. Troncoso, and M. I. Berria. 2006. Regional differences in astrocyte activation in HIV-associated dementia. *Medicina (B Aires)* 66 (2):108–12.

Vargas, M. E., and B. A. Barres. 2007. Why is Wallerian degeneration in the CNS so slow? *Annu Rev Neurosci* 30:153–79.

Verkhratsky, A. 2009. Neuronismo y reticulismo: neuronal-glial circuits unify the reticular and neuronal theories of brain organization. *Acta Physiol (Oxf)* 195 (1):111–22.

Verkhratsky, A. 2010. Physiology of neuronal-glial networking. *Neurochem Int*, in press.

Verkhratsky, A., and F. Kirchhoff. 2007. NMDA receptors in glia. *Neuroscientist* 13 (1):28–37.

Waite, J. J., A. D. Chen, M. L. Wardlow, R. G. Wiley, D. A. Lappi, and L. J. Thal. 1995. 192 immunoglobulin G-saporin produces graded behavioral and biochemical changes accompanying the loss of cholinergic neurons of the basal forebrain and cerebellar Purkinje cells. *Neuroscience* 65 (2):463–76.

Walter, S., M. Letiembre, Y. Liu, H. Heine, B. Penke, W. Hao, B. Bode, N. Manietta, J. Walter, W. Schulz-Schuffer, and K. Fassbender. 2007. Role of the Toll-like receptor 4 in neuroinflammation in Alzheimer's disease. *Cell Physiol Biochem* 20 (6):947–56.

Wellman, C. L., and M. A. Pelleymounter. 1999. Differential effects of nucleus basalis lesions in young adult and aging rats. *Neurobiol Aging* 20 (4):381–93.

Wiley, R. G. 1992. Neural lesioning with ribosome-inactivating proteins: suicide transport and immunolesioning. *Trends Neurosci* 15 (8):285–90.

Wiley, R. G., T. N. Oeltmann, and D. A. Lappi. 1991. Immunolesioning: selective destruction of neurons using immunotoxin to rat NGF receptor. *Brain Res* 562 (1):149–53.

Winkler, J., and L. J. Thal. 1995. Effects of nerve growth factor treatment on rats with lesions of the nucleus basalis magnocellularis produced by ibotenic acid, quisqualic acid, and AMPA. *Exp Neurol* 136 (2):234–50.

Wyss-Coray, T., J. D. Loike, T. C. Brionne, E. Lu, R. Anankov, F. Yan, S. C. Silverstein, and J. Husemann. 2003. Adult mouse astrocytes degrade amyloid-b *in vitro* and *in situ*. *Nat Med* 9 (4):453–57.

Xu, J., S. Chen, S. H. Ahmed, H. Chen, G. Ku, M. P. Goldberg, and C. Y. Hsu. 2001. Amyloid-beta peptides are cytotoxic to oligodendrocytes. *J Neurosci* 21 (1):RC118.

Yamanaka, K., S. J. Chun, S. Boillee, N. Fujimori-Tonou, H. Yamashita, D. H. Gutmann, R. Takahashi, H. Misawa, and D. W. Cleveland. 2008. Astrocytes as determinants of disease progression in inherited amyotrophic lateral sclerosis. *Nat Neurosci* 11 (3):251–53.

Yeh, C. H., M. Olabarria, H. N. Noristani, A. Verkhratsky, and J. J. Rodríguez. 2011. Early astrocytic atrophy in the entorhinal cortex of a triple transgenic animal model of Alzheimer's disease. *ASN Neuro*. Submitted.

Yoshida, M. 2004. Amyotrophic lateral sclerosis with dementia: the clinicopathological spectrum. *Neuropathology* 24 (1):87–102.

Zonta, M., M. C. Angulo, S. Gobbo, B. Rosengarten, K. A. Hossmann, T. Pozzan, and G. Carmignoto. 2003. Neuron-to-astrocyte signaling is central to the dynamic control of brain microcirculation. *Nat Neurosci* 6 (1):43–50.

13 The Role of Glial Pathology in Autism

Rachel E. Kneeland, Stephanie B. Liesch, and Timothy D. Folsom,
Department of Psychiatry, University of Minnesota
Medical School, Minneapolis, Minnesota

S. Hossein Fatemi
Departments of Psychiatry, Pharmacology,
and Neuroscience, University of Minnesota
Medical School, Minneapolis, Minnesota

CONTENTS

13.1 BRAIN PATHOLOGY IN AUTISM

Extensive brain pathology has been documented in subjects with autism (Bauman and Kemper 2005). Multiple findings have been reported, including changes in brain growth, volume, and organization, highlighting the heterogeneous nature of autism. A consistent finding is for postnatal accelerated brain growth, followed by relative growth arrest in childhood (Courchesne et al. 2003, 2007; Hazlett et al. 2005). Indeed, by mid-childhood, approximately 20% of children with autism display macrocephaly (Fombonne et al. 1999). However, macrocephaly has also been observed in the unaffected parents and siblings of autistic children with macrocephaly (Miles et al. 2000) and, as such, may represent a risk factor for autism rather than a pathological aspect of the disorder (Geschwind 2009).

Changes in individual brain regions also help explain deficits associated with autism. In the cerebellum, two recent studies have shown significant reduction in the volume of the vermis in children and adolescents with autism (Scott et al. 2009; Webb et al. 2009). Purkinje cell density has been observed to be reduced (Bauman and Kemper 1994, 2003; Ritvo et al. 1986), and there is also evidence of Purkinje cell atrophy (Fatemi et al. 2000) in cerebella of subjects with autism. Deficits in motor skills and cognition may be partially explained by these anatomical abnormalities (Nayate et al. 2005; Steinlin 2008). Reduction in cell size and increase in cell density have been observed in the hippocampus of subjects with autism (Bauman and Kemper 1994). Moreover, reductions in hippocampal volume in children (Aylward et al. 1999) and adults (Herbert et al. 2003) have been observed in subjects with autism. These changes may impair hippocampal function in memory formation and emotional regulation observed in subjects with autism. In the frontal cortex of subjects with autism, a consistent finding has been mini-columnar abnormalities (Buxhoeveden et al. 2004, 2006; Casanova et al. 2002; Casanova 2007). Additionally, there is evidence of increased white matter in the frontal cortex (Carper et al. 2002; Herbert et al. 2003). These structural changes are likely to impact emotional processing and cognition in subjects with autism. In the parietal cortex, increased gray matter volume has been observed in both subjects with autism and subjects with attention deficit/hyperactivity disorder, which may explain deficits in attention in both disorders (Brieber et al. 2007). While gray matter volume may be increased, other groups have reported an overall decrease in the volume of the parietal cortex (Courchesne et al. 1993; Saitoh and Courchesne 1998), which could potentially impact language and visual processing (Haas et al. 1996; Townsend et al. 1996). Finally, in a stereological study, amygdalae from subjects with autism display a reduced number of neurons when compared with those of control subjects (Schumann and Amaral 2006), potentially contributing to dysregulation of emotion.

Universal changes in brain structure have not yet been identified in autism. However, due to the heterogeneous nature of autism, the disparate findings that have been observed thus far are not entirely unexpected and may reflect different autistic subtypes (Geschwind 2009). Future studies, in which the autistic subtype for each subject is well defined and consistent, and the use of larger sample sizes may yield more consistent results. Regardless, it is clear that brain pathology at both gross and microscopic levels plays a role in the behavioral deficits associated with autism.

13.2   IMMUNE SYSTEM ABNORMALITIES IN AUTISM

Several studies suggest a possible relationship between immune system abnormalities and the etiopathogenesis of autism (Singer et al. 2008; Vargas et al. 2005; Zimmerman et al. 2005). Increased neuroinflammation in autistic patients has been shown by detection of activated microglia and astroglia in autistic brain tissue as well as increased levels of pro-inflammatory cytokines (Vargas et al. 2005; Laurence and Fatemi 2005). Abnormal levels of other markers of immune system activation, such as quinolinic acid, neopterin, and various cytokines and cytokine receptors, have been reported at significantly increased or significantly decreased levels in the cerebral spinal fluid (CSF) and serum of subjects with autism compared to controls (Zimmerman et al. 2005).

Evidence for the association between maternal infection and the development of structural and functional brain abnormalities in offspring has been demonstrated in several rodent models of autism (Fatemi et al. 1999, 2002, 2005, 2008a; Meyer et al. 2006a 2006b, 2007, 2010). The ability of maternal antibodies to cross the placenta and interact with the fetus has been demonstrated (Dalton et al. 2003). Recently, a positive association between the presence of a maternal antibody against fetal brain and the acquisition of autism in offspring has been demonstrated (Singer et al. 2008). The significant presence of several anti-fetal brain antibodies in the mothers of children with autism demonstrates a possible association between maternal immunoreactivity and the development of autism in offspring (Singer et al. 2008).

Significant microglial activation, as defined by increased immunoreactivity for major histocompatibility complex (MHC) class II markers, has been observed in the granular cell layer (GCL) ($p < 0.001$) of the cerebellum and in cerebellar white matter ($p < 0.001$) of subjects with autism (Vargas et al. 2005). Increased glial fibrillary acidic protein (GFAP) immunoreactivity was also observed in the GCL ($p < 0.001$) and cerebellar white matter ($p < 0.007$) in subjects with autism compared to controls showing increased astroglial activation (Vargas et al. 2005). Activated microglia were also found to be associated with areas of degenerating Purkinje cells, granule cells, and axons in these patients (Vargas et al. 2005). In the cerebral spinal fluid (CSF) of children with autism, however, results have shown reductions in two markers of immune activation: quinolinic acid and neopterin (Zimmerman et al. 2005). While a third marker of immune activation, biopterin, was elevated, the authors suggest that this finding does not support an inflammatory response in CSF, but may be consistent with dysregulation of the immune activity in the central nervous system of subjects with autism (Zimmerman et al. 2005).

Abnormal levels of cytokines have also been observed in patients with autism when compared with healthy controls. Zimmerman et al. (2005) found significantly increased levels of soluble tumor necrosis factor receptor II in CSF of patients with autism, while no differences in serum levels of tumor necrosis factor receptor I, interleukin-6 (IL-6), or interleukin-1ra were found (Zimmerman et al. 2005). Vargas et al. (2005) demonstrated increased levels of a number of cytokines in brains from subjects with autism compared to control subjects, including tumor growth factor β-1 and macrophage chemoattractant protein, both of which are derived from neuroglia (Vargas et al. 2005).

13.3 GLIAL CELL LINEAGE IN AUTISM

A number of studies have investigated increases in gray and white matter in children with autism (Schumann et al. 2010) and abnormal development of white matter in adolescents with autism (Cheng et al. 2010). Moreover, Mostofsky et al. (2007) determined that increased white matter in the motor cortex of autistic patients is a predictor of motor impairment. Interestingly, several studies have provided evidence for an association between molecules involved in glial differentiation and psychiatric disorders (Ashwood et al. 2008; Hashimoto et al. 2006; Kwak et al. 2006, 2010; Luo et al. 2004; Okada et al. 2007), suggesting that white matter development may play a role in the etiopathogenesis of autism.

Notch signaling has been shown to be important in the differentiation of embryonic neural stem cells into either glial cells or neurons. Activation of notch signal by the jagged-1 ligand has been shown to commit embryonic stem cells (ESCs) to neural differentiation, while activation of the notch signal by ciliary neurotrophic factor has been demonstrated to promote gliogenesis (Ramasamy and Lenka 2010). This finding builds upon earlier studies demonstrating that the activation of notch signaling was important in gliogenesis (Gaiano and Fishell 2002; Park and Appel 2003). While the only study examining a NOTCH4 polymorphism, conducted among Japanese patients with autism, yielded negative results (Koishi et al. 2004), studies examining an association between the notch signaling pathway and schizophrenia have been inconsistent, with some in support of its association with the disorder (Luo et al. 2004; Prasad et al. 2004; Wang et al. 2005; Wei and Hemmings 2000); others demonstrate no relationship (Carmine et al. 2003; Fan et al. 2002; Kaneko et al. 2004; Ujike et al. 2001).

Brain-derived neurotrophic factor (BDNF) is a molecule involved in axon and dendrite differentiation in neural development that has been shown to increase spine density (Chapleau et al. 2008). BDNF has been shown to be reduced in the serum of patients with autism (Hashimoto et al. 2006). Mutations in genes downstream from BDNF, such as tuberous sclerosis 1 and 2 (TSC1 and TSC2), are associated with tuberous sclerosis, a disease that has been associated with autism, mental retardation, and epilepsy (Meikle et al. 2007). A case study of a patient with tuberous sclerosis has found abnormal differentiation of neurons (Mizuguchi and Takeshima 2001).

Various other molecules supporting gliogenesis have also been shown to be implicated in autism and autism spectrum disorders (ASDs). Amyloid precursor protein (APP) induces glial cell differentiation of neural progenitor cells (Kwak et al. 2006) via the activation of the IL-6/gp-130 pathway (Kwak et al. 2010). Interestingly, this molecule has also been implicated in autism. Children with severe forms of autism have been found to express a 60% increase in the amount of plasma APP than control children (Bailey et al. 2008; Sokol et al. 2006). Additionally, transforming growth factor β-1 (TGFβ1), which promotes astrogliogensis (Stipursky and Gomes 2007), has also shown to be involved in autism. Lower TGFβ1 levels were found in children with autism than in control children and children with delayed developmental problems who did not have ASD (Okada et al. 2007).

While the association between white matter abnormalities and autism is not well understood, evidence does exist for abnormalities in neuron and glial cell differentiation molecules in individuals with psychiatric disorders, most notably autism. Notch signaling may or may not be implicated in psychiatric disorders, while reductions in BDNF have been associated with autism. Additionally, APP and TGFβ1, two molecules involved in gliogenesis, have been found to be higher and lower, respectively, in subjects with autism, providing further support that white matter development may play a role in the pathogenesis of autism.

13.4   GLIAL FIBRILLARY ACIDIC PROTEIN IN AUTISM

Previous studies have shown an elevation of glial fibrillary acidic protein (GFAP), an indicator of astroglial activation, in cerebral spinal fluid (CSF) of autistic children (Ahlsen et al. 1993; Rosengren et al. 1992). This finding suggests that immune dysregulation may play a role in the etiology of autism. GFAP plays a significant role in neuron migration, gliosis, and reactive injury (Fatemi et al. 2002). Based on these prior reports, our laboratory studied GFAP in three regions of the brain (superior frontal cortex, parietal cortex, and cerebellum) to further investigate this protein in subjects with autism vs. matched controls.

We evaluated GFAP protein levels of 16 autistic subjects against 10 control subjects, all matched for age and postmortem interval (PMI). Group differences between the autistic subjects and the control subjects were analyzed using t-tests on all three variables. When compared to controls we found that mean GFAP levels in subjects with autism were elevated by 45% in the frontal cortex, 75% in the parietal cortex, and 49% in the cerebellum (Laurence and Fatemi 2005). We also observed that GFAP/β-actin levels were higher in the autistic subjects vs. control subjects, differing significantly in the frontal cortex ($p < 0.012$) and parietal cortex ($p < 0.043$). There were no confounding effects of medications, mode of death, or presence of seizures on levels of GFAP.

These findings were the first to illustrate that GFAP is significantly increased in three specific brain regions (cerebellum, parietal cortex, and superior frontal cortex) in subjects with autism. The increase may signify reactive astrogliosis and injury to brain. Moreover, a separate study by another group found evidence of activation of microglia and astroglia in the cerebellum and frontal and cingulate cortices of subjects with autism when compared with controls (Vargas et al. 2005). While these findings occurred in brain tissue from postmortem humans, altered GFAP levels can also be seen in an animal model for autism developed by our laboratory (Fatemi et al. 2002).

Our laboratory examined the effects of prenatal viral infection in pregnant mice on GFAP levels. We predicted that there would be altered levels of GFAP in the offspring of pregnant mice who were exposed to the human influenza infection on day 9 of their pregnancy. Specifically, altered expression at three important postnatal time points was examined: postnatal days PN0 (which corresponds to birth), PN14 (which corresponds to childhood), and PN35 (which corresponds to adolescence) (Fatemi et al. 2002). In order to determine the effects of GFAP on postnatal mice, we studied GFAP immunoreactivity levels of astrocytes and ependymal cells (Fatemi et al. 2002). There were significantly increased numbers of GFAP-positive ependymal cells at P14 ($p < 0.039$), cortical cells at P14 ($p < 0.035$) and hippocampal cells at P35 ($p < 0.019$) in the brains of exposed offspring (Fatemi et al. 2002). Additionally, there were also significantly increased cortical (P14, $p < 0.031$) and hippocampal (P14, $p < 0.035$; P35, $p < 0.034$) GFAP-positive astrocytes (Fatemi et al. 2002).

In addition to autism, altered levels of GFAP in the brain have been observed in other psychiatric disorders. Our laboratory has observed significantly reduced GFAP in cerebella of subjects with major depression ($p = 0.015$), while there were no significant changes for subjects with bipolar disorder and schizophrenia (Fatemi et al. 2004). In contrast, a more recent study has also shown a decrease in protein levels of GFAP in the anterior cingulate cortex of subjects with schizophrenia when compared with controls (Steffek et al. 2008), suggesting that altered levels of GFAP may be present in individuals with schizophrenia. Taken together astrocytic pathology appears to occur in multiple psychiatric disorders.

13.5   AQUAPORIN 4 AND CONNEXIN 43 IN AUTISTIC BRAIN

As mentioned in the previous section, increased GFAP levels have been observed in multiple brain regions and in CSF of subjects with autism (Ahlsen et al. 1993; Fatemi et al. 2002; Laurence and Fatemi 2005; Rosengren et al. 1992; Vargas et al. 2005), indicating astroglial activation. Two additional glial proteins, aquaporin 4 (AQP4) and connexin 43 (Cx43), have recently been investigated in brains of subjects with autism (Fatemi et al. 2008b). AQP4 is a transmembrane water channel protein (Baudat et al. 2002). Cx43 is an astrocytic gap junction protein (Giaume et al. 1991; Naus et al. 1991). AQP4 and Cx43 are involved in multiple important processes, including cell signaling, water transport, and transport of nutrients from blood to neurons (Badqut et al. 2002; Nedergaard et al. 2003; Theis et al. 2004), and altered expression of these proteins has been associated with multiple brain pathologies (Badaut et al. 2002; Collignon et al. 2006; Nakase et al. 2006; Papadopoulos et al. 2002; St. Hilarie et al. 2005; Taniguchi et al. 2000).

We compared the protein levels of AQP4 and Cx43 in subjects with autism and matched controls in three brain regions: prefrontal cortex, parietal cortex, and cerebellum. Protein levels were normalized against β-actin, and differences between subjects with autism and matched controls were analyzed using multivariate analysis of covariance (MANCOVA). When significant differences were found, univariate ANCOVAs were performed for each protein. Significant group effects were observed in cerebellum ($p < 0.05$) and prefrontal cortex ($p < 0.017$), while there was no difference in parietal cortex (Fatemi et al. 2008b). ANCOVAs showed that mean Cx43/β-actin values increased significantly in prefrontal cortex of subjects with autism when compared with controls ($p < 0.018$) (Fatemi et al. 2008b). Mean AQP4/β-actin values decreased significantly in cerebella of subjects with autism ($p < 0.040$) (Fatemi et al. 2008b). There were no significant differences in β-actin levels in age or postmortem interval (PMI) between the two groups (Fatemi et al. 2008b).

The reduction in AQP4 is interesting, as it correlates with an animal model of autism developed in our laboratory in which prenatal viral infection of pregnant mice resulted in reduced expression of AQP4 in the brains of the exposed offspring at birth (postnatal day 0 (PN0)) (Fatemi et al. 2005) and at adolescence (P35) (Fatemi et al. 2008a). Interestingly, in astrocytic cell cultures from both humans and rats, AQP4 knockdown via RNA interference displayed altered elongated and stellate morphology, while RNA interference had little effect on morphology of mouse astrocytic cell cultures (Nicchia et al. 2005). It may be the case that the observed reduction in AQP4 in subjects with autism results in similar changes in cell structure, volume, and potentially homeostasis.

Upregulation of Cx43 has been associated with epilepsy/seizure disorders (Rouach et al. 2002), which is of particular importance in autism, as anywhere from 4–44% of subjects with autism also display seizure disorder (Tuchman and Rapin 2002). The increase of Cx43, specifically in the frontal cortex, and potential subsequent increase in cell-to-cell communication and the lack of ability to filter extraneous cellular communication may explain the deficits in sensory processing that occur in subjects with autism (Kern et al. 2006, 2007; Tomchek and Dunn 2007).

Taken together with our previous results regarding GFAP, the altered expression of AQP4 and Cx43 demonstrates dysregulation of three astrocytic markers in brains of subjects with autism. These changes may contribute to pathology associated with autism, including learning deficits, seizure disorder, and immune activation. Additionally, the dysregulation of these proteins in autism may provide insight into the development of future novel therapeutic targets (see Section 1.7 for further discussion).

13.6 ANIMAL MODELS OF AUTISM WITH RELEVANCE TO GLIOGENESIS

Abnormal gliogenesis has been observed in animal models of autism resulting from knockout of genes such as tuberous sclerosis (Onda et al. 2002; Ess et al. 2004), *MECP2*, the neuroligin and neurexin families, *SLC25A12* and *SLC9A6,* and FMR1. Additionally, aberrant gliogenesis has been observed in animal models exhibiting abnormal cytokine levels (Watanabe et al. 2004) and due to prenatal viral infection (Fatemi et al. 2002) (Table 13.1).

Animal models of tuberous sclerosis show altered expression of GFAP (Onda et al. 2002; Uhlmann et al. 2002). Neuroepithelial (NEP) cells from tuberous sclerosis 2 (Tsc2) null mice display increased expression of GFAP, whereas less than 5% of NEP cells from wild type expressed GFAP (Onda et al. 2002). Moreover, cells from Tsc2 null mice display abnormal differentiation into giant cells that express high levels of GFAP and βIII tubulin (Onda et al. 2002). Increased GFAP expression has also been observed in astrocytes in Tsc1^{GFAP} conditional knockout mice in cortex (Uhlmann et al. 2002) (Table 13.1).

Mutations of the methyl-CpG-binding protein 2 (*Mecp2)* gene lead to the development of Rett's syndrome (Amir et al. 1999), a common form of mental retardation that shares many clinical and neurologic features with autism (Lasalle and Yasui 2009). Recently, Maezawa et al. (2009) and Maezawa and Jin (2010) have demonstrated the implications of reduced MeCP2 and coincident abnormal glial cell development. In astrocytes, MeCP2 deficiencies led to reductions in growth rate, dendritic length in hippocampal neurons, and hyperactivation of the p38MAPK pathway (Maezawa et al. 2009). Additionally, MeCP2-deficient microglia have been demonstrated to express toxic levels of glutamate. Hippocampal dendrites treated with media removed from MeCP2-null microglia exhibited stunted growth (Maezawa and Jin 2010), an observation similar to abnormalities seen in a pathological study of Rett's syndrome (Chapleau et al. 2009) (Table 13.1).

The neurexins are presynaptic cell adhesion molecules that interact with the postsynaptic neuroligins, bridging the synaptic cleft (Ichthenko et al. 1996). Both are important for the growth of excitatory glutamatergic and inhibitory GABAergic synapses (Graf et al. 2004; Chih et al. 2005). The neurexin family of genes, most notably the neurexin 1 gene, have been associated with autism (Ching et al. 2010; Abu-Elneel et al. 2008; Kim et al. 2008; Feng et al. 2006). In *Drosophila*, Stork et al. (2009) have demonstrated that neurexin 4 mutants have abnormal glial wrapping around axons. Other studies have also found neurexin 4 to be part of a complex of molecules (neurexin 4/caspr/paranodin) important for the formation of axoglial

TABLE 13.1
Summary of Gene Mutations with Potential Impact on Causation of Autism with Relevance to Glia (included is the gene name, a brief summary of the gene's relevance to glial cells, and the appropriate reference)

Gene	Mechanism with Relevance to Glia	Reference
MeCP2	Mecp2-null microglia express toxic levels of glutamate	Maezawa et al. 2009, Maezawa and Jin 2010
Neuroligin 1	Neuroligin 1KO: ↓ hippocampal dependent spatial memory and LTP	Blundell et al. 2010
Neuroligin 2	Neuroligin 2 KO: ↓ inhibitory synapse transmission	Gutierrez et al. 2009
Neuroligin 3	Neuroligin 3 KI: ↑ inhibitory synapse transmission	Tabuchi et al. 2007, Varoqueaux et al. 2006
Neuroligin 4	Neuroligin 4 R87 mutation (human) corresponded to ↓ spine density	Zhang et al. 2009
Neurexin 1	Neurexin 1 KO: ↓ synaptic activity of GluRδ2	Uemura et al. 2010
	Neurexin 1 mutations (human) associated with ASDs	Ching et al. 2010, Abu-Elneel et al. 2008, Kim et al. 2008
Neurexin 4	Neurexin 4 *Drosophila* mutants exhibit abnormal glial cell wrapping around neurons	Stork et al. 2009
	Neurexin 4 facilitates formation of axoglial junctions	Garcia-Fresco et al. 2006, Bhat et al. 2001
Slc25a12	Slc25a12 encodes AGC1, a mitochondrial protein necessary for myelin formation and other forms of neurodevelopment	Sakurai et al. 2010
Slc9a6	Slc9a6 (human) mutation in x-linked angelman-type mental retardation with observed gliosis pathology	Garbern et al. 2010
FMR1	FMR1 is expressed in astrocytes and oligodendrocytes of pre- and postnatal mouse brain	Jacobs and Doering 2010
	FMR1 expression in astrocytes may play a role in synapse development	Pacey and Doering 2007
	FMRP reduced in adults with autism	Fatemi et al. 2011

Abbreviations: KO = knockout; KI = knock in.

junctions (Garcia-Fresco et al. 2006; Bhat et al. 2001). Neuroligins, located at the postsynaptic end of the neurexin-neuroligin complex, are responsible for maturation of excitatory and inhibitory synapse formation (Chih et al. 2005; Prange et al. 2004; Varoqueaux et al. 2004). While neuroligin abnormalities in autism are well studied (Blundell et al. 2010, 2009; Zhang et al. 2009; Gutierrez et al. 2009; Tabuchi et al. 2007), their relevance to glia is not well documented (Table 13.1).

SLC25A12 is a gene coding for calcium ion (Ca^{2+})-dependent mitochondrial aspartate-glutamate carrier isoform 1 (AGC1). In linkage and association studies, mutations of SLC25A12 have also been identified to be associated with autism (Ramoz et al. 2004). In postmortem studies, SLC25A12 has also been abnormally expressed in individuals with autism spectrum disorders (Lepagnol-Bestel et al. 2008). Recently Sakurai et al. (2010) found that SLC25A12 knockout mice expressed significant reductions (75%) of myelin basic protein compared to controls. It has also been hypothesized that ACG1 is necessary for maintaining adequate levels of aspartate in the cytoplasm, where it is converted to N-acetylaspartate (NAA) (Jalil et al. 2005). Some studies suggest that NAA manufactured in neurons may be transported to oligodendrocytes, where it may be converted to acetyl residues, facilitating the formation of lipids and myelin (Jalil et al. 2005; Ramos et al. 2003) (Table 13.1).

SLC9A6, a gene that encodes the sodium proton exchanger NHE6, has recently been implicated in the etiology of mental retardation (Garbern et al. 2010; Gilfillian et al. 2008; Roxrud et al. 2009). One such study discovered a nine-base-pair mutation of SLC9A6 in a family with an X-linked form of mental retardation with autistic components who also demonstrated prominent gliosis and neurodegeneration in the cerebellum, regions of the midbrain, substantia innominata, and parts of the hippocampus and entorhinal cortex (Garbern et al. 2010). Other studies have highlighted a positive relationship with SLC9A6 mutations and Angelman syndrome (Roxrud et al. 2009), and other forms of X-linked mental retardation (Gilfillan et al. 2008). However, a separate study examining SLC9A6 mutations in patients with similar diagnoses to Angelman syndrome failed to confirm these results (Fichou et al. 2009).

Silencing of the fragile X mental retardation-1 (FMR1) gene leads to fragile X syndrome (FXS) (Oostra and Willemsen 2009). While many FXS patients display overlapping diagnoses of autism (Demark et al. 2003), only a few studies have investigated the dysregulation of FMR1 in autism (Fatemi et al. 2011; Fatemi and Folsom 2011) and other brain disorders (Fatemi et al. 2010). Fragile X mental retardation protein (FMRP) is also present in astrocytes and oligodendrocytes of the pre- and postnatal mouse brain (Pacey and Doering 2007), as well as oligodendroglial progenitor cells and immature oligodendrocytes of the rat brain (Wang et al. 2004). Recently, Jacobs and Doering (2010) demonstrated that wildtype (WT) neurons grown on FMR1 knockout (KO) astrocytes exhibit abnormal dendritic arborization. Moreover, the authors also discovered a reduction in synaptophysin- and PSD-95-immunoreactive puncta per cell on the synapses of WT neurons grown on FMR1 KO astrocytes (Jacobs and Doering 2010). Conversely, FMR1 KO neurons grown on WT astrocytes exhibited normal morphology in addition to significant increases in the number of synaptophysin ($p < 0.003$) and PSD-95 ($p < 0.001$) puncta per cell (Jacobs and Doering 2010). These results suggest a significant role for FMR1 in astrocyte morphology and the development of synaptic protein aggregates (Jacobs and Doering 2010). Since astrocytes are required for synapse development (Pfrieger and Barres 1997; Ullian et al. 2001), FMR1 deficient astrocytes may be unable to properly regulate development of synapses (Jacobs and Doering 2010) (Table 13.1).

Finally, animal models highlighting the association between abnormal cytokine levels and abnormal gliogenesis have been studied. It has been demonstrated that subcutaneous injection of leukemia inhibitory factor in newborn rats resulted in increased GFAP immunoreactivity in neocortex by 451% ($p < 0.001$), while there were no changes in the hippocampus or striatum (Watanabe et al. 2004). These rats also displayed reduced locomotor activity and rearing when compared with saline-injected mice (Watanabe et al. 2004) during the juvenile period (P22–P26) and developed abnormal prepulse inhibition in the acoustic startle test during and after adolescence (Watanabe et al. 2004), a behavioral abnormality observed in subjects with autism (Perry et al. 2007). As discussed previously, prenatal viral infection also leads to abnormal, elevated GFAP expression in mouse brain (Fatemi et al. 2002). GFAP expression increases in the CNS following injury or insult, and the increase in gliogenesis due to cytokine activation and maternal infection may reflect this.

13.7 POTENTIAL IMPACT OF PSYCHOTROPIC MEDICATIONS ON ASTROGLIAL MARKERS IN AUTISM

Astroglial cells are intimately associated with synaptic plasticity and neuroprotection (Chen and Swanson 2003; Danbolt 2001). Psychotropic medications, which may have neuroprotective effects, are currently used to alleviate symptoms associated with autism, such as seizures and learning deficits (Quincozes-Santos et al. 2009). Recent investigations have examined the effects of atypical antipsychotic agents, such as risperidone, on astroglial activity. Risperidone is thought to act via dopamine D2 and serotonin (5-HT (5-hydroxytryptamine)) 2A receptor antagonism (De Deyn and Buitelaar 2006). Past research has shown that when compared to their conventional counterparts, including medications such as haloperidol and chlorpromazine, atypical antipsychotic agents have been demonstrated to have therapeutic advantages in individuals with autism (Owens 1994).

While most studies regarding the pathological abnormalities of autism focus on the neuronal effects of psychotropic medications, a few studies have also been performed to understand the effects of psychotropic medications on glial cells (Quincozes-Santos et al. 2008). A recent study investigated the effects of risperidone on C6 astroglial cell activity, such as glutamate uptake, glutamine synthetase (GS) activity, and glutathione (GSH) levels (Quincozes-Santos et al. 2009). With the use of risperidone, it was observed that glutamate uptake increased ($p < 0.05$), with increases in both GS ($p < 0.05$) and GSH ($p < 0.001$) activity (Quincozes-Santos et al. 2009). These significant increases are important, as they suggest C6 astroglial modifications occur with the use of risperidone.

Similarly, the above-mentioned report investigated the effects of risperidone on C6 astroglial cell morphology as well as S100B secretion (Quincozes-Santos et al. 2008). S100B is an astrocyte-derived neurotrophic cytokine (Quincozes-Santos et al.

2008). Cell morphology and membrane integrity were analyzed under risperidone and control conditions. It was found that risperidone was able to modulate cell morphology and adhesion, while at the same time not interfering with nuclear morphology (Quincozes-Santos et al. 2008).

Finally, our laboratory recently reported the effects of chronic administration of several psychotropic medications on expression of Cx43 and GFAP in two areas of the brain in rats (Fatemi et al. 2008c). Cx43 protein expression in frontal cortex of rats significantly increased following treatment with fluoxetine and clozapine. Additionally, GFAP protein expression decreased significantly following treatment with clozapine and valproic acid (Fatemi et al. 2008c). Both fluoxetine and valproic acid have been used in treatment of some of the symptoms of autism, such as hyperactivity and seizures. These results suggest that the above-mentioned astroglial markers could be potential targets for therapeutic intervention.

13.8   SUMMARY

Autism is a complex, heterogeneous disorder that includes widespread brain pathology. Part of this pathology includes changes in glia. Increased expression of GFAP has been observed by multiple laboratories in both postmortem studies in brains from subjects with autism and in animal models of autism. Increased GFAP is indicative of activation of astroglia and microglia. This activation may be the result of insults such as abnormal levels of cytokines or prenatal viral infection. Other astroglial markers, such as AQP4 and Cx43, also display altered expression both in human brain and in brains of animal models of autism. The altered expression of these markers could potentially contribute to autistic pathology, including seizure disorders, learning deficits, and immune activation. Astroglial markers are also targets for drugs that treat symptoms of autism, including risperidone, fluoxetine, and valproic acid. Further study of astroglial markers has the potential to identify targets for therapeutic interventions for autism.

ACKNOWLEDGMENTS

Dr. S.H. Fatemi's work presented in this chapter has been supported by NICHD R01-HD052074-03 and NICHD R01-HD046589-04.

REFERENCES

Abu-Elneel, K., Liu, T., Gazzaniga, F.S., et al. 2008. Heterogeneous dysregulation of micro-RNAs across the autism spectrum. *Neurogenetics* 9(3):153–161.

Ahlsen, G., Rosengren, L., Belfrage, M., et al. 1993. Glial fibriallary acidic protein in the cerebrospinal fluid of children with autism and other neuropsychiatric disorders. *Biol Psychiatry* 33:734–743.

Amir, R.E., Van den Veyver, I.B., Wan, M., et al.1999. Rett syndrome is caused by mutations in X-linked MECP2, encoding methyl-CpG-binding protein 2. *Nat Genet* 23(2):185–188.

Ashwood, P., Enstrom, A., Krakowiak, P., et al. 2008. Decreased transforming growth factor beta1 in autism: a potential link between immune dysregulation and impairment in clinical behavioral outcomes. *J Neuroimmunology* 204:149–153.

Aylward, E.H., Minshew, N.J., Goldstein, G., et al. 1999. MRI volumes of amygdala and hippocampus in non-mentally retarded autistic adolescents and adults. *Neurology* 53:2145–2150.

Badaut, J., Lasbennes, F., Magistretti, P.J., Regli, L. 2002. Aquaporins in brain: distributions, physiology, and pathophysiology. *Cereb Blood Flow Metab* 22:367–378.

Bailey, A.R., Giunta, B.N., Obregon, D., et al. 2008. Peripheral biomarkers in autism: secreted amyloid precursor protein-alpha as a probable key player in early diagnosis. *Int J Clin Exp Med* 1:338–344.

Bauman, M.L., Kemper, T.L. 1994. Neuroanatomic observations of the brain in autism. In *The neurobiology of autism*, ed. M.L. Bauman and T.L. Kemper, 19–145. Baltimore: Johns Hopkins University Press.

Bauman, M.L., Kemper, T.L. 2003. The neuropathology of autism spectrum disorders: what have we learned? *Novartis Found Symp* 251:112–122; discussion, 22–28, 281–297.

Bauman, M.L., Kemper, T.L. 2005. Structural brain anatomy in autism: what is the evidence? In *The neurobiology of autism*, ed. M.L. Bauman and T.L. Kemper, 121–135. Baltimore: Johns Hopkins University Press.

Bhat, M.A., Rios, J.C., Lu, Y., et al. 2001. Axon-glia interactions and the domain organization of myelinated axons requires neurexin IV/caspr/paranodin. *Neuron* 30(2):369–383.

Blundell, J., Blaiss, C.A., Etherton, M.R., et al. 2010. Neuroligin-1 deletion results in impaired spatial memory and increased repetitive behavior. *J Neurosci* 30(6):2115–2129.

Blundell, J., Tabuchi, K., Bollinger, M.F., et al. 2009. Increased anxiety-like behavior in mice lacking inhibitory synapse cell adhesion molecule neuroligin-2. *Genes Brain Behav* 8:114–126.

Brieber, S., Neufang, S., Bruning, N., et al. 2007. Structural brain abnormalities in adolescents with autism spectrum disorder and patients with attention deficit/hyperactivity disorder. *J Child Psychol Psychiatry* 48:1251–1258.

Buxhoeveden, D.P., Semendeferi, K., Buckwalter, J., Schenker, N., Switzer, R., Courchesne, E. 2006. Reduced minicolumns in frontal cortex of patients with autism. *Neuropathol Appl Neurobiol* 33:720–721.

Buxhoeveden, D.P., Semendeferi, K., Schenker, N., Courchesne, E. 2004. Decreased cell column spacing in autism. Program and abstracts of the Society for Neuroscience 34th Annual Meeting, San Diego, CA, October 23–27, 2004. Abstract 582.6.

Carmine, A., Chheda, M.G., Jonsson, E.G., et al. 2003. Two NOTCH4 polymorphisms and their relation to schizophrenia susceptibility and different personality traits. *Psychiatr Genet* 13:23–28.

Carper, R.A., Moses, P., Tigue, Z.D., Courchesne, E. 2002. Cerebral lobes in autism: early hyperplasia and abnormal age effects. *Neuroimage* 16:1038–1051.

Casanova, M.F. 2007. The neuropathology of autism. *Brain Pathol* 17:422–433.

Casanova, M.F., Buxhoeveden, D.P., Switala, A.E., Roy, E. 2002. Minicolumnar pathology in autism. *Neurology* 58:428–432.

Chapleau, C.A., Calfa, G.D., Lane, M.C. et al. 2009. Dendritic spine pathologies in hippocampal pyramidal neurons from Rett syndrome brain and after expression of Rett-associated MECP2 mutations. *Neurobiol Dis* 35(2):219–233.

Chapleau, C.A., Carlo, M.E., Larimore, J.L., Pozzo-Miller, L. 2008. The actions of BDNF on dendritic spine density and morphology in organotypic slice cultures depend on the presence of serum in culture media. *J Neurosci Methods* 169:182–190.

Chen, Y., Swanson, R.A. 2003. Astrocytes and brain injury. *J Cereb Blood Flow Metab* 23:137–149.

Cheng, Y., Chou, K.H., Chen, I.Y., Fan, Y.T., Decety, J., Lin, C.P. 2010. Atypical development of white matter microstructure in adolescents with autism spectrum disorders. *Neuroimage* 50:873–882.

Chih, B., Engelman, H., Scheiffele, P. 2005. Control of excitatory and inhibitory synapse formation by neuroligins. *Science* 307(5713):1324–1328.

Ching, M.S., Shen, Y., Tan, W.H., et al. 2010. Deletions of NRXN1 (neurexin-1) predispose to a wide spectrum of developmental disorders. *Am J Med Genet B Neuropsychiatr Genet* 153B(4):937–947.

Collignon, F., Wetjen, N.M., Cohen-Gadol, A.A., et al. 2006. Altered expression of connexin subtypes in mesial temporal lobe epilepsy in humans. *J Neurosurg* 1051:77–87.

Courchesne, E., Carper, R., Akshoomoff, N. 2003. Evidence of brain overgrowth in the first year of life in autism. *JAMA* 290:337–344.

Courchesne, E., Pierce, K., Schumann, C.M., et al. 2007. Mapping early brain development in autism. *Neuron* 56:399–413.

Courchesne, E., Press, G.A., Young-Courchesne, R. 1993. Parietal lobe abnormalities detected with MR in patients with infantile autism. *Am J Roentgenol* 160:387–393.

Dalton, P., Deacon, R., Blamir, A., et al. 2003. Maternal neuronal antibodies associated with autism and a language disorder. *Ann Neurol* 53:533–537.

Danbolt, N.C. 2001. Glutamate uptake. *Prog Neurobiol* 65:1–105.

De Deyn, P.P., Buitelaar, J. 2006. Risperidone in the management of agitation and aggression associated with psychiatric disorders. *Eur Psychiatry* 21:21–28.

Demark, J., Feldman, M., Holden, J. 2003. Behavioral relationship between autism and fragile x syndrome. *Am J Ment Retard* 108(1):314–326.

Ess, K.C., Uhlmann, E.J., Li, W., et al. 2004. Expression profiling in tuberous sclerosis complex (TSC) knockout mouse astrocytes to characterize human TSC brain pathology. *Glia* 46(1):28–40.

Fan, J.B., Tang, J.X., Gu, N.F., et al. 2002. A family-based and case-control association study of the NOTCH4 gene and schizophrenia. *Mol Psychiatry* 7:100–103.

Fatemi, S.H., Emamian, E.S., Kist, D., et al. 1999. Defective corticogenesis and reduction in Reelin immunoreactivity in cortex and hippocampus of prenatally infected neonatal mice. *Mol Psychiatry* 4(2):145–54.

Fatemi, S.H., Emamian, E.S., Sidwell, R.W., Kist, D.A., Earle, J., Bailey, K. 2002. Human influenza viral infection *in utero* alters glial fibrillary acidic protein immunoreactivity in the developing brains of neonatal mice. *Mol Psychiatry* 7:633–640.

Fatemi, S.H., Folsom, T.D., Kneeland, R.E., Liesch, S.B. 2011. Metabotropic glutamate receptor 5 upregulation in children with autism is associated with underexpression of both fragile X mental retardation protein and GABA$_A$ receptor beta 3 in adults with autism. *Anatomical Record* (in press).

Fatemi, S.H., Folsom, T.D., Reutiman, T.J., et al. 2008c. Chronic psychotropic drug treatment causes differential expression of connexin 43 and GFAP in frontal cortex of rats. *Schizophr Res* 104(1–3):127–134.

Fatemi, S.H., Folsom, T.D., Reutiman, T.J., Lee, S. 2008b. Expression of astrocytic markers aquaporin 4 and connexin 43 is altered in brains of subjects with autism. *Synapse* 62:501–507.

Fatemi, S.H., Folsom, T.D., Reutiman, T.J., Sidwell, R.W. 2008a. Viral regulation of aquaporin 4, connexin 43, microcephalin and nucleolin. *Schiz Res* 98:163–177.

Fatemi, S.H., Halt, A.R., Earle, J. et al. 2000. Reduced Purkinje cell size in autistic cerebellum. *Biol Psychiatry* 47:S128.

Fatemi, S.H., Kneeland, R.E., Liesch, S.B., Folsom, T.D. 2010. Fragile X mental retardation protein levels are decreased in major psychiatric disorders. *Schizophr Res* 124(1–3): 246–7.

Fatemi, S.H., Laurence, J.A., Araghi-Niknam, M., et al. 2004. Glial fibrillary acidic protein is reduced in cerebellum of subjects with major depression but not schizophrenia. *Schizophr Res* 69:317–323.

Fatemi, S.H., Folsom, T.D. 2011. Dysregulation of Fragile X mental retardation protein and metabotropic glutamate receptor 5 in superior frontal cortex of individuals with antism: a post-mortem brain study. *Mol Autism* (in press)."

Fatemi, S.H., Pearce, D.A., Brooks, A.I., Sidwell, R.W. 2005. Prenatal viral infection in mouse causes differential expression of genes in brains of mouse progeny: a potential animal model for schizophrenia and autism. *Synapse* 57:91–99.

Feng, J., Schroer, R., Yan, J., et al. 2006. High frequency of neurexin 1beta signal peptide structural variants in patients with autism. *Neurosci Lett* 409(1):10–13.

Fichou, Y., Bahi-Buisson, N., Nectoux, J., et al. 2009. Mutation in the SLC9A6 gene is not a frequent cause of sporadic Angelman-like syndrome. *Eur J Hum Genet* 17(11):1378–80.

Fombonne, E., Roge, B., Claverie, J., Courty, S., Fremolle, J. 1999. Microcephaly and macrocephaly in autism. *J Autism Dev Disord* 29:113–119.

Gaiano, N., Fishell, G. 2002. The role of notch in promoting glial and neural stem cell fates. *Annu Rev Neurosci* 25:471–490.

Garbern, J.Y., Neumann, M., Trojanowski, J.Q., et al. 2010. A mutation affecting the sodium/proton exchanger, SLC9A6, causes mental retardation with tau deposition. *Brain* 133(Pt 5):1391–1402.

Garcia-Fresco, G.P., Sousa, A.D., Pillai, A.M., et al. 2006. Disruption of axo-glial junctions causes cytoskeletal disorganization and degeneration of Purkinje neuron axons. *Proc Natl Acad Sci USA* 103(13):5137–5142.

Geschwind, D.H. 2009. Advances in autism. *Ann Rev Med* 60:367–380.

Giaume, C., Fromaget, C., el Aoumari, A., Cordier, J., Glowinski, J., Gros, D. 1991. Gap junctions in cultured astrocytes: single-channel currents and characterization of channel-forming protein. *Neuron* 6:133–143.

Gilfillan, G.D., Selmer, K.K., Roxrud, I., et al. 2008. SLC9AG mutations cause X-linked mental retardation, microcephaly, epilepsy, and ataxia, a phenotype mimicking Angelman syndrome. *Am J Hum Genet* 82(4):1003–10.

Graf, E.R., Zhang, X.Z., Jin, S.X., et al. 2004. Neurexins induce differentiation of GABA and glutamate postsynaptic specializations via neuroligins. *Cell* 119:1013–1026.

Gutierrez, R.C., Hung, J., Zhang, Y., et al. 2009. Altered synchrony and connectivity in neuronal networks expressing an autism-related mutation of neuroligin 3. *Neuroscience* 162:208–221.

Haas, R.H., Townsend, J., Courchesne, E., Lincoln, A.J., Schreibman, L., Young-Courchesne, R. 1996. Neurologic abnormalities in infantile autism. *J Child Neurol* 11:84–92.

Hashimoto, K., Iwata, Y., Nakamura, K., et al. 2006. Reduced serum levels of brain-derived neurotrophic factor in adult male patients with autism. *Prog Neuropsychopharmacol Biol Psychiatry* 30:1529–1531.

Hazlett, H.C., Poe, M., Gerig, G., et al. 2005. Magnetic resonance imaging and head circumference study of brain size in autism: birth through age 2 years. *Arch Gen Psychiatry* 62:1366–1376.

Herbert, M.R., Ziegler, D.A., Makris N., et al. 2003. Localization of white matter volume increase in autism and developmental language disorder. *Ann Neurol* 55:530–540.

Ichtchenko, K., Nguyen, T., Sudhof, T.C. 1996. Structures, alternative splicing, and neurexin binding of multiple *neuroligins*. *J Biol Chem* 271(5):2676–82.

Jacobs, S., Doering, L.C. 2010. Astrocytes prevent abnormal neuronal devleopment in the fragile X mouse. *J Neurosci* 30(12):4508–4514.

Jalil, M.A., Begum, L., Contreras, L., et al. 2005. Reduced *N*-acetylaspartate levels in mice lacking aralar, a brain- and muscle-type mitochondrial aspartate-glutamate carrier. *J Biol Chem* 280:31333–31339.

Kaneko, N., Muratake, T., Amagane, H., et al. 2004. Transmission disequilibrium test and haplotype analysis of the NOTCH4 gene in Japanese patients with schizophrenia. *Psychiatry Clin Neurosci* 58:199–205.

Kern, J.K., Trivedi, M.H., Garver, C.R., et al. 2006. The pattern of sensory processing abnormalities in autism. *Autism* 10:480–494.

Kern, J.K., Trivedi, M.H., Grannemann, B.D., et al. 2007. Sensory correlations in autism. *Autism* 11:123–134.

Kim, H.G., Kishikawa, S., Higgins, A.W., et al. 2008. Disruption of neurexin 1 associated with autism spectrum disorder. *Am J Hum Genet* 82(1):199–207.

Koishi, S., Yamazaki, K., Yamamoto, K., et al. 2004. Notch4 gene polymorphisms are not associated with autism in Japanese population. *Am J Med Genet B Neuropsychiatr Genet* 125B:61–62.

Kwak, Y.D., Brannen, C.L., Qu T., et al. 2006. Amyloid precursor protein regulates differentiation of human neural stem cells. *Stem Cells Dev* 15:381–389.

Kwak, YD., Dantuma, E., Merchant, S., Bushnev, S., Sugaya, K. 2010. Amyloid-beta precursor protein induces glial differentiation of neural progenitor cells by activation of the IL-6/gp130 signaling pathway. *Neurotox Res* 18(3–4): 328–338.

Lasalle, J.M., Yasui, D.H. 2009. Evolving role of MeCP2 in Rett syndrome and autism. *Epigenomics* 1(1):119–130.

Laurence, J.A., Fatemi, S.H. 2005. Glial fibrillary acidic protein is elevated in superior frontal, parietal and cerebellar cortices of autistic subjects. *Cerebellum* 4:206–210.

Lepagnol-Bestel, A.M., Maussion, G., Boda, et al. 2008. *SLC25A12* expression is associated with neurite outgrowth and is upregulated in the prefrontal cortex of autistic subjects. *Mol Psychiatry* 13:385–397.

Luo X., Kemplan, T.A., Lappalainen, J., et al. 2004. NOTCH4 gene haplotype is associated with schizophrenia in African Americans. *Biol Psychiatry* 55:112–117.

Meikle, L., Talos, DM., Onda, H., et al. 2007. A mouse model of tuberous sclerosis: neuronal loss of Tsc1 causes dysplasic and ectopic neurons, reduced myelination, seizure activity, and limited survival. *J Neurosci* 27:5546–5548.

Maezawa, I., Jin, L.W. 2010. Rett syndrome microglia damage dendrites and synapses by the elevated release of glutamate. *J Neurosci* 30(15):5346–5356.

Maezawa, I., Swanberg, S., Harvey, D., et al. 2009. Rett syndrome astrocytes are abnormal and spread MeCP2 deficiency through gap junctions. *J Neurosci* 29(16):5051–5061.

Meyer, U., Feldon, J. 2010. Epidemiology-driven neurodevelopmental animal models of schizophrenia. *Prog Neurobiol* 90:285–326.

Meyer, U., Nyffeler, M., Engler, A., et al. 2006a. The time of prenatal immune challenge determines the specificity of inflammation-mediated brain and behavioral pathology. *J Neurosci* 26:4752–4762.

Meyer, U., Nyffeler, M., Yee, B.K., Knuesel, I., Feldon, J. 2007. Adult brain and behavioral pathological markers of prenatal immune challenge during early/middle and late fetal development in mice. *Brain Behav Immun* 22:469–486.

Meyer, U., Schwendener, S., Feldon, J., Yee, B.K. 2006b. Prenatal and postnatal maternal contributions in the infection model of schizophrenia. *Exp Brain Res* 173:243–257.

Miles, J.H., Hadden, L.L., Takahashi, T.N., Hillman, R.E. 2000. Head circumference is an independent clinical finding associated with autism. *Am J Med Genet* 95:339–350.

Mizuguchi, M., Takashima, S. 2001. Neuropathology of tuberous sclerosis. *Brain Dev* 23:508–515.

Mostofsky, S.H., Burgess, M.P., Gidley Larson, J.C. 2007. Increased motor cortex white matter volume predicts motor impairment in autism. *Brain* 130:2117–2122.

Nakase, T., Yoshida, Y., Nagata, K. 2006. Enhanced connexin 43 immunoreactivity in penumbral areas in the human brain following ischemia. *Glia* 54:369–375.

Naus, C.C., Bechberger, J.F., Caveney, S., Wilson, J.X. 1991. Expression of gap junction genes in astrocytes and C6 glioma cells. *Neurosci Lett* 126:33–36.

Nayate, A., Bradshaw, J.L., Reinhart, N.J. 2005. Autism and Asperger's disorder: are they movement disorders involving the cerebellum and/or basal ganglia? *Brain Res Bull* 67:327–334.

Nedergaard, M., Ransom, B., Goldman, S.A. 2003. New roles for astrocytes: redefining the functional architecture of the brain. *Trends Neurosci* 26:523–530.

Nicchia, G.P., Srinivas, M., Lei, W., Brosnan, C.F., Frigeri, A., Spray, D.C. 2005. New possible roles for aquaporin-4 in astrocytes: cell cytoskeleton and functional relationship with connexin43. *FASEB J* 19:1674–1676.

Okada, K., Hashimoto, K., Iwata, Y., et al. 2007. Decreased serum levels of transforming growth factor-beta1 in patients with autism. *Prog Neuropsychopharmacol Biol Psychiatry* 31:187–190.

Onda, H., Crino, P.B., Zhang, H. 2002. Tsc2 null murine neuroepithelial cells are a model for human tuber giant cells, and show activation of an mTOR pathway. *Mol Cell Neurosci* 21:561–574.

Oostra, B.A., Willemsen, R. 2009. FMR1: a gene with three faces. *Biochim Biophys Acta* 1790:467–477.

Owens, D.G. 1994. Extrapyramidal side effects and tolerability of risperidone: a review. *J Clin Psychiatry* 55:29–35.

Pacey, L.K.K., Doering, L.C. 2007. Developmental expression of FMRP in the astrocyte lineage: implications for fragile X syndrome. *Glia* 55:1601–1609.

Papadopoulos, M.C., Krishna, S., Verkman, A.S. 2002. Aquaporin water channels and brain edema. *Mt Sinai J Med* 69:242–248.

Park, H.C., Appel, B. 2003. Delta-Notch signaling regulates oligodendrocyte specification. *Development* 130:3747–3755.

Perry, W., Minassian, A., Lopez, B., Maron, L., Lincoln, A. 2007. Sensorimotor gating deficits in adults with autism. *Biol Psychiatry* 61:482–486.

Prange, O., Wong, T.P., Gerrow, K. 2004. A balance between excitatory and inhibitory synapses is controlled by PSD-95 and neuroligin. *Proc Natl Acad Sci USA* 101(38):13915–13920.

Prasad, S., Chowdari, K.V., Wood, J., et al. 2004. Association analysis of NOTCH 4 polymorphisms with schizophrenia among two independent family based samples. *Am J Med Genet B Neuropsychiatr Genet* 131B:6–9.

Pfrieger, F.W., Barres, B.A. 1997. Synaptic efficacy enhanced by glial cells *in vitro*. *Science* 277:1684–1687.

Quincozes-Santos, A., Abib, R.T., Leite, M.C., et al. 2008. Effect of the atypical neuroleptic risperidone on morphology and S100B secretion in C astroglial lineage cells. *Mol Cell Biochem* 314:59–63.

Quincozes-Santos, A., Bobermin, L.D., Kleinkauf-Rocha, J., et al. 2009. Atypical neuroleptic risperidone modulates glial functions in C astroglial cells. *Prog Neruopharmacol Biol Psychiatry* 33:11–15.

Ramasamy, S.K., Lenka, N. 2010. Notch exhibits ligand bias and maneuvers stage-specific steering of neural differentiation in embryonic stem cells. *Mol Cell Biol* 30:1946–1957.

Ramos, M., del Arco, A., Pardo, B., et al. 2003. Developmental changes in the Ca2+-regulated mitochondrial aspartate-glutamate carrier aralar1 in brain and prominent expression in the spinal cord. *Brain Res Dev Brain Res* 143:33–46.

Ramoz, N., Reichert, J.G., Smith, C.J., et al. 2004. Linkage and association of the mitochondrial aspartate/glutamate carrier *SLC25A12* gene with autism. *Am J Psychiatry* 161:662–669.

Ritvo, E.R., Freeman, B.J., Scheibel, A.B., et al. 1986. Lower Purkinje cell counts in the cerebella of four autistic subjects: initial findings of the UCLA-NSAC autopsy research report. *Am J Psychiatry* 143:862–866.

Rosengren, L.E., Ahlsen, G., Belfrage, M., Gillberg, C., Haglid, K.G., Hamberger, A. 1992. A sensitive ELISA for glial fibriallary acidic protein: application in CSF of children. *J Neurosci Methods* 44:113–119.

Rouach, N., Avignone, E., Meme, W., et al. 2002. Gap junctions and connexin expression in the normal and pathological central nervous system. *Biol Cell* 94:457–475.

Roxrud, I., Raiborg, C., Gilfillan, G.D., et al. 2009. Dual degradation mechanisms ensure disposal of NHE6 mutant protein associated with neurological disease. *Exp Cell Res* 315(17):3014–3027.

Saitoh, O., Courchesne, E. 1998. Magnetic resonance imaging study of the brain in autism. *Psychiatry Clin Neurosci* 52:S219–S222.

Sakurai, T., Ramoz, N., Barreto, M., et al. 2010. Slc25a12 disruption alters myelination and neurofilaments: a model for hypomyelination syndrome and childhood neurodevelopmental disorders. *Biol Psychiatry* 67(9):887–94.

Schumann, C.M., Amaral, D.G. 2006. Stereological analysis of amygdala neuron number in autism. *J Neurosci* 26:7674–7679.

Schumann, C.M., Bloss, C.S., Barnes, C.C., et al. 2010. Longitudinal magnetic resonance imaging study of cortical development through early childhood in autism. *J Neurosci* 30:4419–4427.

Scott, J.A., Schumann, C.M., Goodlin-Jones, B.L., Amaral, D.G. 2009. A comprehensive volumetric analysis of the cerebellum in children and adolescents with autism spectrum disorder. *Autism Res* 2:246–257.

Singer, H.S., Morris, C.M., Gause, C.D., Gillin, P.K., Crawford, S., Zimmerman, A.W. 2008. Antibodies against fetal brain in sera of mothers with autistic children. *J Neuroimmunol* 194:165–172.

Sokol, D.K., Chen, D., Farlow, M.R., et al. 2006. High levels of Alzheimer beta-amyloid precursor protein (APP) in children with severely autistic behavior and aggression. *J Child Neurol* 21:444–449.

Steffek, A.E., McCullumsmith, R.E., Haroutunian, V., Meador-Woodruff, J.H. 2008. Cortical expression of glial fibrillary acidic protein and glutamine synthetase is decreased in schizophrenia. *Schizophr Res* 103:71–82.

Steinlin, M. 2008. Cerebellar disorders in childhood: cognitive problems. *Cerebellum* 7:607–610.

St. Hilarie, C., Vargas, D., Prado, C.A., et al. 2005. Aquaporin 4 is increased in association with human immunodeficiency virus dementia: implications for disease pathogenesis. *J Neurovirol* 11:535–543.

Stipursky, J., Gomes, F.C. 2007. TGF-beta1/SMAD signaling induces astrocyte fate commitment *in vitro*: implications for radial glia development. *Glia* 55:1023–1033.

Stork, T., Silke, T., Floriano, R., et al. 2009. *Drosophila* Neurexin IV stabilizes neuron-glia interactions at the CNS midline by binding to Wrapper. *Development* 136:1251–1261.

Tabuchi, K., Blundell, J., Etherton, M.R., et al. 2007. A neuroligin-3 mutation implicated in autism increases inhibitory synaptic transmission in mice. *Science* 318:71–76.

Taniguchi, M., Yamashita, T., Kumura, E, et al. 2000. Induction of aquaporin-4 water channel mRNA after focal cerebral ischemia in rat. *Brain Res Mol Brain Res* 78:131–137.

Theis, M., Speidel, D., Willecke, K. 2004. Astrocyte cultures from conditional connexin43-deficient mice. *Glia* 46:130–141.

Tuchman, R., Rapin, I. 2002. Epilepsy in autism. *Lancet Neurol* 1(6):352–358.

Tomchek, S.D., Dunn, W. 2007. Sensory processing in children with and without autism: a comparative study using the short sensory profile. *Am J Occup Ther* 61:190–200.

Townsend, J., Harris, D.L., Courchesne, E. 1996. Visual attention abnormalities in autism: delayed orienting to location. *J Int Neuropsychol Soc* 2:541–550.

Uemura, T., Lee, S.J., Yasumura, M., et al. 2010. Trans-synaptic interaction of GluRdelta2 and Neurexin through Cbln1 mediates synapse formation in the cerebellum. *Cell* 141(6):1068–1079.

Uhlmann, E.J., Wong, M., Baldwin, R.L., et al. 2002. Astrocyte-specific TSC1 conditional knockout mice exhibit abnormal neuronal organization and seizures. *Ann Neurol* 52:285–296.

Ujike, H., Takehisa, Y., Takaki, M., et al. 2001. NOTCH4 gene polymorphism and susceptibility to schizophrenia and schizoaffective disorder. *Neurosci Lett* 301:41–44.

Ullian, E.M., Sapperstein, S.K., Chistopherson, K.S., Barres, B.A. 2001. Control of synapse number by glia. *Science* 291:657–661.

Vargas, D.L., Nascimbene, C., Krishnan, C., Zimmerman, A.W., Pardo, C.A. 2005. Neuroglial activation and neuroinflammation in the brain of patients with autism. *Ann Neurol* 57:67–81.

Varoqueaux, F., Jamain, S., Brose, N. 2004. Neuroligin 2 is exclusively localized to inhibitory synapses. *Eur J Cell Biol* 83(9):449–456.

Varoqueaux, F., Aramuni, G., Rawson, R.L. et al. 2006. Neuroligins determine synapse maturation and function. *Neuron* 51(6):741–754.

Wang, H., Ku, L., Osterhout, D.J., et al. 2004. Developmentally-programmed FMRP expression in oligodendrocytes: a potential role of FMRP in regulating translation in oligodendroglia progenitors. *Hum Mol Genet* 13(1):79–89.

Wang, Z.W., Fang, Y.R., Hong, W., Wang, D.X., Jing, S.D. 2005. Association study of NOTCH4 gene polymorphism with schizophrenia and mood disorders in mixed pedigrees.*Yi Chuan* 27:865–868.

Watanabe, Y., Hashimoto, S., Kakita, A., et al. 2004. Neonatal impact of leukemia inhibitory factor on neurobehavioral development in rats. *Neurosci Res* 48:345–353.

Webb, S.J., Sparks, B.F., Friedman, S.D., et al. 2009. Cerebellar vermal volumes and behavioral correlates in children with autism spectrum disorder. *Psychiatry Res* 172:61–67.

Wei, J., Hemmings, G.P. 2000. The NOTCH4 locus is associated with susceptibility to schizophrenia. *Nat Genet* 25:376–377.

Zhang, C., Milunsky, J.M., Newton, S., et al. 2009. A neuroligin-4 missense mutation associated with autism impairs neuroligin-4 folding and endoplasmic reticulum export. *J Neurosci* 29(35):10843–10854.

Zimmerman, A.W., Jyonouchi, H., Comi, A.M., et al. 2005. Cerebrospinal fluid and serum markers of inflammation in autism. *Pediatr Neurol* 33:195–201.

14 Astrocytes in Major Depressive Disorder and Schizophrenia

Pathophysiological and Therapeutic Aspects

Markus J. Schwarz and Aye-Mu Myint
Klinikum der Universität München, Klinik und
Poliklinik für Psychiatrie und Psychotherapie, Ludwig-
Maximilians Universität München, Munich, Germany

CONTENTS

14.1 ASTROCYTES IN MAJOR DEPRESSION

Major depressive disorders (MDDs) are the commonest psychiatric disorders
and in Europe account for approximately 13% lifetime incidence, with 4% being
diagnosed with major depression within the past 12 months (Alonso et al. 2004).
Depression has a major detrimental impact on the quality of life of patients
regardless of geographical, educational, socioeconomic, and racial back-
grounds. According to the 1990 Global Burden of Disease of the World Health
Organization (WHO), depression has a greater negative impact on the quality of
life than cardiovascular disease and has been projected to be the second most
important cause of disability, as disability adjusted years, by 2030 (Mathers and
Loncar 2005). The pattern of occurrence, increasing severity, and the frequent
resistance to treatment are some of the reasons for the high burden of depression
(Greden 2001a, 2001b). There are different hypotheses upon which antidepressant medications have been developed, but the currently available antidepressants are not adequately effective to treat the disease. Evidence for the role of
astrocytes in pathophysiology of MDD opens the possibility for new mechanisms that are not covered by currently available antidepressants and pave new
ways for new therapeutic strategies.

14.1.1 EVIDENCE OF INVOLVEMENT OF ASTROCYTES IN MAJOR DEPRESSION

14.1.1.1 Morphological Evidences

The first studies reporting evidence for reduced numbers of neuronal and glial cells
in the brains of the subjects with MDD were from Ongur et al. (1998) and Rajkowska
et al. (1999). Altered numbers of astrocytes in different brain regions such as prefrontal cortex (PFC) and cerebellum in MDD were reported soon thereafter (Miguel-
Hidalgo et al. 2000; Si et al. 2004; Fatemi et al. 2004). Miguel-Hidalgo and colleagues
reported that although the mean areal fraction and packing density of glial fibrillary
acidic protein (GFAP)–immunoreactive astrocytes in the dorsolateral PFC were not
significantly different when all MDD patients were compared to controls, there was
a significant strong positive correlation between age and GFAP immunoreactivity
in MDD subjects, such that younger adults (30–45 years old) had values for areal
fractions and packing densities that were lower than the smallest values from the
older (46–86) groups. Based on this finding, the astrogliosis that is the hallmark
of neuronal loss was suspected in pathophysiology of MDD; however, two studies
failed to demonstrate a connection between astrogliosis and MDD (Damadzic et al.
2001; Muller et al. 2001). Damadzic et al. (2001) showed no evidence of astrocytosis

in cortical layers II through VI in the intermediate subarea of the entorhinal cortex in two cohorts, and Muller et al. (2001) demonstrated only a subtle change in astrocyte density in selected portions of the pyramidal cell layer in CA1 and CA2 of the hippocampus in MDD (Damadzic et al. 2001). Instead of astrogliosis, subsequent studies have demonstrated age-dependent loss of GFAP-immunoreactive cells in prefrontal cortex of MDD (Si et al. 2004) and reduction of GFAP-immunoreactive cells in cerebellum of MDD regardless of age (Fatemi et al. 2004).

Regarding the detailed morphological changes of glial cells, the findings can be subdivided with respect to different brain regions: prefrontal cortex (PFC), hippocampus, and amygdala. Furthermore, PFC can be considered as three distinct subregions: dorsolateral PFC, orbitofrontal cortex, and medial PFC. In dorsolateral PFC, two reports (Rajkowska et al. 1999; Cotter et al. 2002) demonstrated lower glial densities in layers III and IV in MDD. However, in detailed analysis, Rajowska and group showed increased astrocytes (glia with large nuclei) in layer III of MDD (Rajkowska et al. 1999), consistent with the results cited above of higher density of GFAP-immunoreactive cells in older MDD brains. In the orbitofrontal area glial cell densities are decreased in MDD brains, particularly for medium-sized and large glial cells in rostral layers IIIa and IV, respectively (Rajkowska et al. 1999). Caudally as well, significant reduction in the glial cell sizes as well as the densities in layers III, V, and VI was reported (Rajkowska et al. 1999). Most morphological and cellular studies in medial PFC have focused on anterior cingulate cortex (ACC), where significant reductions in glial cell number and density have been reported (Ongur et al. 1998; Cotter et al. 2011a; Chana et al. 2003).

Regarding glial changes in hippocampus, most studies have been carried out in animal models and only a few in human. Even though the first study showed reduction in the GFAP-immunoreactive cell density in CA1 and CA2 regions of MDD (Muller et al. 2001), another study reported the increased density of pyramidal neurons and glial cells in CA regions and dentate gyrus of MDD in association with reduction in mean somal size (Stockmeier et al. 2004). This might be due to microgliosis as reported in a later study on S100B-immunoreactive glial cells (Steiner et al. 2008), although this should be further explored. In amygdala, one study has reported reduced glial cell density, especially in the left hemisphere (Bowley et al. 2002), and another study showed only reduced numbers of oligodendrocytes (Hamidi et al. 2004).

14.1.1.2 Biochemical Evidences

The expression of GFAP mRNA in PFC was demonstrated to be reduced in MDD (Johnston-Wilson et al. 2000), although there is a report proposing rebound increase in late-life depression (Davis et al. 2002). Many astrocytes express GFAP, although not all. Most of the astrocytes take up glutamate and convert it to glutamine by the action of glutamine synthetase, which is heavily expressed in grey matter astrocytes (Toro et al. 2006). In one recent research report, the glutamine synthetase levels in the brain of MDD with or without alcoholism were compared to those from brains of normal and alcoholics without MDD. It was observed that glutamine synthetase levels in brains of MDD with alcoholism were lower than normal controls and alcoholics without MDD. Moreover, the area fraction of GFAP was significantly lower in MDD brains (Miguel-Hidalgo et al. 2010). This study also mentioned that excitotoxic amino acid transporter (EAAT) 1 levels in the left Broadmann's area 47, which is part of PFC, in the postmortem brain tissues of MDD patients with or without alcoholism were lower than those in controls.

Unlike the above biochemical, anatomical, or morphological studies in the specific brain areas of postmortem brain tissue, some biochemical studies have been carried out in periphery, and thus only provide indirect evidence for changes. However, since biochemical studies can be carried out in living patients, the effects of medication were better controlled and there was no influence of the cause of death. In terms of biochemical studies, elevated serum S100B, which is an indirect marker of astrocyte damage, was reported in most of the studies (Schroeter et al. 2002; Arolt et al. 2003; Hetzel et al. 2005), although one study failed to demonstrate the elevation of serum S100B in MDD (Jang et al. 2008). However, most of those studies showed association between high serum S100B and better response to treatment (Arolt et al. 2003; Jang et al. 2008). Taken altogether, a question was raised why the increase in serum S100B was associated with better response to treatment if S100B elevation indicated astrocyte damage.

In conclusion, based on the above findings, although it is still not clear whether reduced astroglial cells' density, microgliosis, or astrogliosis is the key pathophysiological mechanism, the involvement of astrocytes in the pathophysiology of MDD is well documented.

14.1.2 Pathophysiological Role of Astrocytes in Major Depression

14.1.2.1 Physiological Roles in Neurotransmission

Before discussing the pathophysiological role of astrocytes in MDD, the physiological roles of astrocytes in relation to neurotransmission and neuronal plasticity should be considered in the context of MDD.

Astrocytes play an important role in brain energy metabolism since glucose in the brain is taken up by astrocytes and either stored as glycogen or metabolized into lactate (Tsacopoulos and Magistretti 1996). Under conditions of high metabolic demand, astrocytic cAMP signaling pathways are activated and the intracellular breakdown of glycogen takes place. The glucose from this breakdown is further metabolized into lactate from which the neurons get energy (Sorg and Magistretti 1991).

Astrocytes also play a major role in glutamate-glutamine metabolism. Astrocyte terminations wrap the glutamatergic synaptic cleft, and gliotransmission between astrocytes and neurons occurs. Each astrocyte may connect about 140,000 synapses in rat hippocampus (Bushong et al. 2002) and 2 million synapses in human cortex (Oberheim et al. 2006). Glutamate released from neurons activates metabotrophic receptors on astrocytes, and astrocytes in turn release glutamate. Astrocytic glutamate release activates both presynaptic (Fiacco and McCarthy 2004; Jourdain et al. 2007; Perea and Araque 2007) and postsynaptic (Angulo et al. 2004; Fellin et al. 2004) glutamate receptors to modulate synaptic transmission. This mechanism contributes to synaptic plasticity and is partly controlled by formation of less toxic glutamine from glutamate by the enzyme glutamine synthetase in astrocytes (Hertz et al. 1999). The glutamate released into extracellular space is transported into astrocytes by EAAT1 and EAAT2 on the astrocytic membrane to prevent excitotoxic neuronal death (Rothstein et al. 1996; Tanaka et al. 1997; Watanabe et al. 1999). The astrocytes also release D-serine-containing vesicles (Mothet et al. 2005), and this form of gliotransmission is hypothesized to regulate N-methyl-D-aspartate (NMDA) receptor function and synaptic plasticity, such as long-term potentiation (LTP) (Panatier et al. 2006).

With regard to NMDA and glutamate receptor activity, astrocytes also play a major role through tryptophan-kynurenine metabolism. In the brain, tryptophan catabolism occurs in astrocytes and microglia (Grant et al. 2000; Grant and Kapoor 1998; Heyes et al. 1996), though 60% of brain kynurenines are contributed from the periphery (Gal and Sherman 1980). Tryptophan is degraded into kynurenine by indoleamine 2,3-dioxygenase (IDO) in most regions of the brain (Heyes et al. 1993; Mellor and Munn 1999), although tryptophan 2,3-dioxygenase 2 (TDO2) immunoreactivity was demonstrated in the frontal cortex of the schizophrenic brain (Miller et al. 2004). The TDO specifically metabolizes only tryptophan (Hayaishi 1980), whereas IDO also metabolizes serotonin and melatonin (Hayaishi 1976). The activity of TDO is enhanced by the tryptophan concentration (Saito et al. 1990; Satyanarayana and Rao 1980; Smith et al. 1980) and by high cortisol (Salter and Pogson 1985), whereas IDO activity is enhanced by the pro-inflammatory cytokines, such as interferon-γ (Carlin et al. 1987, 1989; Hu et al. 1995; Taylor and Feng 1991; Yasui et al. 1986), and inhibited by anti-inflammatory cytokine, IL-4 (Musso et al. 1994).

After tryptophan is first catabolized into kynurenine, there is a further catabolic pathway, the kynurenine pathway, in which the metabolites contribute to the neuroprotective-neurodegenerative changes in the brain. Kynurenine is further catabolized into 3-hydroxykynurenine (3OHK) and the NMDA receptor agonist quinolinic acid (Chiarugi et al. 2001; Bender and McCreanor 1985) (Figure 14.1). The 3OHK causes neuronal apoptosis (Okuda et al. 1998), while quinolinic acid causes excitotoxic neurodegenerative changes (Schwarcz et al. 1983). However, kynurenine can also be catabolized into kynurenic acid (Figure 14.1), which is an NMDA receptor antagonist (Perkins and Stone 1982) and is protective against excitotoxic actions of quinolinic acid (Kim and Choi 1987; Stone and Darlington 2002). Astrocytes were shown to produce mainly kynurenic acid because of the lack of kynurenine 3-monooxygenase enzyme, whereas microglia and macrophages produced mainly quinolinic acid (Guillemin et al. 2000, 2001, 2005). Astrocytes were also demonstrated to metabolize the quinolinic acid produced by the neighboring microglia (Guillemin et al. 2001). The protective effect of kynurenic acid against excitotoxic effects of quinolinic acid has also been detected in neuronal cell cultures (Kim and Choi 1987). Therefore, through the kynurenine pathway, astrocytes play an important neuroprotective role regarding NMDA receptor activity-induced neurotoxicity and glutamate-induced excitotoxicity.

14.1.2.2 Pathophysiological Roles in Immune-Neurochemical Interaction in Astrocytes

The altered glutamatergic neurotransmission in MDD is well documented as part of the pathophysiological mechanisms (Sheline et al. 1998; Auer et al. 2000; Mirza et al. 2004). This might be due to altered glutamatergic pathway itself or through

Simplified schematic presentation of the role of cytokines and kynurenine pathway in
astrocytes and microglia in neuroprotection-neurodegeneration

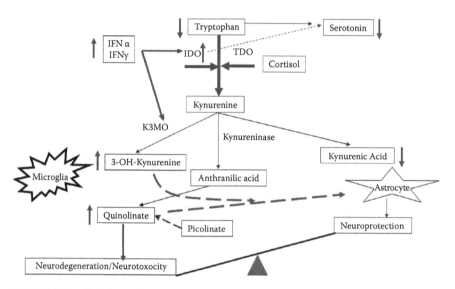

FIGURE 14.1 Principle of the kynurenine pathway. This chart is just a simplified excerpt of
the pathway, showing the most important neurochemically active kynurenine metabolites and
the key enzymes. TDO = tryptophan 2,3-dioxygenase, IDO = indoleamine 2,3-dioxygenase,
KMO = kynurenine 3-monoxygenase, KAT = kynurenine aminotransferase.

astrocyte dysfunction induced by toxic kynurenines. The possible role of imbalanced
kynurenine metabolites in the psychopathology and chronicity of MDD has been
proposed in the neurodegeneration hypothesis of depression (Myint and Kim 2003).
The hypothesis explained the involvement of kynurenine's imbalance in pathophysiology of MDD through immune activation and impaired neuron-astrocytes-microglia
interaction via the enzyme IDO.

As mentioned above, IDO is activated by pro-inflammatory cytokines, and
another enzyme, TDO2, that degrades tryptophan is activated by cortisol. Both
persistent hypercortisolemia (Asnis et al. 1981; Cohen et al. 1984; Lin et al.
1986; Sher et al. 2005) and increased pro-inflammatory cytokines (Kaestner
et al. 2005; Thomas et al. 2005; Mikova et al. 2001; Connor and Leonard 1998;
Anisman et al. 1999; Lanquillon et al. 2000; Myint et al. 2005; Kim et al. 2002)
have been reported in patients with major depression. In the brain, microglia
are the first line of defense, which can produce pro-inflammatory cytokines.
The circulating pro-inflammatory cytokines in patients with MDD may induce
enhanced production of pro-inflammatory cytokines (Miller and O'Callaghan
2005) that might in turn influence the tryptophan-kynurenine pathway and glutamatergic neurotransmission.

The imbalance of neuroprotective vs. neurotoxic kynurenines was also demonstrated in both interferon-induced depressive episodes (Wichers et al. 2005; Raison
et al. 2010) and major depression (Myint et al. 2007). A significant reduction in

plasma kynurenic acid in patients with MDD (Myint et al. 2007) may indirectly reflect impaired production of kynurenic acid due to astrocyte loss. Therefore, it might be concluded that the pro-inflammatory status and hypercortisolemia in MDD could induce increased tryptophan breakdown and serotonin depletion and the shift of kynurenine metabolism toward neurotoxic kynurenines that in turn induces astrocyte loss and impaired neuroprotection and glutamatergic neurotransmission in MDD.

Moreover, since astrocytes are involved in GABA uptake and astrocytes have several antioxidative systems (Aschner 2000; Anderson and Nedergaard 2003), the astrocytes' loss might induce further disturbances in GABA transmission and production of reactive oxygen species. Since reactive oxygen species can activate the IDO enzyme activity, this might result in a vicious cycle of impaired neuroprotection-enhanced neurodegeneration.

14.1.3 THERAPEUTIC ASPECTS

14.1.3.1 Antidepressant Therapy

A modulatory, predominantly inhibitory effect of serotonin reuptake inhibiting drugs on pro-inflammatory immune parameters was demonstrated in animal experiments (Bengtsson et al. 1992; Song and Leonard 1994; Zhu et al. 1994). Different antidepressants reduce the IFN-γ/IL-10 ratio in vitro (Maes et al. 1999), reduce the in vitro production of IFN-γ, IL-2, and sIL-2R (Seidel et al. 1995, 1996), and downregulate the production of IL-6 and TNF-γ (Lanquillon et al. 2000). However, there are also conflicting results, reporting no effect of antidepressants to the in vitro stimulation of cytokines (overview: Kenis and Maes 2002). Since there are different variations in the culture method used in in vitro studies, the methodological issues should be considered regarding those conflicting results.

Regarding serum levels, a reduction of IL-6 during treatment with the serotonin reuptake inhibitor fluoxetine (Sluzewska et al. 1995) or other antidepressant drugs (Frommberger et al. 1997) and a shift of imbalanced IFNγ/IL-4 to a balanced ratio after 6 weeks of antidepressant treatment (Myint et al. 2005) have been observed. A recent in vitro study showed that both tricyclic antidepressants and selective serotonin inhibitors attenuated cytokine-induced PGE$_2$ and nitric oxide production by inflammatory cells (Yaron et al. 1999). These immunological data indicate that antidepressants may also influence the function of astrocytes.

In fact, there are several studies showing the direct impact of antidepressant medication on astrocyte function or cell number. The above-mentioned decreased numbers of astrocytes in depressed patients can be modeled in animals by chronic stress exposure. Interestingly, fluoxetine treatment can prevent this loss of astrocytes, while it has no effect in nonstressed control animals (Czeh et al. 2006). On a cellular level, fluoxetine and citalopram induce delayed and asynchronous calcium transients in astrocytes (Schipke et al. 2010), pointing to the contribution of astrocytes in the cellular action of antidepressants. Chronic fluoxetine treatment also upregulates the expression of calcium-dependent phospholipase A2, which in turn releases arachidonic acid from the cell membrane of astrocytes in vitro (Li et al. 2009). The antidepressants imipramine, venlafaxine, and fluoxetine can even induce astrocytes to express neuronal markers, accompanied by a rapid change of cell morphology toward a neuron-like shape (Cabras et al. 2010). However, an animal study showed the antidepressant-like effects of imipramine in a learned helplessness paradigm were blocked when fluorocitrate, a reversible inhibitor of astrocyte function, was injected into the dentate gyrus or CA3 region (Iwata et al. 2010). This study indicated that the hippocampal astrocytes are essential for the efficacy of antidepressant treatment.

14.1.3.2 Use of Anti-Inflammatory Drugs

In the report on kynurenine-induced changes in MDD, it was stated that 6-week treatment with currently available antidepressants, mainly selective serotonin reuptake inhibitors (SSRIs) and serotonin/norepinephrine reuptake inhibitors (SNRIs), could improve the clinical symptoms but not significantly change the biochemical abnormalities (Myint et al. 2007). Due to the increase of pro-inflammatory cytokines and PGE_2 in depressed patients, and that the kynurenine changes are also directly linked to the pro-inflammatory state, anti-inflammatory treatment would be expected to show antidepressant effects also in depressed patients. In particular, COX-2 inhibitors seem to show advantageous results: animal studies show that COX-2 inhibition can lower the increase of the pro-inflammatory cytokines IL-1, TNF, and PGE_2, but it can also prevent clinical symptoms such as anxiety and cognitive decline, which are associated with this increase of pro-inflammatory cytokines (Casolini et al. 2002). Moreover, treatment with the COX-2 inhibitor celecoxib—but not with a COX-1 inhibitor—prevented the dysregulation of the hypothalamic-pituitary-adrenal (HPA) axis, in particular the increase of cortisol, one of the key biological features associated with depression (Casolini et al. 2002; Hu et al. 2005). This effect might be expected because PGE_2 stimulates the HPA axis in the CNS (Song and Leonard 2000) and PGE_2 is inhibited by COX-2 inhibition. Moreover, the functional effects of IL-1 in the CNS—sickness behavior being one of these effects—were also shown to be antagonized by treatment with a selective COX-2 inhibitor (Cao et al. 1999).

Additionally, COX-2 inhibitors influence the CNS serotonergic system. In a rat model, treatment with rofecoxib was followed by an increase of serotonin in the frontal and temporo-parietal cortex (Sandrini et al. 2002). A possible mechanism of the antidepressant action of COX-2 inhibitors is the inhibition of the release of IL-1 and IL-6. Moreover, COX-2 inhibitors also protect the CNS from effects of quinolinic acid, i.e., from neurotoxicity (Salzberg-Brenhouse et al. 2003). In the depression model of the bulbectomized rat, a decrease of cytokine levels in the hypothalamus and a change in the behavior have been observed after chronic celecoxib treatment (Myint et al. 2007). In another animal model of depression, however, the mixed COX-1/COX-2 inhibitor acetylsalicylic acid showed an additional antidepressant effect by accelerating the antidepressant effect of fluoxetine (Brunello et al. 2006).

The significant therapeutic effect of the COX-2 inhibitor on depressive symptoms was demonstrated in a randomized double-blind pilot add-on study using celecoxib in MDD (Müller et al. 2006). Also in a clinical study, the mixed COX-1/COX-2 inhibitor acetylsalicylic acid accelerated the antidepressant effect of fluoxetine and increased the response rate in depressed nonresponders to a monotherapy with fluoxetine in an open-label pilot study (Mendlewicz et al., 2006).

14.1.3.3 Indirect Manipulation through Kynurenine Pathway

There is evidence from animal experiments that the manipulation of the kynurenine pathway could be a possible therapeutic strategy in MDD. O'Connor and group demonstrated that lipopolysaccharide-induced depressive behavior through the action of enhanced IDO enzyme activity (O'Connor et al. 2009c). This group also demonstrated in the bacille Calmette-Guérin (BCG) mouse model of depression (O'Connor et al. 2009b) that immune activation using BCG could induce depressive behavior and activation of IDO enzyme activity, followed by activation of 3-hydroxyanthranillic acid oxidase (3-HAO) enzyme, which forms neurotoxic quinolinic acid from 3-hydroxyanthranillic acid. Moreover, blockade of IDO could prevent the depressive behavior. In another study, interferon-γ knockout mice did not show the depressive behavior when challenged with BCG since interferon-γ is the inducer of IDO enzyme (O'Connor et al. 2009a). This evidence indicates that manipulating the kynurenine pathway could be a novel therapeutic strategy for counteracting depression. Moreover, this type of manipulation could reduce the production of neurotoxic kynurenines that could induce astrocyte apoptosis, and in turn prevent further loss of astrocytes and enhance response to treatment.

14.2 ASTROCYTES IN SCHIZOPRENIA

Schizophrenia is a devastating psychotic illness characterized by impaired thinking, emotions, and behavior. The key symptoms are auditory hallucinations, paranoid or bizarre delusions, or disorganized speech and thinking with significant social or occupational dysfunction. Earliest manifestation of this disorder typically occurs in young adulthood and affects around 1 to 1.5% of the population worldwide. Schizophrenia belongs to the so-called complex disorders in which both a genetic component and environmental risk factors contribute to the risk to develop this disease. It is a major problem for schizophrenia research that this disease may be a heterogenous disorder, in that several discrete syndromes with different pathoetiological mechanisms may be summarized under the term *schizophrenia*. However, there is general consensus that an alteration of distinct neurotransmitter systems is crucially involved in the pathophysiology of schizophrenia. In the following section, the probable role of astrocytes in the etiopathogenesis of schizophrenia is reviewed.

14.2.1 EVIDENCE FOR THE INVOLVEMENT OF ASTROCYTES IN SCHIZOPHRENIA

14.2.1.1 Altered Numbers

With the emergence of neuropathology and motivated by the successful investigations of Alois Alzheimer in describing brain alterations in demented patients, postmortem histological investigations were among the initial modern biological attempts to investigate schizophrenia. However, most of these investigations produced conflicting results, and also Alois Alzheimer failed to find distinct and disease-specific alteration in schizophrenia. This was 100 years ago. During the following decades, several studies showed gliosis in postmortem brains of schizophrenic patients. Using relatively old methods of staining, increased astrocyte densities have repeatedly been reported

(Stevens 1982; Nasrallah et al. 1983; Bruton et al. 1990). However, using immunohistochemical staining methods, especially against the astrocyte-specific GFAP, most of the studies could not find any indices for increased astrocyte densities (Bernstein et al. 2009). Thus, at this time, there is overall agreement that numbers of astrocytes are not markedly altered in schizophrenia (Bernstein et al. 2009; Cotter et al. 2001).

14.2.1.2 Altered Morphology

There are only very few reports on morphological changes of astrocytes in schizophrenia. The group of Uranova performed ultrastructural investigations of postmortem brains of schizophrenic patients and healthy controls and found no alteration regarding cell size, volume fraction, and area density of mitochondria and lipofuscin granules in the astrocytes. However, they observed a marked reduction of volume fraction and area density of mitochondria in patients with duration of disease more than 21 years, suggesting that there may be a progressive disturbance of astrocyte function due to a disturbed mitochondrial function in schizophrenia (Kolomeets and Uranova 2010).

14.2.1.3 Altered Biochemical Parameters

14.2.1.3.1 S100B

S100B is a calcium, copper, and zinc ion-binding protein that is synthesized in and released from astrocytes (Steiner et al. 2006; Rothermundt et al. 2009). Nanomolar levels stimulate neurite growth and synaptogenesis, while micromolar levels result in opposite effects and can induce neuronal apoptosis (Van Eldik and Wainwright 2003).

The results regarding altered S100B levels in schizophrenia are relatively consistent: elevated levels of S100B have been observed in several studies on peripheral blood and cerebrospinal fluid (CSF) in patients with schizophrenia. Recent meta-analyses confirm the validity of these findings (Schroeter et al. 2009). Elevated blood and cerebrospinal fluid concentrations of S100B are associated with acute exacerbations or deficit symptoms of schizophrenia and have been attributed to schizophrenia-related glial pathology (Steiner et al. 2006; Rothermundt et al. 2004). Indeed, increased glial S100B expression has been observed in a recent postmortem study, especially in the dorsolateral prefrontal cortex of patients with paranoid schizophrenia (Steiner et al. 2008). Cell culture experiments indicate that S100B may exert a modulatory effect on neurotransmission, leading to deficient synaptic glutamate concentrations (via influencing astrocytic glutamate uptake) and enhanced dopaminergic neurotransmission (via binding to the third cytoplasmic loop of the D2 receptor) (Tramontina et al. 2006a, 2006b; Liu et al. 2008).

It has been debated whether elevated S100B levels in schizophrenia indicate brain damage, leakage of the blood–brain barrier, or altered activation state of astrocytes resulting in altered S100B secretion. Recent analyses clearly indicate that the first two explanations can be ruled out, so that an altered function of astrocytes is the most probable explanation (Schroeter et al. 2009).

14.2.1.3.2 GFAP

Glial fibrillary acidic protein is well accepted as a cellular and activation marker of astrocytes, although it is not completely astrocyte specific. The expression of both GFAP and its mRNA are markedly decreased in the white matter of the anterior cigulate cortex of schizophrenic patients (Webster et al. 2005; Steffek et al. 2008). The interpretation of these findings, however, is not clear, because they may point out either a reduction of GFAP-expressing astrocytes in schizophrenia or a functional alteration of still existing GFAP-negative cells (Bernstein et al. 2009). Interestingly, S100B plays an important role in the assembly of components of the cell architecture, and it inhibits the assembly of GFAP (Garbuglia et al. 1999). The elevated levels of S100B could therefore account for the reduced expression of GFAP in schizophrenia.

14.2.1.3.3   EAAT and Glutamine Synthetase

The functional role of astrocytes in glutamatergic neurotransmission is of high importance, especially in the field of schizophrenia research. After glutamate has been released from neurons, it is rapidly taken up into astrocytes, where it is converted into the neurochemically inactive glutamine by the enzyme glutamine synthetase. The excitatory amino acid transporters EAAT1 and EAAT2 are responsible for the uptake (De Keyser et al. 2008). Both expression and function of EAAT2 are increased in the prefrontal cortex of schizophrenic patients (De Keyser et al. 2008). Regarding the enzyme glutamine, synthetase-elevated levels have been described in the thalamus of schizophrenic patients (Bruneau et al. 2005; Toro et al. 2006), while decreased expression or no changes have been reported in cortical areas (Toro et al. 2006; Steffek et al. 2008). Although the findings regarding the expression patterns of this enzyme are not consistent, biochemical investigation clearly demonstrates altered glutamate production in schizophrenia. This leads to the question regarding the neurochemical basis of schizophrenia and the functional role of astrocytes in neurotransmission.

14.2.2   Pathophysiological Role of Astrocytes in Schizophrenia

14.2.2.1   Functional Role in Neurotransmission

14.2.2.1.1   Glutamatergic Neurotransmission in Schizophrenia

Based on the evidence that most drugs ameliorating psychotic symptoms act as dopamine receptor blockers, in particular D2 receptor blockers, a disturbance in dopaminergic neurotransmission is proposed to play a key role in the pathogenesis of schizophrenia (Carlsson 1988; Jentsch and Roth 1999). However, there is general consensus that dopaminergic dysfunction is only part of the pathophysiology of schizophrenia. One important extension was the glutamate hypothesis of schizophrenia, which postulates an equilibrium between inhibiting dopaminergic and inhibiting glutamatergic neurons; the model of a cortico-striato-thalamo-cortical control loop integrates the glutamate hypothesis with neuroanatomical aspects on the pathophysiology of schizophrenia (Carlsson et al. 2001). A hypofunction of the glutamatergic cortico-striatal pathway, i.e., an inhibition of the inhibitory GABA neurons by hyperactivated dopaminergic receptors as well as reduction of the glutamatergic input, is associated with opening of the thalamic filter, which leads to an uncontrolled flow of sensory information to the cortex and to psychotic symptoms (Carlsson 2006).

Hypofunction of the glutamatergic neurotransmitter system as a causal mechanism in schizophrenia was first proposed due to the observation of low concentrations of glutamate in the cerebrospinal fluid (CSF) of schizophrenic patients (Kim et al. 1980). A markedly elevated ratio between glutamine and glutamate has been demonstrated in CSF of schizophrenic patients (Hashimoto et al. 2005). This finding is of special importance since it describes alterations in drug-naive patients during their first episode of schizophrenia; important confounding factors like the effect of antipsychotic medication can therefore be ruled out. Moreover, it underlines the importance of the above-described altered expression of the astrocytic enzyme glutamine synthetase. Since glutamate levels are obviously altered in schizophrenia, experimental manipulation of the glutamate receptors, especially the NMDA receptor, may lead to new insights into the neurochemical basis of schizophrenia.

Dopamine release can be challenged by amphetamine, and the challenge can be blocked by NMDA receptor antagonists. In an animal experiment, treatment with NMDA receptor antagonists leads to a marked, dose-dependent increase of amphetamine-induced dopamine release (Miller and Abercrombie 1996). This observation has been confirmed in humans: blocking of the NMDA receptor by ketamine induced a significant enhancement of amphetamine-induced dopamine release in healthy controls (Kegeles et al. 2000). The magnitude of the increase was comparable to the exaggerated response patients with schizophrenia had to amphetamine alone (Laruelle et al. 2005); in schizophrenics, the amphetamine-induced dopamine release is much higher than in healthy controls (Laruelle et al. 1996). This observation is in accordance with the view that the abnormally elevated dopamine release revealed by the amphetamine challenge in schizophrenia results from a disruption of glutamatergic neuronal systems regulating dopaminergic activity (Laruelle et al. 2005).

Phencyclidine (PCP), ketamine, and MK-801 all block the NMDA receptor complex and are associated with schizophrenia-like symptoms through hypofunction of the glutamatergic neurotransmission (Krystal et al. 1994; Olney and Farber 1995). Other substances, acting as NMDA antagonists, but not at the PCP site, also have psychotogenic properties (CPP, CPP-ene, CGS 19755; Waters et al., 1996). NMDA receptor hypofunction can explain schizophrenic positive and negative symptoms; cognitive deterioration and structural brain changes can be a consequence of NMDA receptor dysfunction (Olney and Farber 1995). Brain imaging offers the chance to selectively investigate those alterations in vivo, and there is evidence from such a study for an NMDA receptor deficit in medication-free schizophrenic patients (Pilowsky et al. 2006).

The reduced glutamatergic neurotransmission may be partly related to the enhanced expression and activity of the glutamate transporter EAAT2 and the glutamate-degrading enzyme glutamine synthetase, as described above. Besides these direct neurochemical effects, there may be an additional link between astrocytes and neurotransmission, which is based on the immune function of astrocytes.

14.2.2.2 Functional Role in Immune Response

Infection during pregnancy in mothers of offspring later developing schizophrenia has been repeatedly described (Brown et al. 2004; Buka et al. 2001; Westergaard et al. 1999) and is discussed as an explanation for the seasonality of schizophrenic births between December and May (Torrey et al. 1997). Results of a Finnish epidemiological study showed that infection of the CNS in childhood increases fivefold the risk of becoming psychotic later on (Koponen et al. 2004). A fivefold increased risk for developing psychoses later on was also observed in Brazil after a (bacterial) meningitis epidemic (Gattaz et al. 2004). Taking into account a sensitization process, an infection during early childhood is in accordance with the assumption that an infection-triggered disturbance in brain development might play a key role in the etiology of schizophrenia.

On the other hand, a persistent (chronic) infection as an etiological factor in schizophrenia has been discussed for many years. Signs of inflammation were observed in schizophrenic brains (Körschenhausen et al. 1996), and the term *mild localized chronic encephalitis* was proposed (Bechter et al. 2003). Following the hypothesis of an ongoing immune process, it would be expected that the brain volume reductions regularly found in schizophrenia are not only due to a neurodevelopmental disturbance (Jakob and Beckmann 1986), but also directly precede the first episode of schizophrenia and are further progressive. In fact, the Edinburgh High Risk Study recently showed that a marked reduction of the inferior temporal gyrus over time preceded the first onset of schizophrenia (Job et al. 2006), and the progressive loss of brain volume has repeatedly been demonstrated (e.g., Chakos et al. 2005).

Due to the characteristics of infectious agents, there are difficulties in proving a localized infection in the brain. A virus or other intracellular infectious agent may be silently hidden in cells of the lymphoid or the nervous system and be activated under certain conditions, such as stress. There is major evidence supporting pre- or perinatal exposure to infection as a risk factor for developing schizophrenia, with the main focus on influenza, rubella, measles, and herpes simplex viruses (Pearce 2001). Moreover, viral infections during childhood (Koponen et al. 2004) and even preceding the onset of the illness have been associated with schizophrenia (Leweke et al. 2004). The levels of the pro-inflammatory cytokine IL-8 were increased during the second trimester of pregnancy in the serum of those mothers whose offspring developed schizophrenia later, i.e., increased IL-8 levels were associated with an increased risk for schizophrenia in the offspring (Brown et al. 2004).

Recent research points out that not one single pathogen, but the immune response of the mother, is related to the increased risk for schizophrenia (Zuckerman and Weiner 2005).

The resident cells responsible for the immune response in the CNS are microglia cells. The role of astrocytes has been underestimated for a long time, but it is now clear that they also contribute markedly to the central nervous innate immune response (Farina et al. 2007). Mediators of astrocytes' function include cytokines like IL-6, IFN-γ, TGF-β, and TNF, and astrocytes express several Toll-like receptors (e.g., Farina et al. 2007). Following the evidence for an ongoing mild infection of the brain in schizophrenia, a chronic overactivation of astrocytes could be proposed. The above-described findings of increased S100B levels in schizophrenia may be interpreted as such an enhanced activation of astrocytes.

The question is now in which way such an ongoing enhanced activation of astrocytes could contribute to the neurotransmitter changes observed in schizophrenia.

14.2.2.3   Relationship between Immune Function and Neurotransmission through Kynurenine Pathway

As already explained in Section 14.1, the kynurenine pathway exhibits several functional links between the immune system and neurotransmission, especially including the glutamatergic, cholinergic, and serotonergic systems.

Astrocytes play a key role in the production of kynurenic acid in the CNS, because they are the main source of KYN-A (Heyes et al. 1997). Since kynurenine 3-monooxygenase (KMO), a critical enzyme in the kynurenine metabolism, is absent in human astrocytes, these cells cannot produce the intermediate 3OHK but are able to produce large amounts of KYN and KYN-A (Guillemin et al. 2001). This supports the observation that inhibition of KMO leads to an increased KYN-A production in the CNS (Chiarugi et al. 1996).

Interestingly, kynurenic acid is the only known endogenous NMDA receptor antagonist (Stone 1993). KYN-A acts both as a blocker of the glycine coagonist site of the NMDA receptor (Kessler et al. 1989) and as a noncompetitive inhibitor of the α7 nicotinic acetylcholine receptor (Hilmas et al. 2001).

The accumulation of KYN-A may lead to schizophrenic symptoms (Erhardt et al. 2003). Accordingly, increased levels of KYN-A have been observed in the CSF of schizophrenic patients (Erhardt et al. 2001a). Since most of the patients in this study were drug-naive first-episode patients, this increase could not be caused by antipsychotic treatment. At any rate, chronic drug treatment with antipsychotics does not result in an increase, but rather in a decrease of KYN-A (Ceresoli-Borroni et al. 2006).

An investigation of CNS tissue specimens in different cortical regions revealed increased KYN-A levels in schizophrenics compared to a control sample, particularly in the prefrontal cortex (Schwarcz et al. 2001). In the amygdala, a small and nonsignificant increase of KYN-A in medicated schizophrenics was observed (Miller et al. 2006). The prefrontal cortex is an area involved in the pathophysiology of schizophrenia (Andreasen et al. 1992).

Besides the effects on the NMDA receptor, KYN-A is also a potent antagonist of α7 nicotinic acetylcholine receptors (Hilmas et al. 2001). This antagonism is associated with cognitive impairment. Compared to other schizophrenic symptoms, cognitive decline is a basic disturbance in schizophrenia (Huber 1983). The effect on acetylcholine receptors is exerted at lower concentrations of KYN-A compared to the antagonism of the NMDA receptor; the affinity of KYN-A to the α7 nicotinic acetylcholine receptor is about twice as high as that to the NMDA receptor (Hilmas et al. 2001). This finding indicates that the impairment of cognitive functions is induced by lower concentrations of KYN-A, while psychotic symptoms appear only at higher concentrations of KYN-A. This view fits with the earlier onset of cognitive disturbance in schizophrenia compared to acute psychotic symptoms.

It was suggested that increased intracerebral KYN-A levels are related to enhanced dopaminergic neurotransmission. A recent study published by Robert Schwarcz's group, however, demonstrated that in the striatum, KYN-A significantly *inhibits* dopamine release (Wu et al. 2006). This effect is mediated through antagonism of the α7 nicotinic ACh receptor, while the NMDA receptor was not involved in this effect. The authors proved this effect for both acute and chronic (knockout mice) modulation of KYN-A levels. Since this effect has specifically been demonstrated for the striatum, it remains to be investigated whether KYN-A has the same effect in other brain regions.

Undoubtedly, the effects of kynurenic acid on glutamatergic and cholinergic neurotransmission are fascinating and may have high relevance for the pathophysiology and future treatment options in schizophrenia. However, the other arm of the kynurenine pathway leading to the NMDA receptor agonist quinolinic has been nearly neglected in schizophrenia research. In our opinion, future research must include this arm, because the indirect and direct neurotoxic effects of 3-OHK and quinolinic acid could be related to negative symptoms and loss of brain volume. Both are major problems in the treatment of schizophrenia, since they are related to lack of response to pharmacological treatment. The described activation of the central nervous immune response in schizophrenia clearly indicates that the "neurotoxic" arm of the kynurenine pathway may also be activated through direct activation of the key enzymes IDO and KMO by pro-inflammatory cytokines.

14.2.3 THERAPEUTIC ASPECTS

14.2.3.1 Manipulation of Neurotransmitter Systems Related to Astrocyte Function

The important role of astrocytes in D-serine metabolism is described above. D-serine and glycine are important endogenous NMDA receptor coagonists. Based on the hypothesis of NMDA receptor hypofunction, modulation of the coagonists came into the focus of schizophrenia research. Decreased plasma levels of glycine in schizophrenic patients and a correlation of glycine levels with the negative symptoms of schizophrenia were found (Sumiyoshi et al. 2004). Baseline glycine levels predicted the treatment outcome of clozapine in negative symptoms (Sumiyoshi et al. 2005). Clinical investigations targeted the glycine coagonist site of the NMDA receptor by administering the amino acids glycine or D-serine, or a glycine pro-drug such as milacemide (Tamminga et al. 1992). Some of these studies have yielded positive results, particularly against the schizophrenic deficit syndrome (Heresco-Levy et al. 1999). In an add-on trial administering D-serine together with olanzapine or risperidone, a beneficial effect on positive, negative, and cognitive symptoms of schizophrenia could be demonstrated (Heresco-Levy et al. 2005). Although these findings give evidence for using the NMDA receptor coagonists as therapeutic agents in schizophrenia, there are some data showing nephrotoxic side effects of the therapeutic administration of D-serine (Carone and Ganote 1975).

The direct effects of antipsychotic drugs on astrocyte function, directly related to neurotransmitter systems, have rarely been investigated. Segnitz and colleagues recently reported that chronic administration of the atypical antipsychotic aripiprazole induces an altered expression of glutamate transporter genes in rats. While the expression of astrocytic transporter EAAT1 was markedly suppressed in several brain regions, EAAT2 expression was dose- and time-dependently modulated by aripiprazole. Those results indicate a marked and distinct effect of antipsychotic drugs on astrocytes, which should be followed up with more detailed investigations.

14.2.3.2 Manipulation of Immune-System-Related Changes

14.2.3.2.1 Effects of Antipsychotic Drugs on the Immune System

In vitro studies show that the blunted IFN-γ production becomes normalized after therapy with neuroleptics (Wilke et al. 1996). Accordingly, numbers of memory cells, which are the main source of IFN-γ, as well as levels of soluble IL-2 receptor as a marker of T cell activation, are increasing during antipsychotic therapy (Müller et al. 1997a, 1997b). Reduced sICAM-1 levels show a significant increase during short-term antipsychotic therapy (Schwarz et al. 2000), and the leukocyte function antigen-1 (LFA-1), the ligand of ICAM-1, shows a significantly increased expression during antipsychotic therapy (Müller et al. 1999). Moreover, the blunted reaction to vaccination with *Salmonella* Typhi was not observed in patients medicated with antipsychotics (Ozek et al. 1971). On the other hand, there are also several studies indicating an anti-inflammatory effect of antipsychotic drugs (for review see Drzyzga et al. 2006). Several studies point out that antipsychotic therapy is accompanied by a functional decrease of the IL-6 system (Maes et al. 1997; Müller et al. 2000). A recent study described the reduction of previously increased IL-6 and TNF levels in schizophrenic patients after a 6-week period of antipsychotic medication, which was accompanied by altered tryptophan breakdown to kynurenine (Kim et al. 2009).

These findings of immune changes in the peripheral immune system may reflect similar changes within the CNS. Regarding the direct effect of antipsychotic drugs on the central nervous immune system and especially on astrocytes, only few studies are available. Some of them have investigated the in vitro effect of antipsychotic drugs on cell lines of astrocytic origin (Steiner et al. 2010; Quincozes-Santos et al. 2009).

14.2.3.2.2 Use of Anti-Inflammatory Drugs

In additional to the above-described immunological mechanism, selective cyclooxygenase-2 (COX-2) inhibitors reduce KYN-A levels by a prostaglandin-mediated mechanism (Schwieler et al. 2005a). COX inhibition provokes differential effects on kynurenine metabolism: while COX-1 inhibitors increase the levels of KYN-A, COX-2 inhibitors decrease them. Therefore, psychotic symptoms and cognitive dysfunctions observed during therapy with COX-1 inhibitors were assigned to the COX-1-mediated increase of KYN-A (Schwieler et al. 2005b).

PGE_2 induces the production of IL-6, a cytokine that is consistently described to be increased in schizophrenia. PGE_2 levels themselves are not well studied in schizophrenia, but increased levels of PGE_2 have been reported (Kaiya et al. 1989). Moreover, increased CNS expression of COX-2 was found in schizophrenia (Das and Khan 1998), although the issue remains controversial (Yokota et al. 2004).

In a prospective, randomized, double-blind study of therapy with the COX-2 inhibitor celecoxib add-on to risperidone in acute exacerbation of schizophrenia, a therapeutic effect of celecoxib was observed (Müller et al. 2002). Immunologically, an increase of the type 1 immune response was found in the celecoxib treatment group (Müller et al. 2004b). The clinical effect of COX-2 inhibition was especially pronounced regarding cognition in schizophrenia

(Müller et al. 2005). The finding of a clinical advantage of COX-2 inhibition, however, could not be replicated in a second study. Further analysis of the data revealed that the outcome depended on the duration of the disease (Müller et al. 2004b). The efficacy of therapy with a COX-2 inhibitor seems most pronounced in the first years of the schizophrenic disease process. This observation is in accordance with results from animal studies showing that the effects of COX-2 inhibition on cytokines, hormones, and particularly on behavioral symptoms are dependent on the duration of the preceding changes and the time point of application of the COX-2 inhibitor (Casolini et al. 2002). Thus, a point of no return for therapeutic effects regarding pathological changes during an inflammatory process has to be postulated.

Regarding the role of the inflammatory process in schizophrenia and possibly other psychiatric disorders, anti-inflammatory therapy should be examined further (Müller et al. 2004a); COX-2 inhibition is one option, among others. Therapeutic research, however, has to consider different levels and different mechanisms for therapeutic targets in the neuroimmune system and the dopaminergic-glutamatergic neurotransmission circuits, including the kynurenine pathway of the tryptophan metabolism.

14.2.3.2.3 *Indirect Manipulation through the Kynurenine Pathway*

In recent years, drugs acting to elevate endogenous KYN-A in the CNS have been identified. One of these substances is PNU 156561A, an inhibitor of KMO. This substance enables studies of the effects of increased endogenous KYN-A levels in animals (Speciale et al. 1996). The effects were similar to the effects observed after administration of MK-801 or PCP: in particular, dopaminergic neurons in the midbrain showed increased activity (Erhardt et al. 2001b). Robert Schwarcz, one of pioneers in this field, proposed to test not only KMO inhibitors, but also inhibitors of the key enzymes TDO and IDO (Wonodi and Schwarcz 2010).

14.3 FUTURE ASPECTS

There is evidence of morphological, biochemical, and functional changes in astrocyte function in both major depression and schizophrenia. It is difficult to assign these changes either causes or effects of psychiatric major depression or schizophrenia, since such changes could be the causes that induce the illnesses and then in turn are induced by the illnesses again. Among the possible neurochemical mechanisms related to astrocytes changes, the role of kynurenines is plausible. The kynurenines are detectable in peripheral blood, and therefore represent possible biomarkers to monitor the response to treatment or to predict the remission and relapse or recurrence. Moreover, changes in those kynurenines follow changes in the immune system, so that correction of pro-inflammatory activity or impaired resolution of inflammation after inflammatory episodes may prevent the negative effects on astrocyte functions and possible psychiatric consequences. Therefore, based on this mechanism, further studies should be carried out to detect early diagnostic or prognostic possibilities and to design new preventive or therapeutic strategies.

REFERENCES

Alonso, J., M. C. Angermeyer, S. Bernert, R. Bruffaerts, T. S. Brugha, H. Bryson, G. de Girolamo, R. Graaf, K. Demyttenaere, I. Gasquet, J. M. Haro, S. J. Katz, R. C. Kessler, V. Kovess, J. P. Lepine, J. Ormel, G. Polidori, L. J. Russo, G. Vilagut, J. Almansa, S. Arbabzadeh-Bouchez, J. Autonell, M. Bernal, M. A. Buist-Bouwman, M. Codony, A. Domingo-Salvany, M. Ferrer, S. S. Joo, M. Martinez-Alonso, H. Matschinger, F. Mazzi, Z. Morgan, P. Morosini, C. Palacin, B. Romera, N. Taub, and W. A. Vollebergh. (2004). Prevalence of mental disorders in Europe: results from the European Study of the Epidemiology of Mental Disorders (ESEMeD) project. *Acta Psychiatr Scand Suppl* 420:21–27.

Anderson, C. M., and M. Nedergaard. (2003). Astrocyte-mediated control of cerebral microcirculation. *Trends Neurosci* 26 (7):340–44; author reply, 344–45.

Andreasen, N. C., K. Rezai, R. Alliger, V. W. Swayze, M. Flaum, P. Kirchner, G. Cohen, and D. S. O'Leary. (1992). Hypofrontality in neuroleptic-naive patients and in patients with chronic schizophrenia. Assessment with xenon 133 single-photon emission computed tomography and the Tower of London. *Arch Gen Psychiatry* 49:943–58.

Angulo, M. C., A. S. Kozlov, S. Charpak, and E. Audinat. (2004). Glutamate released from glial cells synchronizes neuronal activity in the hippocampus. *J Neurosci* 24 (31):6920–27.

Anisman, H., A. V. Ravindran, J. Griffiths, and Z. Merali. (1999). Endocrine and cytokine correlates of major depression and dysthymia with typical or atypical features. *Mol Psychiatry* 4 (2):182–88.

Arolt, V., M. Peters, A. Erfurth, M. Wiesmann, U. Missler, S. Rudolf, H. Kirchner, and M. Rothermundt. (2003). S100B and response to treatment in major depression: a pilot study. *Eur Neuropsychopharmacol* 13 (4):235–39.

Aschner, M. (2000). Neuron-astrocyte interactions: implications for cellular energetics and antioxidant levels. *Neurotoxicology* 21 (6):1101–7.

Asnis, G. M., E. J. Sachar, U. Halbreich, R. S. Nathan, H. Novacenko, and L. C. Ostrow. (1981). Cortisol secretion in relation to age in major depression. *Psychosom Med* 43 (3):235–42.

Auer, D. P., B. Putz, E. Kraft, B. Lipinski, J. Schill, and F. Holsboer. (2000). Reduced glutamate in the anterior cingulate cortex in depression: an *in vivo* proton magnetic resonance spectroscopy study. *Biol Psychiatry* 47 (4):305–13.

Bechter, K., V. Schreiner, S. Herzog, N. Breitinger, K. H. Wollinsky, H. Brinkmeier, P. Aulkemeyer, F. Weber, and R. Schuttler. (2003). [CSF filtration as experimental therapy in therapy resistant psychoses in Borna disease virus-seropositive patients]. *Psychiatr Prax* 30: 216–20.

Bender, D. A., and G. M. McCreanor. (1985). Kynurenine hydroxylase: a potential rate-limiting enzyme in tryptophan metabolism. *Biochem Soc Trans* 13 (2):441–43.

Bengtsson, B.O., J. Zhu, L. H. Thorell, T. Olsson, H. Link, and J. Walinder. (1992). Effects of zimeldine and its metabolites, clomipramine, imipramine and maprotiline in experimental allergic neuritis in Lewis rats. *J. Neuroimmunol.* 39: 109–122.

Bernstein, H. G., J. Steiner, and B. Bogerts. (2009). Glial cells in schizophrenia: pathophysiological significance and possible consequences for therapy. *Expert Rev Neurother* 9:1059–71.

Bowley, M. P., W. C. Drevets, D. Ongur, and J. L. Price. (2002). Low glial numbers in the amygdala in major depressive disorder. *Biol Psychiatry* 52 (5):404–12.

Brown, A. S., M. D. Begg, S. Gravenstein, C. A. Schaefer, R. J. Wyatt, M. Bresnahan, V. P. Babulas, and E. S. Susser. (2004). Serologic evidence of prenatal influenza in the etiology of schizophrenia. *Arch Gen Psychiatry* 61:774–80.

Bruneau, E. G., R. E. McCullumsmith, V. Haroutunian, K. L. Davis, and J. H. Meador-Woodruff. (2005). Increased expression of glutaminase and glutamine synthetase mRNA in the thalamus in schizophrenia. *Schizophr Res* 75:27–34.

Brunello, N., S. Alboni, G. Capone, C. Benatti, J. M. Blom, F. Tascedda, P. Kriwin, and J. Mendlewicz. (2006). Acetylsalicylic acid accelerates the antidepressant effect of fluoxetine in the chronic escape deficit model of depression. *Int. Clin. Psychopharmacol* 21: 219–225.

Bruton, C. J., T. J. Crow, C. D. Frith, E. C. Johnstone, D. G. Owens, and G. W. Roberts. (1990). Schizophrenia and the brain: a prospective clinico-neuropathological study. *Psychol Med* 20:285–304.

Buka, S. L., M. T. Tsuang, E. F. Torrey, M. A. Klebanoff, D. Bernstein, and R. H. Yolken. (2001). Maternal infections and subsequent psychosis among offspring. *Arch Gen Psychiatry* 58:1032–37.

Bushong, E. A., M. E. Martone, Y. Z. Jones, and M. H. Ellisman. (2002). Protoplasmic astrocytes in CA1 stratum radiatum occupy separate anatomical domains. *J Neurosci* 22 (1):183–92.

Cabras, S., F. Saba, C. Reali, M. L. Scorciapino, A. Sirigu, G. Talani, G. Biggio, and V. Sogos. (2010). Antidepressant imipramine induces human astrocytes to differentiate into cells with neuronal phenotype. *Int. J. Neuropsychopharmacol* 13: 603–615.

Cao, C., K. Matsumura, M. Ozaki, and Y. Watanabe. (1999). Lipopolysaccharide injected into the cerebral ventricle evokes fever through induction of cyclooxygenase-2 in brain endothelial cells. *J. Neurosci.* 19: 716–725

Carlin, J. M., E. C. Borden, P. M. Sondel, and G. I. Byrne. (1987). Biologic-response-modifier-induced indoleamine 2,3-dioxygenase activity in human peripheral blood mononuclear cell cultures. *J Immunol* 139 (7):2414–18.

Carlin, J. M., E. C. Borden, P. M. Sondel, and G. I. Byrne. (1989). Interferon-induced indoleamine 2,3-dioxygenase activity in human mononuclear phagocytes. *J Leukoc Biol* 45 (1):29–34.

Carlsson, A. (1988). The current status of the dopamine hypothesis of schizophrenia. *Neuropsychopharmacology* 1:179–86.

Carlsson, A. (2006). The neurochemical circuitry of schizophrenia. *Pharmacopsychiatry* 39 (Suppl 1):S10–S14.

Carlsson, A., N. Waters, S. Holm-Waters, J. Tedroff, M. Nilsson, and M. L. Carlsson. (2001). Interactions between monoamines, glutamate, and GABA in schizophrenia: new evidence. *Annu Rev Pharmacol Toxicol* 41:237–60.

Carone, F. A. and C. E. Ganote. (1975). D-serine nephrotoxicity. The nature of proteinuria, glucosuria, and aminoaciduria in acute tubular necrosis. *Arch Pathol* 99:658–62.

Casolini, P., A. Catalani, A. R. Zuena, and L. Angelucci. (2002). Inhibition of COX-2 reduces the age-dependent increase of hippocampal inflammatory markers, corticosterone secretion, and behavioral impairments in the rat. *J Neurosci Res* 68:337–43.

Ceresoli-Borroni, G., A. Rassoulpour, H. Q. Wu, P. Guidetti, and R. Schwarcz. (2006). Chronic neuroleptic treatment reduces endogenous kynurenic acid levels in rat brain. *J Neural Transm* 113 (10):1355–65.

Chakos, M. H., S. A. Schobel, H. Gu, G. Gerig, D. Bradford, C. Charles, and J. A. Lieberman. (2005). Duration of illness and treatment effects on hippocampal volume in male patients with schizophrenia. *Br J Psychiatry* 186:26–31.

Chana, G., S. Landau, C. Beasley, I. P. Everall, and D. Cotter. (2003). Two-dimensional assessment of cytoarchitecture in the anterior cingulate cortex in major depressive disorder, bipolar disorder, and schizophrenia: evidence for decreased neuronal somal size and increased neuronal density. *Biol Psychiatry* 53 (12):1086–98.

Chiarugi, A., M. Calvani, E. Meli, E. Traggiai, and F. Moroni. (2001). Synthesis and release of neurotoxic kynurenine metabolites by human monocyte-derived macrophages. *J Neuroimmunol* 120 (1–2):190–98.

Chiarugi, A., R. Carpenedo, and F. Moroni. (1996). Kynurenine disposition in blood and brain of mice: effects of selective inhibitors of kynurenine hydroxylase and of kynureninase. *J Neurochem* 67:692–98.

Cohen, M. R., D. Pickar, I. Extein, M. S. Gold, and D. R. Sweeney. (1984). Plasma cortisol and beta-endorphin immunoreactivity in nonmajor and major depression. *Am J Psychiatry* 141 (5):628–32.

Connor, T. J., and B. E. Leonard. (1998). Depression, stress and immunological activation: the role of cytokines in depressive disorders. *Life Sci* 62 (7):583–606.

Cotter, D., D. Mackay, G. Chana, C. Beasley, S. Landau, and I. P. Everall. (2002). Reduced neuronal size and glial cell density in area 9 of the dorsolateral prefrontal cortex in subjects with major depressive disorder. *Cereb Cortex* 12 (4):386–94.

Cotter, D., D. Mackay, S. Landau, R. Kerwin, and I. Everall. (2001a). Reduced glial cell density and neuronal size in the anterior cingulate cortex in major depressive disorder. *Arch Gen Psychiatry* 58 (6):545–53.

Cotter, D. R., C. M. Pariante, and I. P. Everall. (2001b). Glial cell abnormalities in major psychiatric disorders: the evidence and implications. *Brain Res Bull* 55:585–95.

Czeh, B., M. Simon, B. Schmelting, C. Hiemke, and E. Fuchs. (2006). Astroglial plasticity in the hippocampus is affected by chronic psychosocial stress and concomitant fluoxetine treatment. *Neuropsychopharmacology* 31: 1616–1626.

Damadzic, R., L. B. Bigelow, L. S. Krimer, D. A. Goldenson, R. C. Saunders, J. E. Kleinman, and M. M. Herman. (2001). A quantitative immunohistochemical study of astrocytes in the entorhinal cortex in schizophrenia, bipolar disorder and major depression: absence of significant astrocytosis. *Brain Res Bull* 55 (5):611–18.

Das, I., and N. S. Khan. (1998). Increased arachidonic acid induced platelet chemiluminescence indicates cyclooxygenase overactivity in schizophrenic subjects. *Prostaglandins Leukot Essent Fatty Acids* 58:165–68.

Davis, S., A. Thomas, R. Perry, A. Oakley, R. N. Kalaria, and J. T. O'Brien. (2002). Glial fibrillary acidic protein in late life major depressive disorder: an immunocytochemical study. *J Neurol Neurosurg Psychiatry* 73 (5):556–60.

De Keyser J., J. P. Mostert, and M. W. Koch. (2008). Dysfunctional astrocytes as key players in the pathogenesis of central nervous system disorders. *J Neurol Sci* 267:3–16.

Drzyzga, L., E. Obuchowicz, A. Marcinowska, and Z. S. Herman. (2006). Cytokines in schizophrenia and the effects of antipsychotic drugs. *Brain Behav Immun* 20:532–45.

Erhardt, S., K. Blennow, C. Nordin, E. Skogh, L. H. Lindstrom, and G. Engberg. (2001a). Kynurenic acid levels are elevated in the cerebrospinal fluid of patients with schizophrenia. *Neurosci Lett* 313:96–98.

Erhardt, S., H. Oberg, J. M. Mathe, and G. Engberg. (2001b). Pharmacological elevation of endogenous kynurenic acid levels activates nigral dopamine neurons. *Amino Acids* 20:353–62.

Erhardt, S., L. Schwieler, and G. Engberg. (2003). Kynurenic acid and schizophrenia. *Adv Exp Med Biol* 527:155–65.

Farina, C., F. Aloisi, and E. Meinl. (2007). Astrocytes are active players in cerebral innate immunity. *Trends Immunol* 28:138–45.

Fatemi, S. H., J. A. Laurence, M. Araghi-Niknam, J. M. Stary, S. C. Schulz, S. Lee, and Gottesman II. (2004). Glial fibrillary acidic protein is reduced in cerebellum of subjects with major depression, but not schizophrenia. *Schizophr Res* 69 (2–3):317–23.

Fellin, T., O. Pascual, S. Gobbo, T. Pozzan, P. G. Haydon, and G. Carmignoto. (2004). Neuronal synchrony mediated by astrocytic glutamate through activation of extrasynaptic NMDA receptors. *Neuron* 43 (5):729–43.

Fiacco, T. A., and K. D. McCarthy. (2004). Intracellular astrocyte calcium waves *in situ* increase the frequency of spontaneous AMPA receptor currents in CA1 pyramidal neurons. *J Neurosci* 24 (3):722–32.

Frommberger, U.H., J. Bauer, P. Haselbauer, A. Fraulin, D. Riemann, and M. Berger. (1997). Interleukin-6-(IL-6) plasma levels in depression and schizophrenia: comparison between the acute state and after remission. *Eur Arch Psychiatry Clin Neurosci* 247: 228–233.

Gal, E. M., and A. D. Sherman. (1980). L-kynurenine: its synthesis and possible regulatory function in brain. *Neurochem Res* 5 (3):223–39.

Garbuglia, M., M. Verzini, R. R. Rustandi, D. Osterloh, D. J. Weber, V. Gerke, and R. Donato. (1999). Role of the C-terminal extension in the interaction of S100A1 with GFAP, tubulin, the S1. *Biochem Biophys Res Commun* 254:36–41.

Gattaz, W. F., A. L. Abrahao, and R. Foccacia. (2004). Childhood meningitis, brain maturation and the risk of psychosis. *Eur Arch Psychiatry Clin Neurosci* 254:23–26.

Grant, R. S., H. Naif, M. Espinosa, and V. Kapoor. (2000). IDO induction in IFN-gamma activated astroglia: a role in improving cell viability during oxidative stress. *Redox Rep* 5 (2–3):101–4.

Grant, R. S., and V. Kapoor. (1998). Murine glial cells regenerate NAD, after peroxide-induced depletion, using either nicotinic acid, nicotinamide, or quinolinic acid as substrates. *J Neurochem* 70 (4):1759–63.

Greden, J. F. (2001a). The burden of disease for treatment-resistant depression. *J Clin Psychiatry* 62 (Suppl 16):26–31.

Greden, J. F. (2001b). The burden of recurrent depression: causes, consequences, and future prospects. *J Clin Psychiatry* 62 (Suppl 22):5–9.

Guillemin, G. J., S. J. Kerr, G. A. Smythe, D. G. Smith, V. Kapoor, P. J. Armati, J. Croitoru, and B. J. Brew. (2001). Kynurenine pathway metabolism in human astrocytes: a paradox for neuronal protection. *J Neurochem* 78 (4):842–53.

Guillemin, G. J., D. G. Smith, S. J. Kerr, G. A. Smythe, V. Kapoor, P. J. Armati, and B. J. Brew. (2000). Characterisation of kynurenine pathway metabolism in human astrocytes and implications in neuropathogenesis. *Redox Rep* 5 (2–3):108–11.

Guillemin, G. J., G. Smythe, O. Takikawa, and B. J. Brew. (2005). Expression of indoleamine 2,3-dioxygenase and production of quinolinic acid by human microglia, astrocytes, and neurons. *Glia* 49 (1):15–23.

Hamidi, M., W. C. Drevets, and J. L. Price. (2004). Glial reduction in amygdala in major depressive disorder is due to oligodendrocytes. *Biol Psychiatry* 55 (6):563–69.

Hashimoto, K., G. Engberg, E. Shimizu, C. Nordin, L. H. Lindstrom, and M. Iyo. (2005). Elevated glutamine/glutamate ratio in cerebrospinal fluid of first episode and drug naive schizophrenic patients. *BMC Psychiatry* 5:6.

Hayaishi, O. (1976). Properties and function of indoleamine 2,3-dioxygenase. *J Biochem (Tokyo)* 79 (4):13P–21P.

Hayaishi, O. (1980). *Biochemical and medical aspects of tryptophan metabolism*, ed. O. Hayaishi, R. Ishimura, and R. Kideo. Amsterdam: Elsevier/North-Holland Biomedical Press.

Heresco-Levy, U., D. C. Javitt, R. Ebstein, A. Vass, P. Lichtenberg, G. Bar, S. Catinari, and M. Ermilov. (2005). D-serine efficacy as add-on pharmacotherapy to risperidone and olanzapine for treatment-refractory schizophrenia. *Biol Psychiatry* 57:577–85.

Heresco-Levy, U., D. C. Javitt, M. Ermilov, C. Mordel, G. Silipo, and M. Lichtenstein. (1999). Efficacy of high-dose glycine in the treatment of enduring negative symptoms of schizophrenia. *Arch Gen Psychiatry* 56:29–36.

Hertz, L., R. Dringen, A. Schousboe, and S. R. Robinson. (1999). Astrocytes: glutamate producers for neurons. *J Neurosci Res* 57 (4):417–28.

Hetzel, G., O. Moeller, S. Evers, A. Erfurth, G. Ponath, V. Arolt, and M. Rothermundt. (2005). The astroglial protein S100B and visually evoked event-related potentials before and after antidepressant treatment. *Psychopharmacology (Berl)* 178 (2–3):161–66.

Heyes, M. P., C. L. Achim, C. A. Wiley, E. O. Major, K. Saito, and S. P. Markey. (1996). Human microglia convert l-tryptophan into the neurotoxin quinolinic acid. *Biochem J* 320 (Pt 2):595–97.

Heyes, M. P., C. Y. Chen, E. O. Major, and K. Saito. (1997). Different kynurenine pathway enzymes limit quinolinic acid formation by various human cell types. *Biochem J* 326:351–56.

Heyes, M. P., K. Saito, E. O. Major, S. Milstien, S. P. Markey, and J. H. Vickers. (1993). A mechanism of quinolinic acid formation by brain in inflammatory neurological disease. Attenuation of synthesis from L-tryptophan by 6-chlorotryptophan and 4-chloro-3-hydroxyanthranilate. *Brain* 116 (Pt 6):1425–50.

Hilmas, C., E. F. Pereira, M. Alkondon, A. Rassoulpour, R. Schwarcz, and E. X. Albuquerque. (2001). The brain metabolite kynurenic acid inhibits alpha7 nicotinic receptor activity and increases non-alpha7 nicotinic receptor expression: physiopathological implications. *J Neurosci* 21:7463–73.

Hu, B., B. D. Hissong, and J. M. Carlin. (1995). Interleukin-1 enhances indoleamine 2,3-dioxygenase activity by increasing specific mRNA expression in human mononuclear phagocytes. *J Interferon Cytokine Res* 15 (7):617–24.

Hu, F., X. Wang, T. W. Pace, H. Wu, and A. H. Miller. (2005). Inhibition of COX-2 by celecoxib enhances glucocorticoid receptor function. *Mol Psychiatry* 10: 426–428.

Huber, G. (1983). Das Konzept substratnaher Basissymptome und seine Bedeutung für Theorie und Therapie schizophrener Erkrankungen [The concept of substrate-close basic symptoms and its significance for the theory and therapy of schizophrenic diseases]. *Nervenarzt* 54:23–32.

Iwata, M., Y. Shirayama, H. Ishida, G. I. Hazama, and K. Nakagome. (2010). Hippocampal astrocytes are necessary for antidepressant treatment of learned helplessness rats. *Hippocampus*.

Jakob, H., and H. Beckmann. (1986). Prenatal developmental disturbances in the limbic allocortex in schizophrenics. *J Neural Transm* 65:303–26.

Jang, B. S., H. Kim, S. W. Lim, K. W. Jang, and D. K. Kim. (2008). Serum S100B levels and major depressive disorder: its characteristics and role in antidepressant response. *Psychiatry Investig* 5 (3):193–98.

Jentsch, J. D., and R. H. Roth. (1999). The neuropsychopharmacology of phencyclidine: from NMDA receptor hypofunction to the dopamine hypothesis of schizophrenia. *Neuropsychopharmacology* 20:201–25.

Job, D. E., H. C. Whalley, A. M. McIntosh, D. G. Owens, E. C. Johnstone, and S. M. Lawrie. (2006). Grey matter changes can improve the prediction of schizophrenia in subjects at high risk. *BMC Med* 4:29.

Johnston-Wilson, N. L., C. D. Sims, J. P. Hofmann, L. Anderson, A. D. Shore, E. F. Torrey, and R. H. Yolken. (2000). Disease-specific alterations in frontal cortex brain proteins in schizophrenia, bipolar disorder, and major depressive disorder. The Stanley Neuropathology Consortium. *Mol Psychiatry* 5 (2):142–49.

Jourdain, P., L. H. Bergersen, K. Bhaukaurally, P. Bezzi, M. Santello, M. Domercq, C. Matute, F. Tonello, V. Gundersen, and A. Volterra. (2007). Glutamate exocytosis from astrocytes controls synaptic strength. *Nat Neurosci* 10 (3):331–39.

Kaestner, F., M. Hettich, M. Peters, W. Sibrowski, G. Hetzel, G. Ponath, V. Arolt, U. Cassens, and M. Rothermundt. (2005). Different activation patterns of proinflammatory cytokines in melancholic and non-melancholic major depression are associated with HPA axis activity. *J Affect Disord* 87 (2–3):305–11.

Kaiya, H., M. Uematsu, M. Ofuji, A. Nishida, K. Takeuchi, M. Nozaki, and E. Idaka. (1989). Elevated plasma prostaglandin E2 levels in schizophrenia. *J Neural Transm* 77:39–46.

Kegeles, L. S., A. bi-Dargham, Y. Zea-Ponce, J. Rodenhiser-Hill, J. J. Mann, R. L. Van Heertum, T. B. Cooper, A. Carlsson, and M. Laruelle. (2000). Modulation of amphetamine-induced striatal dopamine release by ketamine in humans: implications for schizophrenia. *Biol Psychiatry* 48:627–40.

Kenis, G. and M. Maes. (2002). Effects of antidepressants on the production of cytokines. *Int. J. Neuropsychopharmacol.* 5: 401–412.

Kessler, M., T. Terramani, G. Lynch, and M. Baudry. (1989). A glycine site associated with N-methyl-D-aspartic acid receptors: characterization and identification of a new class of antagonists. *J Neurochem* 52:1319–28.

Kim, J. P., and D. W. Choi. (1987). Quinolinate neurotoxicity in cortical cell culture. *Neuroscience* 23 (2):423–32.

Kim, J. S., H. H. Kornhuber, W. Schmid-Burgk, and B. Holzmuller. (1980). Low cerebrospinal fluid glutamate in schizophrenic patients and a new hypothesis on schizophrenia. *Neurosci Lett* 20:379–82.

Kim, Y. K., A. M. Myint, R. Verkerk, S. Scharpe, H. Steinbusch, and B. Leonard. (2009). Cytokine changes and tryptophan metabolites in medication-naive and medication-free schizophrenic patients. *Neuropsychobiology* 59:123–29.

Kim, Y. K., I. B. Suh, H. Kim, C. S. Han, C. S. Lim, S. H. Choi, and J. Licinio. (2002). The plasma levels of interleukin-12 in schizophrenia, major depression, and bipolar mania: effects of psychotropic drugs. *Mol Psychiatry* 7 (10):1107–14.

Kolomeets, N. S., and N. Uranova. (2010). Ultrastructural abnormalities of astrocytes in the hippocampus in schizophrenia and duration of illness: a postortem morphometric study. *World J Biol Psychiatry* 11:282–92.

Koponen, H., P. Rantakallio, J. Veijola, P. Jones, J. Jokelainen, and M. Isohanni. (2004). Childhood central nervous system infections and risk for schizophrenia. *Eur Arch Psychiatry Clin Neurosci* 254:9–13.

Körschenhausen, D. A., H. J. Hampel, M. Ackenheil, R. Penning, and N. Müller. (1996). Fibrin degradation products in post mortem brain tissue of schizophrenics: a possible marker for underlying inflammatory processes. *Schizophr Res* 19:103–9.

Krystal, J. H., L. P. Karper, J. P. Seibyl, G. K. Freeman, R. Delaney, J. D. Bremner, G. R. Heninger, M. B. Bowers Jr., and D. S. Charney. (1994). Subanesthetic effects of the noncompetitive NMDA antagonist, ketamine, in humans. Psychotomimetic, perceptual, cognitive, and neuroendocrine responses. *Arch Gen Psychiatry* 51:199–214.

Lanquillon, S., J. C. Krieg, U. Bening-Abu-Shach, and H. Vedder. (2000). Cytokine production and treatment response in major depressive disorder. *Neuropsychopharmacology* 22 (4):370–79.

Laruelle, M., A. bi-Dargham, C. H. van Dyck, R. Gil, C. D. D'Souza, J. Erdos, E. McCance, W. Rosenblatt, C. Fingado, S. S. Zoghbi, R. M. Baldwin, J. P. Seibyl, J. H. Krystal, D. S. Charney, and R. B. Innis. (1996). Single photon emission computerized tomography imaging of amphetamine-induced dopamine release in drug-free schizophrenic subjects. *Proc Natl Acad Sci USA* 93:9235–40.

Laruelle, M., W. G. Frankle, R. Narendran, L. S. Kegeles, and A. bi-Dargham. (2005). Mechanism of action of antipsychotic drugs: from dopamine D(2) receptor antagonism to glutamate NMDA facilitation. *Clin Ther* 27 (Suppl A):S16–24.

Leweke, F. M., C. W. Gerth, D. Koethe, J. Klosterkotter, I. Ruslanova, B. Krivogorsky, E. F. Torrey, and R. H. Yolken. (2004). Antibodies to infectious agents in individuals with recent onset schizophrenia. *Eur Arch Psychiatry Clin Neurosci* 254:4–8.

Li, B., S. Zhang, M. Li, L. Hertz, and L. Peng. (2009). Chronic treatment of astrocytes with therapeutically relevant fluoxetine concentrations enhances cPLA2 expression secondary to 5-HT2B-induced, transactivation-mediated ERK1/2 phosphorylation. *Psychopharmacology (Berl)* 207: 1–12.

Lin, S. C., T. Maruta, D. C. Newman, and P. C. Kao. (1986). Plasma levels of cortisol, corticotropin, and beta-endorphin in patients with major depression. *J Clin Psychiatry* 47 (8):413–14.

Liu, Y., D. C. Buck, and K. A. Neve. (2008). Novel interaction of the dopamine D2 receptor and the Ca2+ binding protein S100B: role in D2 receptor function. *Mol Pharmacol* 74:371–78.

Maes, M., E. Bosmans, G. Kenis, R. De Jong, R. S. Smith, and H. Y. Meltzer. (1997). *In vivo* immunomodulatory effects of clozapine in schizophrenia. *Schizophr Res* 26:221–25.

Maes, M., C. Song, A. H. Lin, S. Bonaccorso, G. Kenis, R. De Jongh, E. Bosmans, and S. Scharpe. (1999). Negative immunoregulatory effects of antidepressants: inhibition of interferon-gamma and stimulation of interleukin-10 secretion. *Neuropsychopharmacology* 20: 370–379.

Mathers, C., and Loncar, D. (2005). *Updated projection of global mortality and burden of disease, 2002–2030: data sources, methods and results.* Geneva: WHO.

Mellor, A. L., and D. H. Munn. (1999). Tryptophan catabolism and T-cell tolerance: immunosuppression by starvation? *Immunol Today* 20 (10):469–73.

Mendlewicz J., P. Kriwin, P. Oswald, D. Souery, S. Alboni, N. Brunello. (2006). Shortened onset of action of antidepressants in major depression using acetylsalicylic acid augmentation: a pilot open-label study. *Int Clin Psychopharmacol* 21(4): 227–31.

Miguel-Hidalgo, J. J., C. Baucom, G. Dilley, J. C. Overholser, H. Y. Meltzer, C. A. Stockmeier, and G. Rajkowska. (2000). Glial fibrillary acidic protein immunoreactivity in the prefrontal cortex distinguishes younger from older adults in major depressive disorder. *Biol Psychiatry* 48 (8):861–73.

Miguel-Hidalgo, J. J., R. Waltzer, A. A. Whittom, M. C. Austin, G. Rajkowska, and C. A. Stockmeier. (2010). Glial and glutamatergic markers in depression, alcoholism, and their comorbidity. *J Affect Disord* 127(1–3): 230–240.

Mikova, O., R. Yakimova, E. Bosmans, G. Kenis, and M. Maes. (2001). Increased serum tumor necrosis factor alpha concentrations in major depression and multiple sclerosis. *Eur Neuropsychopharmacol* 11 (3):203–8.

Miller, C. L., I. C. Llenos, J. R. Dulay, M. M. Barillo, R. H. Yolken, and S. Weis. (2004). Expression of the kynurenine pathway enzyme tryptophan 2,3-dioxygenase is increased in the frontal cortex of individuals with schizophrenia. *Neurobiol Dis* 15 (3):618–29.

Miller, C. L., I. C. Llenos, J. R. Dulay, and S. Weis. (2006). Upregulation of the initiating step of the kynurenine pathway in postmortem anterior cingulate cortex from individuals with schizophrenia and bipolar disorder. *Brain Res* 1073–74:25–37.

Miller, D. W., and E. D. Abercrombie. (1996). Effects of MK-801 on spontaneous and amphetamine-stimulated dopamine release in striatum measured with *in vivo* microdialysis in awake rats. *Brain Res Bull* 40:57–62.

Miller, D. B., and J. P. O'Callaghan. (2005). Depression, cytokines, and glial function. *Metabolism* 54 (5 Suppl 1):33–38.

Mirza, Y., J. Tang, A. Russell, S. P. Banerjee, R. Bhandari, J. Ivey, M. Rose, G. J. Moore, and D. R. Rosenberg. (2004). Reduced anterior cingulate cortex glutamatergic concentrations in childhood major depression. *J Am Acad Child Adolesc Psychiatry* 43 (3):341–48.

Mothet, J. P., L. Pollegioni, G. Ouanounou, M. Martineau, P. Fossier, and G. Baux. (2005). Glutamate receptor activation triggers a calcium-dependent and SNARE protein-dependent release of the gliotransmitter D-serine. *Proc Natl Acad Sci USA* 102 (15):5606–11.

Muller, M. B., P. J. Lucassen, A. Yassouridis, W. J. Hoogendijk, F. Holsboer, and D. F. Swaab. (2001). Neither major depression nor glucocorticoid treatment affects the cellular integrity of the human hippocampus. *Eur J Neurosci* 14 (10):1603–12.

Müller, N., M. Empl, M. Riedel, M. Schwarz, and M. Ackenheil. (1997a). Neuroleptic treatment increases soluble IL-2 receptors and decreases soluble IL-6 receptors in schizophrenia. *Eur Arch Psychiatry Clin Neurosci* 247:308–13.

Müller, N., M. Riedel, M. Ackenheil, and M. J. Schwarz. (2000). Cellular and humoral immune system in schizophrenia: a conceptual re-evaluation. *World J Biol Psychiatry* 1:173–79.

Müller, N., M. Riedel, M. Hadjamu, M. J. Schwarz, M. Ackenheil, and R. Gruber. (1999). Increase in expression of adhesion molecule receptors on T helper cells during antipsychotic treatment and relationship to blood-brain barrier permeability in schizophrenia. *Am J Psychiatry* 156:634–36.

Müller, N., M. Riedel, C. Scheppach, B. Brandstätter, S. Sokullu, K. Krampe, M. Ulmschneider, R. R. Engel, H. J. Möller, and M. J. Schwarz. (2002). Beneficial antipsychotic effects of celecoxib add-on therapy compared to risperidone alone in schizophrenia. *Am J Psychiatry* 159:1029–34.

Müller, N., M. Riedel, and M. J. Schwarz. (2004a). Psychotropic effects of COX-2 inhibitors—a possible new approach for the treatment of psychiatric disorders. *Pharmacopsychiatry* 37:266–69.

Müller, N., M. Riedel, M. J. Schwarz, et al. (1997b). Immunomodulatory effects of neuroleptics to the cytokine system and the cellular immune system in schizophrenia. In *Current update in psychoimmunology*, ed. G. Wieselmann, 57–67. New York: Springer.

Müller, N., M. Riedel, M. J. Schwarz, and R. R. Engel. (2005). Clinical effects of COX-2 inhibitors on cognition in schizophrenia. *Eur Arch Psychiatry Clin Neurosci* 255:149–51.

Müller, N., M. Ulmschneider, C. Scheppach, M. J. Schwarz, M. Ackenheil, H. J. Möller, R. Gruber, and M. Riedel. (2004b). COX-2 inhibition as a treatment approach in schizophrenia: immunological considerations and clinical effects of celecoxib add-on therapy. *Eur Arch Psychiatry Clin Neurosci* 254:14–22.

Musso, T., G. L. Gusella, A. Brooks, D. L. Longo, and L. Varesio. (1994). Interleukin-4 inhibits indoleamine 2,3-dioxygenase expression in human monocytes. *Blood* 83 (5):1408–11.

Myint, A. M., and Y. K. Kim. (2003). Cytokine-serotonin interaction through IDO: a neurodegeneration hypothesis of depression. *Med Hypotheses* 61 (5–6):519–25.

Myint, A. M., Y. K. Kim, R. Verkerk, S. Scharpe, H. Steinbusch, and B. Leonard. (2007). Kynurenine pathway in major depression: evidence of impaired neuroprotection. *J Affect Disord* 98 (1–2):143–51.

Myint, A. M., B. E. Leonard, H. W. Steinbusch, and Y. K. Kim. (2005). Th1, Th2, and Th3 cytokine alterations in major depression. *J Affect Disord* 88 (2):167–73.

Nasrallah, H. A., M. Calley-Whitters, L. B. Bigelow, and F. P. Rauscher. (1983). A histological study of the corpus callosum in chronic schizophrenia. *Psychiatry Res* 8:251–60.

Oberheim, N. A., X. Wang, S. Goldman, and M. Nedergaard. (2006). Astrocytic complexity distinguishes the human brain. *Trends Neurosci* 29 (10):547–53.

O'Connor, J. C., C. Andre, Y. Wang, M. A. Lawson, S. S. Szegedi, J. Lestage, N. Castanon, K. W. Kelley, and R. Dantzer. (2009a). Interferon-gamma and tumor necrosis factor-alpha mediate the upregulation of indoleamine 2,3-dioxygenase and the induction of depressive-like behavior in mice in response to bacillus Calmette-Guerin. *J Neurosci* 29 (13):4200–9.

O'Connor, J. C., M. A. Lawson, C. Andre, E. M. Briley, S. S. Szegedi, J. Lestage, N. Castanon, M. Herkenham, R. Dantzer, and K. W. Kelley. (2009b). Induction of IDO by bacille Calmette-Guerin is responsible for development of murine depressive-like behavior. *J Immunol* 182 (5):3202–12.

O'Connor, J. C., M. A. Lawson, C. Andre, M. Moreau, J. Lestage, N. Castanon, K. W. Kelley, and R. Dantzer. (2009c). Lipopolysaccharide-induced depressive-like behavior is mediated by indoleamine 2,3-dioxygenase activation in mice. *Mol Psychiatry* 14 (5):511–22.

Okuda, S., N. Nishiyama, H. Saito, and H. Katsuki. (1998). 3-Hydroxykynurenine, an endogenous oxidative stress generator, causes neuronal cell death with apoptotic features and region selectivity. *J Neurochem* 70 (1):299–307.

Olney, J. W., and N. B. Farber. (1995). Glutamate receptor dysfunction and schizophrenia. *Arch Gen Psychiatry* 52:998–1007.

Ongur, D., W. C. Drevets, and J. L. Price. (1998). Glial reduction in the subgenual prefrontal cortex in mood disorders. *Proc Natl Acad Sci USA* 95 (22):13290–95.

Ozek, M., K. Toreci, I. Akkok, and Z. Guvener. (1971). Influence of therapy on antibody-formation. *Psychopharmacologia* 21:401–12.

Panatier, A., D. T. Theodosis, J. P. Mothet, B. Touquet, L. Pollegioni, D. A. Poulain, and S. H. Oliet. (2006). Glia-derived D-serine controls NMDA receptor activity and synaptic memory. *Cell* 125 (4):775–84.

Pearce, B. D. (2001). Schizophrenia and viral infection during neurodevelopment: a focus on mechanisms. *Mol Psychiatry* 6:634–46.

Perea, G., and A. Araque. (2007). Astrocytes potentiate transmitter release at single hippocampal synapses. *Science* 317 (5841):1083–86.

Perkins, M. N., and T. W. Stone. (1982). An iontophoretic investigation of the actions of convulsant kynurenines and their interaction with the endogenous excitant quinolinic acid. *Brain Res* 247 (1):184–87.

Pilowsky, L. S., R. A. Bressan, J. M. Stone, K. Erlandsson, R. S. Mulligan, J. H. Krystal, and P. J. Ell. (2006). First *in vivo* evidence of an NMDA receptor deficit in medication-free schizophrenic patients. *Mol Psychiatry* 11:118–19.

Quincozes-Santos, A., L. D. Bobermin, J. Kleinkauf-Rocha, D. O. Souza, R. Riesgo, C. A. Goncalves, and C. Gottfried. (2009). Atypical neuroleptic risperidone modulates glial functions in C6 astroglial cells. *Prog Neuropsychopharmacol Biol Psychiatry* 33:11–15.

Raison, C. L., R. Dantzer, K. W. Kelley, M. A. Lawson, B. J. Woolwine, G. Vogt, J. R. Spivey, K. Saito, and A. H. Miller. (2010). CSF concentrations of brain tryptophan and kynurenines during immune stimulation with IFN-alpha: relationship to CNS immune responses and depression. *Mol Psychiatry* 15 (4):393–403.

Rajkowska, G., J. J. Miguel-Hidalgo, J. Wei, G. Dilley, S. D. Pittman, H. Y. Meltzer, J. C. Overholser, B. L. Roth, and C. A. Stockmeier. (1999). Morphometric evidence for neuronal and glial prefrontal cell pathology in major depression. *Biol Psychiatry* 45 (9):1085–98.

Rothermundt, M., J. N. Ahn, and S. Jorgens. (2009). S100B in schizophrenia: an update. *Gen Physiol Biophys* 28 (spec no focus), F76–F81.

Rothermundt, M., P. Falkai, G. Ponath, S. Abel, H. Burkle, M. Diedrich, G. Hetzel, M. Peters, A. Siegmund, A. Pedersen, W. Maier, J. Schramm, T. Suslow, P. Ohrmann, and V. Arolt. (2004). Glial cell dysfunction in schizophrenia indicated by increased S100B in the CSF. *Mol Psychiatry* 9:897–99.

Rothstein, J. D., M. Dykes-Hoberg, C. A. Pardo, L. A. Bristol, L. Jin, R. W. Kuncl, Y. Kanai, M. A. Hediger, Y. Wang, J. P. Schielke, and D. F. Welty. (1996). Knockout of glutamate transporters reveals a major role for astroglial transport in excitotoxicity and clearance of glutamate. *Neuron* 16 (3):675–86.

Saito, K., Y. Ohta, Y. Nagamura, E. Sasaki, and I. Ishiguro. (1990). Relationship between L-tryptophan uptake and L-tryptophan 2,3-dioxygenase activity in rat hepatocytes. *Biochem Int* 20 (1):71–80.

Salter, M., and C. I. Pogson. (1985). The role of tryptophan 2,3-dioxygenase in the hormonal control of tryptophan metabolism in isolated rat liver cells. Effects of glucocorticoids and experimental diabetes. *Biochem J* 229 (2):499–504.

Salzberg-Brenhouse, H.C., E. Y. Chen, D. F. Emerich, S. Baldwin, K. Hogeland, S. Ranelli, D. Lafreniere, B. Perdomo, L. Novak, T. Kladis, K. Fu, A. S. Basile, J. H. Kordower, and R. T. Bartus. (2003). Inhibitors of cyclooxygenase-2, but not cyclooxygenase-1 provide structural and functional protection against quinolinic acid-induced neurodegeneration. *J Pharmacol Exp Ther* 306: 218–228.

Sandrini, M., G. Vitale, and L. A. Pini. (2002). Effect of rofecoxib on nociception and the serotonin system in the rat brain. *Inflamm Res* 51: 154–159.

Satyanarayana, U., and B. S. Rao. (1980). Dietary tryptophan level and the enzymes of tryptophan NAD pathway. *Br J Nutr* 43 (1):107–13.

Schipke, C.G., I. Heuser, and O. Peters (2010). Antidepressants act on glial cells: SSRIs and serotonin elicit astrocyte calcium signaling in the mouse prefrontal cortex. *J Psychiatr Res*

Schroeter, M. L., H. Abdul-Khaliq, A. Diefenbacher, and I. E. Blasig. (2002). S100B is increased in mood disorders and may be reduced by antidepressive treatment. *Neuroreport* 13 (13):1675–78.

Schroeter, M. L., H. Abdul-Khaliq, M. Krebs, A. Diefenbacher, and I. E. Blasig. (2009). Neuron-specific enolase is unaltered whereas S100B is elevated in serum of patients with schizophrenia—original research and meta-analysis. *Psychiatry Res* 167:66–72.

Schwarcz, R., A. Rassoulpour, H. Q. Wu, D. Medoff, C. A. Tamminga, and R. C. Roberts. (2001). Increased cortical kynurenate content in schizophrenia. *Biol Psychiatry* 50:521–30.

Schwarcz, R., W. O. Whetsell Jr., and R. M. Mangano. (1983). Quinolinic acid: an endogenous metabolite that produces axon-sparing lesions in rat brain. *Science* 219 (4582):316–18.

Schwarz, M. J., M. Riedel, M. Ackenheil, and N. Müller. (2000). Decreased levels of soluble intercellular adhesion molecule-1 (sICAM-1) in unmedicated and medicated schizophrenic patients. *Biol Psychiatry* 47:29–33.

Schwieler, L., S. Erhardt, C. Erhardt, and G. Engberg. (2005a). Prostaglandin-mediated control of rat brain kynurenic acid synthesis—opposite actions by COX-1 and COX-2 isoforms. *J Neural Transm* 112:863–72.

Segnitz, N., A. Schmitt, P. J. Gebicke-Harter, and M. Zink, M. (2009). Differential expression of glutamate transporter genes after chronic oral treatment with aripiprazole in rats. *Neurochem Int* 55: 619–628.

Seidel, A., V. Arolt, M. Hunstiger, L. Rink, A. Behnisch, and H. Kirchner. (1995). Cytokine production and serum proteins in depression. *Scand. J. Immunol.* 41: 534–538.

Seidel, A., V. Arolt, M. Hunstiger, L. Rink, A. Behnisch, and H. Kirchner. (1996). Major depressive disorder is associated with elevated monocyte counts. *Acta Psychiatr Scand* 94: 198–204.

Schwieler, L., S. Erhardt, C. Erhardt, and G. Engberg. (2005b). Prostaglandin-mediated control of rat brain kynurenic acid synthesis—opposite actions by COX-1 and COX-2 isoforms. *J Neural Transm* 112:863–72.

Sheline, Y. I., M. H. Gado, and J. L. Price. (1998). Amygdala core nuclei volumes are decreased in recurrent major depression. *Neuroreport* 9 (9):2023–28.

Sher, L., M. A. Oquendo, H. C. Galfalvy, G. Zalsman, T. B. Cooper, and J. J. Mann. (2005). Higher cortisol levels in spring and fall in patients with major depression. *Prog Neuropsychopharmacol Biol Psychiatry* 29 (4):529–34.

Si, X., J. J. Miguel-Hidalgo, G. O'Dwyer, C. A. Stockmeier, and G. Rajkowska. (2004). Age-dependent reductions in the level of glial fibrillary acidic protein in the prefrontal cortex in major depression. *Neuropsychopharmacology* 29 (11):2088–96.

Sluzewska, A., J. K. Rybakowski, M. Laciak, A. Mackiewicz, M. Sobieska, and K. Wiktorowicz. (1995). Interleukin-6 serum levels in depressed patients before and after treatment with fluoxetine. *Ann N Y Acad Sci* 762: 474–6, 474–476.

Smith, S. A., F. P. Carr, and C. I. Pogson. (1980). The metabolism of L-tryptophan by isolated rat liver cells. Quantification of the relative importance of, and the effect of nutritional status on, the individual pathways of tryptophan metabolism. *Biochem J* 192 (2):673–86.

Song, C. and B. E. Leonard. (1994). An acute phase protein response in the olfactory bulbectomised rat: effect of sertraline treatment. *Med Sci Res* 22: 313–314.

Song, C. and B. E. Leonard. (2000). Fundamentals of psychoneuroimmunology. Chichester, New York: J Wiley & Sons.

Sorg, O., and P. J. Magistretti. (1991). Characterization of the glycogenolysis elicited by vasoactive intestinal peptide, noradrenaline and adenosine in primary cultures of mouse cerebral cortical astrocytes. *Brain Res* 563 (1–2):227–33.

Speciale, C., H. Q. Wu, M. Cini, M. Marconi, M. Varasi, and R. Schwarcz. (1996). (R,S)-3,4-dichlorobenzoylalanine (FCE 28833A) causes a large and persistent increase in brain kynurenic acid levels in rats. *Eur J Pharmacol* 315:263–67.

Steffek, A. E., R. E. McCullumsmith, V. Haroutunian, and J. H. Meador-Woodruff. (2008). Cortical expression of glial fibrillary acidic protein and glutamine synthetase is decreased in schizophrenia. *Schizophr Res* 103:71–82.

Steiner, J., H. G. Bernstein, H. Bielau, N. Farkas, J. Winter, H. Dobrowolny, R. Brisch, T. Gos, C. Mawrin, A. M. Myint, and B. Bogerts. (2008). S100B-immunopositive glia is elevated in paranoid as compared to residual schizophrenia: a morphometric study. *J Psychiatr Res* 42 (10):868–76.

Steiner, J., H. Bielau, H. G. Bernstein, B. Bogerts, and M. T. Wunderlich. (2006). Increased cerebrospinal fluid and serum levels of S100B in first-onset schizophrenia are not related to a degenerative release of glial fibrillar acidic protein, myelin basic protein and neurone-specific enolase from glia or neurones. *J Neurol Neurosurg Psychiatry* 77:1284–87.

Steiner, J., M. L. Schroeter, K. Schiltz, H. G. Bernstein, U. J. Muller, C. Richter-Landsberg, W. E. Muller, M. Walter, T. Gos, B. Bogerts, and G. Keilhoff. (2010). Haloperidol and clozapine decrease S100B release from glial cells. *Neuroscience* 167:1025–31.

Stevens, J. R. (1982). The neuropathology of schizophrenia. *Psychol Med* 12:695–700.

Stockmeier, C. A., G. J. Mahajan, L. C. Konick, J. C. Overholser, G. J. Jurjus, H. Y. Meltzer, H. B. Uylings, L. Friedman, and G. Rajkowska. (2004). Cellular changes in the post-mortem hippocampus in major depression. *Biol Psychiatry* 56 (9):640–50.

Stone, T. W. (1993). Neuropharmacology of quinolinic and kynurenic acids. *Pharmacol Rev* 45:309–79.

Stone, T. W., and L. G. Darlington. (2002). Endogenous kynurenines as targets for drug discovery and development. *Nat Rev Drug Discov* 1 (8):609–20.

Sumiyoshi, T., A. E. Anil, D. Jin, K. Jayathilake, M. Lee, and H. Y. Meltzer. (2004). Plasma glycine and serine levels in schizophrenia compared to normal controls and major depression: relation to negative symptoms. *Int J Neuropsychopharmacol* 7:1–8.

Sumiyoshi, T., D. Jin, K. Jayathilake, M. Lee, and H. Y. Meltzer. (2005). Prediction of the ability of clozapine to treat negative symptoms from plasma glycine and serine levels in schizophrenia. *Int J Neuropsychopharmacol* 8:451–55.

Tamminga, C. A., N. Cascella, T. D. Fakouhl, and R. L. Hertin. (1992). Enhancement of NMDA-mediated transmission in schizophrenia: effects of milacemide. In *Novel antipsychotic drugs*, ed. H. Y. Meltzer, 171–77. New York: Raven Press.

Tanaka, K., K. Watase, T. Manabe, K. Yamada, M. Watanabe, K. Takahashi, H. Iwama, T. Nishikawa, N. Ichihara, T. Kikuchi, S. Okuyama, N. Kawashima, S. Hori, M. Takimoto, and K. Wada. (1997). Epilepsy and exacerbation of brain injury in mice lacking the glutamate transporter GLT-1. *Science* 276 (5319):1699–702.

Taylor, M. W., and G. S. Feng. (1991). Relationship between interferon-gamma, indoleamine 2,3-dioxygenase, and tryptophan catabolism. *Faseb J* 5 (11):2516–22.

Thomas, A. J., S. Davis, C. Morris, E. Jackson, R. Harrison, and J. T. O'Brien. (2005). Increase in interleukin-1beta in late-life depression. *Am J Psychiatry* 162 (1):175–77.

Toro, C. T., J. E. Hallak, J. S. Dunham, and J. F. Deakin. (2006). Glial fibrillary acidic protein and glutamine synthetase in subregions of prefrontal cortex in schizophrenia and mood disorder. *Neurosci Lett* 404 (3):276–81.

Torrey, E. F., J. Miller, R. Rawlings, and R. H. Yolken. (1997). Seasonality of births in schizophrenia and bipolar disorder: a review of the literature. *Schizophr Res* 28:1–38.

Tramontina, F., M. C. Leite, D. Goncalves, A. C. Tramontina, D. F. Souza, J. K. Frizzo, P. Nardin, C. Gottfried, S. T. Wofchuk, and C. A. Goncalves. (2006a). High glutamate decreases S100B secretion by a mechanism dependent on the glutamate transporter. *Neurochem Res* 31:815–20.

Tramontina, F., A. C. Tramontina, D. F. Souza, M. C. Leite, C. Gottfried, D. O. Souza, S. T. Wofchuk, and C. A. Goncalves. (2006b). Glutamate uptake is stimulated by extracellular S100B in hippocampal astrocytes. *Cell Mol Neurobiol* 26:81–86.

Tsacopoulos, M., and P. J. Magistretti. (1996). Metabolic coupling between glia and neurons. *J Neurosci* 16 (3):877–85.

Van Eldik, L. J. and M. S. Wainwright. (2003). The Janus face of glial-derived S100B: beneficial and detrimental functions in the brain. *Restor Neurol Neurosci* 21:97–108.

Watanabe, T., K. Morimoto, T. Hirao, H. Suwaki, K. Watase, and K. Tanaka. (1999). Amygdala-kindled and pentylenetetrazole-induced seizures in glutamate transporter GLAST-deficient mice. *Brain Res* 845 (1):92–96.

Waters N., C. Lundgren, L.O. Hansson, M.L. Carlsson. (1996). Concurrent locomotor stimulation and decrease in dopamine release in rats mice after treatment with the competitive NMDA receptor antagonists D-CPPene and CGS 19755. *J Neural Transm* 103(1–2): 117–129.

Webster, M. J., J. O'Grady, J. E. Kleinman, and C. S. Weickert. (2005). Glial fibrillary acidic protein mRNA levels in the cingulate cortex of individuals with depression, bipolar disorder and schizophrenia. *Neuroscience* 133:453–61.

Westergaard, T., P. B. Mortensen, C. B. Pedersen, J. Wohlfahrt, and M. Melbye. (1999). Exposure to prenatal and childhood infections and the risk of schizophrenia: suggestions from a study of sibship characteristics and influenza prevalence. *Arch Gen Psychiatry* 56:993–98.

Wichers, M. C., G. H. Koek, G. Robaeys, R. Verkerk, S. Scharpe, and M. Maes. (2005). IDO and interferon-alpha-induced depressive symptoms: a shift in hypothesis from tryptophan depletion to neurotoxicity. *Mol Psychiatry* 10 (6):538–44.

Wilke, I., V. Arolt, M. Rothermundt, C. Weitzsch, M. Hornberg, and H. Kirchner. (1996). Investigations of cytokine production in whole blood cultures of paranoid and residual schizophrenic patients. *Eur Arch Psychiatry Clin Neurosci* 246:279–84.

Wonodi, I. and R. Schwarcz (2010). Cortical kynurenine pathway metabolism: a novel target for cognitive enhancement in Schizophrenia. *Schizophr Bull* 36: 211–218.

Wu, H. Q., A. Rassoulpour, and R. Schwarcz. (2006). Kynurenic acid leads, dopamine follows: a new case of volume transmission in the brain? *J Neural Transm* 114(1): 33–41.

Yaron, I., I. Shirazi, R. Judovich, D. Levartovsky, D. Caspi, and M. Yaron. (1999). Fluoxetine and amitriptyline inhibit nitric oxide, prostaglandin E2, and hyaluronic acid production in human synovial cells and synovial tissue cultures. *Arthritis Rheum* 42: 2561–2568.

Yokota, O., S. Terada, T. Ishihara, H. Nakashima, A. Kugo, H. Ujike, K. Tsuchiya K. Ikeda, Y. Saito, S. Murayama, H. Ishizu, and S. Kuroda. (2004). Neuronal expression of cyclooxygenase-2, a pro-inflammatory protein, in the hippocampus of patients with schizophrenia. *Prog. Neuropsychopharmacol Biol Psychiatry* 28: 715–721.

Yasui, H., K. Takai, R. Yoshida, and O. Hayaishi. (1986). Interferon enhances tryptophan metabolism by inducing pulmonary indoleamine 2,3-dioxygenase: its possible occurrence in cancer patients. *Proc Natl Acad Sci USA* 83 (17):6622–26.

Zhu, J., B. O. Bengtsson, E. Mix, E., L. H. Thorell, T. Olsson, and H. Link. (1994). Effect of monoamine reuptake inhibiting antidepressants on major histocompatibility complex expression on macrophages in normal rats and rats with experimental allergic neuritis (EAN). *Immunopharmacology* 27: 225–244.

Zuckerman, L., and I. Weiner. (2005). Maternal immune activation leads to behavioral and pharmacological changes in the adult offspring. *J Psychiatr Res* 39:311–23.

Index